ADVANCES IN CLINICAL CHEMISTRY

VOLUME 27

BOARD OF EDITORS

Herbert E. Spiegel
Pierre P. Bourdoux
Margot J. Breidahl
Jui-San Chen
Edward H. Cooper
Leonard K. Dunikoski
M. A. M. Abul-Fadl
Malcolm A. Frazer
Ronald J. Georges

Maurice Green
Mogens Horder
Walter Hordynsky
Olen Hornykiewicz
James E. Logan
Pei-Sheng Mo
Norberto Montalbetti
Elvira De Serratos
Steven J. Soldin

Kihachiro Takahara

COPYRIGHT © 1989 BY ACADEMIC PRESS, INC.
ALL RIGHTS RESERVED.
NO PART OF THIS PUBLICATION MAY BE REPRODUCED OR
TRANSMITTED IN ANY FORM OR BY ANY MEANS, ELECTRONIC
OR MECHANICAL, INCLUDING PHOTOCOPY, RECORDING, OR
ANY INFORMATION STORAGE AND RETRIEVAL SYSTEM, WITHOUT
PERMISSION IN WRITING FROM THE PUBLISHER.

ACADEMIC PRESS, INC.
San Diego, California 92101

United Kingdom Edition published by
ACADEMIC PRESS LIMITED
24-28 Oval Road, London NW1 7DX

LIBRARY OF CONGRESS CATALOG CARD NUMBER: 58-12341

ISBN   0-12-010327-3   (alk. paper)

PRINTED IN THE UNITED STATES OF AMERICA
89 90 91 92   9 8 7 6 5 4 3 2 1

Advances in CLINICAL CHEMISTRY

Edited by

HERBERT E. SPIEGEL

Department of Laboratories
St. Vincent's Hospital
and Medical Center
New York, New York

VOLUME 27

ACADEMIC PRESS, INC.
Harcourt Brace Jovanovich, Publishers
San Diego   New York   Berkeley   Boston
London   Sydney   Tokyo   Toronto

CONTENTS

Contributors ix

Preface xi

Reference Values
Helge Erik Solberg and Ralph Gräsbeck

1. From Normal Values to Reference Values 2
2. Reference Individuals and Specimen Collection 17
3. Assay of Analytes and Quality Control 34
4. Statistical Treatment of Reference Values 36
5. Alternatives to Conventional Reference Values and Intervals 58
6. Presentation of Observed Values Related to Reference Values 61
7. Final Remarks 67
8. Abbreviations and Symbols 68
   References 69

Neopterin as Marker for Activation of Cellular Immunity: Immunologic Basis and Clinical Application
Helmut Wachter, Dietmar Fuchs, Arno Hausen, Gilbert Reibnegger, and Ernst R. Werner

1. Introduction 82
2. Chemistry and Biochemistry of Neopterin 88
3. Methods of Measurement of Neopterin 94
4. Neopterin Concentrations in Healthy Subjects 97
5. Conditions Associated with High Neopterin Levels 99
6. Conclusions 123
   References 125

Biochemical Detection of Hepatitis B Virus Constituents
Hsiang Ju Lin

1. Introduction 143
2. HBV DNA 151
3. HBV Enzymes 174
4. HBV Polypeptides 179
   References 183

Monitoring Acid–Base and Electrolyte Disturbances in Intensive Care
A. KAZDA, A. JABOR, M. ZÁMEČNÍK, AND K. MAŠEK

1. Introduction	202
2. Osmolality	202
3. Sodium	206
4. Potassium	215
5. Renal Function from the Aspect of Water and Ion Balance	217
6. Acid–Base Balance	227
7. Computer Programs for Monitoring Water, Ion, and Acid–Base Metabolism	249
8. Conclusions	261
References	262

Information in the Clinical Laboratory: Computer-Assisted Organization and Management
STANLEY J. ROBBOY AND RONALD TROST

1. Introduction	270
2. Ease of Use	273
3. Hidden Enhancements to Functionality	274
4. Safeguards	275
5. Enhancements for the User	278
6. Common Core Functions to All Lab Modules	278
7. General Chemistry Functions	280
8. Data Presentation and Retrieval	287
9. Management	292
10. Conclusion	301
References	301

Monoclonal Antibodies: Production, Purification, and Technology
DAVID VETTERLEIN

1. Introduction	303
2. Monoclonal Antibody Production	306
3. Monoclonal Antibody Purification	317
4. Monoclonal Antibody Technology	329
References	342

Monoclonal Antibodies: Clinical Applications
RUDOLPH RECKEL

1. Introduction	355
2. Neoplastic Disease	357
3. Hormones and Proteins	376
4. Cell Surface Antigens	381
5. Miscellaneous Clinical Applications	387
6. Conclusions	391
References	391
INDEX	417

CONTRIBUTORS

Numbers in parentheses indicate the pages on which the authors' contributions begin.

DIETMAR FUCHS (81), *Institute of Medical Chemistry and Biochemistry, University of Innsbruck, A-6020 Innsbruck, Austria*

RALPH GRÄSBECK (1), *Minerva Foundation Institute for Medical Research and Laboratory Department, Maria Hospital, Helsinki, Finland*

ARNO HAUSEN (81), *Institute of Medical Chemistry and Biochemistry, University of Innsbruck, A-6020 Innsbruck, Austria*

A. JABOR (201), *Department of Clinical Biochemistry, Postgraduate Medical and Pharmaceutical Institute, 100 05 Prague, Czechoslovakia*

A. KAZDA (201), *Department of Clinical Biochemistry, Postgraduate Medical and Pharmaceutical Institute, 100 05 Prague, Czechoslovakia*

HSIANG JU LIN (143), *Clinical Biochemistry Unit, University of Hong Kong, Hong Kong*

K. MAŠEK (201), *Department of Clinical Biochemistry, Postgraduate Medical and Pharmaceutical Institute, 100 05 Prague, Czechoslovakia*

RUDOLPH RECKEL* (355), *Immunology Development, Immunomedics, Inc., Newark, New Jersey 07103*

GILBERT REIBNEGGER (81), *Institute of Medical Chemistry and Biochemistry, University of Innsbruck, A-6020 Innsbruck, Austria*

*Deceased.

STANLEY J. ROBBOY (269), *Department of Pathology, University of Medicine and Dentistry of New Jersey, Newark, New Jersey 07103*

HELGE ERIK SOLBERG (1), *Department of Clinical Chemistry and Center for Medical Informatics, Rikshospitalet, N-0027 Oslo 1, Norway*

RONALD TROST (269), *Department of Pathology, University of Medicine and Dentistry of New Jersey, Newark, New Jersey 07103*

DAVID VETTERLEIN (303), *Genentech, Recovery Process Research and Development, South San Francisco, California 94080*

HELMUT WACHTER (81), *Institute of Medical Chemistry and Biochemistry, University of Innsbruck, A-6020 Innsbruck, Austria*

ERNST R. WERNER (81), *Institute of Medical Chemistry and Biochemistry, University of Innsbruck, A-6020 Innsbruck, Austria*

M. ZÁMEČNÍK (201), *Department of Clinical Biochemistry, Postgraduate Medical and Pharmaceutical Institute, 100 05 Prague, Czechoslovakia*

PREFACE

Advances in Clinical Chemistry continues the philosophy of expanding the horizon of the clinical chemist. In this volume, several chapters are included which relate to immunology, computer-assisted organization, monitoring electrolyte disturbances, and reference values. The increased emphasis on molecular biology and genetic engineering also reflects the growing importance of these fields to the practice of clinical chemistry. Succeeding volumes of the *Advances* will continue to amplify the scientific and philosophical framework of the clinical chemist both as a practitioner and researcher. The editors welcome any suggestions from the readership for relevant topics for future volumes and, as always, we also welcome any suggestion for improvement.

HERBERT E. SPIEGEL

This volume is dedicated to my son, Jimmy, with my prayers for health and happiness for many years to come.

In Memoriam:
Rudolph Reckel, Ph.D.

On January 11, 1989, Dr. Rudolph Reckel was prematurely taken away from us by an untimely accident. He was my classmate, my friend, and a valued colleague. Rudy was a scientist, a scholar, and a gentleman. He was a man of humor and of kindness. The chapter in this volume entitled "Monoclonal Antibodies: Clinical Applications" was the last professional article Rudy wrote. This series is enriched by his efforts as I was enriched for having known and worked with him.

REFERENCE VALUES

Helge Erik Solberg* and Ralph Gräsbeck†

*Department of Clinical Chemistry and Center for Medical Informatics,
Rikshospitalet, Oslo, Norway
and
†Minerva Foundation Institute for Medical Research and Laboratory
Department,
Maria Hospital,
Helsinki, Finland

1. From Normal Values to Reference Values 2
   1.1. Introduction 2
   1.2. Normal Values 2
   1.3. The Concept of Reference Values 4
   1.4. Different Kinds of Reference Groups 14
2. Reference Individuals and Specimen Collection 17
   2.1. Preanalytical Factors 17
   2.2. Selection of Reference Individuals 22
   2.3. Preparation of Individuals; Collection of Specimens 26
   2.4. Recommended Procedures 28
3. Assay of Analytes and Quality Control 34
   3.1. Analysis and Control 34
   3.2. Transfer of Reference Values 35
4. Statistical Treatment of Reference Values 36
   4.1. Introduction 36
   4.2. Preprocessing of Reference Values 37
   4.3. Estimating Reference Limits 41
5. Alternatives to Conventional Reference Values and Intervals 58
   5.1. The Multivariate Situation 58
   5.2. Time-Specified Reference Values 60
6. Presentation of Observed Values Related to Reference Values 61
   6.1. Desirable Features 61
   6.2. Numerical Methods 62
   6.3. Graphical Presentation 64
   6.4. Avoiding the Conventional Reference Interval 65
   6.5. How to Avoid Being Deluged with Data 66
7. Final Remarks 67
8. Abbreviations and Symbols 68
   References 69

1. From Normal Values to Reference Values

1.1. Introduction

Medicine is an old art, which only during recent centuries has acquired the characteristics of a true science. Clinical medicine still contains some vestiges of its unscientific past, and there is often little documentation of the true value of common procedures, such as confining the patients to bed. Similar conclusions may be drawn from the fact that there are remarkable national differences in the practice of medicine.

The clinical laboratory sciences are no doubt the most objective and exact of the clinical disciplines. However, as exemplified by the most likely unnecessary rite of swabbing the skin before venipuncture (D1), clinical chemistry also has aspects which do not stand scientific critique. An example is the concept of "normal value," which some 20 years ago was found to be too imprecise to satisfy modern demands.

The alternative concept, "reference values," was proposed in 1969 (G15). Since then, the corresponding theory and terminology have been elaborated upon by international, regional, and national organizations and bodies. As a result, there are a number of recommendations representing a coherent philosophy that every clinician and laboratory scientist ought to be acquainted with (A4, A5, H18, I5–I10, S4–S7, S8–S16). As five documents in the series of six recommendations produced by the International Federation of Clinical Chemistry (IFCC) (see Section 8) has recently been internationally approved (I5–I10), it may be valuable to present and critically appraise selected topics covered by these documents. This is one of the purposes of this review, which is written by the two previous chairmen of the Expert Panel on Theory of Reference Values of the IFCC (Gräsbeck, 1975–1978; Solberg, 1979–1987). Furthermore, the review contains an extensive compilation of the literature on reference values and related topics. However, the number of publications in this field is greater than can be included here: in 1980 more than 1000 studies were documented (B8)!

1.2. Normal Values

1.2.1. History

The roots of the modern conception of normality date back to antiquity, particularly to the Platonic doctrine of ideas. Normal essentially means "in accordance with the idea" (V1). The Belgian statistician Quételet (1835), who concerned himself with the "average man," was impressed by the fact that biometric data tended to be distributed as a "vase invertée"

(Q1). This bell-shaped distribution was considered to be adequately described by the Gaussian function. Since the results of laboratory tests are easy to display in the form of histograms and distribution curves, the statistical conception of normality, born in the early decades of the twentieth century (V2), became especially popular in the clinical laboratory sciences.

This simplistic view began to be eroded when it was discovered that the distribution of the concentrations of many serum components were skew and not Gaussian (G17, W12). Research on the relation between blood lipid concentrations and risk for cardiovascular disease revealed that values common among the general population were not necessarily good or desirable. Critical voices were raised against the concepts of normal and normal values (G13, M11), and when the reference value concept was introduced (G15) it found wide acceptance among clinical chemists and hematologists of the world. Thus, around 1970, the Scandinavian, French, and Spanish Societies of Clinical Chemistry, the International Federation of Clinical Chemistry, and the International Committee on Standardization in Haematology founded expert panels on reference values.

1.2.2. *Drawbacks*

The term normal values is ambiguous. It seems to be more precise than is actually the case. Confusion arises because of the varied usages of the word normal (M11), which in laboratory medicine is used with at least three very different connotations:

1. *Statistical sense:* Values are often qualified as normal if they appear to be distributed like the theoretical normal (Gaussian) distribution of statistics. This use of the term normal has often created the erroneous view that the distribution of biological data is symmetrical and bell shaped. To exorcise the "ghost of Gauss," it has been recommended not to use the term normal limits (E2). For a similar reason, the term normal distribution should be replaced by the term Gaussian distribution (M11), as is done in this article.

2. *Epidemiological sense:* An observed value is often declared normal if it is typical of the values found in the general population, and abnormal if the observed value is atypical. When observed values were compared with normal limits, the words normal and abnormal were used. In this context, preferable substitutes for normal are common, frequent, habitual, usual, and typical (M11).

3. *Clinical sense:* The term normal is often used to indicate that values are associated with absence of or low risk for disease. Better terms are healthy, nonpathological, or harmless (M11).

These meanings of the term normal often conflict with each other. For instance, common blood lipid values in the western countries are neither associated with lack of pathology nor distributed in the Gaussian fashion. In the aged, some forms of pathology, e.g., arthrosis, are so common that abnormal is normal. Perusal of clinical–chemical literature reveals titles such as "Blood Chemistry in Normal and Pregnant Women," which implies that pregnancy is abnormal. And only a century ago it was more normal for fertile women to be pregnant than nonpregnant! In fact, a woman who cannot get pregnant could be considered abnormal. Use of such nomenclature is confusing, indeed!

The traditional normal values are usually collected from healthy young ambulant persons such as laboratory personnel and medical students, but are used to judge values observed on bedridden old people. Such values are therefore poorly relevant. Usually nothing is known about the conditions under which they were collected: Were the subjects fasting or sitting and was a tourniquet used during specimen collection? Finally, when a "normal range" is given, say 15–86 mU/liter, it remains obscure whether these figures represent the extreme values observed or a statistically determined interval.

1.3. The Concept of Reference Values

1.3.1. *General Description*

The basic idea is that one should provide values from subjects who are relevant controls for the patients under study and that these controls should be described in detail (G11). One didactic way of expressing this is to state that the diseases may be regarded as experiments of Nature, and that just as in the experimental sciences, one should describe the positive and negative controls in sufficient detail to enable critical evaluation and repetition of the experiments.

Though certain common practices have crystallized (e.g., in the testing of new drugs on animals), there are no strict rules as to how a scientist should design his controls; everyone is free to choose them as he or she finds best or realistic. (Parenthetically, it may be observed that one of the best criteria for judging the quality of a scientific report is to scrutinize its controls.) In the reference value strategy the investigator is also free to choose his own controls.

Another approach to the concept of reference values is by emphasizing the aspect of *comparison*. Clinical data are interpreted and medical decisions made by comparing observations with other data. Every comparison has two operands: the data to be interpreted and those used for comparison, i.e., the *observations* and the *references,* respectively. A medical decision is reached by comparing an observation with one or more references. This is also the case when laboratory results are interpreted. A laboratory result—an *observed value*—is compared with one or more sets of *reference values*.

To interpret a patient's condition, the physician often compares clinical observations with reference data. He makes a diagnosis, estimates a prognosis, or selects a therapy on the basis of empirical data such as those obtained by anamnesis, by physical examination, or by laboratory procedures. If a patient's clinical data resemble those characterizing a particular disease, the doctor may tentatively conclude that the patient is suffering from that illness. This is an example of *positive diagnosis*. This diagnosis is more likely if a set of alternative diagnoses can be excluded by comparison of the observed data with the characteristics of these diseases. That is *diagnosis by elimination*. In both types of diagnosis, the basis is comparison of clinical observations with reference data such as previously collected observations on healthy individuals and on patients having specified diseases.

The process of comparison may be more or less formalized. Sometimes the physician only makes a rough mental matching of observations against his clinical knowledge acquired through personal experience or through scientific communication by education, literature, etc. In other instances he uses scoring or other more formal techniques. At the highest level of formalization, he may apply mathematical models and statistical techniques such as Bayesian procedures, pattern recognition, or discriminant analysis. These methods are often so complex and demand such heavy calculations that the use of a computer is mandatory.

Close examination of all these types of diagnostic decision making discloses the importance of comparison. The approaches are different, but in all cases the basis is the same: to relate, in one way or another, observed data to reference data. But some conditions have to be fulfilled to make a comparison valid. When dealing with clinical laboratory results the following conditions are mandatory (D5):

1. All groups of reference individuals used should be clearly defined.
2. The patient examined should sufficiently resemble the reference individuals (in all groups selected for comparison) in all respects other than those under investigation.
3. The conditions under which the specimens were obtained and processed for analysis should be known.
4. All quantities compared should be of the same type.
5. All laboratory results should be produced by adequately standardized methods under sufficient quality control.

To these general requirements one may add others that become necessary when the advanced techniques for decision making are applied (S43).

6. The stage in the development of each disease that is included into the system should be defined.

7. The diagnostic sensitivity, the diagnostic specificity, the prevalence, and the clinical costs of misclassification should be known for all laboratory tests used.

1.3.2. *The IFCC Recommendations*

The six IFCC recommendations (of which three have also been adopted by the International Committee for Standardization in Haematology) have a common title: "Approved Recommendation on the Theory of Reference Values." The parts are ordered in a logical sequence: definition, production, treatment, and use of reference values.

1. The Concept of Reference Values (I5)
2. Selection of Individuals for the Production of Reference Values (I6)
3. Preparation of Individuals and Collection of Specimens for the Production of Reference Values (I7)
4. Control of Analytical Variation in the Production, Transfer, and Application of Reference Values (I8)
5. Statistical Treatment of Collected Reference Values. Determination of Reference Limits (I9)
6. Presentation of Observed Values Related to Reference Values (I10)

The first document defines reference values and related terms, and it introduces topics treated in greater detail in the later documents: the production, treatment, and use of reference values. Documents two, three, and four describe stages in the production of reference values: a group of adequate reference individuals is selected (second document), the individuals are prepared for specimen collection and specimens for analysis are collected under standardized conditions (third document), and the specimens are analyzed by defined laboratory methods under adequate quality control (fourth document). The statistical analysis of reference values and the parametric and nonparametric estimations of reference limits are presented in the fifth document. The last document concerns how reference values are used for comparison with observed values (clinical values or patient's results).

Flowchart 1 summarizes the production, the statistical analysis, and the use of reference values (the final nonterminating loop in the chart). The present review follows, with some deviations, the order of topics in the six IFCC recommendations. The set of recommendations on reference values published by the Spanish and French societies of clinical chemistry

REFERENCE VALUES

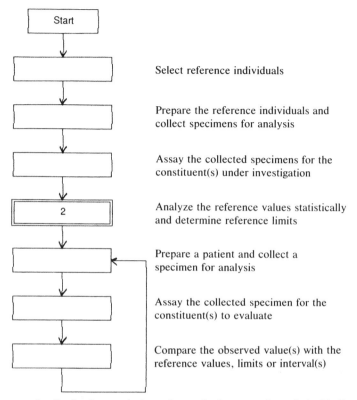

FLOWCHART 1. Production, analysis, and use of reference values. A doubly lined box in this and in the following flowcharts points to the more detailed flowchart indicated in the box.

(S4–S7, S8–S16) is partly based on and conforms with, except for minor differences, the IFCC recommendations.

1.3.3. Definition of Three Basic Terms

The general idea of reference values was presented above. A more formal definition is given in the first IFCC recommendation (I5). This definition and some related terms are presented and discussed below.

The three fundamental terms defined by the IFCC are reference individual, reference value, and observed value:

1. *Reference individual.* A reference individual is an individual selected for comparison using defined criteria. *Note:* It is usually important to define the individual's state of health.

2. *Reference value.* A reference value is the value obtained by observation or measurement of a particular type of quantity on a reference individual. *Note:* The term reference *value* should not be confused with the term reference *limit*.

3. *Observed value* (see Section 1.3.6). An observed value is a value of a particular type of quantity, obtained by observation or measurement and produced to make a medical decision. It can be compared with reference values.

The IFCC actually defines a larger set of terms. The complete sequence is reference individual, reference population, reference sample group, reference value, reference distribution, reference limit, reference interval, and observed values. The relation between the terms is easier to grasp if the three terms defined above are characterized as a basic set supplemented by two secondary sets of terms: (1) one related to reference individuals (reference population and reference sample group) and (2) one related to the reference values as such (reference distribution, reference limit, and reference interval). These two sets of terms are defined and commented on in Sections 1.3.4 and 1.3.5. The definitions of reference value and observed value given above are slightly rephrased. The reader should consult the IFCC recommendation for the exact text (I5).

The concept of *comparison* referred to in the definitions of reference individual and observed value is that presented in Section 1.3.1. There it was asserted that comparison is a central topic of the theory of reference values, and that every observation needs a yardstick to aid in its interpretation.

The concepts *defined criteria* and *state of health* used in the definition of reference individual may seem dull. The IFCC refers to the important and difficult task of selecting suitable individuals for comparison. This problem is discussed in detail in Section 2.2. At this stage, it suffices to mention the following considerations: The characteristics of the group of reference individuals needed depend on the intended use of the reference values. Are they to be used for evaluating the health of other individuals, for classifying patients' diseases, for estimating the prognosis of a patient with a specified stage of a disease, or for selecting a therapy? The physician may need several sets of reference values for a given analyte: health-related reference values, reference values for defined clinical groups, etc. Since a disease represents a deficiency of health, a disease may, in this context, be considered as a "state of health" (I5). To produce relevant sets of reference values, it is thus necessary to define the criteria by which appropriate reference individuals can be selected. *Inclusion criteria* define characteristics of individuals that might enter into the group of reference individuals. *Exclusion criteria* state prohibitive characteristics.

But what is a "value obtained by observation or measurement of a particular type of quantity"? Some find this expression difficult to understand. The problem is that the term *type of quantity,* which is based on an international recommendation (I4), has not prevailed in scientific literature. Dybkær *et al.* (D6) may be consulted for a formal definition. Here is a less formal one: "A *quantity* is one particular instance of the physical reality that can be measured so that a value is obtained. When considering a group of quantities of the same type, for instance, values that concern the concentration of sodium in serum, then this group is called a *type of quantity*" (G12, p. 349). Other common terms that are roughly synonymous, or at least used with a similar meaning, are *analyte, constituent,* and *component.* Because of their frequent usage in laboratory medicine, the latter terms are preferred in this review.

1.3.4. *Definition of Terms Related to Reference Individuals*

Two IFCC-recommended terms supplement that of reference individual (I5):

1. *Reference population.* A reference population consists of all possible reference individuals. *Note 1:* An alternative definition using set theory terms: A reference population is the *set* of all possible reference individuals. *Note 2:* The reference population usually has an unknown number of members and therefore is a hypothetical entity. *Note 3:* The reference "population" may consist of only one member, e.g., an individual may serve as a reference for himself or another individual.

2. *Reference sample group.* A reference sample group is an adequate number of reference individuals taken to represent the reference population. *Note 1:* Alternatively, a reference sample group is a *subset* of the reference population. *Note 2:* The reference individuals in the sample group should preferably be randomly drawn from the reference population.

Two alternative definitions are given for each of the two terms. Those using set theory terms (Note 1 in both cases above) seem to be most frequently used in practice.

The concepts of population and sample are borrowed from statistical theory. The basic idea is that the *population* is the set of all individuals that conform to a stated description (selection criteria; see Section 2.2.2), and the *sample* is the subset of the population actually studied. From a sample of observations one tries to estimate what might have been found if the complete, hypothetical population had been studied. Therefore, when we are going to produce reference values, we select a group of reference individuals (our reference sample group) according to our stated criteria and hope that this group resembles so closely the hypothetical population that our reference values are reliable.

The reference sample group can only be representative of the reference population if all possible reference individuals have the same chance of being included in the sample. A statistician would say that the reference individuals should be drawn from the population by a *random* process. The problem of random sampling is discussed in Section 2.2.1.

The population is often, but not always, *hypothetical* in the sense that one usually does not know its size *a priori*. For instance, one does not know without investigating how many healthy women living in the United States, aged 20–29 years, take at least 10 g of acetylsalicylic acid each month; and such an investigation may possibly never be done.

1.3.5. Definition of Terms Related to Reference Values

The IFCC defined three terms that build upon reference values: reference distribution, reference limit, and reference interval (I5).

1. *Reference distribution.* A reference distribution is the statistical distribution of reference values. *Note 1:* Hypotheses regarding the reference distribution obtained from a reference population may be *tested* using the reference distribution of the reference sample group and adequate statistical methods. *Note 2:* The parameters of the hypothetical reference distribution of the reference population may be *estimated* using the reference distribution of the reference sample group and adequate statistical methods.

It must be admitted that a better definition could have been made; the primary statement does not rise much above mere tautology. During the drafting of this recommendation—a process that lasted for years—much controversy concerning this seemingly harmless definition had to be resolved. Similar difficult problems were often encountered during the preparation of the six IFCC recommendations on reference values. To make an international recommendation is not a totally scientific activity. Sometimes one has to reconcile conflicting opinions and theories in a way that results in a rather vague text. In that sense, the production of international recommendations resembles diplomacy more than science!

The general idea that the definition of the reference distribution should communicate is rather trivial. Every reader of statistical texts knows the concept of a distribution. The problem with the definition above is that it is not explicit enough concerning different connotations of the word distribution. It is used with at least three scopes of information: (1) The topic of the primary statement of the definition is the *empirical sample distribution* of reference values or the distribution actually observed. (2) In addition, the two notes introduce the idea of a *hypothetical population distribution,* i.e., the distribution of values that might have been observed if access to data of the complete reference population was granted. (3) The statements in the two notes also imply the concept of *hypothetical distribution type*. Many of the statistical methods alluded to assume that the hypothetical population distribution is of the Gaussian type. The type is hypothetical in the sense that it is seldom known *a priori*. Therefore, hypotheses regarding the type of distribution underlying the empirical sample distribution may be the subject of statistical testing (see Note 1 of the definition). Problems related to the type of distribution are discussed in more detail in Sections 4.3.5 and 4.3.6.

2. *Reference limit*. A reference limit is derived from the reference distribution and is used for descriptive purposes. *Note 1:* It is a common practice to define a reference limit so that a stated fraction of the reference values would be less than or equal to the limit with a stated probability. *Note 2:* The reference limit is descriptive of the reference values and should be distinguished from various types of decision limits (discrimination values) used for interpretative purposes.

This definition is also rather vague. The general idea of a reference limit is conveyed in two ways. Note 1 refers to the common practice of equating a reference limit with a fractile (or percentile). Note 2 is some sort of negative definition: it tells what a reference limit is not. Various types of reference limits (and intervals) are defined and discussed in Section 4.3.1. The purpose of a reference limit, and of the related reference interval (see below), is data reduction, i.e., to reduce the amount of information contained in the set of reference values (perhaps hundreds or maybe even thousands of values) to something that is manageable in clinical practice: one or two descriptive cutoff values.

Descriptive is a key word here. You cannot make a clinical decision only on the basis of the position an observed value has in relation to a reference limit. But it is possible to say whether the observed value is typical or not of the reference values. Therefore, the reference limit should be clearly distinguished from *decision* or *discrimination limits* (S33, S34, S43), which is not always the case in published studies (e.g., G19). Take, for example, the concentration of glucose in blood collected while the individual is in a fasting state. The clinical decision value for the diagnosis of manifest diabetes mellitus may be quite different from the upper reference limit. The location of the clinical decision limit depends on several factors: (1) the differences in the location, dispersion, and shape of the distributions of glucose values of healthy individuals and of patients with defined stages of diabetes mellitus; (2) the prevalence of diabetes in the general population; and (3) the relative costs of misclassification (is it worse to declare a healthy individual diabetic than the opposite?).

3. *Reference interval.* A reference interval is the interval between, and including, two reference limits.

Thus the definition of the reference interval depends on the definition of reference limit (see Section 4.3.1). Here it may be pertinent to warn against some terms that do not conform with the recommendations of IFCC. Such terms are frequently found in scientific literature, even in journals that state that IFCC recommendations apply. Some use the term reference *values* with the meaning reference *limits* (or reference interval). Others refer to reference *range* where *interval* would be correct. In other contexts range signifies the difference between two values. If, for instance, the upper and lower reference limits of the concentration of albumin in serum are 50 and 35 g/liter respectively, the range, from 50 to 35, is 15 g/liter, while the interval is from 35 to 50 g/liter, both values included. The qualifier *normal* should also be avoided. The IFCC discourages the use of, for example, *normal values* (or related terms) with the signification health-related reference values. Neither should the composite term *normal reference values* (or limits or interval) be used.

1.3.6. *Relation between the Different Terms*

Here is a more complete definition of observed value (see Section 1.3.3):

Observed value. An observed value is a value of a particular type of quantity, obtained by observation or measurement and produced to make a medical decision. It can be compared with reference values, reference distributions, reference limits or reference intervals.

The observed value is thus the result obtained by laboratory analysis of a specimen collected from a patient. Some call such values test values, but the word *test* in this term is ambiguous, since we also use it in other terms, such as laboratory test and statistical test. Some may be misled to believe that the comparison of a test value with reference values is a statistical test in the strict sense (which is usually not the case). Alternatives to the term *observed value* are *clinical value* (or result) or *patient's value* (or result). The following scheme (Fig. 1, modified from Ref. 15) summarizes the relation between the various terms as defined by the IFCC.

1.3.7. *Classes of Reference Values*

It has already been mentioned that the concept of reference values does not automatically imply health. It is therefore a recommendable practice to add qualifying terms relating to health-associated parameters, the name of a disease, or the conditions of the specimen collection (fasting, supine, exercise). Examples of such usage would be health-related reference values, diabetic reference limit, or hospital population reference distribution.

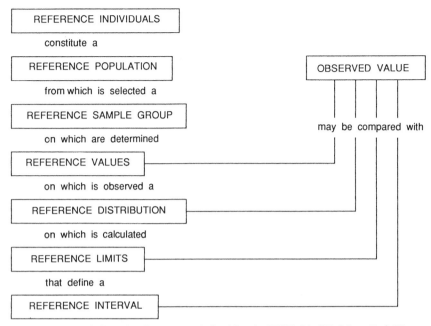

FIG. 1. Relationship of the terms defined by the IFCC. Modified from Ref. I5.

We may further distinguish between group-based (population-based) reference values and subject-based (personal, individual) reference values (G15, H14, I5, W5). Note 3 in the definition of reference population (see Section 1.3.4) alludes to subject-based reference values (see Section 4.2.1).

The reference values may also be univariate or multivariate (A3, B9, I5, W9). If we, for example, produce, treat, and use separately reference values of sodium, potassium, and chloride in serum, we have three sets of univariate reference values. In contrast, we have one set of multivariate reference values if we treat these three sets of reference values in combination (see Section 5.1). Furthermore, reference values may be time unspecified or time specified (see Section 5.2), depending on their relation to biological rhythms or trends (H3, I5).

1.3.8. *Concluding Remarks*

There is a common misunderstanding that the production of reference values is extremely difficult. For instance, it has been stated that reference values "very rarely can be properly used . . . because . . . all statistical requirements [must] be met and all physiological factors [have to be] taken into account—a rarely attainable perfection in this imperfect world" (cited in G16). Let our reply to the editor of *Clinical Chemistry* (G16) be the conclusion to this section:

This is a misunderstanding. Essentially what is required is that a number of reasonable pieces of information are provided . . .:

The characteristics of the population, especially how the subjects were selected and how their health was assessed. It is definitely not required that these criteria be very rigid, only that they be stated. Thus it would be perfectly sufficient to say that the values were derived from the first 100 pretty ladies that Dr. King saw on Broadway, and that his criterion for health was that he thought that they were pretty, and that they were able to walk, and that their age was 'nubile.'

The physiological state of the subjects and the preparation for and the execution of the specimen collection should be stated. It would be compatible with the reference value philosophy to state that the subjects were not asked whether they had eaten, and that the venipunctures were performed on the street corners while the subjects were standing, with the use of a 30-min tourniquet, unsterilized needles, and collection of blood into empty beer cans.

The way the specimen was treated and the analytical procedure used should be stated. Again, it would be compatible with the philosophy to state that the cans were stored on the roof of Empire State Building from July 15 to July 22, 1981, and finally assayed for oxygen according to Lavoisier (1790). (Of course we suspect that the results of this Miss Manhattan study would not be printed by *Clinical Chemistry*).

1.4. DIFFERENT KINDS OF REFERENCE GROUPS

1.4.1. *Introduction*

By extending the concept of the controls to the clinical situation, one can conceive of different sources of reference values. An observed value may be related to reference values from different populations and previous values from the individual himself (Fig. 2; Table 1). The reference individuals are selected depending on the intended use of the reference values and their availability.

FIG. 2. An observed value may be related to reference values from different populations and to previous values from the same individual. From Ref. G8 with permission of John Wiley and Sons.

TABLE 1
TYPES OF REFERENCE POPULATIONS[a]

Nonhospitalized persons
"Healthy"
    Subjectively healthy, unscreened
    Subjectively healthy, health screened
        Superficially health screened
        Thoroughly health screened
    Persons who survived for a specified time
Diseased
    Unselected
    With specified diagnoses
    Unselected, visiting a medical unit

Hospitalized persons
"Healthy"
    Subjectively healthy, unscreened
    Subjectively healthy, health screened
        Superficially health screened
        Thoroughly health screened
Diseased
    Unselected
    With specified diagnoses
    Deemed to be comparable to healthy persons

[a] Modified from Ref. G5 (nonexhaustive list).

Until recently, most reference populations consisted of ambulant, subjectively healthy, usually young and working subjects such as medical students, laboratory staff, blood donors, or soldiers. However, in the clinical situation, many other populations would represent more relevant controls, e.g., healthy hospitalized kidney donors, persons hospitalized for cosmetic or orthopedic operations, or subjects suspected of suffering from a disease (such as myocardial infarction) but subsequently "acquitted." More interest should be paid to the utilization of such reference populations.

In planning the collection of reference values it is good to bear in mind the purposes for which laboratory investigations are ordered: to monitor the patient (follow-up), to make a diagnosis, to select therapy, to evaluate prognosis, for health screening, for forensic purposes, and for research purposes (including testing of drugs). In several of these situations it is necessary to know the kinds of values that occur in healthy subjects. In treating and monitoring the patient the doctor wishes to know whether the values approach or draw away from values typical of health. The use of laboratory tests for diagnosis is based on the observation that the values are dissimilar in healthy and diseased persons, and unalike in different diseases. The same applies to health screening. However, some healthy persons run a higher risk of being afflicted with conditions such as cardiovascular disease and the risk correlates with the results of certain laboratory tests. Low-risk or ideal values may therefore differ from the common values of healthy persons.

Since values from healthy and diseased subjects are of so great importance to the practice of clinical laboratory medicine, it is necessary to subject these concepts to logical analysis. A brief outline is presented here. (For a thorough discussion the reader is referred to Refs. G7 and G9.)

1.4.2. *The Concept of Health*

Medicine has a clear-cut goal, to maintain and restore health. Health is a *goal-oriented* concept. Laboratory values from healthy persons therefore represent goal or target values.

What this goal really is represents a complicated question. The constitution of the World Health Organization gives the following definition (W13): "Health is a state of complete physical, mental and social well-being and not merely the absence of disease or infirmity." This definition is widely recognized as being unrealistic. The fault of the WHO definition is that health is regarded as something absolute. Experience tells us that health is a *relative* concept. What may be considered compatible with health in a developing country may be regarded unhealthy in the affluent parts of the world. Carrying parasites or lacking teeth may be cited as examples. A person may be ill in one respect and well in another; a person whose kidney has been removed may be healthy from the social point of view, but obviously cannot be considered a healthy control for kidney tests. Myopic and color-blind subjects are rarely considered ill. Still their diseases disqualify them from certain jobs.

It is never possible to establish the presence of health positively, only by excluding pathology. Health is thus a *privative* concept. Very thorough medical investigations may be so invasive that they are risky. The subjective feeling of health is not a reliable criterion of health because some states or diseases are associated with increased and erroneous feeling of health, such as mania, frontal lobe brain tumors, and euphoria. Also, some individuals try to conceal their illness, e.g., in order to obtain life insurance. Everybody is probably a heterozygote with respect to a metabolic disorder.

The following definition of health (G7, G9) summarizes the relative, privative, and goal-oriented aspects of health: Health is characterized by a minimum of subjective feelings and objective signs of disease, assessed in relation to the social situation of the subject and the purpose of the medical activity, and it is in the absolute sense an unattainable ideal state.

1.4.3. *Disease*

To define the concept of disease and to categorize subjects into disease groups entails even more difficulties. The names of the diseases and the diagnoses have evolved over a long period of time and have not been created with the same logic. Some diagnoses are based on the pathogenic agent (tuberculosis), some on pathological anatomy (carcinoma and myocarditis), and some on symptoms (stenocardia); some are labels for conditions which may not be homogeneous entities (schizophrenia and rheumatoid arthritis). The categorization of diseases is in a state of continuous flux. Recognized entities tend to break down into smaller subentities; laboratory investigations have greatly contributed to this development. Peculiar names such as pseudo pseudohypoparathyroidism have developed as the result of such subclassification, and several kinds of diabetes mellitus exist today.

A few centuries ago diseases were considered to be comparable to animal and plant species, and attempts were made to produce a taxonomy of diseases; one was made by Linnaeus. Today we do not consider a disease to be an individual entity of its own but rather an interaction between one or many pathogenetic causes and promoting factors and a biological organism. Because individuals have a different genetic and environmental background, different individuals never have exactly the same disease, and thus every diseased individual may be considered to have a unique disease of his own. However, it is necessary to group together relatively similar conditions in different individuals, otherwise it would be impossible to collect and transmit medical experience.

In addition to the individual variability of the diseases, the course of a disease varies over time. Incipient disease is difficult to distinguish from health. It is thus obvious that the laboratory findings in diseases must by necessity overlap with values obtained from healthy subjects. This is one of the main deficiencies of systems for computer-assisted diagnosis. They tend to focus on (often undefined) stages in the evolution of diseases.

2. Reference Individuals and Specimen Collection

Before discussion of this topic it is necessary to discuss some factors that must be taken into account when collecting reference values.

2.1. PREANALYTICAL FACTORS

The rate of production and breakdown of substances in the body and their concentrations in the body fluids are influenced by a multitude of factors, such as intake of meals, posture, and time of the day. The concentrations of substances in the specimens are influenced by the gauge of needles used for venipuncture, whether open or vacuum blood collection tubes are utilized, the additives contained in the specimen collection vessels, etc. Following its collection, the specimen may be treated in various ways, e.g., separation of clot from plasma, centrifugation, and freezing, thawing, and division of the specimen into samples subsequently to be analyzed. The factors influencing the results of a laboratory test between the ordering of the test and the analysis proper are called *preanalytical factors*.

It is convenient to divide the preanalytical factors into *in vivo* or biological factors and *in vitro* factors. The *in vivo* factors act in the subject prior to or during specimen collection; the *in vitro* factors cause variations in the specimen following its collection. Obviously the relevant reference individuals and their specimens should be treated in a manner as similar as possible to the treatment of corresponding patients and their specimens.

A large number of factors cause variation in the composition of body fluids and tissues (Table 2). The quantitative importance of these factors varies, and different components react in different ways (S18, S19, S35). Fortunately, related substances frequently react in approximately the same way, e.g., reaction of macromolecules in plasma to changes in posture and use of tourniquet. In the following sections a few of the more important factors acting *in vivo* on commonly assayed components are briefly dealt with (the *in vitro* factors are discussed in Section 2.3.2).

2.1.1. Fasting and Intake of Meals

Well-known effects of food intake are a rise in blood glucose and insulin concentration. The increase in plasma lipids is evidenced by the appear-

TABLE 2
FACTORS CAUSING BIOLOGICAL VARIATION[a]

Age	Genetic factors
Activity	Pharmacologically active agents
Altitude	Physical fitness
Blood type	Posture
Body mass and surface	Profession
Chronobiological rhythms	Puberty, menstruation, pregnancy, menopause
Diet (type, amount)	Relation to meals
Disease	Sex
Environment (humidity, temperature)	Socioeconomic class
Ethnic group and race	Specimen collection (site, technique)
Exercise	Stress
Exposure to toxic agents, radiation, etc.	Trauma

[a]Nonexhaustive list.

ance of turbid plasma; the turbidity may interfere with analytical procedures. Patients are therefore frequently asked to abstain from food intake before laboratory tests. Digestive enzymes such as serum amylase also react to food ingestion. The nature of the food eaten is often reflected in the composition of plasma and urine. These short-term effects should be distinguished from long-term changes induced by fasting or eating a special diet. To mention a few examples, fasting increases the concentration of bilirubin (S36) and induces acidosis. A purely vegetarian diet may cause low concentrations of serum cobalamin (G14); eating large amounts of oranges or carrots may produce excessively high serum carotene levels; the lipid content of a diet is reflected in the concentrations of plasma cholesterol and lipoproteins.

2.1.2. Posture

Posture is known to influence the concentration of macromolecular and cellular components of blood (F2, S37; see Fig. 3). The serum total protein concentration changes on the average about 8% (in some individuals up to 20%) when going from the supine position to standing (K6); walking

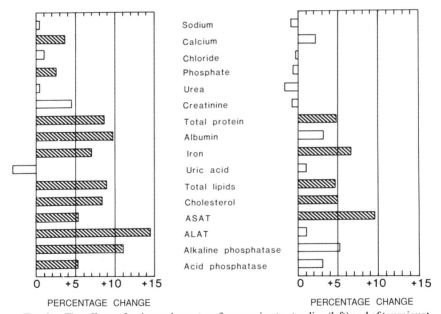

FIG. 3. The effects of a change in posture from supine to standing (left) and of tourniquet (right). The specimens were taken following a 15-min stay in the new posture and following a 3-min versus a 1-min application of tourniquet. ASAT, aspartate aminotransferase; ALAT, alanine aminotransferase. Hatched bars, $p < 0.05$. Modified from Ref. S35 with permission of the authors.

around increases this difference (L3). These changes take place within 15 min. Similar changes have been reported for a large number of plasma proteins and the small molecular substances bound to them. Among the blood cells, at least the erythrocytes and platelets react in this way (L8). In clinical practice the blood hemoglobin concentration has been used to judge whether a patient is fit to be operated upon. Changes in the posture of patients have apparently often caused hemoglobin values to transgress such decision limits and have probably caused unnecessary transfusions and refusals to operate.

2.1.3. Tourniquet

The effects of tourniquet application resemble those of assuming a more upright posture, i.e., the macromolecules and macromolecule-bound substances in plasma increase (K6, S37), see Fig. 3.

2.1.4. Site of Specimen Collection

The composition of the blood varies depending on whether it is obtained by puncturing a vein or an artery or by pricking the skin. Skin puncture blood resembles, but is not identical to, arterial blood. It is more oxygenated and has a higher glucose concentration than does venous blood and its acid–base values resemble those of arterial blood (K6). Skin puncture blood from the ear lobe and finger may differ slightly in composition (W3). The effects of posture also may be slightly different in blood from different sites (E. A. Leppänen, unpublished).

2.1.5. Physical Exercise

Strenuous exercise, such as marching long distances carrying a heavy backpack, marathon runs, and long-distance skiing, may have strong effects, especially on untrained individuals, and may even induce pathological conditions such as march hemoglobinuria. It is well known that serum creatine kinase increases following moderately heavy exercise (S36). However, moderate exercise taken by laboratory technicians on a bicycle ergometer (120–160 W for 15 min) was found to have negligible effects on aspartate and alanine aminotransferases, creatine kinase, lactate dehydrogenase, potassium, and sodium in serum. Very short-term rises were observed in many of the components mentioned, but after 15 min all changes except that of the serum potassium concentration had disappeared (L7).

2.1.6. Chronobiological Rhythms

A large number of components have been shown to exhibit chronobiological rhythms: circadian, weekly, monthly (including menstrual), seasonal, etc. (H3, S38, S39, W11). The strongest circadian variations tend to be exhibited by some hormones such as cortisol and renin. Circadian rhythms have been found in the particle concentration of blood cells, in the serum concentrations of iron, bilirubin, and creatinine, and in the urinary excretion of potassium, corticoids and catecholamines. The circadian rhythm is only partly the result of influences such as light and darkness, sleeping and being awake, change in posture, work and rest, and food intake. Such factors may synchronize hypothetical internal clocks. Circadian and weekly rhythms may be related to work schedules. Seasonal fluctuations have been found for thyroid hormones, immunoglobulin, and cholesterol in serum.

2.1.7. *Activity of Sexual Organs*

Pregnancy has not only strong and specific effects on the production and concentration of several hormones, but also on many other components not directly related to the reproductive function. Especially strong effects have been found on the concentrations of some serum proteins. For instance, lipoproteins and ceruloplasmin increase. Similar changes are induced by the intake of oral contraceptives (see below; L4).

The menstrual cycle is accompanied by, or rather caused by, fluctuations in the secretion of estrogens and progestins. Some indirectly related components such as serum cholesterol also fluctuate during the menstrual cycle (A1). Menopause is accompanied by changes such as a rise in serum alkaline phosphatase.

2.1.8. *Pharmacological Agents*

Drugs and other pharmacological agents have widely varying and frequently strong and specific influences on the constituents of blood and urine (I1). Thorough reviews have been published on this topic (T1, Y2). Among the drugs, the oral contraceptives deserve special mention. A combined estrogen–progestin pill increased α_2-macroglobulin, lipoproteins, transferrin, plasminogen, α_1-antitrypsin, ceruloplasmin, and thyroxine-binding globulin concentrations in serum (L4). Caffeine and ethanol contained in common drinks are also pharmacological agents. They are diuretics, and alcohol intake has been reported to increase serum aspartate and alanine aminotransferases (S35) as well as to induce hypoglycemia (M1). However, alcohol intake at low ("social") levels seems to have negligible effect (see Section 2.2.2.2). Smoking also has a number of pharmacological effects.

2.1.9. *Altitude*

The effect of altitude on hemoglobin concentration is well known. Low air pressure, e.g., during mounting climbing, induces diuresis, and many other effects of mountaineering have been reported (B12, W2).

2.1.10. Age, Sex, Race, Geographical Location, and Socioeconomic Conditions

Compared to the factors enumerated above, these influences are fixed and impossible or difficult to change by exogenous action. Among the best known sex differences are those in the blood hemoglobin and erythrocyte concentrations, but minor differences are noticeable in a multitude of components, for example, serum urate (M7). Great differences, of course, occur in the serum concentrations of sex hormones and sex-specific enzymes (prostatic acid phosphatase).

The concentrations of many components change with age, and relatively great differences are observed for many components during early infancy and puberty and in women after menopause (M10). Genetic differences are naturally observed in genetic markers such as blood group substances and HLA antigens, but there are also differences in genetically less directly controlled components. For instance, American blacks tend to have higher immunoglobulin concentrations in serum than does the white population, and in the United States immigrant Oriental women have been found to have higher plasma estradiol levels than are found in their Caucasian counterparts; the dietary intake of fat and fiber could explain this difference (G3). Average concentrations of serum cholesterol have been observed to be different in different socioeconomic groups, e.g., in South Africa. Dietary habits play a role here, too. Similar factors explain national differences, which may disappear following migration (F1).

2.2. SELECTION OF REFERENCE INDIVIDUALS

Traditionally, values from healthy individuals have beeen used (I6). However, as described in Section 1.4.2, what is meant by health depends on the purpose of the study, and therefore different criteria of health are applied in selecting healthy reference individuals. Finally, values from individuals with defined diseases are needed. According to the relativistic concept of health, diseased individuals are also healthy (I5); their health is decreased but not absent. In preventive medicine one is interested in finding individuals at low risk, whereas in clinical work one is usually content with the aim of restoring the health of the patient to the level he or she enjoyed prior to becoming ill or to a level commonly exhibited by matching individuals.

When selecting the reference individuals one should carefully consider the purpose for which the reference values are to be used. For instance, when it is planned to collect reference values for a test revealing myocardial infarction, the ideal negative controls may be other patients with similar complaints attending the same medical service unit but who are found not to suffer from myocardial infarction. In fact, healthy subjects of the same age and sex category are rather uninteresting controls!

Health-related reference values can be produced longitudinally by excluding values from individuals who died or fell seriously ill within a specified time, say 10 yr. Production of reference values in this way may be especially suited for the higher age groups, where individuals satisfying even mild criteria of health are difficult to find (B5), and for premature babies, for whom neither babies born at term nor unborn fetuses provide suitable controls.

The definition of reference individual (see Section 1.3.3) refers to "defined criteria" to be used in the selection process. The criteria may be of two types: either they include or they exclude individuals from the reference population. An *exclusion criterion* can always be rephrased as an *inclusion criterion*, e.g., "exclude men" equals "include women." The definition of the reference population (see Section 1.3.4) usually consists of a set of inclusion/exclusion criteria. For instance, include males aged 21–40 yr and exclude nonhealthy men and those who take drugs.

In some cases, it is desirable to subclassify the reference values into homogeneous subsets (see Section 4.2.1), e.g., according to age or sex. The distinctions between partitioning criteria and selection criteria are often dependent on the purpose of the study. For instance, both age and sex are selection criteria if one needs a group of male controls aged 21–40 yr; otherwise age and sex are common criteria for partitioning.

2.2.1. Selection Strategies

The selection of reference individuals may be characterized by three pairs of qualifiers: direct or indirect, *a priori* or *a posteriori*, and random or nonrandom. The IFCC accepts only *direct* selection of reference individuals (I5). One selects directly when inclusion/exclusion criteria are applied to individuals of a parent population. For various reasons, mostly related to the costs and labor involved, an *indirect* selection strategy has been suggested (B1, K5, M4, M5, S44). The basis for this strategy is the laboratory results and not the individuals as such. Usually, mathematical or graphical methods are used to identify the "normal" part of the distribution of values obtained by routine analysis of clinical specimens. The indirect method is, however, unreliable and cannot be recommended because the estimated reference limits depend heavily on the particular mathematical method used for the "selection." Even if the same mathematical procedure were used, the resulting estimates would probably vary across hospitals and for the same hospital at different times.

By the *a priori* strategy one selects individuals directly from the parent population by applying the inclusion/exclusion criteria (B5). Because of the amount of work involved, this procedure is best suited for smaller studies. The *a posteriori* method is based on the availability of a large collection of data on medically examined individuals, including their laboratory results (S19). Centers for health screening or preventive medicine may provide such data. The selection criteria are applied to the individuals in the data base, and the laboratory results of those accepted are the required reference values. The *a priori* procedure is the best one, but the *a posteriori* method may provide reliable reference values when correctly used.

Ideally, the reference individuals should be a *random* sample of all individuals fulfilling the selection criteria, i.e., all individuals in the reference population should have the same chance of being selected. This is difficult to achieve in practice. A strict random sampling scheme would imply the examination of and application of inclusion/exclusion criteria to the entire parent population (thousands or millions of persons), and then the random selection, e.g., by raffling, of a subset of individuals from among those accepted. We do not obtain a random sample, in the strict sense, if we start by selecting individuals randomly from the entire population and then apply inclusion/exclusion criteria to select reference individuals. Still this method may be the best approximation to a random sampling scheme we can achieve in practice. Often the situation is even worse: reference values are produced from a sample which is definitely nonrandom, for instance, a group of blood donors or those working in the laboratory.

A few guidelines are recommended: (1) Try to design a sampling scheme that is as random as possible, all practical problems taken into account. (2) The definition and the sampling process should be made known to the user. (3) Interpret the results with due caution, remembering the possible bias introduced by the nonrandomness of the sample.

Among the other strategies mentioned above, the indirect method produces a definitely nonrandom sample, but the *a posteriori* procedure is also prone to nonrandomness.

2.2.2. Selection Criteria

The following sections give examples of some commonly used selection criteria. Some or all of them should be applied, depending on the intended use of the reference values and the analytes assayed.

2.2.2.1. *Disease.* Individuals suffering from diseases and disorders may be excluded by anamnestic data (e.g., data obtained by a questionnaire), clinical examination, and laboratory investigation. The extensive list of possible exclusion criteria prepared by the Scandinavian Committee of Reference Values (A4) may be used as a checklist. However, when these criteria were applied to a general population, a very large proportion of the individuals in the higher age groups were excluded (B5).

2.2.2.2. *Pharmacological Agents.* Individuals taking drugs are usually excluded. Individuals taking oral contraceptives, alcohol, tobacco, and caffeine-containing drinks are also typically excluded. It is usually recommended to exclude individuals who are heavy drinkers and those who have ingested alcoholic beverages later than 24 h prior to blood collection. However, a recent study indicated that social intake of alcohol (0.75 g/kg) on three consecutive evenings had little effect on a number of commonly assayed analytes (L7).

2.2.2.3. *Modified Physiological States.* Pregnant women and persons who recently have ingested food or have exercised heavily are usually excluded. A recent study indicates that moderate exercise has little effect on some commonly assayed serum enzymes (L7). Stress and psychological disorders may also have undesired hormonal and metabolic effects.

2.2.3. Partitioning

The reference sample group may be partitioned in different ways, depending on the analyte and the intended use of the reference values. However, certain criteria must be fulfilled before subgrouping can be considered useful (see Section 4.2.1).

2.2.3.1. *Age and Sex.* Age and sex are the most commonly used criteria for subclassification of the reference values (A7, R3, W1). The age intervals should be chosen in accordance with known age-dependent variation of the component under investigation. Periods where rapid changes are observed, such as early infancy, puberty and menopause, may demand narrower intervals. Bone age, height, and body mass are mostly better indicators for categorizing children than temporal age. One should also attempt to distinguish variations due to deviations from the ideal body mass and differences due to age and sex. This has been shown to be valuable with serum creatinine.

2.2.3.2. *Genetic, Socioeconomic, and Environmental Factors.* For some components it may be of value to subclassify the subjects according to ethnic origin, geographical location, pigmentation, or genetic markers such as blood groups, histocompatibility antigens, and phenotypes of plasma proteins and tissue enzymes.

2.2.3.3. *Biological Criteria.* As described in Section 2.1.2, posture (standing, sitting, or supine) may have strong effects, apparently mediated by hemodynamic factors, kidney perfusion, and hormonal balance. It may therefore be desirable to select separate reference sample groups for ambulant and hospitalized populations in cases where posture has clinically significant influence on the values. In some instances chronobiological data should be used for partitioning (e.g., serum cortisol).

The use of the concept of a *reference state* has been suggested (S19) to facilitate comparison of populations and to enable the transfer of reference data. The idea is to minimize the influence of biological variations. For many components the variation appears to be least in young adults. An individual is considered to be in the reference state if the person is 20–30 yr of age, has an ideal body mass, has fasted for 10 h, does not take any drugs, consumes less than 45 g of alcohol, smokes less than 12 cigarettes per day, and has no apparent illness.

2.3. Preparation of Individuals; Collection of Specimens

It is common practice to attempt to control a number of preanalytical factors when reference values are produced (I7). Great preanalytical variation in reference values diminishes the chance to detect clinically important deviating observed values because of the resulting low signal-to-noise ratio. Failure to control these variations is also incompatible with the basic requirement of the reference value philosophy, that the observed values and reference values (i.e., the experiment and the negative controls) be produced under as similar conditions as possible, which obviously includes the collection of the specimen, its subsequent treatment, and analysis.

Such considerations have led to the standardization of procedures for the preparation of the subject prior to and during specimen collection and for the specimen collection procedure. Most of the current recommendations deal with the collection of blood (A4, A5, H18, M6, N1, N2), but work has been initiated to produce recommendations on the collection of urine.

2.3.1. *Biological Factors*

These factors may be divided into two groups, those that can and those that cannot be influenced (G6). To the latter belong sex, age, race, and similar factors, which are commonly used to partition the reference individuals into groups.

Important factors that one can influence are (1) meals and prolonged fasting, (2) intake of pharmacologically active substances including drugs, hormones, ethanol, caffeine, and tobacco, (3) hemodynamic factors including posture, and (4) cell and tissue damage induced by physical work, muscle massage, venipuncture, etc.

2.3.2. *Methodological Factors*

These factors are related to the specimen collection, the *in vitro* effects of handling of the specimen, and the analysis. During specimen collection there can be interference by blood collection techniques (use of tourniquet, vacuum tubes, etc.), equipment (needle, receptacle, etc.), additives (anticoagulants, aids for separation of serum from plasma, etc.), and the order of filling the tubes. The interference may be caused by cell damage, with release of intracellular components such as potassium (L6), by introduction of contaminants such as heavy metals, or by inhibition of active substances by EDTA.

2.3.3. Handling of Specimen

The conditions prevailing during the storage and transport of the specimen may change the concentration of many substances. The nature of the container and the container temperature may influence the composition of whole blood and serum, etc. The treatment of the specimen prior to analysis may also interfere, e.g., the time and temperature of the storage before analysis, the speed of centrifugation, the techniques used for freezing, thawing, and mixing the specimen, etc. [The difference between the terms specimen and sample should be noted. The *specimen* is collected from the individual, but *sample* is the part of the specimen used for analysis (D6, I2).]

2.3.4. Standardization

Only factors having a clinically meaningful influence are worth controlling. In other words, the magnitude of the effect should be such that there is reason to believe that its elimination influences clinical decision making. However, different components react in different ways, and ideally there should be a separate protocol of standardization for each group of related analytes. For instance, control of body posture is important in studies on particulate elements of blood and the macromolecular substances in plasma, but may be completely unnecessary for many ions and small molecular analytes. On the other hand, it would be highly impractical to devise special protocols for every component to be assayed, because it is common practice to use the same specimen for the assay of several substances. The standard schemes published are therefore based on the philosophy that they are to be used when clinical specimens are taken for the assay of common analytes.

It is relatively easy to arrange standard conditions during a project to produce reference values, but in the clinical situation when specimens are taken from patients it may be difficult. If the specimen has been collected under nonstandardized conditions, the physician needs additional information to be able to interpret the findings in relation to reference values. It is therefore good practice to write relevant information on the forms used to request laboratory tests, e.g., "the patient has not been fasting."

Some components behave so differently during commonly occurring conditions that one should consider producing special sets of reference values for typical conditions, e.g., for ambulant and hospitalized individuals or for supine and sitting subjects. Observed values should also be adjusted to compensate for nonstandard conditions before the interpretation. Quantitative data on the effects of factors such as meals, drug intake, exercise, and posture should be made available to the clinicians to enable such correction. The textbook of Siest *et al.* (S19) is a good source of information, but more studies on such effects are needed, e.g., on the combined effect of two or more sources of variation. Are they additive or multiplicative? Or do they counteract?

Alternatively, reference values could be collected for some typical situations. For instance, traffic accidents tend to occur at night and the victims have often eaten, drunk alcohol, danced, etc. Some such situations may be so common that it would be worthwhile collecting special reference values for them. [This has been called the empirical approach, in contrast to the predictive or corrective approach just described (G6).] At least one such pilot study has been made (G18). Interestingly, it was found that many of the differences in the composition of blood with respect to standard conditions were small and clinically negligible. The probable explanation was the stabilizing effect of a 15-min period of sitting before specimen collection. Recent experience indicate that a 15-min rest before specimen collection is possible to carry out in clinical routine and reduces significantly the disturbing effects of biological variation (G11).

2.4. Recommended Procedures

2.4.1. *General*

Detailed schemes have been published concerning the preparation of individuals before blood collection and the procedures for the collection of the specimen and its subsequent handling. A few examples of such procedures are given below. As emphasized many times above and by the IFCC recommendation (I7), throughout one reference value project and in collecting specimens from patients the preparation of the individuals and the collection and subsequent handling of the specimens should be as similar as possible. Also, when reference values are presented, these standard procedures should be described and made available to the user. Unfortunately, violations of this rule are common.

As discussed in Section 2.3.4, it may sometimes be difficult to follow recommended procedures, e.g., in emergency situations. Therefore, observed values should be interpreted by taking into account the likely effects of deviations from the recommended procedure. When routine specimens are taken from patients, notes of such details should be made and reported to the physician in order to facilitate the interpretation. The relevant IFCC recommendation (I7) enumerates a number of factors which should be annotated and reported when collecting reference values from reference individuals (Table 3).

TABLE 3
CHECKLIST: COLLECTION OF REFERENCE VALUES[a]

Preparation of the individual
Prior diet
    Type (habitual food or defined diet)
    Amount (habitual or restricted, augmented, or supplemented)
    Duration (previous day, previous week, etc.)
Fasting or non-fasting
    Duration (hours, overnight, etc.)
    Extent (food only, both food and drink, only specified foods and drinks, etc.)
Abstinence
    Pharmacologically active substances (alcohol, caffeine-containing drinks, tobacco, dispensable drugs, etc.)
    Duration [hours, previous day(s), previous week(s)]
Drug regimen
    Types of drugs
    Quantity and timing
    Duration
Synchronization in relation to biological rhythm(s)
    Duration (days)
    Sleeping (length and timing)
    Meals (timing)
Physical activity
    Long term (physical training and work-related activity)
    Short term (e.g., walking immediately prior to collection)
Rest period prior to collection
    Posture (sitting, supine)
    Duration (minutes, hours)
Stress
    Emotional stress (including infants crying intensely)
    Fainting during specimen collection
    Noise

Specimen collection
Environmental conditions during collection
    Temperature
    Relative humidity
    Altitude
Time
    Time of day
    Relative (to sleeping, to meals, or to other exogenous factors)
    Season
Body posture
    General (upright, sitting, supine)
    Relative position of collection site (e.g., arm 45° below the horizontal)
Specimen type
    Arterial blood
    Venous blood
    Skin puncture blood
    Other types

(continued)

TABLE 3 (*Continued*)

Collection site
    Dependent on specimen type
Site preparation
    Disinfection
Promoting of blood flow
    Warming or local drugs
    Tourniquet (duration and pressure applied)
    Muscle work of the hand (pumping)
Equipment
    Puncture device (type, dimensions, shape, and material)
    Receptacle (capillary tube or larger tube made of glass, plastic, or other type of material)
    Vacuum tube or not
    Additives (anticoagulant, preservative, silicon, or other type of separation promotor)
Techique
    Puncture
    Collection (free flow or suction)
What to do in case of failure
    Alternative collection sites
    Duration of rest period between attempts

Specimen handling
Transport
    Container, preservatives
    Temperature
    Duration
Clotting
    Time
    Temperature
    Promoting agent
Separation of serum of plasma and particulate elements
    Centrifugation force and time
    Temperature
Storage
    Container
    Preservative, if any
    Temperature (specified in degrees)
    Duration
Preparation for analysis
    Thawing
    Mixing

a This checklist includes factors that should be considered for inclusion and data that should be noted and reported during collection of reference value data (I7).

2.4.2. Specific Recommendations

As examples, summaries of a few recommendations are presented here.

2.4.2.1 Venous Blood from Adults. The Scandinavian Committee on Reference Values has published a recommendation "Standardization of Adult Individuals and the Blood Specimen Collection Procedure for Producing Reference Values." The recommendation, published in 1975 (A4), is now fairly old and is therefore presently under revision. However, its general outlines remain unchanged.

The recommendation envisages the collection of two sets of reference values, one from ambulant subjects, another from supine subjects, corresponding to outpatients and hospitalized patients, respectively. The standardization of ambulant subjects is illustrated in Fig. 4. A very concise summary of the procedure is as follows [NCCLS (N1) has published a less exacting recommendation on venipuncture].

All subjects, the day before specimen collection: Ordinary food intake, not more alcohol than one bottle of beer, after 2200 h no food, no smoking, not more liquid than one glass of water. *Ambulant subjects:* Rise from bed 1–3 h before specimen collection, note time. Not more physical exercise than a short walk. Venipuncture at 0800–1000 h following a 15-min rest in an arm chair. During puncture the arm is kept 45° below the horizontal plane. Use no tourniquet, but the vein may be pressed proximally with a finger. *Afternoon procedure* (controls for afternoon outpatients): Standard morning breakfast ($\sim$1300 kJ = 310 kcal), e.g., two open sandwiches, two cups of tea, coffee, or milk. Venipuncture 1300–1500 h following 15 min of sitting as described above. *Supine subjects:* Remain in bed in the morning (may visit toilet). Specimen collection, 0700–0900 h, with arm in the horizontal plane, no tourniquet.

Wash puncture site mildly with 70% ethanol. Puncture cubital vein with a 20-gauge 1.5 needle using vacuum tube system; note time of puncture. In case of failure, puncture other arm following 15-min rest.

Processing and storage of specimens: To produce serum, keep tube at room temperature for 60–90 min, cover with paraffin film, centrifuge 10 min at 800 g. Loosen clot with plastic rod. To produce plasma, centrifuge at once. For storage, tubes should be at least one–third full; cover with paraffin film. If optimum stability conditions are unknown, analyze specimen within 3 h or store at 4–7°C; store for more than 24 h at $-$20°C and for more than 3 days at $-$70°C or lower. Thaw specimens at room temperature in water bath, invert tubes 20 times to ensure homogeneity; do not refreeze.

2.4.2.2. Skin Puncture Blood from Children. An inquiry demonstrated that widely different procedures were used to take skin puncture blood; the Scandinavian Committee on Reference Values thus recommended the

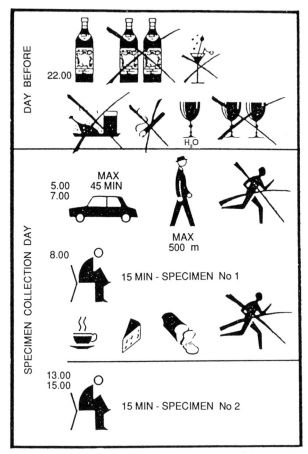

FIG. 4. Standardization of the ambulant subject before specimen collection by venipuncture. From Ref. G10 with permission of Becton Dickinson Vacutainer Systems Europe.

following procedure for children aged 0–15 yr (A5): For neonates note body weight and gestational age. For children approaching 15 yr use posture recommended for adults. Attempt puncture before morning meal; record time of puncture and preceding meal. Keep child calm; if child is not calm, make a note. Children above 1 yr should sit; those above 12 years should sit for 15 min. Select an unused, nonedematous puncture site. Warm puncture site with wet, warm (39°C) cloth (measure temperature!) if the skin is cold or if the child is 0–7 days old, and for hematological and blood gas analyses. Make a note if disinfectant or ointment is used. Use standard lancet (2.4-mm tip for children younger than 3 months).

Figure 5 depicts puncture sites recommended for infants aged 0–3

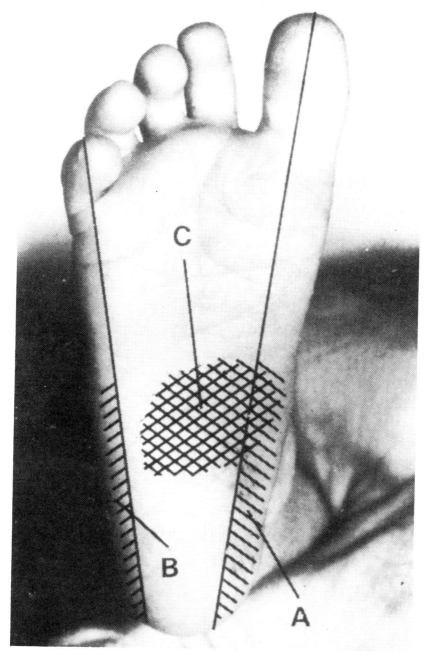

FIG. 5. Recommended puncture sites (hatched areas A, B, and C) of the foot of an infant. From Ref. A5 with permission of *Scand. J. Clin. Lab. Invest.* and Department of Clinical Chemistry, Central Hospital, Eskilstuna, Sweden.

months and for infants aged 3–6 months with birth weights below 2500 g. For bigger children, the specimen is collected from the fourth finger or the ear lobe. Blood from different sites must not be used during the same reference value project. Always make a note of the site.

The recommendation describes the puncture procedure in great detail. Because of the risk for hitting the calcaneus bone with subsequent infection, it is important to observe strictly the instructions for puncturing the skin of the heel. Following the puncture, wipe away the first drops of blood and only use the subsequent drops. Fill the collection vessel maximally without causing venous congestion or massaging the skin. In infants, previous collections may change the composition of the circulating blood.

For every collection of reference values, standardize the *in vitro* procedures and the handling and storage of specimens, e.g., separation of serum from clot. Record these details. Other recommendations on collection of skin puncture blood have also been published (M6, N2).

3. Assay of Analytes and Quality Control

3.1. ANALYSIS AND CONTROL

The analysis of constituents in the collected specimens is briefly described in the fourth IFCC recommendation (I8). As stated in Section 1.3.1, the methods used for the production of reference values and for clinical assay of analytes should produce results that are comparable. Ideally, the analytical methods should be identical. If not, the reference values should be transferable. The problems of transfer of reference values is the topic of Section 3.2.

The analytical method, including the reagents and equipment used, must be described in enough detail to permit adequately trained laboratory personnel to proceed in the same manner and obtain reproducible results. This description is an essential part of the documentation of the reference values produced. In addition, it is necessary to ensure long-term stability of the analytical method both in the period when the reference values were produced and during their subsequent clinical use. This requires quality control to assure both accuracy and precision. A description of quality control is outside the scope of this review, but the reader is referred to the IFCC recommendations (I2, I3, I8) and other relevant literature (D3, S31).

Some claim that the analytical quality, as regards both systematic and random errors, should be better for reference values than for clinical routine assay. This may be true for accuracy. All measures should be taken to eliminate analytical bias or, alternatively, to determine its size and keep it stable. But superprecision is not necessarily needed. To facilitate direct comparison of observed and reference values, it is usually desirable to keep the imprecision during the production of reference values at the same level as that obtained by good-quality routine analysis. Therefore, reference values should preferably be produced under controlled but realistic routine conditions (I8). For the same reason, it is advisable to analyze specimens from reference individuals in several runs to include the between-run component of variation. A safe way to obtain comparability is to insert these specimens among clinical specimens into adequately monitored routine runs.

3.2. Transfer of Reference Values

Production of reliable reference values for all analytes assayed in a laboratory may be a costly exercise, often far beyond the capabilities of a single laboratory. As a result, reference values are often produced by collaborative projects. This is especially true when the availability of specimens is scarce, for instance, when producing pediatric reference values (H17). It is thus necessary to make reference values transferable from one institution to another. The same need arises when analytical methods are changed or modified. Furthermore, comparability of results among laboratories is also important when several laboratories serve the same population, and in multicenter scientific studies.

The problem of transferring reference values obviously emerges only when the laboratories serve the same or sufficiently similar populations. PetitClerc and Kelly (P2) showed that it was difficult to transfer reference values both within the same population in Sherbrooke, Canada, and between two populations in France and Canada. They stated some important prerequisites for successful transfer:

1. The populations should be adequately described and matching.
2. In a pilot study, small data sets from both populations should be compared to check for systematic differences arising mainly from preanalytical and analytical factors. Preferably, a well-defined and stable subclass of the population may be selected for this study. Healthy males aged 20–24 yr, who are considered to be in a "reference state" (S19), may be used for this purpose.
3. Quality control specimens should be exchanged to ensure close agreement in analytical performance in both laboratories.
4. The preparation of reference individuals and the collection of specimens should follow the same scheme of standardization.

5. Mathematical correction can compensate for discrepancies. One possibility is to adjust by linear regression.

Other methods for mathematical adjustment during the transfer of reference values have been suggested. Some procedures (S3) are probably too simple to be valid. It is necessary to adjust for the effects of differences in both bias and imprecision (S25). Among the methods suggested, that of Strike *et al.* (S42) is probably best one published to date. Their method was developed and tested in a recent theoretical and empirical study. They eliminated the interlaboratory bias and accounted for differences in analytical imprecision by an extension of the well-known SD unit transform (G20).

4. Statistical Treatment of Reference Values

4.1. INTRODUCTION

Reference values may be analyzed statistically in many ways, depending on the project (H9). Typically, one may wish to test the hypothesis that two samples of reference values are drawn from the same parent population, i.e., to test that there are no statistically significant differences between two sets of reference values. This is, for instance, relevant if a laboratory method has been changed or substituted by another method. Then it is necessary to test whether the methodological change causes a need for new reference limits or not. Another example concerns whether it is necessary to make available different sets of reference values for an analyte depending on the conditions during specimen collection (sitting/supine, fasting/nonfasting, etc.). Most of these problems may be solved by standard parametric or nonparametric statistical methods (B11, H4, L5, S1, S20).

Another main problem is reliable determination of reference limits (H9, I9, M4, S21, S22, S25, S27, W7). Many of the statistical methods applied in this context are standard methods, but—as always—it is necessary to use them correctly. Some of these methods are not found in standard textbooks on statistics. A third use of statistics is to permit presentation of observed values in relation to reference values (see Section 6.2).

The statistical analysis of reference values and the estimation of reference limits and their confidence intervals may be done without a computer. But computing facilities may be convenient if the size of the set of reference values is large. The iterative transformation routines described in Section 4.3.6 require a computer. Standard statistical software packages

(B6, S2, S30, S32) contain most of the routines needed for such work. The additional computation required can usually be done with a calculator.

The two-stage iterative transformation method is not available in standard statistical programs. Therefore, the special-purpose statistical program REFVAL was designed to solve this problem (S24). It treats reference values as recommended by the IFCC (I9). Two versions are available: (1) a FORTRAN subroutine package that can be incorporated into a user's own program, and (2) a complete program (written in TurboPascal) for IBM PCs or look-alikes running PC/MS-DOS.

Flowchart 2 shows the recommended order of the different steps (I9). This flowchart also describes the sequence of routines in the REFVAL program (S24). Tests used to detect the need for partition of reference values into homogeneous subsets (see Section 4.2.1) were omitted from REFVAL because of their general availability in standard statistical programs (B6, S2, S30, S32). REFVAL is thus not designed to be an alternative to statistical software packages. It is a supplement.

Here it is assumed that *quantitative data* are basically of the continuous type, which usually is the case with most analytes assayed in clinical laboratories. The real distributions are, however, often stepwise, because of rounding of the values to a specified least significant digit for the analysis. The effects of rounding may demand special treatment of data (see Section 4.3.5.3). Laboratory data may, however, be of other types. *Qualitative data* are either binary (yes/no, present/absent, positive/negative, male/female, etc.) or polytomous (i.e., several values or categories are possible such as blood groups or types of blood cells). Some types of quantitative data have a set of discrete values. Such *semiquantitative data* may be coded as $-$, $+$, $++$, ..., by integers (0, 1, 2, ...), or by descriptive labels such as infant, child, teenager, and adult. The treatment of qualitative or semiquantitative data is outside the scope of this review.

4.2. Preprocessing of Reference Values

The routines shown in the top five boxes in Flowchart 2 may be grouped as data preprocessing, that is, preliminary steps before the estimation of reference limits. Most of the techniques applied here are conventional statistical procedures. It is thus not necessary to discuss these steps in great detail in this review.

4.2.1. *Partitioning of Data; Subject-Based Reference Values*

Many biochemical and physiological quantities are dependent on sex, age, and other characteristics (A7, R3, W1). *Partitioning* of the set of

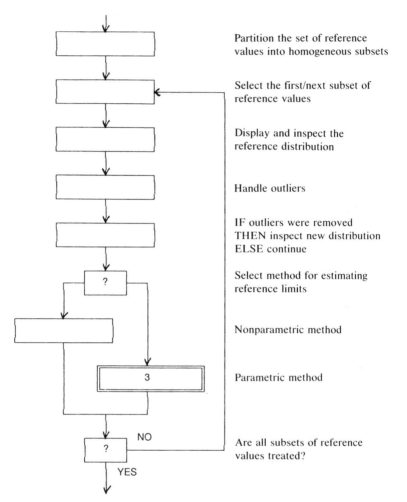

FLOWCHART 2. Statistical treatment of reference values.

reference values into more homogeneous subsets may therefore reduce variation among subsets and result in narrower reference intervals (19). Such reference intervals are then qualified as age-specific, sex-specific, age- and sex-specific, etc. The process of dividing the set of reference values according some criterion or criteria (see Section 2.2.3) is called, in addition to partitioning, stratification, categorization, subgrouping, and subclassification. The resulting subsets are also called partitions, strata, categories, subgroups, and subclasses.

Though apparently simple in principle, partitioning is often a rather complex affair. The most difficult problem is to determine when partitioning is really needed:

1. Excessive subgrouping should be avoided because it necessarily results in small subsets. Since each subset is analyzed separately (Flowchart 2), partitioning reduces the precision of the estimated fractiles.

2. In some cases, the data at hand are already subclassified according to some criteria. Then combination of the subsets into one big set may be considered and tested for validity (B4).

3. Statistically significant differences in location or dispersion of the reference values in possible subclasses may indicate that partitioning should be considered. Standard statistical parametric or nonparametric tests are appropriate here (B11, H4, L5, S1, S20).

4. Significantly different means or variances, however, do not justify partitioning without additional evidence. If the intended purpose of a reference interval is to detect individual biochemical changes, differences among means of class-specific reference values may be statistically significant and still too small to warrant replacing a single reference interval with several class-specific reference intervals, e.g., for age and sex.

Harris (H5, H6) has studied this problem extensively. To simplify, he assumed Gaussian distributions for both individual and group-based variation. He showed that the average ratio r of intra- to interindividual variation was decisive. More formally, the average ratio r is defined as follows: the average value of individual variances divided by the variance of mean values. The population-based reference interval has less than the desired sensitivity to changes in the biochemical status of an individual if $r \leq 0.6$, while this interval may be a more trustworthy reference when $r > 1.4$, at least for individuals whose intraindividual variations are close to average. Even if $r > 1.4$, the reference interval remains less sensitive than desired in persons whose intraindividual variations are smaller than average and too sensitive in those whose variations are greater than average.

The r ratio has consistently shown the same variation among commonly assayed analytes in several studies (H13, P3, W6, Y1). As an example, the study of Winkel and Statland (W6) shall be cited. They determined the average ratio r of 39 analytes in blood. Their results fall into four groups, within which the analytes are listed according to increasing r values:

1. $r \leq 0.20$: IgM, IgA, haptoglobin, alkaline phosphatase, IgG, γ-glutamyltransferase, complement C3, complement C4, α_1-antitrypsin, and α_2-macroglobulin.

2. $r = 0.21–0.40$: basophils, cholesterol, transferrin, orosomucoid, eosinophils, hemoglobin, hematocrit, platelets, lymphocytes, monocytes, and triglycerides.

3. $r = 0.41–1.00$: lactate dehydrogenase, alanine aminotransferase, creatinine, total WBC, cortisol, creatine kinase, neutrophils, thyroxine, urea, aspartate aminotransferase, and uric acid.

4. $r > 1.00$: total protein, chloride, iron, potassium, sodium, calcium, and albumin.

Of these blood analytes, 27 (69%) had $r \leq 0.6$ (all up to and including creatine kinase), while only 3 (8%) had $r > 1.4$ (sodium, calcium, and albumin). Thus, the group-based reference intervals for the majority of these analytes would be bad referents for assessing a change in individual biochemical status, and they would probably be reliable for only 3 analytes. An unanswered question in the study (W6) is whether partitioning of the reference values into subsets would produce a more sensitive reference interval.

The desired goal of partitioning is to increase the ratio r to at least 1.4 in the subsets. The required percentage reduction of the standard deviation (from set to subset) depends on the prior value of the ratio in the set (H6): $r = 0.4$, 54%; $r = 0.6$, 37%; $r = 0.8$, 23%; $r = 1.0$, 13%; and $r = 1.2$, 6%. This reduction, for instance 37% when the prior ratio is 0.6, can be difficult to achieve in practice. The alternative is to compare observed values with individual or *subject-based reference values*. The theory of individual reference values is either based on time-series models (H5–H10, H12, H14, W10) or on physiological models (W5, W6, W8). The reader is directed to the referenced literature on this topic. It has also been suggested to determine age- and sex-related reference intervals by a linear regression model (G1). In the following discussion it is assumed that a homogeneous reference distribution (either consisting of the complete set of reference values or a subset obtained by partitioning) is available for the statistical treatment (Flowchart 2).

4.2.2. *Visual Inspection of Data; Handling of Outliers*

Visual inspection of the reference distribution, e.g., in a conventional histogram, is a safeguard against the misapplication of statistical methods and the misinterpretation of their outcome. It is difficult to identify and eliminate erroneous reference values. Statistical methods usually can only detect incorrect values that are unusual in some respect. If each analyte is treated separately, erroneous values can only be identified as *outliers*, i.e., values far from the other values (B3, H15). Incorrect values hidden in the body of the reference distribution pass undetected. In contrast, multivariate methods can also identify a hidden single value as a possible error because it causes the multivariate pattern to be atypical (B3, H15, K1).

Visual inspection of histograms is a simple and reliable method for identification of outliers (I9). If the reference distribution has a long right tail (positive skewness), a preliminary logarithmic transformation may facilitate the visual identification. Many numerical or statistical methods for outlier detection have been suggested (B3, H15), but none are applicable in all circumstances. Some fail because they assume a Gaussian (or other type of) distribution. The type is, however, not known *a priori,* and it may be impossible to test in the presence of outliers. Other methods fail when more than a single outlier exists.

Some recommend iterative trimming of extreme values by excluding those outside, say, three standard deviations in each pass. A preliminary logarithmic transformation may be necessary (G4). The IFCC recommendation (I9) advocates a modified Dixon test (D4) which identifies an extremely low or high value as a possible outlier if its distance to the next value in the distribution exceeds one-third of the total range of values. This method, which is implemented in the REFVAL program (S24), has been criticized (L11) because of its fixed criterion (one-third of the range). However, because the type of distribution is generally not known *a priori,* the method is only a rough rule of thumb and a single, approximate criterion is adequate.

Values identified as possible outliers should not be discarded automatically (as is done with some statistical systems). The complete procedure for the production of the dubious value (Flowchart 1) should be back-traced to detect any errors. It may be necessary to reanalyze the specimen. The value should be corrected, kept, or discarded according to best judgment. The reference distribution should be reinspected if one of more outliers have been removed (Flowchart 2).

4.3. ESTIMATING REFERENCE LIMITS

4.3.1. *Types of Reference Limits; Confidence Intervals*

A rather imprecise IFCC definition of the term *reference limit* was presented in Section 1.3.5. One reason for its vagueness is that at least three kinds of group-based reference intervals have been proposed in the literature (I9, S21, S25, W7):

1. The *tolerance interval* is defined as an interval which can be claimed to contain at least a specified fraction of the population's values with a

stated degree of confidence (H1, H2, L5, W7). The size of the tolerance interval is dependent upon the number of reference values and upon the specified confidence, being wider with smaller samples and with higher confidence. It is also possible to define a tolerance interval without specified confidence (L5). Then it will contain on the average a stated proportion of the reference values. Since the tolerance interval is an estimate of an interval of the population's hypothetical reference distribution, the use of the tolerance interval is strictly limited to situations where the assumption of random sampling is fulfilled (see Section 2.2.1).

2. The *prediction interval* is bounded by upper and lower limits, between which a future observed value is expected to fall with a specified probability (H1, H2, W7). Like the tolerance interval, this kind of reference interval is strictly limited to situations where the sampling process was completely random. It further presupposes that the future observation is a true member of the reference population from which the sample was obtained. The latter assumption may be false for observations obtained at later times.

3. The *interfractile interval* (or interpercentile interval) is the most frequently used kind of reference interval and is recommended by the IFCC (19). It is defined as an interval bounded by two fractiles (or percentiles) of the reference distribution. The terms fractiles and percentiles mean the same, only the scale is different. The α-*fractile* is the value below and including that in which lies a specified fraction α of the cumulative distribution (D6, H4, I5). The *percentile* is the fractile expressed with 100 (instead of one) as the basis (K4). The preferred terms herein are fractile and interfractile interval.

The distinction between these three types of reference intervals is mainly theoretical. When the sample size is at least 100 reference values, the differences between the numerical values of these intervals are too small to be practically important. The following discussion of the estimation of reference intervals is based on the interfractile interval in accordance with the IFCC recommendation (I9).

The size of the reference interval should be adjusted according to clinical requirements or usual practice. The use of the *central 0.95 interval* defined by the 0.025 and 0.975 fractiles (Fig. 6) is so common that, in the absence of a specification, this size is usually assumed. As other sizes and asymmetric locations of the fractiles are sometimes used, authors should always state what kind of interval they report. To simplify the following presentation, the lower and upper reference limits are assumed to be the 0.025 and 0.975 fractiles, respectively. The central 0.95 interval is obtained by cutting off 2.5% of the values in each tail of the reference distribution.

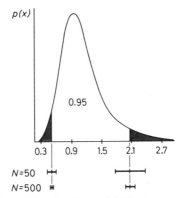

FIG. 6. The 0.025 and 0.975 fractiles (central 0.95 reference interval) and the effect of sample size on their 0.90 confidence intervals. From Ref. 15 with permission of the IFCC.

The techniques presented are general and may easily be adapted to other sizes or locations of the interval.

Fractiles, determined on a sample reference distribution, are estimates of the true fractiles of the population. The smaller the sample the greater is their imprecision. The IFCC therefore recommends to specify the *confidence interval* of each reference limit (Fig. 6). A confidence interval is bounded by two limits, within which the true population value is found with the stated confidence β. Because of its common usage, the confidence $\beta = 0.90$ is assumed here although another β value may be chosen.

Figure 6 shows also that the width of these confidence intervals are unequal if the reference distribution is asymmetric (the interval in the long tail is wider). This fact has obvious, but often ignored, consequences for the interpretation of an observed value close to a reference limit. For instance, a value slightly smaller than the lower reference limit in Fig. 6 may be highly unusual, but a value considerably above the upper reference limit may be relatively typical.

4.3.2. *Parametric or Nonparametric Method?*

Reference limits (fractiles) and their confidence intervals may be determined by parametric or nonparametric statistical methods. The parametric types assume a specified (usually Gaussian) distribution and use estimates of parameters (e.g., mean, standard deviation, and coefficients that determine shape). In contrast, nonparametric methods do not make any assumptions of distribution type nor do they use estimates of distribution parameters. Fractiles may also be determined visually on a graph of the cumulative reference distribution. Some assume a Gaussian or log-

Gaussian distribution (linear plot on Gaussian or log-Gaussian probability paper), but the graphical method is applicable without this hypothesis (L2).

The confidence intervals obtained by parametric methods are seemingly narrower than those of the nonparametric estimates. Theoretical and empirical evidence indicates that the true precision is fairly similar (I9, L11). The formulas used to compute the parametric confidence intervals of fractiles (see Section 4.3.7) assume a Gaussian reference distribution. But the true reference distribution is not known *a priori*. Therefore, one has to test how well the sample distribution conforms with the hypothetical distribution type. If the test declares that the distribution is non-Gaussian, it must be transformed by mathematical functions. These two steps, testing and transforming, are sources of variation that add to the calculated imprecision of the fractiles. A study based on computer simulation seems to indicate that the calculated parametric 0.90 confidence intervals of fractiles should be expanded by 19–33% (mean, 25%) (L11). For instance, it was found that a better estimate for a confidence interval 222–253 of a upper reference limit of 236 was 217–256, i.e., 26% wider.

The confidence intervals described above give limits for true fractiles in the population. Another type of confidence interval of fractiles has also been described (H9, H16): confidence limits for the true fraction of a population included within the sample-based reference interval. In this case both parametric and nonparametric methods are also available for estimation (H9).

There are several studies wherein the parametric and nonparametric methods for estimating fractiles have been compared on simulated or real data (B10, L11, R1, R2, S17, S21). The results of these studies may seem somewhat confusing. Some claim the superiority of the nonparametric method (R1, R2, S17), while others recommend the parametric method (B10), and a third group finds no hard evidence in favor of either method (L11, S21). There may be several reasons for the conflicting conclusions: (1) these studies did not apply the same parametric methods; (2) different nonparametric methods were also used; and (3) even more important, different types of distributions were employed, some of these being very atypical of those found with biological data. A critical evaluation of these comparative studies implies that, with the types of distributions usually encountered in clinical laboratory medicine, both methods, parametric and nonparametric, perform well. Therefore, one can usually select between the two methods (see Flowchart 2) according to personal preferences or computing facilities at hand. When successful, both methods usually give very similar estimates of reference limits. The precision of their estimates is probably not too dissimilar (see above) to be of importance in the selection.

4.3.3. *The Nonparametric Method*

The IFCC (I9) recommends a simple method based on ranked data (M8, R2) because it may be applied manually. A computer implementation (REFVAL) also exists (S24). The procedure is as follows:

1. Sort the N reference values in an ascending order: $x_1 \leq x_2 \leq \cdots \leq x_N$.
2. Assign rank numbers to the values: 1, 2, ..., N. Consecutive rank numbers should be given to values that are equal ("ties").
3. Compute the rank number of the α fractile (lower reference limit): $\alpha(N + 1)$. *Note:* It has been claimed (R4) that the rank number of the α fractile is more accurately determined by the expression $\alpha(N + 0.2) + 0.4$.
4. Find the original reference value corresponding to the fractile's rank number if it is an integer. Otherwise, interpolate between the two reference values between which the noninteger rank value lies.
5. Look up in Table 4 the rank numbers of the fractile's $\beta = 0.90$ confidence interval. Alternatively, find the ranks a and b of the two β confidence limits by a formula based on the binomial distribution (M8, R2):

$$\sum_{i=a}^{b-1} \binom{N}{i} \alpha^i (1 - \alpha)^{N-i} \geq \beta$$

This numerical method has been implemented in the REFVAL computer program (S24). An approximation based on the Gaussian distribution has also been published (H9). The nonparametric 0.90 confidence interval is only defined for $N \geq 119$ (I9, R2).

6. Repeat steps 3–5 for the $1 - \alpha$ fractile (upper reference limit).

More refined methods for nonparametric estimates of fractiles and their confidence interval have been published. In some cases, the tails of the reference distribution are smoothed by fitting a mathematical function (R4, S17). A weighted fractile method has also been described (S17). These methods are probably somewhat more accurate and precise, but they usually require the assistance of computer programs.

4.3.4. *The Parametric Method: Introduction*

Flowchart 3 outlines the parametric method. The reference limits (fractiles) and their confidence intervals can be estimated directly if the sample reference distribution is found to be nonsignificantly different from the Gaussian distribution. Otherwise, the reference distribution should be transformed to a Gaussian shape before making the estimates. In that case

TABLE 4
CONFIDENCE INTERVALS OF REFERENCE LIMITS (NONPARAMETRIC)[a]

Sample size	Rank numbers		Sample size	Rank numbers	
	Lower	Upper		Lower	Upper
119–132	1	7	566–574	8	22
133–160	1	8	575–598	9	22
161–187	1	9	599–624	9	23
188–189	2	9	625–631	10	23
190–218	2	10	632–665	10	24
219–248	2	11	666–674	10	25
249–249	2	12	675–698	11	25
250–279	3	12	699–724	11	26
280–307	3	13	725–732	12	26
308–309	4	13	733–765	12	27
310–340	4	14	766–773	12	28
341–363	4	15	774–799	13	28
364–372	5	15	800–822	13	29
373–403	5	16	823–833	14	29
404–417	5	17	834–867	14	30
418–435	6	17	868–871	14	31
436–468	6	18	872–901	15	31
469–470	6	19	902–919	15	32
471–500	7	19	920–935	16	32
501–522	7	20	936–967	16	33
523–533	8	20	968–970	17	33
534–565	8	21	971–1000	17	34

[a] The table shows the rank numbers of the 0.90 confidence interval of the 0.025 fractile for samples with 119–1000 values. To obtain the corresponding rank numbers of the 0.975 fractile, subtract the rank number in the table from $N + 1$, where N is the sample size. (From Ref. I9 with permission of the IFCC.)

the estimated reference limits and confidence intervals must be transformed back to the original data scale.

When the values are distributed in a non-Gaussian fashion, the calculation of reference limits by a method that assumes Gaussianity may result in useless or even bizarre estimates. It is very common that distributions of laboratory data exhibit positive skewness, that is, they have a heavy lower part and a long tail toward high values. The common practice of calculating the reference limits as the values located two standard deviations below and above the arithmetic mean leads to serious bias. In some cases, the lower limit may even be negative, which obviously is nonsense if the reference values are all positive. Still such "estimates" of fractiles may be found in the literature (although, for cosmetic reasons, the negative

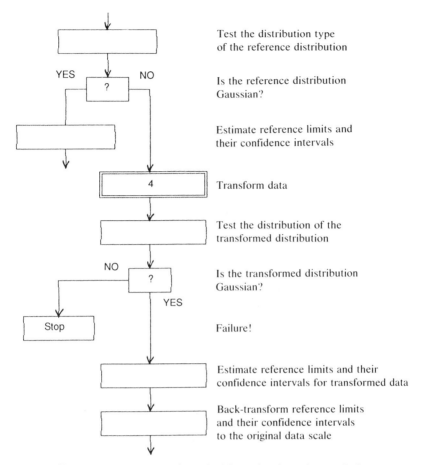

FLOWCHART 3. Parametric method for estimating reference limits.

lower limits may have been adjusted to zero). The three essential parts of the parametric method are therefore as follows:

1. *Tests of Gaussian distribution,* i.e., statistical tests of how well the reference distribution conforms with the Gaussian distribution (see Section 4.3.5).

2. *Transformations.* A transformation is a mathematical function that is applied to all reference values to obtain a new distribution closer to Gaussian form than the original one. For instance, the logarithms of the reference values may adjust a positively skewed distribution to symmetry. Other transformations are presented in Section 4.3.6.

3. *Estimation* of the fractiles (reference limits) and their confidence intervals are straightforward when the distribution is Gaussian—originally or by transformation (see Section 4.3.7). When necessary, the estimates are back-transformed to the original data scale by applying the inverses of the transforming function(s).

4.3.5. Tests of Gaussian Distribution

Many tests of Gaussianity have been proposed. Mardia's extensive review may be recommended as a source of information on both univariate and multivariate tests (M3). In the following discussion only univariate tests that have proved useful in the statistical treatment of reference values will be presented (I9).

The need to test the Gaussian hypothesis arises in two situations of the parametric method: (1) *prior to estimating reference limits*, as shown in Flowchart 3 (at the top and following the transformation box), and (2) *during the transformation*, for monitoring the convergence toward Gaussian shape (see Flowchart 4). To test the success with the same test as used during the transformation may give a too optimistic picture. Therefore, the two situations demand two classes of distribution tests based on different principles. Broadly, tests of Gaussian distribution may be divided into two categories:

1. *Goodness-of-fit tests* are based on the differences between the individual values in the sample distribution and the corresponding values of the hypothetical Gaussian distribution. The differences between the sample and hypothetical distributions are nonsignificant if these differences are all small or if test statistics calculated from them are within critical limits. Examples of this category of distribution test are the graphical test, the Kolmogorov–Smirnov test, the Anderson–Darling test, and the Cramér–von Mises test. These tests are presented in detail below. The χ^2 test (L5, S1, S20) belongs to the same family, and it was previously a popular test. It is not discussed here as it has been shown to be less effective than the alternatives (M3, S40).

2. *Coefficient-based tests* use parametric measures of the shape of the distribution. With these tests the hypothesis of Gaussian distribution is rejected if the calculated coefficient exceeds critical limits. The most frequently used tests in this category are based on the coefficient of skewness (a measure of asymmetry of the distribution) and on the coefficient of kurtosis (a measure of tail heaviness).

A note on nomenclature: There is no generally accepted standard nomenclature for tests of Gaussian distribution, which explains the confusing use of terms in scientific literature. Some call all such tests, of both cat-

REFERENCE VALUES

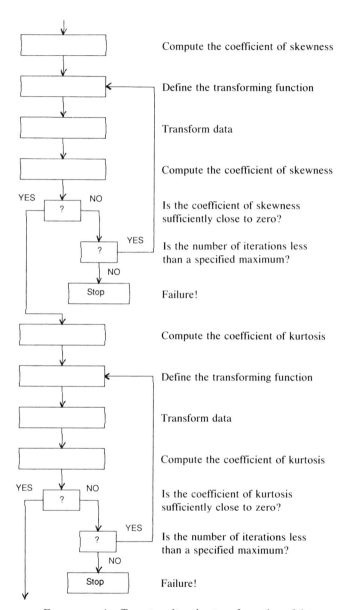

FLOWCHART 4. Two-stage iterative transformation of data.

egories mentioned above, goodness-of-fit tests. Here it is preferred to reserve that term to the group of tests that evaluate the deviations of the individual data points from the hypothesis. This use of the word "fit" is similar to its usage in other domains of statistics, e.g., in regression analysis. Some use the term "empirical distribution function" (EDF) tests for those called goodness-of-fit tests here (S40).

The two categories of tests are based on different principles. This may provide dissimilar tests for the two test situations described above. The IFCC recommends to apply a goodness-of-fit test prior to estimating the reference limits and to use coefficient-based tests for monitoring during the transformation (I9).

The REFVAL computer program (S24, S26) performs all tests described below except the graphical test. Its iterative two-stage transformation to Gaussianity is monitored by the coefficients of skewness and kurtosis, while the Anderson–Darling test is applied for testing the distribution before the estimation of fractiles and their confidence intervals.

4.3.5.1. *Goodness-of-Fit Tests.* The graphical test on Gaussian probability paper has frequently been used in the past, but it should be avoided because the nonlinear ordinate makes the visual evaluation unreliable (I9). The reliable Kolmogorov–Smirnov test is the numerical analog of the graphical test (I9). The procedure is as follows (I9, S26, S40):

1. Sort the reference values x_i according to numerical values: $x_1 \leq x_2 \leq \cdots \leq x_N$.

2. Compute the standardized differences: $v_i = (x_i - AV_x)/SD_x$, where AV_x is the arithmetic mean (average) and SD_x is the standard deviation.

3. Transform the differences v_i to the corresponding cumulative probabilities w_i of the standard Gaussian distribution. These probabilities may be found in statistical tables (L5, P1, S1, S20) or by numerical approximation on a computer (C2, S24).

4. Find the maximum difference D_{max} between the cumulative reference distribution and the hypothetical Gaussian distribution and compute the sample size-adjusted test criterion $*D_{max}$:

$$D^+ = \max(i/N - w_i) \quad (i = 1, \ldots, N)$$
$$D^- = \max[w_i - (i - 1)/N] \quad (i = 1, \ldots, N)$$
$$D_{max} = \max(D^+, D^-)$$
$$*D_{max} = D_{max}(\sqrt{N} - 0.01 + 0.85/\sqrt{N})$$

5. Evaluate the test criterion $*D_{max}$ by comparison with tabulated critical values (S40) or by using a computed approximation (S24, S26). The Gaussian hypothesis is rejected at the 0.05 significance level if the criterion exceeds the critical value 0.895. The critical value at the 0.01 significance level is 1.035.

The Anderson–Darling test and the Cramér–von Mises test have many properties in common with the Kolmogorov–Smirnov test, and their procedures share the first three steps of the latter test (see above).

The Anderson–Darling test statistic A^2 and the sample size-adjusted criterion $*A^2$ are computed from the cumulative Gaussian probabilities w_i for $i = 1, \ldots, N$ (S40):

$$A^2 = -\{\sum (2i - 1)[\ln(w_i) + \ln(1 - w_{N+1-i})]\}/N - N$$

$$*A^2 = A^2(1 + 4/N - 25/N^2)$$

The Gaussian hypothesis is rejected at the 0.05 significance level if the $*A^2$ criterion exceeds the critical value 0.787. The critical value at the 0.01 significance level is 1.092.

In a later paper, Stephens (S41) published a revised size-adjusted test criterion $*A^2 = A^2(1.0 + 0.75/N + 2.25/N^2)$ and the critical values 0.752 and 1.035 for the significance levels 0.05 and 0.01, respectively. The IFCC recommendation (I9) and the REFVAL program (S24), however, use the earlier version. The same is the case with some other published studies (L11, S26). Therefore, the discussion below is based on the earlier version of the Anderson–Darling test.

The Cramér–von Mises test statistic W^2 and the sample size-adjusted criterion $*W^2$ are computed from the cumulative Gaussian probabilities w_i for $i = 1, \ldots, N$ (S40):

$$W^2 = 1/(12N) + \sum [w_i - (2i - 1)/2N]^2$$

$$*W^2 = W^2(1 + 0.5/N)$$

The Gaussian hypothesis is rejected at the 0.05 significance level if the $*W^2$ criterion exceeds the critical value 0.126. The critical value at the 0.01 significance level is 0.178.

More details of critical values for the Kolmogorov–Smirnov test, the Anderson–Darling test, and the Cramér–von Mises test have been published (L11, L12, S40, S41) and computer approximations, using third-degree polynomials, have been developed for the REFVAL program (S24, S26).

4.3.5.2. *Coefficient-Based Tests.* These tests of distribution type use approaches other than the goodness-of-fit tests. Rather than comparing each value in the sample cumulative distribution with the corresponding cumulative Gaussian distribution, the coefficient-based tests rely on estimated parameters that carry information about the shape of the distribution. The two coefficient-based tests presented here use measures of (1) the asymmetry (coefficient of skewness) and (2) the tail heaviness (coefficient of kurtosis) of the distribution. Both coefficients are computed from statistics called *moments about the mean* (C3, I9, M3, S20). These moments are simply average powers of the differences between the sample values x_i and the sample arithmetic mean AV_x and thus carry information about the dispersion of the values around the mean. The coefficients utilize the second, third, and fourth moments about the mean (m_2, m_3, and m_4, respectively):

$$m_r = \sum (x_i - AV_x)^r/N \qquad (r = 2, 3, \text{ or } 4; \quad i = 1, \ldots, N)$$

The second moment about the mean is closely related to the sample variance: $SD_x^2 = \Sigma(x_i - AV_x)^2/(N - 1)$.

The coefficient of skewness g_s carries information about the asymmetry (skewness) of the distribution. Its value is zero when the distribution is symmetric. The sign of a nonzero coefficient indicates the type of skewness (Fig. 7). The distributions found with clinical chemical and hematological data are most frequently positively skewed or near-symmetric, with negative skewness being relatively rare (D7, S29). The coefficient of skewness computed from the second and third moments about the mean (SD_s is its standard deviation):

$$g_s = m_3/(m_2 \sqrt{m_2})$$
$$SD_s = \sqrt{6/N}$$

The coefficient should preferably be corrected for bias (C3, D2).

The coefficient of kurtosis g_k carries information about the relative weight of the tails of a distribution. The Gaussian distribution has zero kurtosis. The sign of a nonzero coefficient indicates whether more (positive) or less (negative) values are located in the tails of the distribution as compared with the Gaussian distribution (Fig. 7). This coefficient is most useful when the distribution is symmetric. The coefficient of kurtosis g_k is computed from the second and fourth moments about the mean (SD_k is its standard deviation):

$$g_k = (m_4/m_2^2) - 3$$
$$SD_k = 2\sqrt{6/N}$$

The coefficient should preferably be corrected for bias (C3, D2).

Very approximate two-sided tests at the 0.01 level of significance are obtained by rejecting the Gaussian hypothesis if one or both of the coefficients are greater than 2.6 times their corresponding standard deviation. The reason for the inaccuracy is that the sampling distributions of the coefficients are far from Gaussian when the sample sizes are small (M3, S20). Although both distributions approach the Gaussian type with in-

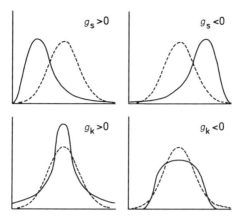

FIG. 7. Skewed and kurtotic distributions (continuous curves) compared with the Gaussian distribution (dotted curves). The g_s and g_k are the coefficients of skewness and kurtosis, respectively. From Ref. S25 with permission of W. B. Saunders Co.

creasing sample size, the distribution of the coefficient of kurtosis is still definitely non-Gaussian with a sample as large as 1000 values (M3). More accurate critical values have been published (P1, S20). Solberg (S26) has developed computer approximations for critical values of both coefficients. They are available in the REFVAL program (S24). These approximations were based upon reshaping of the sampling distributions of the two coefficients to near-Gaussianity by Johnson's S_U transformation (J2, J3), a method that results in rather accurate critical values of the two coefficients (S26).

4.3.5.3. *Correction for Rounding of Data.* Such correction may be necessary because the numerical goodness-of-fit tests assume a continuous distribution, whereas the rounding produces a stepwise distribution. Reference values are often rounded to the nearest significant digit (e.g., serum sodium concentration measured as 141.6 mmol/liter may be reported as 142 mmol/liter. The tests therefore become more reliable when corrected for the rounding. A convenient method is to add random noise to each reference value before performing the test (I9):

$$x_i + L(r_i - 0.5)$$

where L is the step size of the value scale (least significant digit; e.g., $L = 0.1$ when data are reported to one decimal place) and r_i is a computer-generated random number in the interval 0–1. Though not explicitly recommended by the IFCC (I9), the same correction for rounding may also be used with the coefficient-based tests. The REFVAL computer program applies this correction prior to all tests of the distribution type (S24).

4.3.5.4. Distribution Tests: Concluding Remarks. The IFCC recommendation on the statistical treatment of reference values presents all the tests described above except the Cramér–von Mises test (I9). The graphical test should not be used, according to the IFCC, but its recommendation does not give any further guidance on the selection of tests. There are numerous comparative studies on the relative merits of the distribution-type tests [see Mardia (M3), for an extensive review]. In the context of reference values in laboratory medicine, the recent studies of Linnet (L10, L11) and Solberg (S26) contain useful information.

Solberg (S26) developed computer approximations for the critical values of the test statistics of all the tests included in the REFVAL program (S24), i.e., all the numerical tests described above. The methods and their computer implementation were studied by extensive computer simulations. It was verified that the empirical probabilities of false rejections when the underlying distribution was Gaussian were close to the stated significance levels. The two coefficient-based tests were recommended for use during the mathematical transformation of a reference distribution to Gaussian shape. The Anderson–Darling test was a good all-around test for Gaussianity.

Linnet (L10, L11) also used computer simulations and applied routines from the REFVAL computer program (S24). He discovered that goodness-of-fit tests used with transformed data tended to give a too optimistic picture of the situation and proposed empirical correction factors for the Anderson–Darling and Kolmogorov–Smirnov tests.

4.3.6. Transformation of the Reference Distribution

4.3.6.1. Rigid and Elastic Transformations. If, as is often the case, the initial test of distribution type (see Flowchart 3) rejects the hypothesis that the distribution is Gaussian, then data must be transformed to approximate such a distribution. To transform data one applies a mathematical function to each value. The type of mathematical function determines the effect on the distribution.

For example, if the original distribution is positively skewed it is often observed that the distribution of the logarithms of the values is close to symmetry. In fact, the simple logarithmic transformation $y_i = \ln(x_i)$ has frequently been used to "normalize" the reference distribution. Its advantage is that it can be calculated with an ordinary calculator. An equally simple alternative is the square root transformation $y_i = \sqrt{x_i}$, which also may reshape a positively skewed distribution to symmetry. Still, there are several problems with these two transformations:

1. They can only transform positively skewed distributions. However, it is possible to alter a distribution with negative skewness into a positively skewed mirror distribution at the expense of extra work.
2. All reference values must be positive. A possible solution is to shift the location of the distribution above zero by adding a constant to all reference values.
3. These functions are rigid. If none of them produce a symmetrical distribution, more flexible functions are needed.

Table 5 shows a selection of elastic transforming functions, i.e., functions whose properties may be changed continuously by adjusting parameter(s). They are able to reshape the reference distribution as if it was made of rubber. It is usually possible with the types of distributions encountered in clinical laboratory medicine to stretch or relax the reference distribution to produce a new distribution of transformed values that are close to symmetry.

4.3.6.2. *Two-Stage Transformation.* Although symmetry often is obtained by the application of a suitable function, non-Gaussian kurtosis may remain. Then it is necessary to use a second function that compresses or expands the distribution symmetrically around the mean (Table 5). Also for this purpose there are many alternative functions and many of them provide elastic solutions.

To obtain a complete reshaping of the reference distribution to Gaussian form, one thus often needs a two-stage transformation procedure, as was first proposed by Harris and DeMets (H11). The first step is transformation of a skewed distribution to symmetry, and the second step removes residual non-Gaussian kurtosis (see Flowchart 4.) Table 5 lists some functions that may be used for the two-stage transformation. Other transforming functions have also been suggested (K3, M4, M8). Several two-stage transformation systems that utilize different combinations of the functions shown in Table 5 have been described and evaluated (B10, D2, H11, L11, R1, S21, S22, S24–S27, U1).

To overcome the deficiencies of previously described procedures, Solberg (S24–S27) developed a two-stage transformation system for the final version of the IFCC recommendation (I9). The exponential function of Manly (M2) was selected as the first-stage transformation to correct for skewness in either direction. This function can handle values of any magnitude and sign. For numerical reasons a preliminary standardization of the reference distribution to zero mean and unit standard deviation is rec-

TABLE 5
TRANSFORMING FUNCTIONS[a]

Stage 1: Transformation to symmetry
In the following functions, Q is a function parameter to be estimated (for example, by an iterative procedure)
    Logarithmic function, elastic type (H11)
        $y_i = \ln(x_i + Q)$
        Valid data: $x_i > -Q$
        May correct positive skewness (see, however, the text)
    Power function (B7, S21)
        $y_i = (x_i^Q - 1)/Q$ (if $Q \neq 0$)
        $y_i = \ln(x_i)$ (if $Q = 0$)
        Valid data: $x_i > 0$
        May correct negative and positive skewness
    Exponential function (I9, M2)
        $y_i = [\exp(Qx_i) - 1]/Q$ (if $Q \neq 0$)
        $y_i = x_i$ (if $Q = 0$)
        Valid data: no restriction (see, however, the text).
        May correct negative and postive skewness

Stage 2: Transformation to Gaussian kurtosis
In the following functions, R and S are function parameters to be estimated analytically or by iterative procedures. It is assumed that the input data have a symmetric distribution.
    Hyperbolic sine function[b] (B10)
        $z_i = R \sinh(y_i/S)$
        May correct negative kurtosis
    Inverse hyperbolic sine function[b] (H11, J2)
        $z_i = R \sinh^{-1}(y_i/S)$
        May correct positive kurtosis
    Power function (B10)
        $z_i = |y_i|^R$ (if $y \geq 0$)
        $z_i = -|y_i|^R$ (if $y < 0$)
        May correct negative and positive kurtosis
    Modulus function (I9, J1)
        $z_i = S\{[(|y_i| + 1)^R - 1]/R\}$ (if $R \neq 0$)
        $z_i = S[\ln(|y_i| + 1)]$ (if $R = 0$)
        Here S denotes a sign factor (-1 or $+1$) corresponding to the sign of the input value y_i
        May correct negative and positive kurtosis

[a] The table shows selected functions that have been suggested for the two-stage transformation of reference distributions to Gaussian shape. Only elastic functions are shown (see the text). The references to the literature point to the first description of a function and/or its first use with reference values. More references are given in the text.

[b] Definition of the two hyperbolic functions: the hyperbolic sine function is $\sinh(y) = [\exp(y) - \exp(-y)]/2$; the inverse hyperbolic sine function is $\sinh^{-1}(y) = \ln[y + \sqrt{(y^2 + 1)}]$.

ommended: $^*x_i = (x_i - AV_x)/SD_x$, where AV_x is the arithmetic mean (average) and SD_x is the standard deviation. Among several possible functions for correction of negative or positive kurtosis, the modulus function of John and Draper (J1) was chosen without actually claiming its superiority over the alternatives. It is also recommended, for numerical reasons, to standardize the input distribution to the modulus transform to zero mean and unit standard deviation: $^*y_i = (y_i - AV_y)/SD_y$. The two-stage parametric procedure based on the exponential and modulus functions is implemented in the REFVAL computer program (S24), which estimates the function parameters by iterations guided by the coefficients of skewness and kurtosis (S27), as shown in Flowchart 4. The Anderson–Darling test was applied for the final affirmation of Gaussianity (Flowchart 3).

Linnet (L11) studied the IFCC-recommended system as implemented in the REFVAL program by Monte Carlo methods. He found that, in some cases, the system was not able to transform successfully data to Gaussianity. The frequency of failure is fairly low with most real reference distributions studied (H. E. Solberg, unpublished data).

4.3.7. Calculation of Reference Limits and Their Confidence Intervals

At this step in the parametric method a Gaussian distribution or a close approximation to it is assumed. Otherwise, reliable estimates cannot be produced. Then the only alternative is to estimate reference limits and their confidence intervals by the nonparametric method. Here the individual values are symbolized z_i, that is, values resulting from the two-stage transformation procedure described in the preceding section (see also Table 5). If the original distribution was Gaussian or if only a single transformation was used, we use the following dummy transformation: $z_i = x_i$ (no transform), $y_i = x_i$ (no first-stage transform), or $z_i = y_i$ (no second-stage transform).

The reference limits, e.g., the α fractile and the $1 - \alpha$ fractile, are estimated from the arithmetic mean (average) AV_z and the standard deviation SD_z of the sample of z values (I9):

$$AV_z \pm c_{1-\alpha} SD_z$$

Here $c_{1-\alpha}$ ($= c_\alpha$) denotes a standard Gaussian deviate as may be read from statistical tables (L5, P1, S1, S20) or obtained by a numerical approximation on a computer (C2, S24). The relevant c value for the central 0.95 interfractile reference interval ($\alpha = 0.025$) is 1.960, i.e., the 0.025 and 0.975 fractiles are approximately equal to the mean minus/plus two standard deviations. Other c values have to be used when another interfractile interval is required. It is not necessary to estimate a symmetric reference interval. Two different c values are needed when the interval has an asymmetric location.

The common method for estimating the confidence interval of a fractile is based on its standard error SE_f (I9):

$$SE_f = \sqrt{(2 + c_{1-\alpha}^2) SD_z^2/2N}$$

Here $c_{1-\alpha}$ and SD_z denote the same values as above. The β confidence interval of the fractile f is then

$$f \pm u_{(1 + \beta)/2} SE_f$$

Here $u_{(1 + \beta)/2}$ denotes a Gaussian deviate.

For the $\beta = 0.90$ confidence interval of the central 0.95 interfractile reference interval ($\alpha = 0.025$), the u value is 1.645. By simple algebra, the following expression for the 0.90 confidence interval is obtained (I9):

$$f \pm 2.81 \cdot SD_z/\sqrt{N}$$

The confidence interval of a parametrically determined fractile estimated as shown above is probably too narrow because of sampling variation of transformation parameters (L11) (see Section 4.3.2).

The final step in the parametric method (Flowchart 3) is back-transformation of the estimates to the original data scale. This step can be omitted if all $z_i = x_i$ (i.e., the estimation was done with original data). Otherwise, the inverse transformation functions should be applied. If, for instance, $z_i = \ln(x_i)$, then the inverse function is $x_i = \exp(z_i)$. Simple algebra suffices to define the inverses of the transforming functions listed in Table 5.

5. Alternatives to Conventional Reference Values and Intervals

5.1. The Multivariate Situation

The commonly used reference interval for a single analyte is designed to contain the central 95% of the reference values. If observed values of several analytes measured in the same individual are compared with their corresponding 0.95 interfractile intervals, more than the expected average of 5% of the values are eventually located outside reference intervals. Some statistical reasoning may show that the expected frequency of location of one or more observed values outside the interval is $100(1 - 0.95^k)$. Here k is the number of variables (in this context, observed values of different analytes). We would, for instance, expect to find $100(1 - 0.95^{10}) = 40\%$ false positives with 10 analytes assayed in each individual.

One possibility is to increase each reference interval to keep the expected frequency of at least one random false positive at 5% when several analytes are measured in the same individual (H9, H16). Another possibility is to consider the set of simultaneously determined observed values as one multivariate observation; that is, as a point in a k-dimensional coordinate system. Then, instead of k individual reference distributions, we have a single k-variate distribution. Analogous to the univariate 0.95 reference interval, it is possible to estimate a 0.95 k-variate reference region (B9, H9, W4, W9). The bivariate reference region is an ellipse in the plane of the two axes, more or less slim depending on the correlation between the two variables (E1). The trivariate reference region is an ellipsoid body in three-dimensional space (K1). The region is an ellipsoid hyperbody in hyperspace if $k > 3$. The estimation of the reference region is straightforward when the distribution is multivariate Gaussian or is transformed to that type by application of mathematical functions (see Section 4.3.6) (B9, H9, K1, N3). The basis is multivariate statistical theory (A6, M9). In practice, an index (Mahalanobis' squared distance) is calculated for each multivariate observation and compared with a critical value for the index determined on the multivariate reference distribution (A3, B9, K1). These calculations are easily done by computer programs.

It has been claimed that comparison of multivariate observations with the corresponding k-dimensional 0.95 reference region may hide single values outside univariate reference intervals (H9). This is particularly true for batteries of highly correlated analytes. This has been confirmed by Boyd and Lacher (B9) in an empirical study of a 20-dimensional clinical chemistry profile: the multivariate reference region could be quite insensitive to highly deviating results for a single analyte. They also confirmed the high frequency of false positives with multiple univariate comparisons: 68% found against 64% theoretically expected (calculated by the formula given above). By contrast, the expected 5% of the multivariate observations were outside the 0.95 20-dimensional reference region. Both theoretical considerations and the results of the latter study also indicate that an observation may be distinctly unusual in the multivariate sense, even though each individual result is within its proper univariate reference interval.

To avoid hiding atypical single results, it has been suggested to limit multivariate reference regions to two or three dimensions (N3). Kågedal et al. (K2) evaluated both a multivariate reference region and the univariate reference intervals for three thyroid hormones on 510 routine patients. Of these, 108 patients were misclassified by multiple univariate comparison while only a single patient was misclassified by the multivariate method.

In conclusion, the multivariate method has several advantages when compared with the multiple univariate method. It should be the method of choice in situations where one or a few multiple test profiles are used, e.g., in health-screening programs. The method has, however, doubtful value in the common clinical routine where many different test combinations are requested or where the result of a single analyte may be decisive.

5.2. Time-Specified Reference Values

Many biological variables such as the analytes assayed in clinical laboratories show more or less systematic time-dependent variation. The nature of this variation may be a trend or a rhythm or a mixture of both types. A *trend* is a systematic decrease or increase of the mean value of the analyte over time (F3, G2). The trend may be linear or of another type. It may be observed by follow-up of a single individual and it may be reflected as an average trend in a population. Some analytes exhibit *rhythmic variation* with a single frequency or a mixture of several frequencies with a period measured in hours (W11), days, months (L9), or years. This cyclic variation seems to be caused by biological processes (internal clocks), although external factors such as meals, activity, posture, sleeping, and psychological and climatic factors may modulate or synchronize the rhythms.

The science dealing with these phenomena is called *chronobiology*. Halberg and Montalbetti (H3) have published a theoretical and practical introduction to the specification of reference values which takes into account these types of time-dependent variation. Although the determination of time-specified reference intervals, which usually are narrower and more specific than their static counterparts, has a solid theoretical basis, such intervals are yet relatively rarely used in practical laboratory medicine. One reason may be the unusual terminology used (H3). Concepts such as mesor, acrophase, chronodesm, mondesm, and merodesm may constitute an intellectual barrier for many laboratory scientists. The theory as such, however, is not too difficult to understand, and its practical implications are important. For instance, it is well known that the concentration of cortisone in plasma exhibits a cyclic variation throughout a 24-h period: high concentrations in the morning and low levels in the evening. One may therefore estimate separate reference intervals for plasma cortisol assayed in samples collected in the morning and in the evening. Alternatively, one may determine a reference band covering the complete 24-h period using empirical data collected at selected times and appropriate mathematical methods (H3). Frequently an underlying cosine function is assumed (Fig. 8). Subject-based (personal) or group-based time-specified reference intervals or reference intervals or reference bands with appropriate periods may greatly reduce the frequency of false positives.

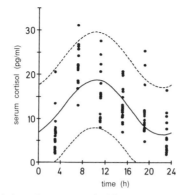

FIG. 8. Circadian variation of serum cortisol in 17- to 20-yr-old women in Como, Italy (May, 1961; $n = 16$). The curves for the mean value and reference limits were obtained by fitting a cosine function. From Ref. H3 with permission of the authors and *Bull. Mol. Biol. Med.*

6. Presentation of Observed Values Related to Reference Values

6.1. DESIRABLE FEATURES

A satisfactory laboratory report should not only mention patient identification data, nature of specimen (serum, urine), time when specimen was taken, type of quantity (component, analyte, constituent), and analytical results, but should also help the clinician to interpret the result. This can be done by indicating the position of the observed value in relation to reference values (I10). In addition, further statements could be made, such as "clearly pathological" or "rather low for third trimester."

Ideally, observed values should be related to several sets of reference values, such as from healthy persons, from the general hospital population, from patients with typical diseases, and from ambulant persons, and such as ideal values and previous values from the same subject (Fig. 2). If comparison is made with values from populations, they should be from relevant age and sex categories (see Sections 2.2.3 and 4.2.1). The specimens should have been collected under the same conditions; in cases where posture has an effect, values from sitting subjects should be compared with reference values from sitting populations, and values from supine persons should be compared with values from supine populations, etc.

It is desirable to automatically supply the clinician with many sets of reference values because the laboratory staff usually does not know which reference values are the most relevant for the patient, who may be lying in bed on one day and ambulant on another. The doctor would then have the option of selecting the reference values he considers most relevant.

6.2. NUMERICAL METHODS

The sixth document of the IFCC recommendations (I10) deals with numerical methods for presentation of observed values (clinical results) in relation to reference values. Such methods may be univariate (D5, I10) or multivariate (A3). Univariate methods, i.e., the traditional ones, relate each observed value to reference values of the same analyte, while multivariate methods compare a pattern, consisting of results of several analytes, with their corresponding joint reference distribution.

The numerical comparisons of observations with the relevant reference distributions are usually not statistical tests in the strict sense. It is not always obvious that the patients match the reference individuals. Furthermore, clinical specimens may be collected under conditions different from those when the reference values were produced. Still, methods from statistical theory may be applied as tools, provided due reservations are made when clinical data are interpreted (I10).

Irrespective of the method utilized, the original observed value of a given analyte must be reported to allow further uses of the result, e.g., comparison with other analytes and metabolic calculations. The minimum requirement is thus to report the observation as such and make reference information available to the user in the form of graphs or tables of the reference distribution(s) or reference interval(s). Reference intervals may either be published separately in laboratory guidebooks or be printed on report sheets (preprinted or printed by the laboratory computer system). To facilitate comparison of observed values with reference values, numerical indices may be calculated and reported together with the results (A3, D5, S23). Some frequently used methods are reviewed briefly below.

An observed value may be flagged on the report with an appropriate symbol according to its locations relative to the reference interval. It is important that such signaling is not interpreted as the observed value being normal, abnormal, pathological, etc. (see Section 1.2.2). The comparison is only descriptive in the sense that the value is declared typical or not of the reference values.

The classification into three classes (below, within, and above the reference interval) is wasteful of the information contained in the reference distribution. A subclassification into more than three categories only partly solves this problem. Another method is to calculate a measure of relative distance of the observed value from the location of the reference values (D5, I10, S23). A popular index, which ambiguously has been called the

SD unit or the normal equivalent deviate (G20), is $*x = (x - AV_x)/SD_x$, where x is the observed value, and AV_x and SD_x are the arithmetic mean and standard deviation, respectively, of the reference distribution. For instance, if the mean and standard deviation of triglycerides in serum are 1.13 and 0.41 mmol/liter, respectively, the distance of an observed value of 0.51 mmol/liter is $(0.51 - 1.13)/0.41 = -1.51$. The serum triglyceride concentration may accordingly be reported as 0.51 mmol/liter (-1.51). Several varieties of this distance measure have been suggested (D5), for instance, by substituting the mean with the median, and the standard deviation with other measures of the dispersion of the reference values.

Such measures of the relative distance are, however, very misleading if the reference distribution is far from the Gaussian type, skewness being particularly disastrous. The distance measure described above (the SD unit) would be -1.96 and $+1.96$ for observed values located at the limits of the central 0.95 reference interval if the reference distribution was Gaussian. This is not true when the values have a skewed distribution. Solberg (S23) therefore suggested a method to make the relative distance measure more informative in computer-based reporting systems. It is based on storage of a parameter of the mathematical function that may transform the reference distribution to approximate symmetry (Table 5) together with the mean and the standard deviation of the transformed distribution. Then the distance relative to this transformed reference distribution is found by applying the mathematical transformation prior to calculation of the index. If this method is applied to the serum triglyceride result (0.51 mmol/liter; see above), the distance measure is -1.96, i.e., the value is located at the lower reference limit. A computer subroutine for the calculation of distances with compensation for skewness, utilizing a power transformation (B7), has been published (S23). It may easily be modified for other transformations (Table 5), for example, the exponential function recommended by the IFCC (I9, M2).

The location of the observed value may also be expressed as a fractile (0.0–1.0, or, as a percentile, 0–100) of the reference distribution (D5, I10). This method is restricted to observed values located within the interval of registered reference values. An observed value located at the conventional lower or upper reference limit has a fractile value of 0.025 or 0.975, respectively.

The index of atypicality carries essentially the same information and has the same range restriction. This index may take values from 1.0 − to 1.0 + and is defined as the "probability" of finding a result closer to the mean than the observed value (the sign suffixed to the value indicates location below or above the mean). An observed value located at the lower or upper limit of the 0.95 reference interval has an index value of 0.95 −

or 0.95+, respectively. Computer routines are available for the calculation of fractiles (R4, S23) and the index of atypicality (A2). The method suggested by Solberg (S23) applies a prior transformation to symmetry as described above.

The distance, fractile, and index of atypicality methods are easily extended to the multivariate situation (A3). The relevant distance measure is Mahalanobis' squared distance, which is a multivariate analog to the square of the univariate relative distance (SD unit; see above). It is an index that locates the multivariate observation in relation to the corresponding reference region (see Section 5.1), and it may be expressed as a fractile (A3, K1). The multivariate index of atypicality is analogous to the univariate index. It indicates the "probability" of finding a result pattern closer to the common mean of the multivariate reference distribution than the observation (A2, A3).

6.3. Graphical Presentation

As described in the previous section, one possibility is to supplement the result with a distance, a fractile, or an index of the relevant reference population. A more sophisticated system would give several such measures corresponding to different reference sample groups, e.g., ambulant, supine, healthy, and diseased. As described in Ref. G8, the doctors could ask the computer to provide such comparisons. The intervals can also be presented as figures; such graphs are produced by some automatic analyzers (Fig. 9). One variant is the polygon given in Fig. 10. In these systems it is also possible to provide confidence limits for the reference limits.

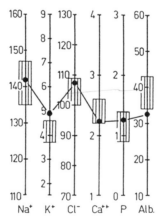

FIG. 9. The result from multichannel analysis of a specimen displayed on a chart providing reference intervals. From Ref. G8 with permission of John Wiley and Sons.

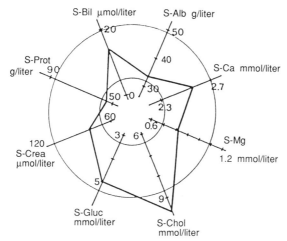

FIG. 10. Radial plot modified from a system used by B.T. Williams, Regional Health Center, Urbana, Illinois. The circles correspond to mean ± 2 SD. All analytes are from serum (S). From Ref. G8 with permission of John Wiley and Sons.

6.4. AVOIDING THE CONVENTIONAL REFERENCE INTERVAL

Today there seems to be an unwritten convention that the reference interval should include the central 95% of the reference values. Thus 5% of the data are cut off and considered to be rare or even pathological. When several tests are performed on the same subject the chance is rapidly increased that one or several of the results fall outside the respective reference intervals. One way of avoiding the problem of such false positives is to utilize multivariate analysis, i.e., to compare packages of tests with reference intervals (space figures) for such packages (see Section 5.1).

Another alternative is to abstain from reporting reference intervals, as they tend to lead the physician to use only two verdicts in judging observed values, viz. pathological or nonpathological (G10). The fractile or index of atypicality system described in Section 6.2 avoids this coarse binary evaluation of results.

One can also use clinical decision limits (S34). Clinicians utilize many limits, but frequently subconsciously. For instance, in judging hemoglobin values that are somewhat low, the doctor first sets a limit where he notes a suspect value and perhaps decides to check the result next time. When the value is lower he makes the diagnosis anemia and at this point, or when the value is still lower, he decides to treat the patient. When the value is very low and transgresses the emergency limit, the patient may be given a transfusion. A computerized laboratory report could warn about the crossing of such limits.

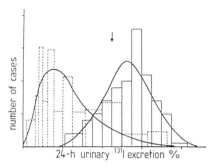

FIG. 11. The location of an observed value (arrow) related to the real (histograms) and hypothetical (curves) distributions of reference values. Results from urinary radioiodine excretion tests. From Lamberg et al. (L1), with permission of the author and *Acta Endocrinol. (Copenhagen)*.

To avoid the use of limits, a result may be presented as an arrow above one or several distribution curves or histograms describing relevant reference populations, such as euthyroid or hyperthyroid ones (Fig. 11). Figure 12 represents the same system but uses the cumulative curve. This system corresponds to the fractile system discussed above.

6.5. How to Avoid Being Deluged with Data

As stated, it is desirable to relate an observed result to several collections of reference values. However, that means that the laboratory report contains numerous figures or pictures. When several tests are made on the same patient and results of previous tests are also reported, huge quantities of data tend to accumulate. Cumulative reports are therefore needed. A way of preventing the accumulation of too much paper is to place a display terminal in the doctor's office. In a "laboratory round" the doctor would

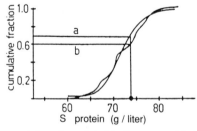

FIG. 12. A serum (S) protein value related to the parametric (a) and nonparametric (b) cumulative reference distributions. From Ref. G8 with permission of John Wiley and Sons.

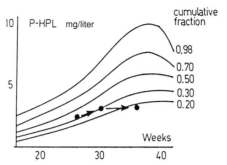

FIG. 13. Three plasma human placental lactogen (P-HPL) values (connected by arrows) at different times of pregnancy in the same individual, plotted on a chart giving the fractiles observed at different times of uneventful pregnancies. From Ref. G8 with permission of John Wiley and Sons.

inspect the graphs on the screen (C1, S28), then ask the computer some questions, possibly noting some unusual results, and finally ordering printouts of interesting figures (G8).

It is also possible to present cumulative reports graphically. Figure 13 provides reference values from a population and gives simultaneously the current observed value and population-based and individual reference data. Most of the display systems discussed require a computer. The reference value field therefore presents interesting challenges to computer programmers.

7. Final Remarks

The reference value field is vast, and one may wonder whether optimally good data and utilization procedures will ever be produced, especially since the analytical procedures change and therefore new reference values are constantly needed. On the other hand, we clearly have to improve the production and use of our reference data. And now that good data are still lacking, teaching of the biological and mathematical principles governing this field is needed to produce enlightened physicians and laboratory scientists who in due course hopefully will produce good reference data and understand how to use them.

It is also important to know what should not be done. A selection of a few rules concludes this review:

1. Do not use reference value nomenclature if you do not describe how you obtained and treated the reference values.

2. Do not call the reference interval for reference values or "normal reference range," etc.

3. Do not compute the reference limits as mean ± 2SD regardless of the shape of the distribution.

4. Do not supply reference values collected in other places or with other methods without checking their relevance in your laboratory and cautioning the user.

8. Abbreviations and Symbols

A^2	Anderson–Darling test statistic
AV_x, AV_y, AV_z	Arithmetic mean (average) of x, y, and z, respectively
α	Fraction defining a fractile
β	Degree of confidence
c	Standard Gaussian deviate
D_{max}	Kolmogorov–Smirnov test statistic
exp	$\exp(x) = e^x$, i.e., the xth power of e (natural base number)
f	Fractile (reference limit)
g_k, g_s	Coefficients of kurtosis and skewness, respectively
IFCC	International Federation of Clinical Chemistry
k	Number of dimensions in a multivariate system
ln	Natural (Napierian) logarithm
L	Least significant digit
m_r	rth moment about the mean
N	Number of reference values (size of reference sample group)
NCCLS	National Committee for Clinical Laboratory Standards
Q, R, S	Parameters of transforming functions.
r	Ratio of intra- to interindividual variation
SD_k, SD_s	Standard deviations of the coefficients g_k and g_s, respectively (see above)
SD_x, SD_y, SD_z	Standard deviations of x, y, and z, respectively
SE_f	Standard error of the fractile f
v_i	Standardized difference between the ith reference value and the mean
w_i	Cumulative Gaussian probability corresponding to the ith reference value
W^2	Cramér–von Mises test statistic
x_i	ith reference value
Y_i	ith transformed reference value (in the two-stage transformation it denotes the value after transformation to correct for skewness)
z_i	ith transformed reference value after the application of both transformations in the two-stage procedure

Acknowledgment

We thank the Nordic Clinical Chemistry Project (NORDKEM) and the Sigrid Jusélius Foundation for support.

REFERENCES

A1. Adlercreutz, H., and Tallqvist, G., Variations in the serum total cholesterol and hematocrit values in normal women during the menstrual cycle. *Scand. J. Clin. Lab. Invest.* **11,** 1–9 (1959).
A2. Albert, A., Atypicality indices as reference values for laboratory data. *Am. J. Clin. Pathol.* **76,** 421–425 (1981).
A3. Albert, A., and Heusghem, C., Relating observed values to reference values: The multivariate approach. *In* "Reference Values in Laboratory Medicine. The Current State of the Art" (R. Gräsbeck and T. Alström, eds.), pp. 289–296. Wiley, Chichester, 1981.
A4. Alström, T., Gräsbeck, R., Hjelm, M., and Skandsen, S., Recommendations concerning the collection of reference values in clinical chemistry and activity report. *Scand. J. Clin. Lab. Invest.* **35,** Suppl. 144 (1975).
A5. Alström, T., Dahl, M., Gräsbeck, R., Hagenfeldt, L., Hertz, H., Hjelm, M., Järvenpää, A.-L., Kantero, R., Larsson, A., Leskinen, E. E. A., Lindblad, B., Moe, P. J., Nyberg, A. P. W., Näntö, V., Olesen, H., Siimes, M., Solberg, H. E., Strandvik, B., Wimberley, P. D., and Winkel, P., Recommendation for collection of skin puncture blood from children, with special reference to production of reference values. *Scand. J. Clin. Lab. Invest.* **47,** 199–205 (1987).
A6. Anderson, T. W., "An Introduction to Multivariate Statistical Analysis." Wiley, New York, 1958.
A7. Ash, K. O., Clark, S. J., Sandberg, L. B., Hunter E., and Woodward, S. C., The influences of sample distribution and age on reference intervals for adult males. *Am. J. Clin. Pathol.* **79,** 574–581 (1983).
B1. Baadenhuijsen, H., and Smit, J. C., Indirect estimation of clinical chemical reference intervals from total hospital patient data: Application of a modified *Bhattacharya* procedure. *J. Clin. Chem. Clin. Biochem.* **23,** 829–839 (1985).
B2. Baadenhuijsen, H., Arts, J., Somers, L., and Smit, J. C., "REFVALUE. A Software Package to Calculate Reference Intervals from Total Hospital Patient Laboratory Data." Elsevier, Amsterdam, 1984. (See Ref. S24 concerning a different program with a similar name: REFVAL.)
B3. Barnett, V., and Lewis, T., "Outliers in Statistical Data." Wiley, Chichester, 1978.
B4. Behr, W., Schlimok, G., Firchau, V., and Paul, H. A., Determination of reference intervals for 10 serum proteins measured by rate nephelometry, taking into consideration different sample groups and different distribution functions. *J. Clin. Chem. Clin. Biochem.* **23,** 157–166 (1985).
B5. Berg, B., Nilsson, J.-E., Solberg, H. E., and Tryding, N., Practical experience in the selection and preparation of reference individuals: Empirical testing of the provisional Scandinavian recommendations. *In* "Reference Values in Laboratory Medicine. The Current State of the Art" (R. Gräsbeck and T. Alström, eds.), pp. 55–64. Wiley, Chichester, 1981.
B6. "BMDP" (for Mainframes) and "BMDPC" (for IBM PC/XT/AT or Compatibles). BMDP Statistical Software, Inc., Los Angeles, California.
B7. Box, G. E. P., and Cox, D. R., An analysis of transformations. *J. R. Stat. Soc., Ser. B* **26,** 211–252 (1964).
B8. Boyd, J. C., Perspectives on the use of chemometrics in laboratory medicine. *Clin. Chem. (Winston-Salem, N.C.)* **32,** 1726–1733 (1986).
B9. Boyd, J. C., and Lacher, D. A., The multivariate reference range: An alternative interpretation of multi-test profiles. *Clin. Chem. (Winston-Salem, N.C.)* **28,** 259–265 (1982).

B10. Boyd, J. C., and Lacher, D. A., A multi-stage Gaussian transformation algorithm for clinical laboratory data. *Clin. Chem. (Winston-Salem, N.C.)* **28,** 1735–1741 (1982).
B11. Bradley, J. V., "Distribution-Free Statistical Tests." Prentice-Hall, Englewood Cliffs, New Jersey, 1968.
B12. Brendel, W., and Zink, R. A., eds., "High Altitude Physiology and Medicine." Springer-Verlag, New York, 1982.
C1. Connelly, D. P., Lasky, L. C., Keller, R. M., and Morrison, D. S., A system for graphical display of clinical laboratory data. *Am. J. Clin. Pathol.* **78,** 729–737 (1982).
C2. Cooke, D., Craven, A. H., and Clarke, G. M., "Statistical Computing in Pascal." Edward Arnold, London, 1985.
C3. Cramér, H., "Mathematical Methods of Statistics," pp. 184, 352, and 361. Princeton Univ. Press, Princeton, New Jersey, 1946.
D1. Dann, T.C., Routine skin preparation before injection. *Lancet* **1,** 96–98 (1969).
D2. DeMets, D. L., and Harris, E. K., "Transformation of Observed Distributions to Gaussian Form," Tech. Rep. No. 8. Laboratory of Applied Studies, Division of Computer Research and Technology, National Institutes of Health, Bethesda, Maryland, 1972.
D3. De Verdier, C.-H., Aronsson, T., and Nyberg, A., eds., Quality control in clinical chemistry—Efforts to find an efficient strategy. *Scand. J. Clin. Lab. Invest.* **44,** Suppl. 172 (1984).
D4. Dixon, W. J., Processing data for outliers. *Biometrics* **9,** 74–89 (1953).
D5. Dybkær, R., Observed value related to reference values. *In* "Reference Values in Laboratory Medicine. The Current State of the Art" (R. Gräsbeck and T. Alström, eds.), pp. 263–278. Wiley, Chichester, 1981.
D6. Dybkær, R., Jörgensen, K., and Nyboe, J., Statistical terminology in clinical chemistry reference values. *Scand. J. Clin. Lab. Invest.* **35,** Suppl. 144, 45–74 (1975).
D7. Dybkær, R., Lauritzen, M., and Krakauer, R., Relative reference values for clinical chemical and haematological quantities in "healthy" elderly people. *Acta Med. Scand.* **209,** 1–9 (1981).
E1. Edén, S., Jagenburg, R., Lindstedt, G., Lundberg, P.-A., and Mellström, D., Interrelationships among body mass, thyrotropin, thyroid hormones, and thyroid-hormone binding proteins in healthy 70-year-old men. *Clin. Chem. (Winston-Salem, N.C.)* **30,** 681–686 (1984).
E2. Elveback, L. R., Guillier, C. L., and Keating, F. R. Health, normality, and the ghost of Gauss. *JAMA, J. Am. Med. Assoc.* **211,** 69–75 (1970).
F1. Fejfar, Z., WHO's work in cardiovascular diseases: A brief review. *WHO Chron.* **25,** 354–362 (1971).
F2. Felding, P., Tryding, N., Petersen, P. H., and Hörder, M., Effects of posture on concentrations of blood constituents in healthy adults: Practical application of blood specimen collection procedures recommended by the Scandinavian Committee on Reference Values. *Scand. J. Clin. Lab. Invest.* **40,** 615–621 (1980).
F3. Forest, J.-C., Garrido-Russo, M., Lemay, A., Carrier, R., and Dube, J.-L., Reference values for the oral glucose test at each trimester of pregnancy. *Am. J. Clin. Pathol.* **80,** 828–831 (1983).
G1. Gallo, C., Truco, M. C., Sanzi, G., La Rocca, S., Sacchetti, L., and Salvatore, F., Application of a statistical approach to determining reference intervals in clinical chemistry. *Clin. Chem. (Winston-Salem, N.C.)* **31,** 1367–1370 (1985).
G2. Gerver, W. J. M., Laan, C. G. v. d., Drayer, N. M., and Schaafsma, W., Smoothing techniques for obtaining reference values for measurements of children. *Int. J. Bio-Med. Comput.* **16,** 29–34 (1985).

G3. Goldin, B. R., Adlercreutz, H., Gorbach, S. L., Woods, M. N., Dwyer, J. T., Conlon, T., Bohn, E., and Gershoff, S. N., The relationship between estrogen levels and diets of Caucasian American and Oriental immigrant women. *Am. J. Clin. Nutr.* **44**, 945–953 (1986).
G4. Gómez, P., Coca, C., Vargas, C., Acebillo, J., and Martinez, A., Normal reference-intervals for 20 biochemical variables in healthy infants, children, and adolescents. *Clin. Chem. (Winston-Salem, N.C.)* **30**, 407–412 (1984).
G5. Gräsbeck, R., Types of reference groups. *Scand. J. Clin. Lab. Invest.* **29**, Suppl. 126, 19.2a-c (1972).
G6. Gräsbeck, R., Terminology and biological aspects of reference values. *In* "Logic and Economics of Clinical Laboratory Use" (E. S. Benson and M. Rubin, eds.), pp. 77–90. Am. Elsevier, New York, 1978.
G7. Gräsbeck, R., Health as seen from the laboratory. *In* "Reference Values in Laboratory Medicine. The Current State of the Art" (R. Gräsbeck and T. Alström, eds.), pp. 17–24. Wiley, Chichester, 1981.
G8. Gräsbeck, R., Display of results with reference values. *In* "Reference Values in Laboratory Medicine. The Current State of the Art" (R. Gräsbeck and T. Alström, eds.), pp. 279–288. Wiley, Chichester, 1981.
G9. Gräsbeck, R., Health and disease from the point of view of the clinical laboratory. *In* "Health, Disease, and Causal Explanations in Medicine" (L. Nordenfelt and B. I. B. Lindahl, eds.), pp. 47–60. Reidel Publ., New York, 1984.
G10. Gräsbeck, R., Considerations in determination of reference intervals. *In* "Clinical Laboratory Forum" (J. Homolko, ed.), pp. 9–13. Becton Dickinson Vacutainer Systems Europe, Meylan, 1985.
G11. Gräsbeck, R., Reference values: Relevant controls for clinical decision-making. *Eur. J. Haematol.* **40**, 1–6 (1988).
G12. Gräsbeck, R., and Alström, T., eds., "Reference Values in Laboratory Medicine. The Current State of the Art." Wiley, Chichester, 1981.
G13. Gräsbeck, R., and Fellman, J., Normal values and statistics. *Scand. J. Clin. Lab. Invest.* **21**, 193–195 (1968).
G14. Gräsbeck, R., and Salonen, E.-M., Vitamin B12. *Prog. Food Nutr. Sci.* **2**, 193–231 (1976).
G15. Gräsbeck, R., and Saris, N.-E., Establishment and use of normal values. *Scand. J. Clin. Lab. Invest.* **26**, Suppl. 110, 62–63 (1969).
G16. Gräsbeck, R., and Solberg, H. E., Can reference values be properly used? *Clin. Chem. (Winston-Salem, N.C.)* **27**, 1795–1796 (1981).
G17. Gräsbeck, R., Nyberg, W., Saarni, M., and von Bonsdorff, B., Lognormal distribution of serum vitamin B12 levels and dependence of blood values on the B12 level in a large population heavily infected with *Diphyllobothrium latum*. *J. Lab. Clin. Med.* **59**, 419–429, 1962.
G18. Gräsbeck, R., Tikkanen, I., Yki-Järvinen, H., Ojala, K., and Weber, T., Common types of quantities in laboratory personnel under standardized conditions, in the afternoon (1430 h) and following an evening party. *In* "Reference Values in Laboratory Medicine. The Current State of the Art" (R. Gräsbeck and T. Alström, eds.), pp. 369–375. Wiley, Chichester, 1981.
G19. Gruemer, H.-D., Miller, W. G., Chinchilli, V. M., Leshner, R. T., Hassler, C. R., Blasco, P. A., Nance, W. E., and Goldsmith, B. M., Are reference limits for serum creatine kinase valid in detection of the carrier state for Duchenne muscular dystrophy? *Clin. Chem. (Winston-Salem, N.C.)* **30**, 724–730 (1984).
G20. Gullick, H. D., and Schauble, M. K., SD unit system for standardizing reporting and interpretation of laboratory data. *Am. J. Clin. Pathol.* **57**, 517–525 (1972).

H1. Hahn, G. J., Statistical intervals for a normal population, Part I. Tables, examples and applications. *J. Qual. Technol.* **2**, 115–125 (1970).
H2. Hahn, G. J., Statistical intervals for a normal population. Part II. Formulas, assumptions, some derivations. *J. Qual. Technol.* **2**, 195–206 (1970).
H3. Halberg, F., and Montalbetti, N., Chronobiologic specification of reference values. *Bull. Mol. Biol. Med.* **8**, 75–103 (1983).
H4. Hald, A., "Statistical Theory with Engineering Applications." Wiley, New York, 1952.
H5. Harris, E. K., Effects of intra- and interindividual variation on the appropriate use of normal ranges. *Clin. Chem. (Winston-Salem, N.C.)* **20**, 1535–1542 (1974).
H6. Harris, E. K., Some theory of reference values. I. Stratified (categorized) normal ranges and a method for following an individual's clinical laboratory values. *Clin. Chem. (Winston-Salem, N.C.)* **21**, 1457–1464 (1975).
H7. Harris, E. K., Some theory of reference values. II. Comparison of some statistical models of intraindividual variation in blood constituents. *Clin. Chem. (Winston-Salem, N.C.)* **22**, 1343–1350 (1976).
H8. Harris, E. K., Step-by-step algorithm for computing critical range of next observation based on previous time series. *Clin. Chem. (Winston-Salem, N.C.)* **23**, 2179–2180 (1977).
H9. Harris, E. K., Statistical aspects of reference values in clinical pathology. *Prog. Clin. Pathol.* **8**, 45–66 (1981).
H10. Harris, E. K., Further applications of time series analysis to short series of biochemical measurements. *In* "Reference Values in Laboratory Medicine. The Current State of the Art" (R. Gräsbeck and T. Alström, eds.), pp. 167–176. Wiley, Chichester, 1981.
H11. Harris, E. K., and DeMets, D. L., Estimation of normal ranges and cumulative proportions by transforming observed distributions to Gaussian form. *Clin. Chem. (Winston-Salem, N.C.)* **18**, 605–612 (1972).
H12. Harris, E. K., and Yasaka, T., On the calculation of a "reference change" for comparing two consecutive measurements. *Clin. Chem. (Winston Salem, N.C.)* **29**, 25–30 (1983), with an addendum in *ibid.*, p. 997.
H13. Harris, E. K., Kanofsky, P., Shakarji, G., and Cotlove, E., Biological and analytic components of variation in long-term studies of serum constituents in normal subjects. *Clin. Chem. (Winston-Salem. N.C.)* **16**, 1022–1027 (1970).
H14. Harris, E. K., Yasaka, T., Horton, M. R., and Shakarji, G., Comparing multivariate and univariate subject-specific reference regions for blood constituents in healthy persons. *Clin. Chem. (Winston-Salem, N.C.)* **28**, 422–426 (1982).
H15. Hawkins, D. M., "Identification of Outliers." Chapman & Hall, London, 1980.
H16. Healy, M. J. R., Normal values from a statistical viewpoint. *Bull. Acad. R. Med. Belg.* **9**, 703–718 (1969).
H17. Hicks, J. M., Hammond, K., and Boeckx, R. L., Pediatric reference values. *In* "Reference Values in Laboratory Medicine. The Current State of the Art" (R. Gräsbeck and T. Alström, eds.), pp. 297–309. Wiley, Chichester, 1981.
H18. Hjelm, M., Leyssen, M. H. T., Tentori, L., Verwilghen, R. L., Solberg, H. E., PetitClerc, C., Stamm, D., and Wilding, P., Standardization of blood specimen collection procedure for reference values. *Clin. Lab. Haematol.* **4**, 83–86 (1982).
I1. International Federation of Clinical Chemistry (Expert Panel on Drug Effects in Clinical Chemistry), Drug effects in clinical chemistry. Part 1. The basic concepts (a proposal for an IFCC recommendation). *J. Clin. Chem. Clin. Biochem.* **22**, 271–274 (1984); *Clin. Chim. Acta* **139**, 215F–221F (1984).
I2. International Federation of Clinical Chemistry (Expert Panel on Nomenclature and Principles of Quality Control in Clinical Chemistry), Approved recommendation (1978)

on quality control in clinical chemistry. Part 1. General principles and terminology. *Clin. Chim. Acta* **98**, 129F–143F (1979); *J. Clin. Chem. Clin. Biochem.* **18**, 69–77 (1980).

I3. International Federation of Clinical Chemistry (Expert Panel on Nomenclature and Principles of Quality Control in Clinical Chemistry), Approved recommendation (1983) on quality control in clinical chemistry. Part 4. Internal quality control. *J. Clin. Chem. Clin. Biochem.* **21**, 877–884 (1983).

I4. International Federation of Clinical Chemistry (Expert Panel on Quantities and Units in Clinical Chemistry) and International Union of Pure and Applied Chemistry (Commission on Quantities and Units in Clinical Chemistry), Approved recommendation (1978), quantities and units in clinical chemistry. *Clin. Chim. Acta* **96**, 155F–183F (1979); *J. Clin. Chem. Clin. Biochem.* **17**, 807–821 (1979); *Pure Appl. Chem.* **51**, 2451–2479 (1979).

I5. International Federation of Clinical Chemistry (Expert Panel on Theory of Reference Values) and International Committee for Standardization in Haematology, Approved recommendation (1986) on the theory of reference values. Part 1. The concept of reference values. *J. Clin. Chem. Clin. Biochem.* **25**, 337–342 (1987); *Clin. Chim. Acta* **165**, 111–118 (1987); *Labmedica* **4**, 27–31 (1987); *Ann. Biol. Clin.* **45**, 237–241 (1987).

I6. International Federation of Clinical Chemistry (Expert Panel on Theory of Reference Values), Approved recommendation (1987) on the theory of reference values. Part 2. Selection of individuals for the production of reference values. *J. Clin. Chem. Clin. Biochem.* **25**, 639–644 (1987); *Clin. Chim. Acta* **170**, S1–S12 (1987).

I7. International Federation of Clinical Chemistry (Expert Panel on Theory of Reference Values), Approved recommendation (1988) on the theory of reference values. Part 3. Preparation of individuals and collection of specimens for the production of reference values. *J. Clin. Chem. Clin. Biochem.* **26**, 593–598 (1988); *Clin. Chem. Acta* **177**, S1–S12 (1988).

I8. International Federation of Clinical Chemistry (Expert Panel on Theory of Reference Values), The theory of reference values. Part 4. Control of analytical variation in the production, transfer and application of reference values. In preparation.

I9. International Federation of Clinical Chemistry (Expert Panel on Theory of Reference Values) and International Committee for Standardization in Haematology, Approved recommendation (1987) on the theory of reference values. Part 5. Statistical treatment of collected reference values. Determination of reference limits. *J. Clin. Chem. Clin. Biochem.* **25**, 645–656 (1987); *Clin. Chim. Acta* **170**, S13–S32 (1987).

I10. International Federation of Clinical Chemistry (Expert Panel on Theory of Reference Values) and International Committee for Standardization in Haematology, Approved recommendation (1987) on the theory of reference values. Part 6. Presentation of observed values related to reference values. *J. Clin. Chem. Clin. Biochem.* **25**, 657–662 (1987); *Clin. Chim. Acta* **170**, S33–S42 (1987); *Labmedia* **5**, 27–30 (1988).

J1. John, J. A., and Draper, N. R., An alternative family of transformations. *Appl. Stat.* **29**, 190–197 (1980).

J2. Johnson, N. L., Systems of frequency curves generated by methods of translation. *Biometrika* **36**, 149–176 (1949).

J3. Johnson, N. L., Tables to facilitate fitting S_u frequency curves. *Biometrika* **52**, 547–558 (1965).

K1. Kågedal, B., Sandström, A., and Tibbling, G., Determination of a trivariate reference region for free thyroxine index, free triiodothyronine index, and thyrotropin from results obtained in a health survey of middle-aged women. *Clin. Chem. (Winston-Salem, N.C.)* **24**, 1744–1750 (1978).

K2. Kågedal, B., Larsson, L., Norr, A., and Toss, G., Trivariate evaluation of a thyroid hormone panel in clinical practice compared with multiple univariate evaluation. *Scand. J. Clin. Lab. Invest.* **42**, 177–180 (1982).

K3. Kaskey, G., Krishnaiah, P. R., Kolman, B., and Steinberg, L., Transformations to normality. *In* "Handbook of Statistics" (P. R. Krishnaiah, ed.), Vol. 1, pp. 321–341. North-Holland Publ., Amsterdam, 1980.

K4. Kendall, M. G., and Buckland, W. R., "A Dictionary of Statistical Terms." Longmans, Green, London, 1971.

K5. Kieviet, W. de, Slaats, E. H., and Abeling, N. G. G. M., Pediatric reference values for calcium, magnesium and inorganic phosphorus in serum obtained from *Bhattacharya* plots for data from unselected patients. *J. Clin. Chem. Clin. Biochem.* **24**, 233–242 (1986).

K6. Kreutz, F. H., Auswirkungen der Probenahme auf klinisch-chemische Untersuchungsergebnisse. *In* "Optimierung der Diagnostik" (H. Lang, W. Rick, and L. Róka, eds.), pp. 149–163. Springer-Verlag, Berlin, 1973.

L1. Lamberg, B.-A., Heinonen, O. P., Liewendahl, K., Kvist, G., Viherkoski, M., Aro, A., Laitinen, O., and Knekt, P., Laboratory tests on thyroid function in hyperthyroidism. II. A statistical evaluation of 13 different variables. *Acta Endocrinol.* Suppl. **146**, 23–25 (1970).

L2. Landahl, S., Jagenburg, R., and Svanborg, A., Blood components in a 70-year-old population. *Clin. Chim. Acta* **112**, 301–304 (1981).

L3. Lange, H. F., The normal plasma protein values and their relative variations. *Acta Med. Scand.,* Suppl. **176** (1946).

L4. Laurell, C. B., Kullander, S., and Thorell, J., Effect of administration of a combined estrogen-progestin contraceptive on the level of individual plasma proteins. *Scand. J. Clin. Lab. Invest.* **21**, 337–343 (1967).

L5. Lentner, C., ed., "Geigy Scientific Tables," Vol. 2. Ciba-Geigy Ltd., Basle, 1982.

L6. Leppänen, E. A., and Gräsbeck, R., The effect of the order of filling tubes after venipuncture on serum potassium, total protein, and aspartate and alanine aminotransferases. *Scand. J. Clin. Lab. Invest.* **46**, 189–191 (1986).

L7. Leppänen, E. A., and Gräsbeck, R., Experimental basis of standardized specimen collection: Effects of moderate alcohol consumption and exercise on some serum components. *Scand. J. Clin. Lab. Invest.* **46**, Suppl. 185, 191 (1986).

L8. Leppänen, E. A., and Gräsbeck, R., Experimental basis of standardized specimen collection: The effect of posture on the blood picture. *Eur. J. Haematol.* **40**, 222–226 (1988).

L9. Letellier, G., and Desjarlais, F., Study of seasonal variations for eighteen biochemical parameters over a four-year period. *Clin. Biochem. (Ottawa)* **15**, 206–211 (1982).

L10. Linnet, K., "Goodness-of-Fit Tests on Distributions of Transformed Sample Values," Tech. Rep. Department of Clinical Chemistry KK, Rigshospitalet, Copenhagen, Denmark, (1986).

L11. Linnet, K., Two-stage transformation systems for normalization of reference distributions evaluated. *Clin. Chem. (Winston-Salem, N.C.)* **33**, 381–386 (1987).

L12. Linnett, K., Testing normality of transformed data. *Appl. Stat.* **37**, 180–186 (1988).

M1. Madison, L. L., Ethanol-induced hypoglycemia. *Adv. Metab. Disord.* **3**, 85–109 (1968).

M2. Manly, B. F. J., Exponential data transformations. *Statistician* **25**, 37–42 (1976).

M3. Mardia, K. V., Tests of univariate and multivariate normality. *In* "Handbook of Statistics" (P. R. Krishnaiah, ed.), Vol. 1, pp. 279–320. North-Holland Publ., Amsterdam, 1980.

M4. Martin, H. F., Gudzinowicz, B. J., and Fanger, H., "Normal Values in Clinical Chemistry. A Guide to Statistical Analysis of Laboratory Data." Dekker, New York, 1975.

M5. Martin, H. F., Hologgitas, J. V., Driscoll, J., Fanger, H., Gudzinowicz, B. J., Barnett, R., and Weisbrot, I., Reference values based on populations accessible to hospitals. *In* "Reference Values in Laboratory Medicine. The Current State of the Art" (R. Gräsbeck and T. Alström, eds.), pp. 233–262. Wiley, Chichester, 1981.

M6. Meites, S., and Levitt, M. J., Skin-puncture and blood-collecting techniques for infants. *Clin. Chem. (Winston-Salem, N. C.)* **25,** 183–189 (1979).

M7. Mikkelsen, W. M., Dodge, H. J., and Valkenburg, H., The distribution of serum uric acid values in a population unselected as to gout or hyperuricemia. *Am. J. Med.* **39,** 242–251 (1965).

M8. Miller, W. G., Chinchilli, V. M., Gruemer, H.-D., and Nance, W. E., Sampling from a skewed population distribution as exemplified by estimation of the creatine kinase upper reference limit. *Clin. Chem. (Winston-Salem, N. C.)* **30,** 18–23 (1984).

M9. Morrison, D. F., "Multivariate Statistical Methods," 2nd ed. McGraw-Hill, New York, 1976.

M10. Munan, L., Population based reference data. *In* "Logic and Economics of Clinical Laboratory Use" (E. S. Benson and M. Rubin, eds.), pp. 117–128. Am. Elsevier, New York, 1978.

M11. Murphy, E. A., The normal, and the perils of the sylleptic argument. *Perspect. Biol. Med.* **15,** 566–582 (1972).

N1. National Committee for Clinical Laboratory Standards, "Standard Procedures for the Collection of Diagnostic Blood Specimens by Venipuncture." NCCLS, Villanova, 1977.

N2. National Committee for Clinical Laboratory Standards, "Approved Standard Procedures for the Collection of Diagnostic Blood Specimens by Skin Puncture." NCCLS, Villanova, 1982.

N3. Naus, A. J., Borst, A., and Kuppens, P. S., Determination of n-dimensional reference ellipsoids using patient data. *J. Clin. Chem. Clin. Biochem.* **20,** 75–80 (1982).

P1. Pearson, E. S., and Hartley, H. O., eds., "Biometrika Tables for Statisticians," 3rd ed., Vol. 1. Cambridge Univ. Press, London and New York, 1966.

P2. PetitClerc, C., and Kelly, A., Transferability of reference data. *In* "Reference Values in Laboratory Medicine. The Current State of the Art" (R. Gräsbeck and T. Alström, eds.), pp. 177–192. Wiley, Chichester, 1981.

P3. Pickup, J. F., Harris, E. K., Kearns, M., and Brown, S. S., Intra-individual variation of some serum constituents and its relevance to population-based reference ranges. *Clin. Chem. (Winston-Salem, N. C.)* **23,** 842–850 (1977).

Q1. Quételet, A., "Sur l'homme et le développement de ses facultés, ou Essai de physique sociale." Bachelier, Imprimeur-Libraire, Paris, 1835.

R1. Reed, A. H., and Wu, G. T., Evaluation of a transformation method for estimation of normal range. *Clin. Chem. (Winston-Salem, N. C.)* **20,** 576–581 (1974).

R2. Reed, A. H., Henry, R. J., and Mason, W. B., Influence of statistical method used on the resulting estimate of normal range. *Clin. Chem. (Winston-Salem, N. C.)* **17,** 275–284 (1971).

R3. Reed, A. H., Cannon, D. C., Winkelman, J. W., Bhasin, Y. P., Henry, R. J., and Pileggi, V. J., Estimation of normal ranges from a controlled sample survey. I. Sex- and age-related influence on the SMA 12/60 screening group of tests. *Clin. Chem. (Winston-Salem, N. C.)* **18,** 57–66 (1972).

R4. Rossing, R. G., and Hatcher, W. E., A computer program for estimation of reference percentile values in laboratory data. *Comput. Programs Biomed.* **9**, 69–74 (1979).
S1. Sachs, L., "Applied Statistics. A Handbook of Techniques." Springer-Verlag, New York, 1982.
S2. "SAS" (Statistical Analysis System), Version 5 (for Mainframe Computers) and Version 6 (for IBM PC/XT/AT, IBM PS/2, or Compatibles). SAS Institute Inc., Cary, North Carolina.
S3. Schauble, M. K., Bectel, J. M., Gullick, H. D., and Kaplow, L. S., Pooled normal values, a useful technic. *Am. J. Clin. Pathol.* **67**, 386–392 (1977).
S4. SEQC (Sociedad Española de Química Clínica) (Comisión Valores de Referencia), Documento A: Concepto de valores de referencia en química clínica. *Quím. Clín.* **2**, 39–41 (1983).
S5. SEQC (Sociedad Española de Química Clínica) (Comisión Valores de Referencia), Documento B: Bases estadísticas de la theoría de valores de referencia. *Quím. Clín.* **2**, 95–105 (1983).
S6. SEQC (Sociedad Española de Química Clínica) (Comisión Valores de Referencia), Documento C: Variaciones analíticas y extraanalíticas en la producción de valores de referencia. *Quím. Clín.* **3**, 43–49 (1984).
S7. SEQC (Sociedad Española de Química Clínica) (Comisión Valores de Referencia), Documento D: Producción y utilización de valores de referencia. *Bol. Soc. Esp. Quím. Clín.* **31** (1985).
S8. SFBC (Société Française de Biologie Clinique) (Commission "Valeurs de référence"), Document A: Le concept de valeurs de référence en biologie clinique. *Ann. Biol. Clin. (Paris)* **39**, 381–384 (1981).
S9. SFBC (Société Française de Biologie Clinique) (Commission "Valeurs de référence"), Document B: Langage et principes statistiques pour les valeurs de référence. *Ann. Biol. Clin. (Paris)* **43**, 297–305 (1985).
S10. SFBC (Société Française de Biologie Clinique) (Commission "Valeurs de référence"), Document C: Influence des facteurs analytiques sur les valeurs de référence. *Ann. Biol. Clin. (Paris)* **43**, 306–309 (1985).
S11. SFBC (Société Française de Biologie Clinique) (Commission "Valeurs de référence"), Document D: Les variations biologiques des examens de laboratoire. *Ann. Biol. Clin. (Paris)* **37**, 229–239 (1979).
S12. SFBC (Société Française de Biologie Clinique) (Commission "Valeurs de référence"), Document E: Facteurs à prendre en considération pour le prélèvement sanguin en vue de l'établissement des valeurs de référence. *Ann. Biol. Clin. (Paris)* **38**, 251–265 (1980).
S13. SFBC (Société Française de Biologie Clinique) (Commission "Valeurs de référence"), Document G: Production des valeurs de référence de sujets sains. *Ann. Biol. Clin. (Paris)* **39**, 235–244 (1981); appendices in *ibid.* **42**, 243–265 (1984).
S14. SFBC (Société Française de Biologie Clinique) (Commission "Valeurs de référence"), Document H: Traitement des valeurs de référence et détermination de l'intervalle de référence. *Ann. Biol. Clin. (Paris)* **41**, 63–79 (1983).
S15. SFBC (Société Française de Biologie Clinique) (Commission "Valeurs de référence"), Document I: Présentation des valeurs observées par rapport aux valeurs de référence. *Ann. Biol. Clin. (Paris)* **41**, 225–231 (1983).
S16. SFBC (Société Française de Biologie Clinique) (Commission "Valeurs de référence"), Document J: Utilisation des valeurs de référence. *Ann. Biol. Clin. (Paris)* **40**, 697–708 (1982).

S17. Shultz, E. K., Willard, K. E., Rich, S. S., Connelly, D. P., and Critchfield, G. C., Improved reference-interval estimation. *Clin. Chem. (Winston-Salem, N. C.* **31,** 1974–1978 (1985).
S18. Siest, G., Strategy for the establishment of healthy population reference values. *In* "Reference Values in Laboratory Medicine. The Current State of the Art" (R. Gräsbeck and T. Alström, eds.), pp. 45–53. Wiley, Chichester, 1981.
S19. Siest, G., Henny, J., Schiele, F., and Young, D. S., eds., "Interpretation of Clinical Laboratory Tests. Reference Values and their Biological Variation." Biomedical Publications, Foster City, California, 1985.
S20. Snedecor, G. W., and Cochran, W. G., "Statistical Methods," 6th ed. Iowa State Univ. Press, Ames, 1967.
S21. Solberg, H. E., Statistical treatment of collected reference values and determination of reference limits. *In* "Reference Values in Laboratory Medicine. The Current State of the Art" (R. Gräsbeck and T. Alström, eds.), pp. 193–205. Wiley, Chichester, 1981.
S22. Solberg, H. E., Statistical treatment of reference values. *Bull. Mol. Biol. Med.* **8,** 13–19 (1983).
S23. Solberg, H. E., Presentation of observed values in relation to reference values. *Bull. Mol. Biol. Med.* **8,** 21–26 (1983).
S24. Solberg, H. E., "REFVAL," Tech. Rep. Department of Clinical Chemistry, Rikshospitalet, Oslo, Norway, 1983. (The program exists both as a complete program, written in TurboPascal, for PCs running under PC/MS-DOS and as a general FORTRAN subroutine library. The programs are available upon request to the author. See Ref. B2 concerining a different program with a similar name: REFVALUE).
S25. Solberg, H. E., Establishment and use of reference values. *In* "Textbook of Clinical Chemistry" (N. W. Tietz, ed.), pp. 356–386. Saunders, Philadelphia, Pennsylvania, 1986.
S26. Solberg, H. E., Statistical treatment of reference values in laboratory medicine: Testing the goodness-of-fit of an observed distribution to the Gaussian distribution. *Scand. J. Clin. Lab. Med.* **46,** Suppl. 184, 125–132 (1986).
S27. Solberg, H. E., Establishment and use of reference values. With an introduction to statistical techniques. *In* "Fundamentals of Clinical Chemistry" (N. W. Tietz, ed.), 3rd ed., pp. 197–212. Saunders, Philadelphia, Pennsylvania, 1987.
S28. Solberg, H. E., and Hansen, E., CLDS (Clinical Laboratory Display System): Clinical laboratory data presented by computer colour graphics. *In* "Computing in Clinical Laboratories. Databases, Data Presentation, Expected Developments" (C. Trendelenburg, ed.), pp. 61–62. Stuttgart, 1985.
S29. Soliman, S. A., Abdel-Hay, M. H., Tayeb, O. S., and Sulaiman, M. I., Application of transformation algorithm and nonparametric calculation in determining the reference intervals of some urire constituents and characteristics. *Clin. Chim. Acta* **166,** 9–16 (1987).
S30. "SPSS" (Statistical Package for Social Sciences; exists both in mainframe and PC versions). SPSS Inc., Chicago, Illinois.
S31. Stamm, D., A new concept for quality control of clinical laboratory investigations in the light of clinical requirements and based on reference method values. *J. Clin. Chem. Clin. Biochem.* **20,** 817–824 (1982).
S32. "STATGRAPHICS" (for IBM PC/XT/AT or Compatibles). STSC Software Publishing Group, Lockville, Maryland.
S33. Statland, B. E., Establishing decision levels in clinical chemistry. *In* "Reference Values in Laboratory Medicine. The Current State of the Art" (R. Gräsbeck and T. Alström, eds.), pp. 207–221. Wiley, Chichester, 1981.

S34. Statland, B. E., "Clinical Decision Levels for Laboratory Tests." Medical Economics Books, Oradell, New Jersey, 1983.
S35. Statland, B. E., and Winkel, P., Effects of preanalytical factors on the intraindividual variation of analytes in the blood of healthy subjects: Consideration of preparation of the subject and time of venipuncture. *CRC Crit. Rev. Clin. Lab. Sci.* **8,** 105–144 (1977).
S36. Statland, B. E., and Winkel, P., Selected pre-analytical sources of variation. *In* "Reference Values in Laboratory Medicine. The Current State of the Art" (R. Gräsbeck and T. Alström, eds.), pp. 127–137. Wiley, Chichester, 1981.
S37. Statland, B. E., Bokelund, H., and Winkel, P., Factors contributing to intra-individual variation of serum constituents. 4. Effects of posture and tourniquet application on variation of serum constituents in healthy subjects. *Clin. Chem. (Winston-Salem, N. C.)* **20,** 1513–1519 (1974).
S38. Statland, B. E., Winkel, P., and Bokelund, H., Factors contributing to intra-individual variation of serum constituents. 1. Within-day variation of serum constituents in healthy subjects. *Clin. Chem. (Winston-Salem, N. C.)* **19,** 1374–1379 (1973).
S39. Statland, B. E., Winkel, P., and Killingsworth, L. M., Factors contributing to intraindividual variation of serum constituents. 6. Physiological day-to-day variation in concentrations of 10 specific proteins in sera of healthy subjects. *Clin. Chem. (Winston-Salem, N. C.)* **22,** 1635–1638 (1976).
S40. Stephens, M. A., EDF statistics for goodness of fit and some comparisons. *J. Am. Stat. Assoc.* **69,** 730–737 (1974).
S41. Stephens, M. A., Anderson-Darling test for goodness of fit. *In* "Encyclopedia of Statistical Sciences" (S. Kotz, N. L. Johnson, and C. B. Read, eds.), Vol. 1, pp. 81–85. Wiley, New York, 1982.
S42. Strike, P. W., Michaeloudis, A, and Green, A. J., Standardizing clinical laboratory data for the development of transferable computer-based diagnostic programs. *Clin. Chem. (Winston-Salem, N. C.)* **32,** 22–29 (1986).
S43. Sunderman, F. W., Current concepts of "normal values," "reference values," and "discrimination values" in clinical chemistry. *Clin. Chem. (Winston-Salem, N. C.* **21,** 1873–1877 (1975).
S44. Swaanenburg, J. C. J. M., Rutten W. P. F., Holdrinet, A. C. J. M., and Strik, R., The determination of reference values for hematologic parameters using results obtained from patient populations. *Am. J. Clin. Pathol.* **88,** 182–191 (1987).
T1. Tryding, N., and Roos, K.-.A., "Drug Interferences and Effects in Clinical Chemistry," 4th ed. Apoteksbolaget, Stockholm, 1986.
U1. Uesaka, H., and Goto, M., Statistical analysis of clinical chemistry data based on power normal distribution. *In* "MEDINFO 80" (D. B. Lindberg and S. Kaihara, eds.), pp. 1014–1018. North-Holland Publ., Amsterdam, 1980.
V1. Vácha, J., Biology and the problem of normality. *Scientia (Milan)* **113,** 823–846 (1978).
V2. Vácha, J., German constitutional doctrine in the 1920s and 1930s and pitfalls of the contemporary conception of normality in biology and medicine. *J. Med. Philos.* **10,** 339–367 (1985).
W1. Werner, M., Tolls, R. E., Hultin, J. V., and Mellecker, J., Influence of sex and age on the normal range of eleven serum constituents. *J. Clin. Chem. Clin. Biochem.* **8,** 105–115 (1970).
W2. West, J. B., and Lahiri, S., eds., "High Altitude and Man." Williams & Wilkins, Baltimore, Maryland, 1984.
W3. Wimberley, P. D., Hansen, E., Bucher, D., and Olesen, H., Ear, finger and vein blood: Analytical variation. *Abstr., Scand. Congr. Clin. Chem., 19th, 1984* (1984).

W4. Winkel, P., Patterns and clusters—Multivariate approach for interpreting clinical chemistry results. *Clin. Chem. (Winston-Salem, N. C.)* **19,** 1329–1338 (1973).

W5. Winkel, P., The use of the subject as his own referent. In "Reference Values in Laboratory Medicine. The Current State of the ART" (R. Gräsbeck and T. Alström, eds.), pp. 65–78. Wiley, Chichester, 1981.

W6. Winkel, P., and Statland, B. E., Using the subject as his own referent in assessing day-to-day changes of laboratory test results. *Contemp. Top. Anal. Clin. Chem.* **1,** 287–317 (1977).

W7. Winkel, P., and Statland, B. E., Reference values. In "Clinical Diagnosis and Management by Laboratory Methods" (J. B. Henry, ed.), 16th ed., pp. 29–52. Saunders, Philadelphia, Pennsylvania, 1979.

W8. Winkel, P., Gaede, P., and Lyngbye, J., Method for monitoring plasma progesterone concentrations in pregnancy. *Clin. Chem. (Winston-Salem, N. C.)* **22,** 422–428 (1976).

W9. Winkel, P., Lyngbye, J., and Jörgensen, K., The normal region—A multivariate problem. *Scand. J. Clin. Lab. Invest.* **30,** 339–344 (1972).

W10. Winkel, P., Bentzon, M. W., Statland, B. E., Mouridsen, H., and Sheike, O., Predicting recurrence in patients with breast cancer from cumulative laboratory results: A new technique for the application of time series analysis. *Clin. Chem. (Winston-Salem, N. C.)* **28,** 2057–2067 (1982).

W11. Wisser, H., and Breuer, H., Circadian changes of clinical chemical and endocrinological parameters. *J. Clin. Chem. Clin. Biochem.* **19,** 323–337 (1981).

W12. Wootton, J. D. P., and King, E. J., Normal values for blood constituents. Interhospital differences. *Lancet* **1,** 470–471 (1953).

W13. World Health Organization, "The First Ten Years of the World Health Organization," p. 459. WHO, Geneva (1958).

Y1. Young, D. S., Harris, E. K., and Cotlove, E., Biological and analytic components of variation in long-term studies of serum constituents in normal subjects. IV. Results of a study designed to eliminate long-term analytic deviations. *Clin. Chem. (Winston-Salem, N. C.)* **17,** 403–410 (1971).

Y2. Young, D. S., Pestaner, L. C., and Gibberman, V., Effects of drugs on clinical laboratory tests. *Clin. Chem. (Winston-Salem, N. C.)* **21,** 1D–432D (1975).

NEOPTERIN AS MARKER FOR ACTIVATION OF CELLULAR IMMUNITY: IMMUNOLOGIC BASIS AND CLINICAL APPLICATION

Helmut Wachter, Dietmar Fuchs, Arno Hausen, Gilbert Reibnegger, and Ernst R. Werner

Institute of Medical Chemistry and Biochemistry,
University of Innsbruck,
Innsbruck, Austria

1. Introduction	82
1.1. Nomenclature	84
1.2. Discovery of Neopterin	84
1.3. Early Synthesis of Pteridines	85
1.4. Biological Function and Occurrence of Pteridines	85
2. Chemistry and Biochemistry of Neopterin	88
2.1. Chemistry	88
2.2. Biochemistry	90
2.3. Catabolism	91
2.4. Cellular Source	92
2.5. Induction Signal	93
2.6. Physiological Role	94
3. Methods of Measurement of Neopterin	94
3.1. Measurement by HPLC	95
3.2. Measurement by RIA	96
4. Neopterin Concentrations in Healthy Subjects	97
4.1. Urinary Excretion Levels	97
4.2. Levels in Serum	97
4.3. Circadian Variations	98
4.4. Long-Term Stability	99
5. Conditions Associated with High Neopterin Levels	99
5.1. Allograft Rejections	100
5.2. Viral Infections	104
5.3. Infections by Intracellular Protozoa	109
5.4. Infections by Bacteria	110
5.5. Autoimmune Diseases	112
5.6. Other Inflammatory Diseases	116
5.7. Malignant Diseases Involving Activation of Cell-Mediated Immunity	118
6. Conclusions	123
References	125

1. Introduction

Neopterin was isolated from larvae of bee (R19), from worker bees, and from royal jelly (R20) in 1963. Oxidation of the isolated component with permanganate yielded 6-pterincarboxylic acid, demonstrating the presence of a 6-substituted pterin moiety. The structure was characterized by comparison with small amounts of synthesized material. By condensation of 2,4,5-triamino-6-hydroxypyrimidine with D-ribose in the presence of hydrazine, the side chain was identified as a 1,2,3-trihydroxypropyl residue. The configuration of the side chain was shown to be the D-*erythro* form by comparison with the four synthetic isomers (R19). Originally, H. Rembold intended to term the new compound, 2-amino-4-hydroxy-(erythro-1′,2′,3′-trihydroxypropyl)-pteridine, "novapterin," to indicate that it was a new (from Latin, *nov*um) molecule isolated from honey bees (Latin, *Apis*) and with a *pterin* structure. When discussing the name with A. Butenandt, however, they felt that this new compound might be more than just a new pterin of the honey bee. Due to its occurrence also in royal jelly, they speculated that it might open a new horizon of biologically important pteridines. Thus, the compound finally was termed "neopterin" to denote that it might start a new (Greek, *neo*) epoch in pteridine research. And indeed, 25 years later, we now feel their vision turns out to be correct. Figure 1 shows the chemical structure of neopterin. Neopterin can be synthesized with high yield and purity starting from 2,4,5-triamino-6-oxo-1,6-dihydropyrimidine and D-arabinosephenylhydrazone, with subsequent oxidation of the condensed product (V5).

In 1967, Sakurai and Goto isolated 25 mg of neopterin from 500 liters of human urine (S1). They did not examine urinary neopterin excretion during disease states. Our laboratory studied enhanced excretion of neopterin by comparing fluorescent components in urinary specimens from healthy subjects and from patients with neoplastic diseases. Analyses were performed by paper electrophoresis (W7) and by high-voltage electrophoresis (W5) with fluorescence detection. Differences in the then-unidentified

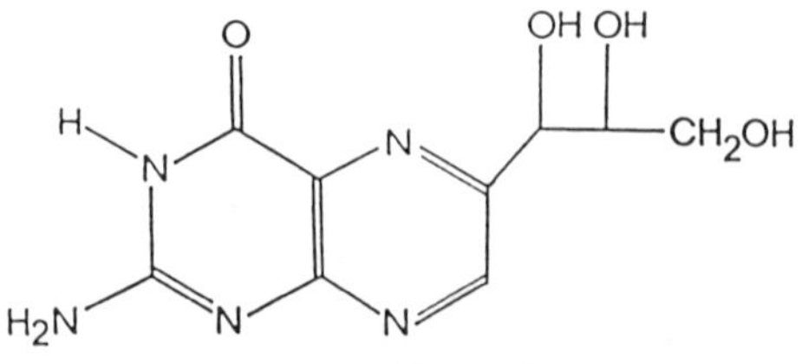

FIG. 1. Neopterin.

fluorescent urinary excretion products were observed. Enhanced excretion of a fluorescent component with similar electrophoretic mobility was also found by *in situ* fluorometry after high-voltage electrophoresis of urine of mice with Ehrlich ascites tumor, when compared to healthy animals (G5). Urinary excretion of this fluorescent compound (related to indoxyl sulfate as the invariate) was four times higher for mice with Ehrlich ascites tumor ($n = 120$) 5 days after transplantation of tumor, when compared to healthy mice ($n = 110$). This fluorescent component was identified as 6-hydroxylumazine; it originated during the analytical process by autoxidation from 7,8-dihydro-6-hydroxylumazine (W1). Excretion of this product increased on the day subsequent to inoculation of the tumor until the seventh day, in parallel to the increase in number of tumor cells (H8).

Following the identification of a pteridine as the fluorescent component that was elevated in mice with Ehrlich ascites tumor, the corresponding substance from human urine was isolated and characterized. It was found that the fluorescent component previously observed in urine of patients with malignant diseases was neopterin. In the first report of this observation, Wachter and co-workers found elevated rates of neopterin excretion in a group of patients with various malignant disorders, as well as in patients with viral diseases (W2). Soon after publication of this report, these results were confirmed in several other laboratories (D5, R28, S19).

These studies stimulated work in examining the correlation of neopterin with tumor stage and its potency for prognosis and monitoring treatment of cancer. The initial data obtained on homogeneous patient groups with hematological and gynecological neoplasias suggested that the neopterin assay conveys information on extent, progression, and outcome in these diseases (B5, H11). Particularly, the close correlation of neopterin excretion with hepatosplenomegaly, with hematological criteria, and with concomitant viral diseases directed interest to the biological significance of neopterin excretion. In 1981, it was suggested that neopterin originated from the immune response of the host directed against tumor cells or virally transformed cells (H11). This hypothesis was then tested using *in vitro* experiments in mixed lymphocyte cultures (F14). Further *in vitro* studies (H21,H22,N4) revealed that human monocytes/macrophages produce neopterin when stimulated by interferon-γ. This lymphokine is released from activated T cells. Other cell types do not produce measurable amounts of neopterin following various stimuli. Therefore, neopterin production appears to be closely associated with activation of the cellular immune system. These *in vitro* experiments are consistent with numerous clinical studies of urinary and serum neopterin levels.

This review will focus on the immunological basis and clinical experiences with neopterin as an indicator for activation of the cell-mediated immune system.

1.1. NOMENCLATURE

The bicyclic nitrogenous ring system pyrazino-(2,3-d)-pyrimidine is now termed pteridine according to the International Union of Pure and Applied Chemistry. As proposed by Pfleiderer, the term pterin denotes derivatives of 2-amino-4-oxo-3,4-dihydropteridine and the term lumazine denotes derivatives of 2,4-dioxo-1,2,3,4-tetrahydropteridine (P8) (Fig. 2).

In contrast to these "unconjugated" pteridines, which bear comparatively small substituents, derivatives with larger residues are termed "conjugated" pteridines, e.g., folic acid (S21), because of the presence of the three components, 6-hydroxymethylpteridine, p-aminobenzoic acid, and glutaminic acid, in the molecules.

1.2. DISCOVERY OF NEOPTERIN

In 1889, Hopkins, who had discovered tryptophan and glutathione, isolated a pigment from the wings of lepidoptera (H20). This work was continued by Wieland and Schöpf (W18), who in 1936 named these pigments pteridines (S10), a term which has its origin in the Greek word for wing, *pteron*. However, attempts to elucidate the structures of these compounds were unsuccessful until Purrmann (P21–P23) showed that three insect pig-

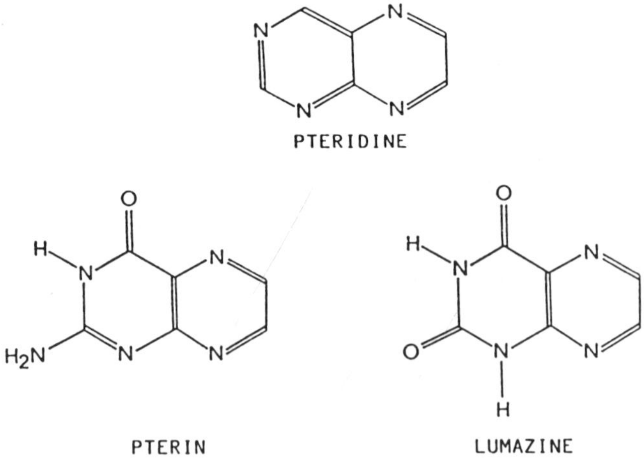

FIG. 2. Structures of parent compounds of unconjugated pterins.

ments, xanthopterin, isoxanthopterin, and leucopterin, contain the bicyclic nitrogenous ring system pyrazino-(2,3,-d)-pyrimidine.

1.3. Early Synthesis of Pteridines

It should be noted that chemists in Berlin synthesized pteridines before Hopkins did, but no connection to the pigments of butterflies was suggested. In 1884, Kühling (K17) oxidized the benzene ring of tolualloxazine to a dicarboxylic acid that then decarboxylated to lumazine. This compound was made also by Gabriel and Sonn in 1907 starting from pyrazino-2,3-dicarboxamide (G2). Gabriel and Colman (in 1901) (G1) and Isay (in 1906) (I1) condensed 5,6-diaminopyrimidine with a 1,2-dicarbonyl compound. This Gabriel–Isay reaction is currently used. Another commonly applicable synthesis, also published before the work of Purrmann, developed by Kuhn and Cook from Heidelberg (K19), is the principle of preparing lumazines from 4,5-diamino-2,6-dioxopyrimidine and glyoxal.

1.4. Biological Function and Occurrence of Pteridines

Unconjugated pteridines are ubiquitous in nature. However, they occur in comparatively high concentrations only as pigments of insects, amphibia, reptiles, and fish (B7,F6,Z1). Because of their low concentrations in other species, the presence of pteridines was only recently reported in mammalian species. Koschara (K15) demonstrated the occurrence of xanthopterin in human urine by comparison of the urinary component with the material isolated by Schöpf from wings of the common English brimstone butterfly. Patterson *et al.* first identified biopterin from human urine by analyzing its structure (P5).

Today, several biological functions of pteridines are established. Biopterin stimulates the mitosis of the trypanosome flagellate *Crithidia fasciculata* (B12,P6). The first metabolic role for an unconjugated pterin was shown by Kaufman (K2), who demonstrated that 5,6,7,8-tetrahydrobiopterin serves as the cofactor for mammalian aromatic amino acid monooxygenases, enzymes that hydroxylate phenylalanine, tyrosine, and tryptophan and thus control the biosynthesis of the neurotransmitters dopamine, norepinephrine, and serotonin. Therefore, a deficiency of tetrahydrobiopterin causes severe neurological illness by accumulation of phenylalanine and depletion of neurotransmitters. The enzyme deficiencies that result in reduced levels of phenylalanine, tyrosine, and tryptophan hydroxylases because of tetrahydrobiopterin deficiency are termed atypical phenylketonuria (PKU). The classical PKU is caused by deficiency of the enzyme. About 1–3% of phenylketonuria patients (D2) suffer from the atypical form, the tetrahydrobiopterin deficiency that can be caused by dihydropteridine reductase deficiency (K3), by dihydrobiopterin synthetase deficiency, and by GTP cyclohydrolase deficiency.

In dihydropteridine reductase deficiency, levels of biopterin in body fluids are high (C6, W10). In dihydrobiopterin synthetase deficiency the most frequent defect is a low or absent activity of 6-pyruvoyltetrahydropterin synthetase, the first enzyme leading from dihydroneopterin triphosphate to tetrahydrobiopterin (N8). 6-Pyruvoyltetrahydropteridine synthetase catalyses the elimination of the triphosphate group from dihydroneopterin triphosphate. Dihydropteridine synthetase deficiency is characterized by high neopterin, dihydroneopterin, and 3'-hydroxysepiapterin concentrations and low biopterin concentrations. In the third deficiency, the GTP cyclohydrolase deficiency (N7), levels of all pterins are low.

A tetrahydropterin derivative is also cofactor for the enzymatic oxidation of glyceryl ethers to alcohols (T4). Recently, the pterin component, termed molybdopterin, of the molybdenum cofactor that is present in nitrate reductase, sulfite oxidase, and xanthine oxidase was structurally characterized (J3). While the presence of a pterin nucleus in molybdopterin is unequivocally established, the exact nature of the labile molecule is not presently known. The cofactor is converted *in vitro* by oxidation into two stable, fluorescent products of molybdopterin. *In vivo*, urothione, the structure of which was established by Goto *et al.* (G9), is the excreted metabolic product. This is based on the observation that patients with a deficiency of molybdenum cofactor do not excrete detectable urothione (J2).

Because of the occurrence of unconjugated pteridines in a diversity of biological sources, the involvement of pteridines in further oxygenase reactions and in other biological processes has been suggested (F6). Association between pteridines and tryptophan was demonstrated (F21). A possible interaction between isoxanthopterin and DNA was reported (F6,L2). Tetrahydrobiopterin accumulates in the regeneration bud of *Triturus* species after tail amputation and disappears with redifferentiation (K12), whereas the contrary is the case for isoxanthopterin. High levels of tetrahydrobiopterin were found in human squamous carcinoma cells (K14) and in blood of certain tumor patients (K13).

It was claimed that a degradation product of folic acid, 6-pterinaldehyde, is characteristic for cancer cells in tissue culture and for urine of cancer patients in 1977 (H5). In 1978, it was shown using a thin-layer chromatographic system that 6-pterinaldehyde is present in healthy subjects as well as in patients with cancer (D6). This folate catabolite was later reidentified as 6-hydroxymethylpterin (S20). Stea reported in 1979 in his dissertation (S18) that in malignant cells, but not in normal cells, the folate catabolite 6-hydroxymethylpterin was formed from radioactively labeled folic acid as well as from 6-pterinaldehyde. In a patient with ovarian carcinomatosis and abdominal and thoracic metastases, Stea found elevated levels of 6-pterinaldehyde and 6-hydroxymethylpterin in peritoneal fluid but not in urine. However, he noted that this patient also excreted higher urinary levels of neopterin and biopterin when compared to healthy subjects. The observation of elevated neopterin levels in 20 patients with malignant diseases was published in 1981 by Stea *et al.* (S19). However, this report showed also that the levels of 6-hydroxymethylpterin in cancer patients did not differ significantly from those in healthy subjects.

Various prokaryotic and eukaryotic organisms secrete pteridines. *Escherichia coli* excretes monapterin (L-threo-6-trihydroxypropylpterin). The release of monapterin experiences a burst at the switch from the exponential growth phase to the stationary phase and then ceases (W4). The slime mold *Physarum polycephalum*, the nuclear division of which occurs synchronously in macroplasmodia, excretes isoxanthopterin when undergoing exponential growth (L11). The release of isoxanthopterin shows a pronounced maximum in the early G_2 phase (4 h before mitosis) and declines at entry into the stationary phase.

Furthermore, during sclerotization, a differentiation process from proliferation to quiescence, the excretion of isoxanthopterin rapidly increases with a concomitant decrease of intracellular isoxanthopterin levels.

The cellular slime mold *Dictyostelium discoideum* releases lumazines into the growth medium (G7). Folic acid (P3), monapterin (T6), and other pterins (V3) act as chemotactic signals involved in cellular locomotion, as is also the case for cyclic AMP and glorin (S12). These chemoattractants serve as chemotactic signals that cause the predatory amebae to move toward their bacterial food source. The chemotaxis and motility require formation of filopodia and thus reorganization of the cytoskeleton (M3,M6,R24). The chemotactic molecules bind to specific receptors on the cell surface (M7) and are inactivated by enzymes that have been detected in extracellular, intracellular, and particulate fractions obtained from three strains of *D. discoideum* (P2). These enzymes catalyze the hydrolytic deamination of folic acid and pterin derivatives to chemotactically inactive D-deamino folic acid and lumazine.

The most significant roles of conjugated pteridines are as tetrahydrofolate cofactors involved in thymine synthesis and in the transfer of one-carbon groups in a variety of reactions in purine, pyrimidine, and amino acid metabolism (D4,S21).

Dihydroneopterin triphosphate functions biosynthetically as a direct precursor of the 6-hydroxymethylpterin component of folic acid and riboflavin in microorganisms, but there is no biosynthetic or degradative pathway in mammals to or from these vitamins that leads to pterins with a propyl side chain in position six, as has been shown by Pabst and Rembold (P1).

Methanopterin has been characterized in methanogenic bacteria (V2). Its exact chemical structure is not known. Methanopterin serves as single-carbon carrier at the oxidation level of methenyl, methylene, and methyl in the conversion of CO_2 to methane, and thus resembles the cofactor function of folic acid (E5).

2. Chemistry and Biochemistry of Neopterin

2.1. CHEMISTRY

Comprehensive information on the chemistry of pteridines has been provided by Pfleiderer (P9). Synthetic methods for some natural pteridines have been reviewed by Taylor (T2) and the reactivity of pteridines have been examined by Pfleiderer et al. (P10). Here, some reactions are discussed that are important for measurement of neopterin in biological specimens. Particularly important are the sensitivity of pteridines to photodecomposition and the lability of reduced pterins to oxidizing conditions.

Determination of pteridines in biological material is complicated by the fact that they exist in three oxidation states, fully oxidized (aromatic) pterins, 7,8-dihydropterins and 5,8-dihydro(quinonoid)pterins, and 5,6,7,8-tetrahydropterins. Because 5,6,7,8-tetrahydrobiopterin participates as cofactor in biological hydroxylations of aromatic amino acids, only the reduced pteridines have been considered as biologically significant. On the other hand, the fully oxidized pteridines are fluorescent compounds and can therefore be determined with high sensitivity. Neopterin and the dihydro and tetrahydro forms have been measured as "total neopterins" by fluorescence following oxidation with iodine (F22), with ferricyanide (H3), or with manganese dioxide (N6). Fukushima and Nixon (F22) have shown that 7,8-dihydroneopterin and 5,6,7,8-tetrahydroneopterin were oxidized almost quantitatively by iodine in 0.1 mol/liter HCl. Oxidation in 0.1 mol/liter NaOH of tetrahydroneopterin yields only a trace of neopterin but 80% pterin. The same authors report that autoxidation of tetrahydrobiopterin in 0.1 mol/liter sodium bicarbonate (pH 7.5) produced biopterin, xanthopterin, and pterin, but autoxidation of dihydrobiopterin under the same conditions yielded only biopterin and xanthopterin (F22). The oxidation reactions of reduced biopterin and neopterin are considered to be very similar except that dihydroneopterin appears to be more labile to alkaline oxidation (F23). Consistent with this view are the studies of Armarego and Randles (A6), who reported that aerobic oxidation at pH 7.6 of 5,6,7,8-tetrahydroneopterin produces quinonoid 7,8-(6H)-dihydroneopterin, which rapidly loses the side chain and forms 7,8-dihydropterin. This intermediate adds water across the 5,6-double bond. Further aerobic oxidation forms quinonoid 6-hydroxy-7,8-(6H)-dihydropterin, which rearranges to 7,8-dihydroxanthopterin (A6) (Fig. 3).

Chemical reactivities indicate that both neopterin and its acid-oxidizable reduced forms (total neopterin) can be measured with sufficient accuracy. Clinical urine and serum specimens, however, sometimes have to be shipped from the clinician or physician to the laboratory, and sometimes have to be stored for days. In this case, the acid-oxidizable reduced forms of neopterin are converted to a variable extent into dihydroxanthopterin, xanthopterin, and pterin, resulting in an erroneous reflection of the true concentration of neopterin. Therefore, it must be recommended to determine only the concentration of neat neopterin (i.e., not the acid-oxidizable forms) in serum or urine, since storage for 2 days at 4°C (or even longer at − 20°C) does not influence its concentration.

Furthermore, a study using freshly collected and uniformly handled samples (L7) demonstrates that the ratio of neat neopterin to total neopterin has a fairly constant value for both urine (seven patients; 0.45) and serum (five patients; 0.43), with coefficients of variation of 14.4 and 21.5%, respectively. Therefore, both methods should give comparable results if bias caused by handling and storage is excluded.

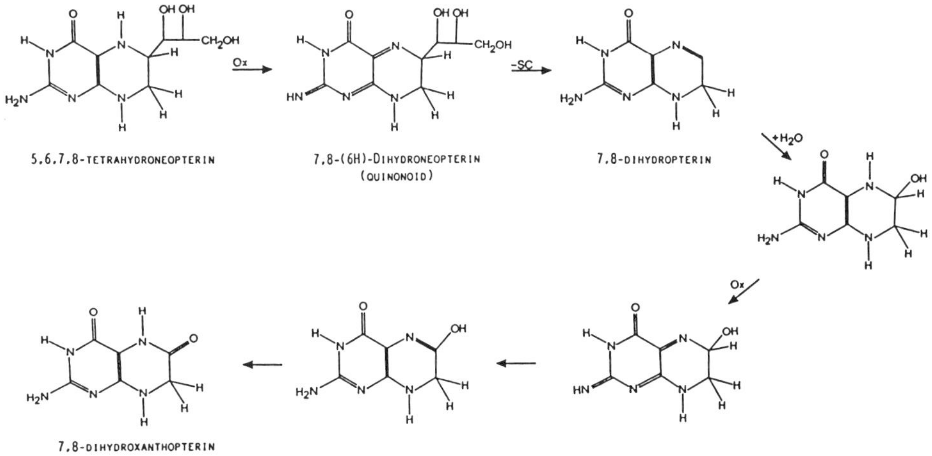

FIG. 3. Aerobic oxidation of 5,6,7,8-tetrahydroneopterin; Ox, oxidation; −SC, cleavage of side chain.

2.2. BIOCHEMISTRY

It was suggested by Albert (A3) in 1957 that purines may represent the biological precursors of pteridines; he observed that purines, treated with glyoxal-type compounds produce pterins. Weygand *et al.* (W16) provided evidence in insects that radioactive precursors were incorporated into leucopterin. Jones and Brown showed that dihydroneopterin is formed enzymatically from guanosine triphosphate in *E. coli* (J4). This and further studies (R22,S14) have led to a biosynthetic pathway (Fig. 4) according to which all naturally occurring pterins, including the pteridine moiety of folic acid and riboflavin, are formed.

The biosynthesis of pterins begins with guanosine triphosphate. The initial enzyme is guanosine triphosphate cyclohydrolase. Opening of the imidazole ring of the purine is followed by removal of the original C-8 of the starting compound. By Amadori rearrangement, the ribosyl residue is converted to a 1-deoxypentulose. The pteridin ring is then formed, yielding 7,8-dihydroneopterin triphosphate as first intermediate (this compound is an intermediate in the synthesis of other natural pterins.

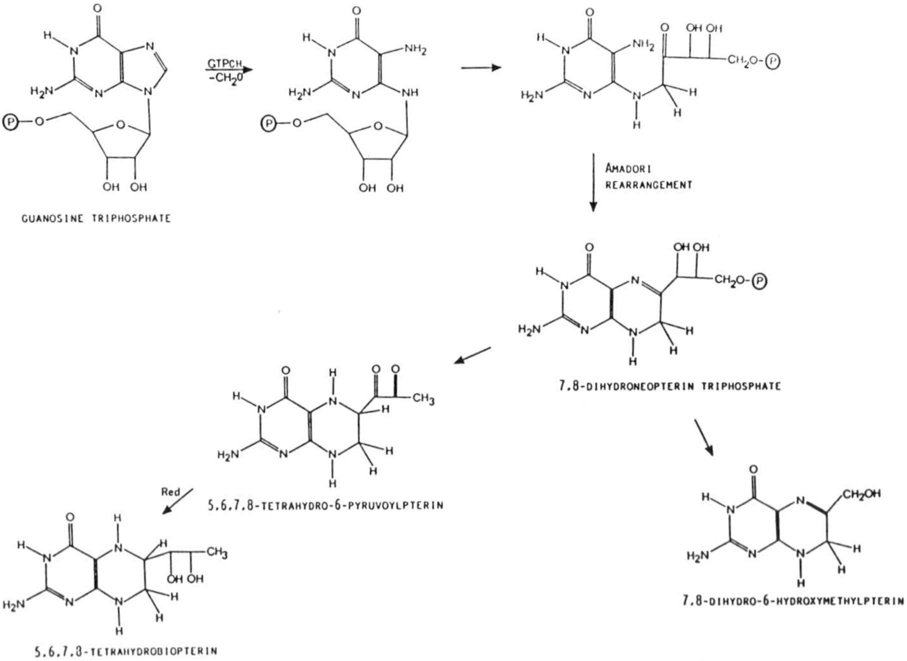

FIG. 4. Biosynthetic pathway for conversion of GTP into the unconjugated pterins: 7,8-dihydroneopterin triphosphate, 7,8-dihydro-6-hydroxymethylpterin, and 5,6,7,8-tetrahydrobiopterin; GTPCH, guanosine triphosphate cyclohydrolase; Red, reduction.

Unconjugated pteridines but not folic acid and riboflavin can be synthesized *de novo* by mammals and other higher animals, as has been shown by Kraut *et al.* (K16) and by Pabst and Rembold (P1). The biosynthesis of 7,8-dihydroneopterin and neopterin by macrophages is considered to result from the same pathway (S8). However, increased dihydroneopterin triphosphate in macrophages has not been detected (S8). Therefore, the possibility that an alternate pathway leads to synthesis of neopterin in macrophages cannot be ruled out with certainty.

2.3. CATABOLISM

All naturally occurring pterins, including folic acid and riboflavin, are biosynthesized from guanosine triphosphate. Dihydroneopterin triphosphate has been established as an intermediate in all pathway experiments. However, there is no reversibility of the metabolic pathway from folic acid and riboflavin back to dihydroneopterin. Folic acid and riboflavin were in fact excluded as a source of biopterin because rats and chicken that were fed a diet lacking these vitamins but containing sulfonamide to suppress production of these vitamins by the intestinal flora continued to excrete biopterin in normal amounts (F4,K16,P1). The major metabolites of folic acid that are excreted in urine of the rat were identified as 5-methyltetrahydrofolic acid, 10-formyltetrahydrofolic acid, and 4α-hydroxy-5-methyltetrahydrofolic acid (B2). The theoretically possible degradation products, tetrahydrofolic acid, xanthopterin, and pterin, were not found in rat urine. The main excretion species of riboflavin in human urine are 7α- and 8α-hydroxyriboflavin (O1).

The catabolism of neopterin in humans has not been investigated, but data are available from studies using rats; in contrast to humans, a pterin deaminase is present in rats. The catabolism in rats of biopterin, tetrahydrobiopterin, and tetrahydroneopterin has been studied by Rembold (R16,R17). Biopterin was recovered mainly unchanged, partly as 6-pterincarboxylic acid and, as a minor fraction, as pterin and isoxanthopterin. The main degradation products of tetrahydrobiopterin and of tetrahydroneopterin in rat were 6-hydroxylumazine, lumazine, and xanthopterin. The degradation pathways involve three steps: cleavage of the side chain, deamination of the pterin ring, and introduction of an oxygen function into the C-6 position. Two enzymes, a pterin deaminase and a xanthine oxidase, were responsible for the two latter reactions. The catabolic reaction sequences according to Rembold are shown in Fig. 5. Due to the presence of a pterin deaminase, the main excretory products in rats and mice are derivatives of lumazine. In humans and primates only pterins but no lumazines are found (D7). Further, humans and primates are the only species in which neopterin can be detected in body fluids.

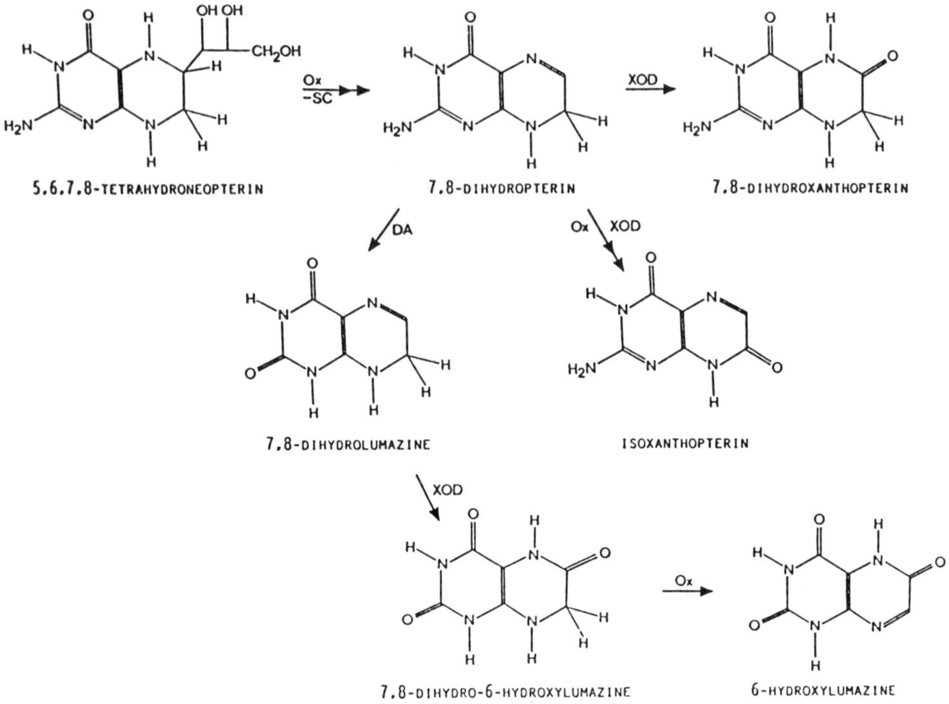

FIG. 5. Catabolic reaction sequences in degradation of 5,6,7,8-tetrahydroneopterin leading to the metabolites 7,8-dihydroxanthopterin, isoxanthopterin, and 7,8-dihydro-6-hydroxylumazine; Ox, oxidation; −SC, cleavage of side chain; XOD, xanthine oxidase; DA, pterin deaminase.

2.4. Cellular Source

Whereas pteridines other than neopterin are released from metabolically active cells into the culture supernatants when cells cease to proliferate or start to differentiate (W3), cellular excretion of neopterin has not been observed. Since it was suggested that high urinary neopterin levels detected in patients with viral diseases and with malignancies (W2) might reflect a host response (H11), *in vitro* studies on activated human immune cells were performed. Neopterin was detected for the first time in cell culture media of human mixed peripheral blood mononuclear cells that were stimulated by allogeneic cells, by virally or chemically modified autologous cells, and by mitogens (F14,H21). Human peripheral blood mononuclear cells not only secrete neopterin when adequately activated but also synthesize it from guanosine triphosphate. This was concluded because the activity of the enzyme GTP cyclohydrolase I increased 5–10 times after stimulation by phytohemagglutinin (B8). Human peripheral blood mononuclear cells include T lymphocytes, B lymphocytes, large granular lymphocytes, and monocytes. It has been demonstrated that secretion of neopterin is associated with proliferation of T cells, but cloned T cells do not release neopterin upon stimulation. Therefore, it was assumed that products of activated T cells induce neopterin production by other components of human peripheral blood mononuclear cells.

Studies directed at the cellular source of neopterin excretion demonstrated that human monocytes/macrophages release neopterin (H22). *In vivo*, only patients with extremely high neopterin levels showed a moderate increase of monapterin levels. Neopterin secretion was not detected from other hemopoietic cells, such as natural killer cells, B lymphocytes, granulocytes, nor from fibroblasts or a variety of tumor cell lines (H22,N4), except that certain subclones of the U937 monocytic cell line produce neopterin in small amounts (R29). Interestingly, no other pterin derivatives were released from macrophages upon stimulation with interferon-γ. *In vivo*, only patients with extremely high neopterin levels showed a moderate increase of monapterin levels.

2.5. INDUCTION SIGNAL

Highly purified monocytes/macrophages released neopterin into culture media when stimulated by supernatants from activated T cells (H22,N4). These supernatants contain, among other immunomodulators, interferon-γ, which is produced during early steps of T cell activation. Apart from its immunoregulatory and antiviral action, interferon-γ regulates oxidative metabolism of macrophages and primes them for defense against intracellular pathogens, for antiviral resistance, and for antitumor activity (N5). Interferon-γ and other lymphokines were examined as possible inducers of neopterin secretion.

Maximal release of neopterin was achieved by doses of recombinant interferon-γ ranging from 10 to 100 U/ml. Approximately 10^3 times higher doses of interferon-α were required for detectable production of neopterin. Experiments with monoclonal antibodies against human recombinant interferon-γ showed complete suppression of the neopterin-inducing activity in supernatants from activated T cells, providing further evidence for interferon-γ as specific inducer (H22). Monocytes/ macrophages do not secrete detectable neopterin *in vitro* when exposed to other inducers of macrophage activity, e.g., zymosan, phorbol myristate acetate, colony-stimulating factor, and interferon-β (H22,N4). Neopterin production from macrophages exposed to interferon-α was either not observed (N4) or required extremely high concentrations (H22), hence being of questionable physiological relevance. Lipopolysaccharide variably stimulated neopterin secretion in small amounts from macrophages *in vitro* (N4). Thus the *in vitro* experiments are compatible with the view that interferon-γ represents the only apparent inducer of neopterin secretion from macrophages. This is supported by numerous *in vivo* observations.

The question whether macrophages synthesize neopterin upon activation or secrete it from a presynthesized pool was recently resolved. *In vitro* studies in macrophages showed that interferon-γ augments the intracellular concentration of guanosine triphosphate (GTP) as well as the conversion of GTP to neopterin (S8). The conversion rates correlated well with enhanced neopterin levels (dihydroneopterin and neopterin were measured as total neopterin) in macrophages (S7). Most likely, conversion of GTP is due to activation of GTP cyclohydrolase I; however, elevated levels of dihydroneopterin triphosphate were not detected.

2.6. Physiological Role

Questions about possible physiological functions of neopterin or of biosynthetic precursors cannot be answered at present. The secretion of neopterin correlated with the capacity of monocytes/macrophages to secrete H_2O_2. It should be noted that the secretion of H_2O_2 by macrophages is a two-step process: the activation by interferon-γ must be followed by a second stimulus, e.g., phorbol myristate acetate. Whereas the first step leads to synthesis of neopterin, the second step, the release of H_2O_2, is not paralleled by further neopterin production (N4).

Since the main physiological role of interferon-γ may be the induction of antibacterial, antiprotozoal, and antifungal states in parasitized cells, it has been suggested that neopterin might act as an endogenous inhibitor of folate synthesis by intracellular pathogenic microorganisms (N4).

A further study suggests a connection of neopterin and the tryptophan metabolism of macrophages. Interferon-γ stimulates macrophages to degrade tryptophan, closely correlated with production of kynurenine, 3-hydroxyanthranilic acid, anthranilic acid, and neopterin (W13). Other cells that degrade tryptophan following stimulation with interferon-γ (P7,W12) formed neither neopterin nor 3-hydroxyanthranilic acid (W12). It was therefore proposed that tetrahydroneopterin might serve as coenzyme for anthranilic acid 3-monooxygenase. Anthranilic acid 3-monooxygenase has been isolated from the tropical bush *Tecoma stans* and required a tetrahydropteridine as cofactor (N3).

3. Methods of Measurement of Neopterin

Isolation and measurement of pteridines has in general posed problems caused by their presence in trace amounts in biological material. Due to photolability and susceptibility of reduced forms to aerobic oxidation, it is recommended that all manipulations of pteridines be conducted in dim light or in darkness.

In addition to microbiological assay techniques for total biopterin, sepiapterin, monapterin, and their reduced forms, using the growth response of *C. fasciculata* (P6), determinations of pteridines have been performed using paper chromatography (H4,K21). More suitable methods are separation by column chromatography and measurement by fluorescence or UV absorption techniques (F23,R18,S1). Advances in liquid chromatography have led to the development of high-performance liquid chromatographic (HPLC) methods for measurement of pteridines in biological materials (B1,F22,K11,S17), with fluorescence detection. Only oxidized pterins are fluorescent and can be measured directly by a fluorescence detector. The dihydro- and tetrahydropterins require oxidation steps before measurement by fluorescence. Nonfluorescent dihydro- and tetrahydroneopterins can be measured subsequent to HPLC separation by electrochemical detection. Several pterin species can be measured in various oxidation states by liquid chromatography/electrochemistry using a dual-electrode detector (B1,L12). The extensive sample purification required by this method was avoided using ion-pair reversed-phase HPLC for separation and sequential detection steps. Tetrahydropterins were measured electrochemically, whereas dihydropterins were measured by fluorescence following postcolumn oxidation and oxidized pterins were measured by native fluorescence (H24).

A sensitive method for identification of natural compounds is gas chromatographic separation of derivatized pteridines (G4); this technique enables elucidation of structure when combined with mass fragmentography (H7,K20,R30) and can serve as a reference method due to the high precision.

Due to growing interest in analysis of pteridines in urine and serum, radioimmunoassays (RIAs) have become available for measurement of biopterin (N2) and for neopterin (R26,S24). To date, determination of neopterin for clinical use is done by HPLC for urine and serum or by RIA for serum, since these methods are suited for application in clinical laboratories due to their simplicity and accuracy.

3.1. Measurement by HPLC

Pteridines, including biopterin and neopterin, have been separated by cation exchange (S17) and by reversed-phase HPLC techniques (F23,K11). The procedures involve time-consuming chemical oxidation of dihydro- and tetrahydro forms. Furthermore, dihydro- and tetrahydro forms are partly converted into pterins other than the corresponding oxidized compounds, depending on pH value, when not immediately analyzed.

Therefore, a method was developed for rapid separation and sensitive quantitation of urinary neat oxidized neopterin by reversed-phase HPLC on a 10-μm octadecylsilica column (H9). The analytes are eluted with 15 mmol/liter potassium phosphate buffer at pH 6.4 and at a flow rate of 0.8 ml/min. Urinary neopterin can be measured by fluorescence and related to creatinine determined by ultraviolet absorption in order to account for fluctuating concentrations of urine. The method has good performance characteristics and is easy to handle. The procedure was modified for routine laboratory automated analysis without any pretreatment except dilution of samples with aqueous potassium phosphate buffer using guard columns (F10).

Determination of neopterin in serum by HPLC is more difficult to perform than in urine due to the presence of protein and due to the about 200-fold lower concentration of neopterin. Particularly, during the precipitation of protein as proposed by Fukushima and Nixon (F22), a variable amount of neopterin might be coprecipitated. Precipitation is not necessary and further pretreatment steps are avoided if solid-phase extraction is combined with on-line elution of the solid-phase cartridge directly onto the HPLC column (W11). Acidified but not deproteinized serum is injected to a cartridge with silica sorbent modified by 4-propylbenzenesulfonic acid, which quantitatively retains the analytes but not the serum proteins. Then the analytes are eluted from the cartridge by a pulse of more concentrated potassium phosphate buffer (0.4 mol/liter; pH 6.8). The final separation is achieved by isocratic elution on octadecyl silica column with potassium phosphate buffer (15 mmol/liter; pH 6.0). Neopterin or other pterins are measurable by fluorescence. Creatinine can be simultaneously determined by ultraviolet absorption.

The glomerular filtration rate of neopterin was found to average 1.8 times that of creatinine in healthy subjects (F22,W14) as well as in renal allograft recipients with variably impaired kidney function (R7). Therefore, neopterin and creatinine rise in a dependent manner and relating serum neopterin to serum creatinine allows discrimination of the influence of renal impairment on serum neopterin levels. This is of particular importance for the interpretation of neopterin levels in serum of patients with impaired kidney function, such as renal allograft recipients.

3.2. Measurement by RIA

The particular advantage of RIA compared to HPLC is its suitability for large-scale applications. When measurement of neopterin was recognized to be useful for diagnostic purposes, radioimmunoassay procedures were quickly developed (N1,R26).

Antisera were obtained in rabbits immunized with thyreoglobulin and hemocyanin conjugates of N-(3-carboxy)propylneopterin. The ω-carboxyl alkyl derivatives were coupled to tyramine by the mixed anhydride method and then radioiodinated by the chloramine T method, yielding ^{125}I-labeled tracers (R26).

In rabbits, specific antibodies against neopterin have been prepared with a conjugate of neopterin to bovine serum albumin (neopterincaproyl–bovine serum albumin). With this specific antiserum a conjugate of neopterin with tyramine was synthesized and labeled with ^{125}I as ligand for the radioimmunoassay (N1).

Measurement of neopterin by RIA in serum is sensitive and specific. The results of serum neopterin determination correlate with those of measurement by HPLC (W14). In contrast, the results of both methods did not correlate for urinary neopterin values (W14), possibly due to the cross-reactivity of the antisera (R26).

4. Neopterin Concentrations in Healthy Subjects

4.1. URINARY EXCRETION LEVELS

Results of urinary neopterin levels in four healthy subjects have been published by Fukushima and Shiota (F24). A distinct dependency of neopterin on sex was later shown, the mean urinary neopterin/creatinine ratio in males ($n = 12$) being lower than that in females ($n = 43$) (W2). Mean neopterin/creatinine ratios and corresponding upper normal limits for different sex and age groups obtained from 417 healthy adults and from 211 healthy children were age dependent (H12,R5,S13) (Table 1). The highest urinary neopterin/creatinine ratios were observed in neonates (S13). The values in children decreased gradually with increasing age toward those in adults. The lowest values were found in males aged 26–45 yr and in females aged 18–35 yr. The values increased slightly for the older age groups, being lower in general in males than in females ($p <$ 0.01). Daily neopterin excretion levels reveal that the sex dependence and, at least in part, the age dependence are caused by variations of urinary creatinine levels. In fact, daily neopterin excretion of males was found to be higher than of females (F10).

4.2. LEVELS IN SERUM

Neopterin levels were determined by radioimmunoassay in serum specimens from 662 healthy subjects, aged 1–98 yr. Of these, 263 were children,

TABLE 1
URINARY NEOPTERIN LEVELS IN HEALTHY INDIVIDUALS IN DIFFERENT SEX AND AGE GROUPS[a]

Age	Sex[b]	Number of subjects	Neopterin mean[c] (SD)	Upper normal limits[c]
0–3 days	m,f	13	972 (661)	—
4 days	m,f	21	1510 (641)	—
5 days	m,f	15	1602 (657)	—
1 month	m,f	9	906 (527)	—
3–8 months	m,f	4	560 (53)	—
1–3^{11}/12 yr	m,f	13	267 (94)	432
4–6^{11}/12 yr	m,f	25	226 (76)	405
7–11^{11}/12 yr	m,f	55	181 (73)	374
12–14^{11}/12 yr	m,f	45	171 (73)	343
15–18 yr	m,f	11	144 (65)	320
18–25 yr	m	42	123 (30)	195
26–35 yr	m	29	101 (33)	182
36–45 yr	m	41	109 (28)	176
46–55 yr	m	32	105 (36)	197
56–65 yr	m	31	119 (39)	218
>65 yr	m	33	133 (38)	229
18–25 yr	f	55	128 (33)	208
26–35 yr	f	28	124 (33)	209
36–45 yr	f	31	140 (39)	239
46–55 yr	f	28	147 (32)	229
56–65 yr	f	26	156 (35)	249
>65 yr	f	41	151 (40)	251

[a] From Shintaku et al. (S13), Reibnegger et al. (R5), and Hausen et al. (H12).
[b] m, male; f, female.
[c] Values in μmol/mol creatinine.

aged 1–18 yr (W14) (Table 2). There was no statistically significant sex dependence (Kruskal–Wallis test, $p > 0.05$). The neopterin levels in serum of subjects between 18 and 75 yr old did not differ significantly, but subjects younger than 18 yr or older than 75 yr had significantly higher neopterin levels than the former group. The serum neopterin levels of the group aged 18–75 yr agree with data of 1837 blood donors aged 18–67 yr (K5) and of 518 blood donors older than 20 yr (H1).

4.3. CIRCADIAN VARIATIONS

Several laboratories examined neopterin excretion or serum levels during a 24-h period. Neopterin was excreted in constant amounts from one in-

TABLE 2
Neopterin Levels in Serum of Healthy Individuals in Different Age Groups[a]

Age (years)	Number of subjects	Neopterin mean (SE)	Percentile	
			5th	95th
0–18	263	6.78 (0.22)	3.5	13.5
19–75	359	5.34 (0.14)	2.6	8.7
> 75	40	9.67 (0.79)	3.7	19.0

[a]Values in nmol/liter. Three age groups showed statistically significant different neopterin levels (Kruskal–Wallis test, $p < 0.0001$), but no statistically significant sex dependence (W14). The 95th percentiles were chosen as upper normal limits.

dividual during 24 h (S17). The urinary neopterin/creatinine ratios of three subjects exhibited a mean coefficient of variation of 15.7% (H9). The excretion of neopterin was calculated per hour using spontaneous urine specimens from nine subjects. The highest neopterin excretion level was observed during night and early morning (F10). This finding corresponds with serum neopterin levels obtained from a healthy individual (R27) and higher total neopterin excretion in the morning (P11).

Due to the observed physiological circadian variations of neopterin/creatinine ratios, we recommend collection of first-morning urine specimens for diagnostic purposes.

4.4. Long-Term Stability

Neopterin/creatinine ratios were determined in 25 individuals (10 males, 10 females, and 5 children) every 2 weeks during 1 yr (H. Wachter, unpublished). Two representative determinations are depicted in Fig. 6. The ratios vary little and remain within the normal range except as shown in subject B, when a sharp peak appears during a viral infection. Another study showed that neopterin levels in serum obtained daily from six healthy females were within the normal range during one complete estrous cycle (H2).

5. Conditions Associated with High Neopterin Levels

Elevated neopterin levels in general are caused by activation of cell-mediated immunity, as will be discussed below. However, there are two further conditions known to lead to elevated neopterin levels. One con-

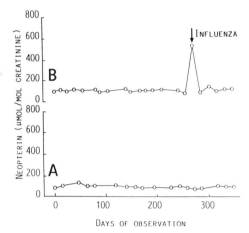

FIG. 6. Neopterin/creatinine ratios of two adults during 1 yr. The ratios were determined every 2 weeks. Subject B experienced a viral infection (influenza) during the observation period.

dition is impaired renal function, which leads to elevated neopterin levels but to normal neopterin/creatinine ratios in serum and urine. The other condition is atypical phenylketonuria. This rare metabolic defect (incidence 1/1,000,000 of newborns) is caused by dihydrobiopterin synthetase deficiency (N8), consisting in most cases of deficiency of 6-pyruvoyl-tetrahydropterin synthetase, an enzyme eliminating the triphosphate group from dihydroneopterin triphosphate. Patients are commonly detected by postnatal screening for concentrations of serum phenylalanine.

Elevated neopterin levels in urine and serum are associated with conditions involving activation of cell-mediated immunity. High neopterin levels in patients correlate with *in vitro* studies which have demonstrated that neopterin release by monocytes/macrophages depends on their activation by supernatants from activated T lymphocytes or by interferon-γ. Thus, high neopterin levels are not specific for a certain disease but indicate the presence of at least one of several conditions involving activation of cellular immunity. The presently known disease states related to neopterin levels are listed in Table 3.

5.1. ALLOGRAFT REJECTIONS

Allograft rejections and infections represent a major problem in the follow-up of allograft recipients. Release of lymphokines is one of the earliest events following stimulation of T cells by alloantigens. Interleukin-2 and interferon-γ are produced locally during rejections or systemically follow-

TABLE 3
DISEASE STATES ASSOCIATED WITH
ACTIVATION OF CELLULUAR
IMMUNITY[a]

Allograft rejections
Graft versus host disease
Infections
  By viruses (including HIVs)
  By intracellular protozoa
  By intracellular bacteria

Autoimmune diseases
  Rheumatoid arthritis
  Ulcerative colitis
  Crohn's disease
  Autoimmune thyroiditis
  Diabetes mellitus type I

Other inflammatory diseases
  Sarcoidosis
  Celiac disease
  Multiple sclerosis
  Aseptic meningoencephalitis

Certain malignant diseases
  Hematological neoplasias
  Gynecological cancer
  Tumors of the genitourinary tract
  Lung cancer
  Gastrointestinal carcinoma
  Pancreatic carcinoma

[a] As measured by high frequency of elevated neopterin levels.

ing infections. Interferon-γ stimulates macrophages to release neopterin. Determination of neopterin levels has been found useful for early assessment of major clinical complications of allograft recipients. It was observed that monitoring changes in daily neopterin levels is most significant in detection of these complications, whereas the actual level of neopterin has less diagnostic potency.

The use of daily urinary neopterin measurement to monitor the postoperative course was tested in recipients of 96 cadaver kidneys, three livers, and one pancreas (M5). Posttransplant courses without acute rejection episodes or viral infections ($n = 29$) were characterized by stable and/or low neopterin levels. Acute rejection episodes ($n = 38$), viral infections ($n = 17$), or both complications ($n = 8$) were preceded by increasing neopterin levels, on the average by 1 day. Increased neopterin levels also followed withdrawal of cyclosporine A, since this drug inhibits secretion of interferon-γ by T cells. Two characteristic courses of urinary neopterin levels, one obtained from a patient with an uncomplicated posttransplant course and one from a patient with a rejection episode, are shown in Fig. 7. Assessment of urinary neopterin levels enabled prediction of acute rejections and viral infections in 95 and 100% of patients, respectively. Therefore, neopterin appears to be a useful marker for detection of immunological complications in allograft recipients.

These results were confirmed by several reports on the clinical relevance of urinary and of serum or plasma neopterin levels. Serial plasma samples

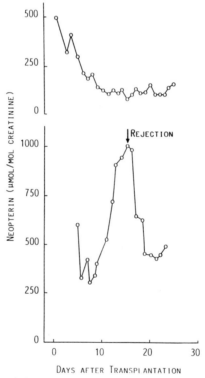

FIG. 7. Two characteristic posttransplant urinary neopterin levels of two patients having received renal allografts. The upper curve, obtained from a patient with uncomplicated posttransplant recovery, is characterized by decreasing neopterin levels during about 2 weeks. The levels then remained nearly unchanged. The neopterin levels of a patient with rejection episode (lower curve) show a pronounced maximum that peaks at the day the rejection was clinically diagnosed (arrow).

of 172 kidney transplant recipients were evaluated retrospectively by radioimmunoassay (S5). Of a total of 169 rejection episodes, a significant rise in neopterin levels was observed 1 day prior to initiation of bolus steroid therapy due to clinical diagnosis of acute rejection. The sensitivity of neopterin measurement was 95% in biopsy-proved rejections.

Levels of neopterin and of interferon-γ were simultaneously assessed daily in 36 patients having received cadaveric renal allografts; 26 acute rejection episodes and 9 serious infectious complications were observed (W22). The appearance of small amounts of interferon-γ in serum was invariably paralleled by increasing neopterin levels. Rejection episodes were associated by increasing neopterin levels even when interferon-γ could not be detected. It has been assumed that interferon-γ is produced in peripheral tissues but rarely enters the bloodstream. In contrast, neopterin enters the circulation due to its small size and chemical stability.

These studies were extended by determining serum levels of interferon-γ and neopterin concomitantly in 63 kidney transplant patients and in 6 heart transplant patients (W21). Distinct increases of neopterin levels accompanied both rejections and infections. In all patients with severe infectious episodes, distinct peaks of interferon-γ levels were also observed, followed 1–2 days later by increases of neopterin levels. In contrast, interferon-γ levels did not rise or only slightly rose during rejections, whereas neopterin levels were clearly elevated.

Neopterin was determined in serum and urine by RIA in 33 renal allograft recipients who were treated with cyclosporine A maintenance therapy (W19). Patients without clinical complications showed slightly elevated but stable urinary and serum neopterin levels. The highest neopterin levels in serum were observed in patients with allograft rejections, while the highest urinary neopterin levels were found in patients with viral infections. Among the studied parameters, the ratio of neopterin clearance and serum neopterin was found most useful for early diagnosis of allograft rejection.

In these studies, daily determination of neopterin in serum and/or urine was found to be useful for detection of immunologic complications, particularly for early diagnosis of acute rejections and viral infections during the posttransplant period.

In addition, the clinical value of neopterin levels was compared to that of $β_2$-microglobulin for monitoring the course of disease in 116 renal allograft recipients (S4). The data of these studies indicated that elevations of neopterin levels clearly preceded those of serum creatinine, but $β_2$-microglobulin levels remained nearly constant 4 days prior to and 4 days after bolus steroid therapy.

A further study that was conducted on 117 patients with kidney, 17 with liver, 8 with heart and 18 with pancreas allografts evaluated the application of urinary neopterin levels for diagnosis and prognosis of cytomegalovirus infection (T5). Measurement of urinary neopterin levels has provided useful information because levels increased approximately 1 week earlier than antibody titers, because the magnitude of the rise was related to the severity of infection, and because levels normalized concomitantly with resolution of infection.

A similar correlation between serial urinary neopterin levels and clinical course was observed in monitoring of 16 (N10), 6 (L6), and 25 (V8) bone marrow transplant recipients. The excretion of neopterin was lowest simultaneously with the nadir of peripheral blood leukocyte counts during the bone marrow aplasia. The engraftment was associated with rising urinary neopterin levels. Activation of cellular immune responses caused either by graft-versus-host disease or by infection was indicated by an increase of neopterin excretion that preceded the clinical signs by 2 days (N10), or by 2.9 ± 1.6 days (V8). Successful treatment of graft-versus-host disease with steroids cut the rapid increase of neopterin levels, but when in addition to this disease infections occurred, corticoids did not depress the neopterin excretion (V8). Therefore, it was concluded that measurement of neopterin excretion appears to be useful for the early prediction of developing immunologic complications in recipients of bone marrow transplants.

5.2. Viral Infections

Interferon-γ is produced whenever T lymphocytes are activated by specific antigens, including viral antigens (K8,L8). Hence, viral infections are typical clinical conditions associated with highly elevated neopterin levels. Elevated neopterin levels were found in urine from patients with various viral infections, as was reported in the very first publication (1979) describing altered neopterin excretion for a variety of clinical conditions (W2).

This important observation was confirmed by various authors. For example, in four children with chicken pox, serial urinary neopterin levels were determined during the viral incubation period, during the period of clinical symptoms, and during convalescence (R4). Invariably, neopterin levels began to rise 1 to 2 days before the first exanthem was observed. Sharp peaks of neopterin levels appeared within a few days. Subsequently, neopterin levels decreased rapidly and normalized. The decrease of neopterin levels coincided with the period when specific antibodies are measurable. Thus, the elimination of the pathogen by a functioning immune system in these children was indicated by normalizing neopterin levels. Importantly, there was no correlation between increased body temperature and neopterin levels in these children.

Significantly increased serum neopterin levels were seen in 9 patients with mononucleosis due to infection with Epstein–Barr virus, in 10 patients with cytomegalovirus infection, and in 20 patients with acute virus-induced hepatitis (6 cases with hepatitis A, 11 with hepatitis B, and 3 with non-A, non-B hepatitis) (K6).

Serial urinary neopterin determinations were performed daily in 18 patients with rubella, in 12 with chicken pox, and in 15 with influenza, from infection until recovery (M11). A slow increase in neopterin level was observed during the incubation period, and, with onset of clinical symptoms, high peak levels were reached in all cases (between 1000 and 2000 μmol neopterin per mol creatinine). The peak height corresponded with severity of diseases. At the maximum of clinical symptoms, neopterin levels showed a rapid decline and finally normalized.

In patients with virus-induced hepatitis, several additonal studies were undertaken (K6). In a study on 95 patients with hepatitis B, urinary neopterin levels were determined and compared with clinical status (F1). There were 50 patients with acute, 20 with chronic aggressive, and 25 with chronic persistent hepatitis. Mean neopterin excretion was significantly elevated in these patients. There was a correlation with status, patients with acute hepatitis having higher levels than those with chronic disease, and patients with chronic aggressive hepatitis having higher levels than those with chronic persistent hepatitis.

Urinary neopterin levels were also assessed in 51 patients with acute hepatitis and were related with the type of the hepatitis (R13). There were 13 patients with hepatitis A, 26 with hepatitis B, and 12 with non-A, non-B hepatitis. A good correlation was found with markers of clinical activity, such as serum levels of aminotransferases in individual patients. In this study, patients with various nonviral pancreatic and biliary disorders as well those with alcohol-toxic hepatitis were included for control. In all these patients, neopterin levels were normal or only slightly elevated, but were well below the levels found in virus-mediated hepatitis (serologically proved in the case of hepatitis A and B). Further, in the control subjects there was no correlation between levels of neopterin and other biochemical tests for liver function. As neopterin levels were invariably elevated in proved viral hepatitis, it was concluded that neopterin might be a valuable marker for suspected non-A, non-B hepatitis.

Another study underlined the potential of neopterin determination to aid in the discrimination between chronic persistent non-A, non-B hepatitis and steatosis hepatitis (P19). Patients with these disease entities usually present with very similar signs, including mild elevations of serum levels of transaminases, but rarely show clinical disease. The prognosis and the clinical impact of both diseases, however, are completely different. In a comparison of neopterin levels and other biochemical tests for liver function, neopterin levels were shown to have the greatest potential for differential diagnosis, because neopterin levels were elevated in the chronic persistent non-A, non-B hepatitis cases but were normal in fatty liver patients.

In addition to these studies, essentially the same observations were made in children after vaccination with a live measles–mumps vaccine. The typical pattern of sharply increasing levels was observed, with a peak at the time when viremia is known to be highest in wild-type measles infection (12 to 15 days after vaccination). Subsequently, neopterin levels rapidly declined and normalized. This decline coincided with the period in which specific antibodies become detectable. It is important to note that in none of these children were clinical symptoms apparent (R4).

The typical pattern of peaking neoptern levels preceding the appearence of specific antibodies in serum, and the decline associated with increasing antibody titers, was also observed in kidney allograft recipients with cytomegalovirus infection (T5).

Summarizing the available evidence, one can conclude that neopterin is a highly sensitive indicator for the presence of activation of cell-mediated immune responses by viral pathogens. Neopterin elevations in viral infection do not appear to depend heavily on the type of the virus, and hence, neopterin determination should be kept in mind as a potent general marker for known viral infection, and also for new viruses against which no antigen–antibody system exists.

In this context, it should be considered that a fast and generally applicable method to identify viral infections is not yet available. Virus serology and culture methods cannot be used routinely in clinical practice because they are time consuming, expensive, and often the pathogenic agent cannot be identified. Furthermore, viral and bacterial infections require different therapeutic modalities. Therefore, it seems reasonable to predict that a sensitive marker indicating viral infections and activity of disease would be useful in management and therapy control of viral infections.

The large number of diseases indicated by elevation of serum neopterin concentrations suggests the application of neopterin for routine screening of blood donors. One of the most serious hazards of blood transfusion is the potential transmission of malignant and infectious diseases, including acquired immunodeficiency syndrome. Neopterin levels were determined in parallel with established laboratory tests in 6968 voluntary blood donors during 3 months (S9). The observed donor loss due to elevated neopterin levels was comparable to that due to elevated alanine aminotransferase and was therefore within tolerable limits. It can be expected that the average donor loss due to elevated neopterin levels will not exceed 2% over a period of 1 yr. Because a panel of hazardous diseases can be detected due to its association with activated cellular immunity, measurement of neopterin to minimize transfusion hazards can be recommended.

5.2.1. *Infections by Human Immunodeficiency Virus*

Acquired immunodeficiency syndrome (AIDS) was first described in 1981 in the United States. Human immunodeficiency virus (HIV) has been identified as the causative agent of AIDS in humans. At present, HIV infection is a major worldwide health problem. HIV selectively infects and destroys the helper/inducer subpopulation of T lymphocytes. The progressive depletion of these cells leads to the wide spectrum of clinical manifestations characterized by high susceptibility to opportunistic infections, which are generally fatal, and to rare malignancies. The participation of T cell activation in the pathogenesis of the disease is of increasing concern.

In 1983, when our laboratory, collaborating with T. J. Spira from the Centers for Disease Control, Atlanta, Georgia, had begun to study urinary neopterin levels in patients with acquired immunodeficiency syndrome and with prodromal stages (W8), the pathogenic agent of AIDS was not known. Surprisingly, it was concluded in part from the extremely high neopterin levels of these patients that their cellular immune system was strongly activated. Due to the severely immunocompromised condition of patients with advanced disease, the results presented provoked much debate or even denial. Later on, it was observed that cultivation of human immunodeficiency virus strictly depended on activation of target cells. Thus, the data became reasonable. Particularly, both infection of T cells and replication of HIV *in vitro* require activation of cells, which is usually achieved by phytohemagglutinin and interleukin-2 (B3,P14). A pivotal role of T cell activation *in vivo* has been concluded from neopterin data. Neopterin levels increase early during the course of HIV infection and correlate positively with progressive disease according to the Walter Reed staging classification. Neopterin is a valuable marker for AIDS patients due to its predictive significance.

The preceding pilot study (W8) was conducted using 5 patients with AIDS and 12 patients in prodromal stages. The results of this report were extended to 38 patients with AIDS and 64 patients with persistent generalized lymphadenopathy (F15); results confirmed by another group using 36 patients with AIDS and 48 patients with AIDS-related complex (ARC) (A1). This study showed that almost all patients had significantly elevated neopterin levels but biopterin levels were within the normal range.

Detailed investigations of neopterin levels in groups with a high incidence of AIDS were of significance for the recognition and confirmation of the role the T cell activation plays in HIV infection. Increased urinary and/

or serum neopterin levels were found in more than 90% of 59 HIV-seropositive parenteral drug addicts (F19), in approximately 85% of seropositive homosexuals (F13,G8), and in 60–80% of seropositive hemophilia patients (D1,F17). However, in addition to most of the HIV-infected individuals, a substantial fraction of seronegative symptomless members of groups at high risk for AIDS showed elevated neopterin levels, for instance, 15% of seronegative homosexuals and 50% of seronegative parenteral drug addicts and hemophilia patients (D1,F13,F19). This observation possibly identifies a special disposition in high-risk groups. The role of T cell activation for HIV infections appears to be reflected by the observation that the magnitude of neopterin levels was correlated with behavioral factors of risk-group members: with the number of partners of receptive anal intercourse in homosexuals (F18) and with the parenteral mode of drug administration in drug addicts (F20). Infections, drugs, and adulterants that are often mixed with drugs likely stimulate T cells, leading to enhanced production of neopterin (W9). The presence of at least transient immune stimulation was also shown in recipients of blood transfusions (W9).

These data have indicated that activation of cellular immunity represents the often-suggested cofactor for increased susceptibility for seroconversion and progressive disease, in addition to increased risk of infection following exposure to HIV. Thus, those individuals who are at risk for exposure to HIV (homosexuals, parenteral drug abusers, and hemophiliacs) and recipients of blood transfusions who are exposed to HIV are likely to be effectively infected if a preexisting T cell activation promotes production of HIV (F16,W9). In East Africa activation of the cell-mediated immune system occurs frequently because of infections with parasites or fungi, thus AIDS spreads rapidly, regardless of sex, among African communities (W6).

According to this proposed role of T cell activation, immunostimulatory treatment such as application of interleukin-2 will exacerbate the course of the disease in HIV-infected patients (H14,H15). In fact, failure of treatment with interleukin-2 has been reported (F5,V1,V7). Immunosuppressive therapeutic regimens, however, have been shown to be beneficial (A4,M1).

Neopterin levels follow closely the course of HIV infection. In one patient, neopterin levels increased during acute infection before antibodies against HIV became detectable (E1). Progressive infection is paralleled by increasing neopterin levels, as was reported first for a single patient (A1) then for groups of patients (F13,K10). Whereas there exists no difference in neopterin levels among asymptomatic HIV seropositives and those with persistent generalized lymphadenopathy, the correlation to the Walter Reed staging classification was found to be highly significant (F13).

Neopterin levels and $CD4^+/CD8^+$ T cell ratios correlate inversely in ARC and AIDS patients. In asymptomatic seropositives the ratios were usually only slightly below the level found in seronegative risk-group members (H23,K10). Similar conclusions were drawn regarding the correlation of neopterin and β_2-microglobulin, because neopterin levels begin to rise earlier than do β_2-microglobulin levels (K10,L3).

Neopterin has predictive potential in HIV-infected individuals. HIV-seropositive subjects with high neopterin levels developed symptoms earlier than did those with lower levels (U1). High neopterin levels precede progression within the Walter Reed staging classification (B9). AIDS patients with the highest neopterin levels were observed to die earlier than others (H23).

Neopterin levels in risk groups and HIV-infected individuals can serve as a valuable marker because neopterin levels increase early in course of HIV infection with high frequency and because they continue to rise steadily when the disease progresses, and thus are of predictive significance.

5.3. Infections by Intracellular Protozoa

It is well established that cell-mediated immunity is involved in defense against malaria parasites, but antibodies also play a role particularly in the case of continued challenge by the parasites (F3,H17,R23). Mediators released from T cells and macrophages such as interferon-γ, tumor necrosis factor, and reactive forms of oxygen are thought to play an important role (C1). Therefore, it is not unexpected that neopterin levels, indicating activation states of T cells and macrophages, increase during infection by *Plasmodium* species.

Extremely high urinary neopterin levels were reported in a group of 27 malarial patients from Bangkok, Thailand, where malaria is semiendemic (R8).

A study of 55 patients with malaria showed that the serum neopterin levels were almost invariably highly elevated during acute malarial attacks (K4). In 17 patients with *Plasmodium falciparum* malaria, a weak correlation between degree of parasitemia and neopterin levels was observed. Neopterin levels of 41 nonimmune patients were higher than were those of 14 semiimmune individuals. When clinical disease resolved within 3–7 days of treatment, neopterin levels normalized rapidly.

Urinary neopterin levels were further measured in 128 infected individuals from Tanzania, where the disease is holoendemic. Neopterin levels of a subgroup of 117 previously untreated patients were compared with those of 19 untreated malarial patients from Bangkok (R9). The levels of the Thai patients were considerably higher than those of the Tanzanian subjects of similar age. An overwhelming age dependency was detected among the Tanzanian patients (children showed by far the highest levels), but other variables did not influence neopterin levels significantly. The observed influence of age on neopterin levels likely reflects a switch from cellular to humoral immune mechanisms against *Plasmodium*. In holoendemic regions, malarial infection provokes predominantly cellular immune phenomena in young children. With increasing age of subjects, antibodies against the circumsporozoite protein are thought to become more important for acquired immunity in endemic areas where the prevalence and severity of *P. falciparum* infection decreases (H17). It should be noted that urinary neopterin levels were also reported to be significantly elevated in four children with low-grade parasitemia without clinical symptoms.

The studies on neopterin levels in malaria patients suggest that neopterin is an early and sensitive marker in this disease. Since neopterin levels decrease toward normal with recovery, their measurement might also be useful for follow-up treatment.

5.4. Infections by Bacteria

5.4.1. *Intracellular Bacteria*

Intracellular bacteria provoke cellular immune responses of the infected host. Interferon-γ is regarded as the primary lymphokine responsible for defense against intracellular pathogens (N5). Although direct measurements of interferon-γ levels have not yet been performed, there is indirect evidence that this lymphokine is produced endogenously in diseases caused by intracellular pathogens (C2). For example, injection of monoclonal antibody to interferon-γ prevented mice from recovering from listeriosis (B13). Due to the importance of cell-mediated immunity in resistance to infection caused by intracellular pathogens, it appeared appropriate to study neopterin levels as a marker for activation of cellular immunity in patients with such infections. So far, studies on neopterin have been performed in patients with pulmonary tuberculosis and leprosy.

A study conducted on 55 patients with pulmonary tuberculosis demonstrated that urinary neopterin levels were elevated in 83.6% of patients with active disease (F12). There was a correlation between mean neopterin levels and extent of disease as well as activity of disease. Serial measurement of neopterin levels showed that neopterin levels reflect accurately the clinical activity of tuberculosis under treatment. These findings appear to be of practical value because there is no reliable parameter reflecting rapid changes of activity in this disease.

Urinary excretion of neopterin was studied in 77 patients with leprosy (S6). Neopterin levels were increased in 75% of patients. No difference was apparent between patients with tuberculous, with borderline lepromatous, and with lepromatous leprosy. Also there was no correlation between neopterin levels and duration of disease or prior medication. The value of neopterin levels for assessment of activity or of efficacy of therapy in this disease remains to be examined.

5.4.2. *Other Bacteria*

The question as to which bacteria induce interferon-γ and which do not cannot be answered systematically by experimental studies from the literature (K9). However, it can be concluded from the high susceptibility to viral, fungal, protozoan, and intracellular bacterial infections of individuals with deficient T cell function that immune responses against these pathogens are mediated by immune cells (T cells and macrophages). During this immune response, $CD4^+$ and $CD8^+$ cells secrete interferon-γ (M9), inducing neopterin production by macrophages.

The host defenses against other (nonintracellular) bacteria are thought to be mainly mediated by antibody and complement factors. Therefore, interferon-γ and neopterin should not be produced to any great extent in infections by such bacteria. Indeed, bacterial infections are not commonly associated with high neopterin levels (W2), particularly when infections are locally restricted. However, increased neopterin levels are observed in a substantial fraction of patients with staphylococcal pneumonia (N11, and H. Wachter, unpublished results).

Possibly, high neopterin levels reflect a combined viral–bacterial pneumonia. It is thought that virus infection in the respiratory tract favors growth conditions for bacteria. In addition, it was reported that coinfection with *Staphylococcus aureus* enhanced greatly pathogenicity of influenza virus due to virus activation by a protease produced by *S. aureus* (T1).

Alternatively, production of interferon-γ and of neopterin might be found in advanced stages of bacterial infections such as bacteriemia. This view is supported by a study on plasma neopterin levels in intensive care patients with and without septic complications (S23). Twenty-one patients were classified according to their clinical course and outcome. Highly significant differences of neopterin levels were observed among survivors and nonsurvivors. It was concluded that estimation of neopterin might become a helpful tool for assessment of the clinical course in septic patients. This is also supported by a study on neopterin levels in 42 patients with bronchial asthma (M8). Patients with exacerbated asthma showed significantly higher neopterin levels in urine and blood than did patients with stable asthma ($p < 0.01$).

New information about bacterially stimulated release of interferon-γ could lead to insights on human immune responses and to recognition of neopterin elevations in other bacterial infections.

5.5. Autoimmune Diseases

The participation of cellular immune events in the pathogenesis of autoimmune diseases is now generally accepted. Activation of T cells was determined by expression of HLA-DR antigen (B10) and by significantly elevated levels of interferons in patients with autoimmune diseases (H18,H19,P16). It was concluded that interferon found in sera from patients with systemic lupus erythematosus was an unusual acid-labile form of interferon-α but that interferon from patients with autoimmune vasculitis was predominantly interferon-γ (P15). Furthermore, it has been hypothesized that interferon has a role in the development of autoimmune disorders. Indeed, accelerated progression of disease was observed in several independent studies in experimental animals (A2,E4,H16). As a consequence of the presence of interferon-γ, it was suspected that neopterin levels might be of interest in autoimmune diseases. Neopterin levels are high in acute phases and reflect extent and activity of disease in rheumatoid arthritis, Crohn's disease, ulcerative colitis, autoimmune thyroiditis, and early-onset diabetes mellitus type I.

5.5.1. *Rheumatoid Arthritis*

Activated lymphocytes and lymphokines are found in synovial membrane and synovial fluid of patients with rheumatoid arthritis during clinically active periods of disease (J1). The clinical course of rheumatoid arthritis fluctuates. The patients require medication during periods of high clinical activity but not during quiescence of the disease. Apart from clinical criteria, a panel of laboratory data is commonly used to assess clinical activity of the disease. However, a reliable, sensitive, and reproducible method to quantitate activity of rheumatoid arthritis would be particularly valuable.

A pilot study conducted in 30 patients with the classical form of rheumatoid arthritis appeared promising because a correlation between urinary neopterin levels and stage and particularly activity of disease was observed (H13).

Investigations of urinary neopterin levels and other commonly used laboratory variables were conducted in 106 patients with rheumatoid arthritis and compared with the data of 45 patients with osteoarthritis as an example of a degenerative rheumatic disorder (R11). Neopterin levels were significantly associated with stage as well as with activity of rheumatoid arthritis. Significantly higher neopterin levels were found even in patients with stage I rheumatoid arthritis compared to patients with osteoarthritis. Correlations with erythrocyte sedimentation rate, hemoglobin level, white cell blood count, and presence or absence of C-reactive protein and of rheumatoid factor were weak. Therefore, measurement of neopterin confers independent information. This conclusion was corroborated by multivariate analysis of the data, showing that levels of urinary neopterin, C-reactive protein, and erythrocyte sedimentation rate were the best independent combined predictors of clinical activity. There was a dominant contribution of neopterin to this combination. The ability of neopterin levels to reflect clinical activity of disease was also apparent in follow-up studies, which showed that neopterin levels decreased if therapy effectively lowered the activity of the autoimmune process.

A close association between urinary neopterin levels and clinical activity of rheumatoid arthritis was also shown by another group (H6). Neopterin levels, serum levels of C-reactive protein, and activity of disease were correlated. Patients with rheumatoid arthritis ($n = 67$) had significantly higher levels than did a healthy control group ($n = 67$) or patients with arthralgia or arthrosis ($n = 24$).

Because it is assumed that activated T cells produce interferon-γ within inflamed joints, it seems likely that this lymphokine locally induces macrophages for neopterin production.

A recent study demonstrated that neopterin levels in synovial fluid are higher than in serum of patients with inflammatory rheumatic diseases ($n = 30$) (M2). Neopterin levels in synovial fluids from inflammatory joints ($n = 30$) were significantly elevated compared to those from noninflammatory joints ($n = 30$). It was concluded that neopterin is locally produced in synovial fluid of patients with inflammatory rheumatoid diseases but not with noninflammatory disease; then neopterin appears in circulating blood and is finally excreted in urine. There was no correlation between neopterin levels and immunoglobulin levels, acute-phase proteins or albumin in synovial fluids of inflammatory joints.

It might be expected from the published results that neopterin levels can advantageously complement the typically employed clinical and laboratory variables to discriminate between rheumatoid arthritis and osteoarthritis and to assess the clinical activity of rheumatoid arthritis, particularly during treatment. Application of neopterin in other inflammatory rheumatoid diseases, for instance, systemic lupus erythematosus and juvenile rheumatoid arthritis, might be of importance for insights into pathogenetic backgrounds.

5.5.2. Ulcerative Colitis

In addition to phenomena usually ascribed to the activity of humoral immunity in the pathogenesis of ulcerative colitis, participation of cell-mediated immune events is generally supposed. Therefore, it can be inferred that production of interferon-γ by activated T cells and of neopterin by induced macrophages should be observed in active periods of ulcerative colitis. Because assessment of clinical activity of ulcerative colitis is based predominantly on secondary clinical features, such as local irritation and systemic symptoms, a sensitive marker for activity of critical immune events is desirable.

Urinary neopterin levels were measured in 25 patients with ulcerative colitis (N9). A correlation between neopterin levels and severity of clinical symptoms was observed. The closest correlation of clinical activity was obtained with extent of bowel involvement. Neopterin levels were the second best index, followed by erythrocyte sedimentation rate, by number of bowel movements per day, and by body temperature. C-Reactive protein and orosomucoid correlated less than neopterin with clinical activity of disease. Therefore, determination of neopterin was proposed to be useful for monitoring activity of ulcerative colitis.

In further studies, the results were confirmed in 11 patients with colitis ulcerosa. In addition, a significant correlation between the number of peripheral activated lymphocytes and neopterin levels was observed (R1).

5.5.3. Crohn's Disease

There are several unresolved questions in the pathogenesis and assessment of clinical activity of Crohn's disease. However, there is no debate that both humoral and cellular immune phenomena play a role in this disease. Similar to other autoimmune disorders, Crohn's disease is characterized by alternating acute (active) and quiescent periods. Therefore, criteria for assessment of clinical activity are sought. Neopterin levels as indicators of activation of cellular immunity were found appropriate to characterize the activity of Crohn's disease.

Urinary neopterin levels were measured in 34 patients with Crohn's disease (P20). Multiple stepwise regression analyses were used to evaluate a possible association of neopterin with 15 other clinical and laboratory variables. Neopterin was significantly correlated with clinical activity, with disease duration (inverse correlation), and with index of body weight. Extent and localization of the disease did not influence neopterin excretion.

For assessment of the course of Crohn's disease, several indices have been suggested, e.g., the Crohn's Disease Activity Index (CDAI) includes one laboratory parameter (hematocrit) and seven other somewhat subjective factors. On the basis of the correlation of neopterin and activity of disease, a simple triple-parametric index for clinical activity of Crohn's disease was proposed (R14). This index consists of two laboratory variables, hematocrit and urinary neopterin levels, and one clinical parameter, the frequency of liquid or very soft stools. The results obtained with the latter index were at least equal to those based on the eight parametric CDAI.

A study of 51 patients with Crohn's disease revealed significant association of neopterin concentrations with clinical activity, with CDAI, and with the number of peripheral activated lymphocytes (R1). Therefore, neopterin levels might be considered valuable assays in Crohn's disease.

5.5.4. Autoimmune Thyroiditis

Excessive expression of HLA-DR antigen was observed in T cell clones from autoimmune thyroid glands (B10). Because interferon-γ represents the best known inducer of DR antigen expression, it has been suggested that this lymphokine is produced locally and within the environment of thyroid glands (T7). Neopterin levels were examined in patients with different thyroid diseases (S11). Neopterin levels in serum were elevated in 11 of 13 patients with autoimmune thyrotoxicosis, in 2 of 8 patients with disseminated autonomy, in 5 of 6 patients with autoimmune hypothyroidism, and in 0 of 4 patients with goiter. From these results, it was suggested that neopterin might be helpful to differentiate autoimmune from nonautoimmune thyroid diseases and in the management of autoimmune-induced thyrotoxicosis. During treatment, neopterin levels normalized in hyperthyroid patients when thyroid hormones normalized (5 patients) but remained elevated when the course of disease was unchanged (1 patient). Further work on autoimmune thyroiditis may help to consolidate the value of the neopterin assay in this field.

5.5.5. Diabetes Mellitus Type I

Type I diabetes mellitus (insulin-dependent diabetes mellitus) is considered to be an autoimmune disease (E2). Increased $CD4^+/CD8^+$ ratio and activated T cells were found in peripheral blood, with high numbers of $CD8^+$ T cells positive for HLA-DR antigen in the pancreas of patients with type I diabetes mellitus (B11). There is evidence for involvement of cell-mediated immunity.

A pilot study showed that urinary neopterin levels were significantly increased in newly diagnosed patients with early onset of type I diabetes ($n = 8$) and with long-lasting diabetes with anticytoplasmic islet cell antibodies ($n = 3$), but not in long-lasting diabetic patients without anticytoplasmic islet cell antibodies ($n = 13$) (M4). It was proposed that neopterin determination would be useful in monitoring patients with type I diabetes mellitus. Additional investigations might show the predictive value of neopterin in patients who are genetically predisposed to diabetes mellitus type I.

5.6. Other Inflammatory Diseases

This group comprises diseases with recognized involvement of cell-mediated immunity as an essential element in the pathology of disease. At part, they are considered as autoimmune-like (celiac disease, multiple sclerosis). However, the precise mechanisms of these diseases are not yet defined and remain a matter of debate.

5.6.1. *Sarcoidosis*

Pulmonary sarcoidosis, a chronic disorder of unknown etiology, is characterized by accumulation of T lymphocytes and macrophages at the site of disease activity (V4). Activated immune cells produce, among other lymphokines, interferon-γ (R25, T3).

In order to test the hypothesis that neopterin levels reflect activity of pulmonary sarcoidosis, urinary neopterin levels were measured concomitantly with clinical status and ^{67}Ga scans in 40 patients with biopsy-proved disease (L1). It was shown that neopterin levels were significantly elevated in the patients with sarcoidosis. Patients with active sarcoidosis had higher neopterin levels than did those who did not meet the clinical criteria for active disease. Furthermore, neopterin levels of patients with extended alveolar lympholysis (more than 30%) were higher than was found in those with low alveolar lympholysis (less than 30%). A significant correlation between urinary neopterin levels and results of ^{67}Ga scans, but not with serum angiotensin-converting enzyme levels and urinary calcium excretion, was observed.

The results of further studies were in accord with the former. (E3, P18, W20). Also, serum neopterin levels were significantly elevated in patients with sarcoidosis. Serum and urinary neopterin levels increased in parallel to clinical activity of the disease and were correlated to ^{67}Ga scans.

The results of these studies suggest that measurement of neopterin might be valuable in determining the degree disease activity and in follow-up, and justify further investigations.

5.6.2. *Celiac Disease*

Celiac disease is a malabsorptive disorder defined as permanent intolerance against wheat gluten A-gliadin and similar proteins of other cereals. Apart from genetic and environmental factors, pronounced immunological events of both humoral and cellular types are important in the pathogenesis of celiac disease (F2, K1). Oral intake of gliadin is known to activate celiac disease and leads to lesions, termed villous atrophy, of the small bowel mucosa. Activation of T lymphocytes and production of lymphokines play a crucial role (S22).

A pilot study on children with active celiac disease demonstrated that urinary neopterin levels were elevated when the children were fed with gluten-containing diet. Introducing a gluten-free diet was accompanied by rapid fall of neopterin levels (F11).

Urinary neopterin levels were measured in 52 children with celiac disease, as reported in an extended work (G10). A highly significant difference between children with active celiac disease and patients without clinical symptoms or patients on gluten-free diet was observed. There was a close correlation between neopterin levels and titers of antigliadin antibodies in serum (an indicator of specific humoral immune response) when samples were collected on the same day ($n = 16$).

Because neopterin levels reflect activity of celiac disease, neopterin can be regarded as useful, independent marker for follow-up of childhood celiac disease.

5.6.3. *Multiple Sclerosis and Aseptic Meningoencephalitis*

Multiple sclerosis is presently an incurable and often debilitating neurological disease in which a number of immunological phenomena have been reported. The presence of mononuclear cell infiltrates in demyelinating lesions and of oligoclonal immunoglobulin G in the cerebrospinal fluid strongly suggests the involvement of an immunopathological process (P4). The involvement of the immune system in the pathology of multiple sclerosis is generally accepted, but it is not known which antigens trigger the immune response. It has been suggested that infectious agents or autoimmune responses against components of the central nervous system are implicated. Certainly, there is a substantial role of cell-mediated immunity in the pathogenesis of multiple sclerosis (B4). Analogously, intrathecal cellular immune mechanisms are demonstrated to be involved in the pathogenesis of aseptic meningoencephalitis (L10).

A recent study revealed elevated neopterin levels in cerebrospinal fluid (CSF) of 10 of 12 patients with multiple sclerosis during exacerbation when compared to periods of remission (F8). High neopterin levels in CSF were not paralleled in the serum of the patients studied.

Neopterin levels were further measured in CSF of patients with acute-phase aseptic meningoencephalitis, with active noninflammatory neurological disease, and with tension headache or psychoneurosis as controls (F7). Elevated neopterin levels were found in CSF in most patients during acute-phase aseptic meningoencephalitis. The neopterin levels in CSF were strongly correlated to the number of mononuclear cells in CSF during the course of disease and normalized with recovery. Neopterin levels in CSF of patients with active noninflammatory neurological disease and of controls were lower than those of a majority of patients with aseptic meningoencephalitis. The observed elevation of neopterin levels in CSF argues for high-level activation of these immune cells.

The results of neopterin levels in CSF of patients with inflammatory neurological diseases are promising and further examinations would be of considerable interest.

5.7. Malignant Diseases Involving Activation of Cell-Mediated Immunity

Immunogenicity of tumors was convincingly demonstrated in the 1950s (N12). The interactions between tumors and immune system, however, are complex and therefore remain a matter of much debate. The original concept of immune surveillance, whereby actions of the immune system are invariably beneficial to the host of a tumor, clearly can be regarded as an oversimplification (P13,P17). It has been strongly suggested that activated macrophages even promote tumor growth (M10,S15). Extensive investigations leave little doubt that a substantial number of tumors evoke cellular immune responses in the host whether the growth of the tumor is restricted or not.

In several malignant diseases, elevated levels of neopterin in urine and serum were observed. The extent of elevation of neopterin levels was correlated with stage and prognosis of disease in several tumor types. The data support the view that persistent immunoactivation is not necessarily of benefit to tumor patients (R6).

5.7.1. *Hematological Neoplasias*

Several studies describe neopterin levels in patients with hematological neoplasias (H12,L5,M11,R5,S2,S3,Z2). Patients with active malignancies (chronic lymphocytic leukemia, chronic myelocytic leukemia, multiple myeloma, Hodgkin's lymphoma, non-Hodgkin's lymphoma, and polycythemia vera) ($n = 135$) showed significantly elevated urinary neopterin levels with high frequency, except that only 29% of patients with multiple myeloma in stage I exhibited neopterin levels above the upper limit of normal (H12). It was of particular interest that more than 90% of patients with non-Hodgkin's lymphoma and with leukemia showed elevated neopterin levels. A pronounced influence of stage was observed in patients with chronic myelocytic leukemia, with multiple myeloma, and with non-

Hodgkin's lymphoma. The majority of patients with hematological neoplasms in remission had neopterin levels within the normal range.

Several further studies underline these results. Urinary and/or serum neopterin levels were examined in 14 children with active hematological neoplasia and in 25 children with hematological neoplasia in remission (R5), in 19 adults with active hematological neoplasia (S2), in 64 patients with hematological neoplasias, untreated or in stable remission (L5), in 28 patients with active neoplasia, in 2 patients with disease in remission (S3), in 28 patients with Hodgkin's disease (Z2), and in 21 patients with multiple myeloma (M11). The latter study showed a significant correlation between urinary neopterin levels and the total estimated mass of tumor cells that was even more pronounced than the correlation between β_2-microglobulin and mass of tumor cells. These data support the view that neopterin levels might be useful complements to the hematological parameters commonly employed in these diseases.

Several other studies should be mentioned. In 1981, a simple method was published to determine 6-hydroxymethylpterin (R2). The described assay claims to absorb 6-hydroxymethylpterin from urine on charcoal, followed by elution and measurement by fluorescence. This report was followed by two further studies from the same laboratory and by one from another laboratory (A5,R3,T8). The results were excellent: 120 patients with various types of cancer excreted significantly higher levels of 6-hydroxymethylpterin (R3). The levels correlated with the percentage of blasts in patients with acute myeloid leukemia (T8) and normalized when remission was achieved. An analogous correlation was observed in patients with acute lymphoblastic leukemia (R3). However, the method used for quantification of 6-hydroxymethylpterin in patients with tumors was questioned (D6,D8), and according to our experience, only neopterin levels are consistently increased in patients with malignancies, particularly in hematological neoplasias. Examination of the HPLC method proposed by Rao and co-workers showed that most of fluorescent pterins were absorbed on and eluted from charcoal. The major fluorescence measured by Rao and associates was caused by neopterin, not by 6-hydroxymethylpterin (H10). From this view, the findings are consistent with other reports on neopterin.

5.7.2. Gynecological Tumors

In 287 women with carcinomas of the genital tract pretherapeutical urinary neopterin levels were measured (B5,B6,R12,R15). The percentage of patients with elevated neopterin levels depended on tumor type and tumor stage. The highest frequencies of elevated levels were observed in patients with ovarian carcinoma and generally in patients with the most advanced tumor stages. In contrast, in patients with benign gynecological tumors ($n = 53$) and in women with precancerous lesions of the cervix and endometrium ($n = 24$), neopterin levels below the upper limits of normal were commonly observed (B6).

The relation between pretherapeutically and serially measured neopterin levels and the outcome of disease was examined in 186 women with cervical cancer (R15) and in 74 women with ovarian cancer (R12). High pretherapeutic neopterin levels were significantly predictive for a more unfavorable course of disease and earlier death in both groups of patients.

The possibility that the predictive significance of neopterin depends on other factors, such as tumor stage or surgery, was examined by multivariate statistical analysis with the Cox regression procedure and stratification according to tumor stage. Laboratory variables chosen for inclusion in these analyses were urinary neopterin levels, white blood cell count, thrombocyte number, hemoglobin, hematocrit, erythrocyte sedimentation rate, serum urea, and liver function tests. Levels of hemoglobin, leukocyte number, and neopterin were independent significant predictors of survival time in cervical cancer even when adjusted for tumor stage (R15). In ovarian cancer, neopterin was the unique significant laboratory test (R12).

A potential association of neopterin levels measured serially during follow-up with risk of death, recurrence, or metastasis was statistically assessed by a variant of the Cox technique with a time-dependent covariate. This model is particularly suited for analysis of profiles of tumor markers (G3). A highly significant association of elevated neopterin levels and a higher risk for death, recurrence, or metastasis were detected in cervical and ovarian cancer. In cervical cancer, a significant rise of neopterin levels was observed up to 5 months prior to death, 1 month prior to recurrence, and 2 months prior to metastasis. Significantly elevated neopterin levels were found 7 months before death in ovarian cancer.

There was a significant association between serial neopterin levels and histological diagnosis of patients with ovarian carcinoma at second-look laparotomy ($p = 0.016$). This observation might be of particular value because a substantial number of patients commonly refuse second-look surgery. Furthermore, patients with evidence of disease at second look and with normal neopterin levels were found to have a better outcome than women with evidence of tumor and with elevated neopterin levels (R12).

5.7.3. *Tumors of the Genitourinary Tract*

Urinary neopterin levels were measured in patients with bladder tumor ($n = 54$), with carcinoma of prostate ($n = 34$), with tumors of testicle (n

= 32), and with renal cell carcinoma ($n = 11$) (A7,F9). Except for the latter group, a substantial correlation between stage of tumor and urinary neopterin excretion was found. All patients with bladder tumor stage T1 had normal neopterin levels, but 89% of 35 patients with higher stages had elevated neopterin levels. Similarly, all patients with prostatic carcinoma stage A or B had normal levels but 84% of 19 patients in stage C or D had significantly elevated neopterin levels.

Neopterin levels were determined in urinary samples from 71 testicular tumor patients during a 4-yr follow-up. Of these patients, 38 had seminomas and 33 had nonseminomatous tumors (J5). The observed correlation between urinary neopterin levels and clinical condition demonstrates that neopterin represents a valuable marker for monitoring the course of disease in patients with testicular tumor.

A further study of 93 patients with prostatic carcinoma revealed a predictive significance of serum neopterin levels (L9). Patients with initially elevated neopterin levels had significantly worse survival rates than did those with normal neopterin levels ($p = 0.001$).

The therapeutical modalities of patients with prostatic carcinoma stages A and B differ from those with stages C and D, and with bladder tumor stage T1 from those with higher stages. Due to its predictive significance in patients with prostatic carcinoma, the value of neopterin determination for these tumor types is obvious.

5.7.4. Lung Cancer

Of patients with lung cancer ($n = 103$), 58% showed elevated urinary neopterin levels (C3). There was no association between neopterin levels and histological type of lung cancer or tumor stage. However, there was a significant difference of neopterin levels between limited and extensive disease in small cell carcinoma ($p = 0.0117$; Kruskal–Wallis test). Furthermore, 25 of 36 patients (69%) with tumor recurrence had elevated neopterin levels.

The predictive value of tumor stage, of therapy, and of pretherapeutic neopterin level was examined by multivariate analyses according to the Proportional Hazards model (C5). The analysis was stratified by tumor histology to correct for possible nonproportional effects of histology on base-line survival curves.

There was no significant difference in survival of patients with normal and elevated neopterin levels in squamous cell carcinoma ($n = 41$, $p = 0.24$; Breslow test) but there was a significant difference in adenocarcinoma ($n = 21$, $p = 0.018$) and in small cell lung cancer ($n = 24$, $p = 0.025$).

The multivariate stepwise regression analysis showed significant influence of tumor stage ($p = 0.0001$), therapy ($p = 0.0056$), and neopterin levels ($p = 0.0055$) on survival time of patients with lung cancer. The relative risk of death of patients with elevated neopterin levels was 2.1 times higher than that of patients with neopterin levels below the upper normal limit, even when corrected for effects of stage and therapy. Thus, the predictive value of neopterin levels was observed independently from studies of tumor stage and presence or absence of specific therapy.

Due to the limited number of patients in the various histologic types and stages, these investigations of lung cancer patients should be extended. However, the results of neopterin determinations in patients with lung cancer are promising.

5.7.5. Gastrointestinal and Pancreatic Carcinoma

Urinary neopterin levels were elevated in approximately 50% of patients with gastric carcinoma ($n = 42$) or with carcinoma of colon or of rectum ($n = 25$) (C4). Patients with pancreatic carcinoma ($n = 26$; 69%) and patients with carcinoma of biliary tract (8 of 9) showed elevated urinary neopterin levels. For these tumor localizations, no correlation of neopterin with tumor stage could be demonstrated, probably because of the limited numbers of patients investigated.

5.7.6. Malignant Diseases with Low Frequencies of Elevated Neopterin Levels

It should be noted that certain tumor types show elevated neopterin levels only in low percentages (below 25%): melanomas (Z2), breast tumors (D5, W17), and head and neck cancers (D5, R10). However, a possible predictive value of neopterin has not been investigated.

5.7.7. Activation of Cellular Immunity and Prognosis

The observed correlation between elevated neopterin levels and activation of cell-mediated immunity might be considered surprising. Due to the close association between neopterin release by macrophages and production of interferon-γ by activated T cells, elevated neopterin levels in patients with malignant disease are considered as indirect evidence that these types of neoplastic cells evoke at least one or several first steps of the cell-mediated immune response.

From *in vitro* results it is recognized that interferon-γ induces macrophages to secrete neopterin and concomitantly primes them for capacity to secrete H_2O_2. As already mentioned, the secretion of H_2O_2 by macrophages is a two-step process: activation by interferon-γ (paralleled by secretion of neopterin) must be followed by a second step for H_2O_2 production. Thus, it is possible that this second and/or further steps are not initiated or are suppressed despite a high degree of neopterin release.

There is some evidence that tumor-specific immunity, particularly actions of macrophages, might support tumor growth. Activated macrophages might fuse with malignant cells and the resulting hybrids could be responsible for metastatic spreading of tumors (M10). In addition, angiogenesis of tumors was reported to be facilitated by secretory products of macrophages (P13). It was also found that tumor-specific immune reactions supported tumor development in experimental tumors of mice (P17). On the basis of these studies, the observed association of high activation of macrophages and poor prognosis appears more understandable.

Finally, it is certainly conceivable that measurement of neopterin as an index for production of interferon-γ and activation of T cells might be of some importance in drawing conclusions about features of cellular immune reactions in the complex field of tumor immunology.

5.7.8. Follow-Up during Therapeutic Immunomodulation

Due to its special source from interferon-γ-induced macrophages, neopterin is particularly suited to assess treatment modalities aimed at modifying immune responses. Therapy with interferons (D3,L4), tumor necrosis factor (D3), interleukin-2 (K7), and immunoactivating antitumoral peptides (V6) was accompanied by elevation of neopterin levels in serum and urine. In this way, the minimal therapeutic doses of interferon-α were determined for treatment of hairy cell leukemia (G6). Evidence was obtained that doses ranging from 5 to 10% of those commonly applied yield clinical results comparable to conventional doses, with greatly reduced side effects. The minimal dose was defined as the dose of interferon-α inducing maximal neopterin release in serum and urine.

6. Conclusions

The low-weight metabolite neopterin is biosynthetically derived from guanosine triphosphate via 7,8-dihydroneopterin triphosphate. *In vitro*, human monocytes/macrophages produce neopterin when stimulated by interferon-γ released from activated T cells. Other cell types do not produce measureable amounts of neopterin following various stimuli. Therefore, neopterin production appears to be closely associated with activation of the cellular immune system.

Numerous clinical studies have demonstrated that neopterin represents an early, specific, and sensitive marker for *in vivo* T cell activation, interferon-γ production, and monocyte/macrophage activation. High neopterin levels were observed in clinical settings recognized or supposed to involve activation of cell-mediated immunity in acute allograft rejections, viral infections, infections by intracellular parasites and bacteria, autoimmune diseases, and certain malignancies.

Several studies focus on the application of neopterin for monitoring the posttransplant course in allograft recipients. Acute rejections and viral complications are rapidly and accurately indicated. Increased neopterin production is a very early and sensitive response to most forms of infection, with exception of localized infections by extracellular bacteria. The rise of neopterin levels precedes the appearance of specific antibodies. Neopterin levels decrease toward normal rapidly with recovery. Neopterin can, therefore, be used as an objective index of disease activity and response to therapy.

Studies of neopterin levels in groups at high risk for AIDS and in patients infected with HIV have demonstrated that activation of T lymphocytes and macrophages (analogous to activation *in vitro*) represents a crucial event for HIV production. HIV attaches preferentially to stimulated T cells and is produced by such cells. Consistent with these data, neopterin levels are of prognostic significance, identifying among healthy subjects those with high susceptibility for HIV infection and among HIV-infected individuals those whose disease is more likely to progress to AIDS.

In autoimmune diseases, neopterin determination provides useful information because neopterin levels reflect accurately the clinical activity and the extent of disease.

In malignant diseases, the degree of neopterin elevation is a measure of the clinical activity, and in some tumor types serves as a measure of the extent of disease. In ovarian cancer, in cancer of the uterine cervix, in prostatic tumor, and in carcinoma of the lung, neopterin is of predictive value particularly for follow-up studies.

In several clinical situations, neopterin as an early, specific, and sensitive marker for activation of the cellular immune system is well suited to discriminate between disorders otherwise hardly distinguishable by laboratory assays, e.g., between chronic persistent non-A, non-B hepatitis and steatosis hepatitis, or between osteoarthritis and rheumatoid arthritis at an early stage.

Finally, it should be noted that neopterin measurement can be easily performed by RIA in serum or by HPLC in serum or urine. Neopterin determination provides information on the *in vivo* situation of the cellular immune system and is more simply performed than, for instance, determination of T lymphocyte activation markers. Neopterin is diffusible and is not metabolized subsequent to *in vivo* formation, enabling ease of measurement compared to interferon-γ, which enters the circulation minimally when locally produced and which has a comparatively short biological half-life.

That neopterin is not specific for a special disease or pathogen does not disqualify it as a useful measure. The wide application range of neopterin resembles, for instance, that of the erythrocyte sedimentation rate and of C-reactive protein. However, it should be emphasized that neopterin carries completely independent and unique information.

REFERENCES

A1. Abita, J. P., Cost, H., Milstien, S., Kaufman, S., and Saimot, G., Urinary neopterin and biopterin levels in patients with AIDS and AIDS related complex. *Lancet* **2**, 51–52 (1985).

A2. Adam, C., Thoua, Y., Ronco, P., Verroust, P., and Tovey, M., The effect of exogenous interferon: Acceleration of autoimmune and renal diseases in (NZB/W) F_1 mice. *Clin. Exp. Immunol.* **40**, 373–382 (1980).

A3. Albert, A., Transformation of purines into pteridines. *Biochem. J.* **65**, 5310–5314 (1957).

A4. Andrieu, J. M., Even, P., Venet, A., Audroin, C., Stern, M., Israel-Biet, D., Tourani, J. M., and Lowenstein, W., Preliminary results of cyclosporin A in AIDS and related disorders. *Abstr., Int. Conf. AIDS, 2nd, 1986*, (1986).

A5. Andrysek, O., and Gregora, V., Urinary excretion of pterins in tumor-bearing patients. *Neoplasma* **30**, 497–507 (1983).

A6. Armarego, W. L. F., and Randles, D., Aerobic oxidation of 5,6,7,8-tetrahydroneopterin. *In* "Biochemical and Clinical Aspects of Pteridines" (H. C. Curtius, W. Pfleiderer, and H. Wachter, eds.), pp. 423–427. de Gruyter, Berlin and New York, 1983.

A7. Aulitzky, W., Frick, J., Fuchs, D., Hausen, A., Reibnegger, G., and Wachter, H., Significance of urinary neopterin in patients with malignant tumors of the genitourinary tract. *Cancer (Philadelphia)* **55**, 1052–1055 (1985).

B1. Bailey, S. W., and Ayling, J. E., High pressure liquid chromatography of substituted pteridines and tetrahydropteridines. *In* "Chemistry and Biology of Pteridines" (W. Pfleiderer, ed.), pp. 633–643. de Gruyter, Berlin and New York, 1975.

B2. Barford, P. A., and Blair, J. A., Novel urinary metabolites of folic acid in the rat. *In* "Chemistry and Biology of Pteridines" (W. Pfleiderer, ed.), pp. 413–427. de Gruyter, Berlin and New York, 1975.

B3. Barré-Sinoussi, F., Chermann, J. C., Rey, F., Nugeyre, M. T., Chamaret, S., Gruest, J., Dauguet, C., and Axler-Blin, C., Isolation of a T-lymphotropic retrovirus from a patient at risk for acquired immune deficiency syndrome (AIDS). *Science* **220**, 868–871 (1983).

B4. Bellamy, A. S., Calder, V. L., Feldmann, M., and Davison, A. N., The distribution of interleukin-2 receptor bearing lymphocytes in multiple sclerosis: Evidence for a key role of activated lymphocytes. *Clin. Exp. Immunol.* **61**, 248–256 (1985).

B5. Bichler, A., Fuchs, D., Hausen, A., Hetzel, H., König, K., and Wachter, H., Urinary neopterin excretion in patients with genital cancer. *Clin. Biochem. (Ottawa)* **15**, 38–40 (1982).

B6. Bichler, A., Fuchs, D., Hausen, A., Hetzel, H., Reibnegger, G., and Wachter, H., Measurement of urinary neopterin in normal pregnant and non-pregnant women and in women with benign and malignant genital tract neoplasms. *Arch. Gynecol.* **233**, 121–130 (1983).

B7. Blakley, R. L., Natural occurrence of pterins and folate derivatives. *In* "The Biochemistry of Folic Acid and Related Pteridines" (W. Pfleiderer and E. C. Taylor, eds.), pp. 8–57. North-Holland Publ., Amsterdam, 1969.

B8. Blau, N., Joller, P., Atarés, M., Cardesa-Garcia, J., and Niederwieser, A., Increase of GTP cyclohydrolase I activity in mononuclear blood cells by stimulation: Detection of heterozygotes of GTP cyclohydrolase I deficiency. *Clin. Chim. Acta* **148**, 47–52 (1985).

B9. Bogner, R. J., Goebel, F. D., Kronawitter, U., and Keller, S., Prognostic value of laboratory parameters in the clinical course of HIV-infection. *Abstr., Int. AIDS Conf., 3rd, 1987*, WP165 (1987).

B10. Bottazzo, G. F., Pujol-Borell, R., Hanafusa, T., and Feldmann, M., Role of aberrant HLA-DR expression and antigen presentation in induction of endocrine autoimmunity. *Lancet* **2**, 1115–1119 (1983).

B11. Bottazzo, G. F., Dean, B. M., McNally, J. M., MacKay, E. H., Swift, P. G. F., and Gamble, D. R., *In situ* characterization of autoimmune phenomena and expression of HLA molecules in the diabetic insulitis. *N. Engl. J. Med.* **313**, 353–360 (1985).

B12. Broquist, H. P., and Albrecht, A. M., Pteridines and the nutrition of the protozoan *Crithidia fasciculata*. *Proc. Soc. Biol. Med.* **89**, 178–180 (1955).

B13. Buchmeier, N. A., and Schreiber, R. D., Requirement of endogenous interferon-gamma production for resolution of *Listeria monocytogenes* infection. *Proc. Natl. Acad. Sci. U.S.A.* **82**, 7404–7408 (1985).

B14. Byrne, G. I., Lehmann, L. K., and Landry, G. J., Induction of tryptophan catabolism is the mechanism for gamma-interferon-mediated inhibition of intracellular *Chlamydia psittaci* replication in T24 cells. *Infect. Immun.* **53**, 347–351 (1986).

C1. Clark, I. A., Cell-mediated immunity in protection and pathology of malaria. *Parasitol. Today* **3**. 300–305 (1987).

C2. Collins, F. M., The immunology of tuberculosis. *Am. Rev. Respir. Dis.* **125**, 42–49 (1982).

C3. Conrad, F., Salzer, G. M., Fuchs, D., Hausen, A., Reibnegger, G., and Wachter, H., Neopterin—A marker of the cellular immune response in patients with lung cancer. *In* "Biochemical and Clinical Aspects of Pteridines" (H. Wachter, H. C. Curtius, and W. Pfleiderer, eds.), pp. 535–544. de Gruyter, Berlin and New York, 1985.

C4. Conrad, F., Bodner, E., Fuchs, D., Grubauer, G., Hausen, A., Reibnegger, G., and Wachter, H., Determination of neopterin—A marker of cellular immunity in gastrointestinal and pancreatic carcinoma. *In* "Biochemical and Clinical Aspects of Pteridines" (W. Pfleiderer, H. Wachter, and H. C. Curtius, eds.), pp. 357–366. de Gruyter, Berlin and New York, 1984.

C5. Conrad, F., Fuchs, D., Hausen, A., Reibnegger, G., Werner, E. R., Salzer, G. M., and Wachter, H., Prognostic value of neopterin in patients with lung cancer. *In* "Biochemical and Clinical Aspects of Pteridines" (W. Pfleiderer, H. Wachter, and J. A. Blair, eds.), pp. 233–241. de Gruyter, Berlin and New York, 1987.

C6. Curtius, H. C., Niederwieser, A., Viscontini, M., Otten, A., Schaub, J., Scheibenreiter, S., and Schmidt, H., Atypical phenylketonuria due to tetrahydrobiopterin deficiency. Diagnosis and treatment with tetrahydrobiopterin, dihydrobiopterin and sepiapterin. *Clin. Chim. Acta* **93**, 251–262 (1979).

D1. Daniel, V., Opelz, G., Schäfer, A., Schimpf, K., Wendler, I., and Hunsmann, G., Correlation of immune defects in hemophilia with HTLV-III antibody titers. *Vox Sang.* **51**, 35–39 (1986).

D2. Danks, D. M., Cotton, R. G. H., and Schlesinger, P., Variant forms of phenylketonuria. *Lancet* **1**, 1236–1237 (1976).

D3. Datta, S. P., Brown, R. R., Borden, E. C., Sondel, P. M., and Trump, D. L., Interferon and interleukin-2 induced changes in tryptophan and neopterin metabolism: Possible markers for biologically effective doses. *Proc. Am. Assoc. Cancer Res.* **28**, 338 (1987).

D4. Davis, R. E., Clinical chemistry of folic acid. *Adv. Clin. Chem.* **25**, 233–294 (1986).
D5. Dhondt, J. L., Hayte, J. M., Bonneterre, J., Adenis, L., Demaille, A., Arduin, P., and Farriaux, J. P., Pteridines in urine and serum from cancer patients. *In* "Biochemical and Clinical Aspects of Pteridines" (H. Wachter, H. C. Curtius, and W. Pfleiderer, eds.), pp. 133–140. de Gruyter, Berlin and New York, 1982.
D6. Dinovo, E. C., Lynn, J. K., McIntosh, M. E., and Johnson, M., Is pterin-6-aldehyde diagnostic for cancer? *Clin. Chem. (Winston-Salem, N. C.)* **24**, 1002–1003 (1978).
D7. Duch, D. S., Bowers, S. W., Woolf, J. H., and Nichol, C. A., Biopterin cofactor biosynthesis: GTP cyclohydrolase, neopterin and biopterin in tissues and body fluids of mammalian species. *Life Sci.* **35**, 1895–1901 (1984).
D8. Duvall, S., Ridgway, H., and Speer, R. J., Measurement of urinary pteridines for the detection of cancer: Evaluation of the Rao *et al.* method. *J. Clin. Hematol. Oncol.* **12**, 49–54 (1982).
E1. Eberhartinger, C., and Simader, R., Akute AIDS—Retrovirus Infektion. Klinisches Krankheitsbild bei einem Patienten mit HTLV-III Serokonversion. *Wien. Klin. Wochenschr.* **99**, 18–20 (1987).
E2. Eisenbarth, G. S., Type I diabetes mellitus. A chronic autoimmune disease. *N. Engl. J. Med.* **314**, 1360–1368 (1986).
E3. Eklund, A., and Blaschke, E., Elevated serum neopterin levels in sarcoidosis. *Lung* **164**, 325–332 (1986).
E4. Engleman, E., Sonnenfeld, G., Dauphinee, M., Greenspan, J., Talal, N., McDevitt, H. O., and Merigan, T. C., Treatment of NZB/NZW Fo1 hybrid mice with *Mycobacterium bovis* strain BCG or type II interferon preparations accelerates autoimmune disease. *Arthritis Rheum.* **24**, 1396–1402 (1981).
E5. Escalante-Semerena, J. C., Leigh, J. A., Rinehart, K. L., and Wolfe, R. S., Formaldehyde activation factor, tetrahydromethanopterin, a coenzyme of methanogenesis. *Proc. Natl. Acad. Sci. U.S.A.* **81**, 1976–1980 (1984).
F1. Farci, A. M. G., Laconi, R., Cabras, F., Loviselli, A., Cappai, G., Balestrieri, A., Garau, V. L., Tocco, M., and Casula, D., Urinary neopterin in acute and chronic liver disease. *In* "Biochemical and Clinical Aspects of Pteridines" (H. Wachter, H. C. Curtius, and W. Pfleiderer, eds.), pp. 453–460. de Gruyter, Berlin and New York, 1985.
F2. O'Farrely, C., Feighery, C. F., Whelan, C. A., and Weir, D. G., Suppressor-cell activity in coeliac disease induced by alpha-gliadin, a dietary antigen. *Lancet* **2**, 1305–1307 (1984).
F3. Ferreira, A., Schofield, L., Enea, V., Schellekens, H., Van der Meide, P., Collins, W. E., Nussenzweig, R. S., and Nussenzweig, V., Inhibition of development of exo-erythrocytic forms of malaria parasites by gamma-interferon. *Science* **232**, 881–884 (1986).
F4. Fleming, A. F., and Broquist, H. P., Biopterin and folic acid deficiency. *Am. J. Clin. Nutr.* **20**, 613–621 (1967).
F5. Flomenberg, N., Welte, K., Mertelsmann, R., Kernan, N., Ciobanu, N., Venuta, S., Feldman, S., Kruger, G., Kirkpatrick, D., Dupont, B., and O'Reilly, R., Immunologic effects of interleukin-2 in primary immunodeficiency diseases. *J. Immunol.* **130**, 2644 (1983).
F6. Forrest, H., and VanBaalen, C., Microbiology of unconjugated pteridines. *Annu. Rev. Microbiol.* **24**, 91–108 (1970).
F7. Fredrikson, S., Eneroth, P., and Link, H. Intrathecal production of neopterin in aseptic meningo-encephalitis and multiple sclerosis. *Clin. Exp. Immunol.* **67**, 76–81 (1987).

F8. Fredrikson, S., Link, H., and Eneroth, P., CSF neopterin as marker of disease activity in multiple sclerosis. *Acta Neurol. Scand.* **75**, 352–355 (1987).

F9. Frick, J., Aulitzky, W., Fuchs, D., Hausen, A., Joos, H., Reibnegger, G., and Wachter, H., The value of urinary neopterin as an immunological parameter in patients with malignant tumors of the genitourinary tract. *Urol. Int.* **40**, 155–159 (1985).

F10. Fuchs, D., Hausen, A., Reibnegger, G., and Wachter, H., Automatized routine estimation of neopterin in human urine by HPLC on reversed phase. *In* "Biochemical and Clinical Aspects of Pteridines" H. Wachter, H. C. Curtius, and W. Pfleiderer, eds.), pp. 67–79. de Gruyter, Berlin and New York, 1982.

F11. Fuchs, D., Granditsch, G., Hausen, A., Reibnegger, G., and Wachter, H., Urinary neopterin excretion in coeliac disease. *Lancet* **2**, 463 (1983).

F12. Fuchs, D., Hausen, A., Kofler, M., Kosanowski, H., Reibnegger, G., and Wachter, H., Neopterin as an index of immune response in patients with tuberculosis. *Lung* **162**, 337–346 (1984).

F13. Fuchs, D., Reibnegger, G., Wachter, H., Jaeger, H., Popescu, M., and Kaboth, W., Neopterin levels correlating with Walter Reed staging classification in HIV infection. *Ann. Intern. Med.* **107**, 784–785 (1987).

F14. Fuchs, D., Hausen, A., Huber, C., Margreiter, R., Reibnegger, G., Spielberger, M., and Wachter, H., Pteridinausscheidung als Marker für alloantigen-induzierte Lymphozytenproliferation. *Hoppe-Seyler's Z. Physiol. Chem.* **363**, 661–664 (1982).

F15. Fuchs, D., Hausen, A., Reibnegger, G., Reissigl, H., Schönitzer, D., Spira, T., and Wachter, H., Urinary neopterin in the diagnosis of acquired immune deficiency syndrome. *Eur. J. Clin. Microbiol.* **3**, 70–71 (1984).

F16. Fuchs, D., Hausen, A., Hengster, P., Reibnegger, G., Schulz, T., Werner, E. R., Dierich, M. P., and Wachter, H., In vivo activation of $CD4^+$ cells in AIDS. *Science* **235**, 356 (1987).

F17. Fuchs, D., Dierich, M. P., Hausen, A., Hengster, P., Reibnegger, G., Reissigl, H., Schönitzer, D., Vinazzer, H., Werner, E. R., and Wachter, H., Änderungen immunologischer Parameter und Serokonversion bei massiver Substitution von Hämophilen. *In* "16.Hämophilie-Symposium, Hamburg 1985" (G. Landbeck and R. Marx, eds.) pp. 63–70. Springer-Verlag, Berlin and New York, 1987.

F18. Fuchs, D., Hausen, A., Reibnegger, G., Wachter, H., Roessler, H., Hinterhuber, H., Krahnke, H., Kathke, N., Spira, T. J., and Vinazzer, H., Urinary neopterin evaluation in risk groups for the acquired immunodeficiency syndrome (AIDS). *In* "Biochemical and Clinical Aspects of Pteridines" (W. Pfleiderer, H. Wachter, and H. C. Curtius, eds.), pp. 457–467, de Gruyter, Berlin and New York, 1984.

F19. Fuchs, D., Hausen, A., Reibnegger, G., Werner, E. R., Wachter, H., Blecha, H. G., Roessler, H., Unterweger, B., Hinterhuber, H., Hengster, P., Schulz, T., Dierich, M. P., and Renner, D., Importance of neopterin determination in individuals at risk of AIDS. *In* "Chemistry and Biology of Pteridines" (B. A. Cooper and V. M. Whitehead, eds.), pp. 427–430. de Gruyter, Berlin and New York, 1986.

F20. Fuchs, D., Blecha, H. G., Dierich, M. P., Hausen, A., Hengster, P., Hinterhuber, H., Reibnegger, G., Rössler, H., Schauenstein, K., Schönitzer, D., Schulz, T., Traill, K., Unterweger, B., Werner, E. R., and Wachter, H., Immune status of drug abusers. *Cancer Detect. Prev., Suppl.* **1**, 535–541 (1987).

F21. Fujimori, E., Interaction between pteridines and tryptophan. *Proc. Natl. Acad. Sci. U. S. A.* **45**, 133–136 (1959).

F22. Fukushima, T., and Nixon, J. C., Oxidation and conversion of reduced forms of biopterin. *In* "Chemistry and Biology of Pteridines" (R. L. Kisliuk and G. M. Brown, eds.) pp. 31–34. Elsevier/North-Holland, Amsterdam, 1979.

F23. Fukushima, T., and Nixon, J. C., Analysis of reduced forms of biopterin in biological tissues and fluids. *Anal. Biochem.* **102**, 176–188 (1980).

F24. Fukushima, T., and Shiota, T., Pterins in human urine. *J. Biol. Chem.* **247**, 4549–4556 (1972).

G1. Gabriel, S., and Colman, J., Synthesen in der Purinreihe. *Ber. Dtsch. Chem. Ges.* **34**, 1234–1257 (1901).

G2. Gabriel, S., and Sonn, A., Übergang von der Chinoxalin zur Pyrazinreihe. *Ber. Dtsch. Chem. Ges.* **40**, 4850–4860 (1907).

G3. Gail, M. H., Evaluating serial cancer marker studies in patients at risk of recurrent disease. *Biometrics* **37**, 67–78 (1981).

G4. Gál, E. M., and Sherman, A. D., Rapid isolation and quantitation of biopterin, neopterin, and their guanine ribotide precursor from biological samples. *Prep. Biochem.* **7**, 155–164 (1977).

G5. Gantner, G., Grassmayr, K., Hausen, A., and Wachter, H., Erhöhte Ausscheidung einer fluoreszierenden Substanz im Harn Ehrlich-Ascites-Tumor tragender Mäuse. *Mikrochim. Acta* **2**, 33–38 (1977).

G6. Gastl, G., Aulitzky, W., Tilg, H., Nachbaur, K., Troppmair, J., Flener, R., and Huber, C., A biological approach to optimize interferon treatment in hairy cell leukemia. *Immunobiology* **172**, 262–268 (1986).

G7. Gerisch, G., Blank, G., Schweiger, M., Fuchs, D., Hausen, A., Reibnegger, G., and Wachter, H., Pteridines released from *Dictyostelium discoideum* cells into the extracellular medium. *In* "Biochemical and Clinical Aspects of Pteridines" (H. Wachter, H. C. Curtius, and W. Pfleiderer, eds.) pp. 253–256. de Gruyter, Berlin and New York, 1982.

G8. Goebel, F. D., Erfle, V., Piechowiak, H., Hien, P., Schloz, H., and Hehlmann, R., The relations of HTLV-III antibodies to neopterin and β_2-microglobulin in the serum of patients with AIDS or persons at risk. *In* "Biochemical and Clinical Aspects of Pteridines" (H. Wachter, H. C. Curtius, and W. Pfleiderer, eds.), pp. 319–334. de Gruyter, Berlin and New York, 1985.

G9. Goto, M., Sakurai, A., Ohta, K., and Yamakami, H., Die Struktur des Urothions. *J. Biochem. (Tokyo)* **65**, 611–620 (1969).

G10. Granditsch, G., Fuchs, D., Hausen, A., Reibnegger, G., and Wachter, H., Urinary neopterin excretion in childhood coeliac disease. *In* "Biochemical and Clinical Aspects of Pteridines" (W. Pfleiderer, H. Wachter, and H. C. Curtius, eds.), pp. 335–343. de Gruyter, Berlin and New York, 1984.

H1. Haas, R., and Gerstner, L., Radioimmunologische Bestimmung von Neopterin bei Blutspendern. *Allerg. Immunol.* **31**, 179–182 (1985).

H2. Haas, R., and Gerstner, L., Neopterinschwankungen während des weiblichen Zyklus. *Allerg. Immunol.* **32**, 169–173, (1986).

H3. Haavik, J., and Flatmark, T., Isolation and characterization of quinonoid dihydropterins by high-performance liquid chromatography. *J. Chromatogr.* **257**, 361–372 (1983).

H4. Hadorn, E., and Mitchell, H. K., Properties of mutants of *Drosophila melanogaster* and changes during development as revealed by paper chromatography. *Proc. Natl. Acad. Sci. U.S.A.* **37**, 650–665 (1951).

H5. Halpern, R., Halpern, B. C., Stea, B., Dunlap, A., Conklin, K., Clark, B., Ashe, H., Sperling, L., Halpern, J. A., Hardy, D., and Smith, R. A., Pterin-6-aldehyde, a cancer cell catabolite: Identification and application in diagnosis and treatment of human cancer. *Proc. Natl. Acad. Sci. U.S.A.* **74**, 587–591 (1977).

H6. Hannonen, P., Tikanoja, S., Hakola, M., Möttönen, T., Viinikka, L., and Oka, M., Urinary neopterin index as a measure of rheumatoid activity. *Scand. J. Rheumatol.* **15**, 148–152 (1986).

H7. Haug, P., Mass spectral fragmentation of trimethylsilyl derivatives of 2-amino-4-hydroxypteridines. *Anal. Biochem.* **37,** 285–292 (1970).
H8. Hausen, A., and Wachter, H., Pteridines in the assessment of neoplasia. *J. Clin. Chem. Clin. Biochem.* **20,** 593–602 (1982).
H9. Hausen, A., Fuchs, D., König, K., and Wachter, H., Determination of neopterin in human urine by reversed-phase high-performance liquid chromatography. *J. Chromatogr.* **227,** 61–70 (1982).
H10. Hausen, A., Fuchs, D., Reibnegger, G., and Wachter, H., Urinary pteridines in patients suffering from cancer. *Cancer (Philadelphia)* **53,** 1634–1636 (1984).
H11. Hausen, A., Fuchs, D., Grünewald, K., Huber, H., König, K., and Wachter, H., Urinary neopterin as marker for haematological neoplasias. *Clin. Chim. Acta* **117,** 297–305 (1981).
H12. Hausen, A., Fuchs, D., Grünewald, K., Huber, H., König, K., and Wachter, H., Urinary neopterin in the assessment of lymphoid and myeloid neoplasia, and neopterin levels in haemolytic anaemia and benign monoclonal gammopathy. *Clin. Biochem. (Ottawa)* **15,** 34–37 (1982).
H13. Hausen, A., Fuchs, D., Reibnegger, G., Wachter, H., Egg, D., and Günther, R., Neopterin as index for activity of disease in patients with rheumatoid arthritis. *In* "Biochemical and Clinical Aspects of Pteridines" (H. C. Curtius, W. Pfleiderer, and H. Wachter, eds.), pp. 245–254. de Gruyter, Berlin and New York, 1983.
H14. Hausen, A., Fuchs, D., Reibnegger, G., Werner, E. R., Dierich, M. P., and Wachter, H., Immunosuppressants in treatment of patients with AIDS. *Lancet* **2,** 214–215 (1987).
H15. Hausen, A., Dierich, M. P., Fuchs, D., Hengster, P., Reibnegger, G., Schulz, T., Werner, E. R., and Wachter, H., Immunosuppressant in patients with AIDS. *Nature (London)* **320,** 114 (1986).
H16. Heremans, H., Biliau, A., Columbatti, A., Hilgers, J., and de Somer, P., Interferon treatment of NZB mice. Accelerated progression of autoimmune disease. *Infect. Immun.* **21,** 925–928 (1978).
H17. Hoffman, S. L., Wistar, R., Ballou, W. R., Hollingdale, M. R., Wirtz, R. A., Schneider, I., Marwoto, H. A., and Hockmeyer, W., Immunity to malaria and naturally acquired antibodies to the circumsporozoite protein of *Plasmodium falciparum*. *N. Engl. J. Med.* **315,** 601–606 (1986).
H18. Hooks, J. J., Jordan, G. W., Cupps, T., Moutsopoulos, H. M., Fauci, A. S., and Notkins, A. L., Multiple interferons in the circulation of patients with systemic lupus erythematosus and vasculitis. *Arthritis Rheum.* **25,** 396–400 (1982).
H19. Hooks, J. J., Moutsopoulos, H. M., Geis, S. A., Stahl, N. I., Decker, J. L., and Notkins, A. L., Immune interferon in the circulation of patients with autoimmune disease. *N. Engl. J. Med.* **301,** 5–8 (1979).
H20. Hopkins, F. G., Note on a yellow pigment in butterflies. *Nature (London)* **40,** 335 (1889).
H21. Huber, C., Fuchs, D., Hausen, A., Margreiter, R., Reibnegger, G., Spielberger, M., and Wachter, H., Pteridines as a new marker to detect human T cells activated by allogeneic or modified self major histocompatibility complex (MHC) determinants. *J. Immunol.* **130,** 1047–1050 (1983).
H22. Huber, C., Batchelor, J. R., Fuchs, D., Hausen, A., Lang, A., Niederwieser, D., Reibnegger, G., Swetly, P., Troppmair, J., and Wachter, H., Immune response-associated production of neopterin. Release from macrophages primarily under control of interferon-gamma. *J. Exp. Med.* **160,** 310–316 (1984).
H23. Hutterer, J., Fuchs, D., Eder, G., Hausen, A., Knapp, W., Köller, U., Reibnegger, G., Werner, E. R., Wachter, H., Stingl, G., and Wolff, K., Neopterin as discriminating and prognostic parameter in healthy homosexuals, ARC and AIDS patients. *Wien. Klin. Wochenschr.* **99,** 531–535 (1987).

H24. Hyland, K., Estimation of tetrahydro-, dihydro- and fully oxidised pterins by high-performance liquid chromatography using sequential electrochemical and fluorometric detection. *J. Chromatogr.* **343**, 35–41 (1985).
I1. Isay, O., Eine Synthese des Purins. *Ber. Dtsch. Chem. Ges.* **39**, 250–265 (1906).
J1. Janossy, G., Duke, O., Poulter, L. W., Panayi, G., Bofill, M., and Goldstein, G., Rheumatoid arthritis: A disease of T-lymphocyte/macrophage immunoregulation. *Lancet* **2**, 839–842 (1981).
J2. Johnson, J. L., and Rajagopalan, K. V., Structural and metabolic relationship between the molybdenum cofactor and urothione. *Proc. Natl. Acad. Sci. U.S.A.* **79**, 6856–6860 (1982).
J3. Johnson, J. L., Hainline, B. E., Rajagopalan, K. V., and Arison, B. H. The pterin component of the molybdenum cofactor. *J. Biol. Chem.* **259**, 5414–5422 (1984).
J4. Jones, T. H. D., and Brown, G. M., The biosynthesis of folic acid. VII. Enzymatic synthesis of pteridines from guanosine triphosphate. *J. Biol. Chem.* **242**, 3989–3997 (1967).
J5. Joos, H., Aulitzky, W., Frick J., Fuchs, D., Hausen, A., Reibnegger, G., and Wachter, H., Urinary neopterin excretion in testicular cancer long term follow up. *In* "Biochemical and Clinical Aspects of Pteridines" (W. Pfleiderer, H. Wachter, and J. A. Blair, eds.), pp. 251–260. de Gruyter, Berlin and New York, 1987.
K1. Kagnoff, M. F., Austin, R. K., Hubert, J. J., Bernardin, J. E., and Kasarda, D. D., Possible role for a human adenovirus in the pathogenesis of celiac disease. *J. Exp. Med.* **160**, 1544–1557 (1984).
K2. Kaufman, S., The structure of phenylalanine hydroxylation cofactor. *Proc. Natl. Acad. Sci. U.S.A.* **50**, 1085–1093 (1963).
K3. Kaufman, S., Holtzman, N. A., Milstien, S., Butler, I. J., and Krumholz, A. Phenylketonuria due to a deficiency of dihydropteridine reductase. *N. Engl. J. Med.* **293**, 785–790 (1975).
K4. Kern, P., Horstmann, R. D., and Dietrich, M., Clinical resolvement of malaria infection parallels decrease of highly elevated serum neopterin levels. *In* "Biochemical and Clinical Aspects of Pteridines" (H. Wachter, H. C. Curtius, and W. Pfleiderer, eds.), pp. 341–345. de Gruyter, Berlin and New York, 1984.
K5. Kern, P., Krebs, H. J., and Dietrich, M., Serum neopterin: Screening test to exclude transfusion hazards by blood donors with cytomegalovirus infection, AIDS, etc. *Abstr., Congr. Int. Soc. Blood Transfus., 18th, 1984,* (1984).
K6. Kern, P., Rokos, H., and Dietrich, M., Raised serum neopterin levels and imbalances of T-lymphocyte subsets in viral diseases, acquired immune deficiency and related lymphadenopathy syndromes. *Biomed. Pharmacother.* **38**, 407–411 (1984).
K7. Kern, P., Toy, J., and Dietrich, M., Preliminary clinical observations with recombinant interleukin-2 in patients with AIDS or LAS. *Blut* **50**, 1–6 (1985).
K8. Kirchner, H., Interferon gamma. *Prog. Clin. Biochem. Med.* **1**, 169–203 (1984).
K9. Kirchner, H., Digel, W., and Storch, E., Interferons and bacterial infections. *Klin. Wochenschr.* **60**, 740–742 (1982).
K10. Kofler, H., Fuchs, D., Hintner, H., Wachter, H., and Fritsch, P., Urinary neopterin: An early marker for HIV infection. *Eur. J. Clin. Microbiol.* **6**, 698–699 (1987).
K11. Kohashi, M., Tomita, K., and Iwai, K., Analysis of unconjugated pterins in food resources and human urine. *Agric. Biol. Chem.* **44**, 2089–2094 (1980).
K12. Kokolis, N., and Ziegler, I., Wiedererscheinen von Tetrahydrobiopterin in der Regenerationsknospe von Triturus-Arten. *Z. Naturforsch., B: Anorg. Chem., Org. Chem., Biochem., Biophys., Biol.* **23B**, 860–865 (1968).
K13. Kokolis, N., and Ziegler, I., On the levels of phenylalanine, tyrosine and tetrahy-

drobiopterin in the blood of tumor-bearing organisms. *Cancer Biochem. Biophys.* **2**, 79–85 (1977).

K14. Kokolis, N., Mylonas, N., and Ziegler, I., Pteridine and riboflavin in tumor tissue and the effect of chloramphenicol and isoxanthopterin. *Z. Naturforsch., B: Anorg. Chem., Org. Chem., Biochem., Biophys., Biol.* **27B**, 292–295 (1972).

K15. Koschara, W., Isolierung eines gelben Farbstoffs (Uropterin) aus Menschenharn. *Hoppe-Seyler's Z. Physiol. Chem.* **240**, 127–151 (1936).

K16. Kraut, H., Pabst, W., Rembold, H., and Wildemann, L., Über das Verhalten des Biopterin im Säugetier-Organismus. I. Bilanz- und Wachstumsversuche an Ratte. *Hoppe-Seyler's Z. Physiol. Chem.* **332**, 101–110 (1963).

K17. Kühling, O., Über die Oxidation des Tolualloxazins. I. *Ber. Dtsch. Chem. Ges.* **27**, 2116–2119 (1894).

K18. Kühling, O., Über die Oxidation des Tolualloxazins. II. *Ber. Dtsch. Chem. Ges.* **28**, 1968–1971 (1895).

K19. Kuhn, R., and Cook, A. H., Über Lumazine und Alloxazine. *Ber. Dtsch. Chem. Ges.* **70**, 761–768 (1937).

K20. Kuster, T., Matasovic, A., and Niederwieser, A., Application of gas chromatography–mass spectrometry to the study of biopterin metabolism in man. *J. Chromatogr.* **290**, 303–310 (1984).

K21. Kwietny, H., and Bergmann, F., Separation and identification of pteridines by paper chromatography. *J. Chromatogr.* **2**, 162–172 (1959).

L1. Lacronique, J., Auzeby, A., Barbosa, M. A., Valeyre, D., Soler, P., Marsac, J., Battesti, J. P., and Touitou, Y., Urinary neopterin as a new marker of lymphocytic alveolitis in pulmonary sarcoidosis. *Am. Rev. Respir. Dis.* **133**, A24 (1986).

L2. Lagowski, J. M., and Forrest, H. S., Interaction *in vitro* between isoxanthopterin and DNA. *Proc. Natl. Acad. Sci. U.S.A* **58**, 1541–1547 (1967).

L3. Lambin, P., Desjobert, H., Debbia, M., Fine, J. M., and Muller, J. Y., Serum neopterin and β_2-microglobulin in anti-HIV positive blood donors. *Lancet* **2**, 1216 (1986).

L4. Lang, A., Niederwieser, D., Huber, C., Swetly, P., Fuchs, D., Hausen, A., Reibnegger, G., and Wachter, H., Treatment with human recombinant interferon-alpha-2 induces increase of in vivo neopterin excretion. *In* "Biochemical and Clinical Aspects of Pteridines" (W. Pfleiderer, H. Wachter, and H. C. Curtius, eds.), pp. 251–254. de Gruyter, Berlin and New York, 1984.

L5. Lang, H. R. M., Linkesch, W., and Huber, J. F. K., Determination of serum neopterin in patients with acute leukemia and malignant lymphomas. *In* "Biochemical and Clinical Aspects of Pteridines" (H. Wachter, H. C. Curtius, and W. Pfleiderer, eds.), pp. 525–533. de Gruyter, Berlin and New York, 1985.

L6. Lang, H. R. M., Schmiedmeier, W., and Huber, J. F. K., Serum neopterin determination in bone marrow transplantation in childhood. *In* "Biochemical and Clinical Aspects of Pteridines" (H. Wachter, H. C. Curtius, and W. Pfleiderer, eds.), pp. 405–410. de Gruyter, Berlin and New York, 1985.

L7. Levine, R. A., and Milstien, S., The ratio of reduced to oxidized neopterin and biopterin in human fluids: Significance to the study of human disease. *In* "Biochemical and Clinical Aspects of Pteridines" (W. Pfleiderer, H. Wachter, and H. C. Curtius, eds.), pp. 277–284. de Gruyter, Berlin and New York, 1984.

L8. Levine, S., and Hahn, T., Evaluation of the human interferon system in viral disease. *Clin. Exp. Immunol.* **46**, 475–483 (1981).

L9. Lewenhaupt, A., Ekman, P., Eneroth, P., Eriksson, A., Nilsson, B., and Nordström, N., Serum levels of neopterin as related to the prognosis of human prostatic carcinoma. *Eur. Urol.* **12**, 422–425 (1986).

L10. Link, H., Kam-Hansen, S., Forsberg, P., and Henrikson, A., Humoral and cellular immunity in patients with acute aseptic meningitis. In "Immunology of Nervous System Infections" (P. O. Behan, V. ter Meulen, and F. Clifford Rose, eds.), pp. 29. Elsevier, Amsterdam, 1983.

L11. Loidl, P., Fuchs, D., Gröbner, P., Hausen, A., Reibnegger, G., and Wachter, H., Pteridines during growth and differentiation of *Physarum polycephalum*. In "Biochemical and Clinical Aspects of Pteridines" (H. Wachter, H. C. Curtius, and W. Pfleiderer, eds.), pp. 257–266. de Gruyter, Berlin and New York, 1982.

L12. Lunte, C. E., and Kissinger, P. T., Determination of pterins in biological samples by liquid chromatography/electrochemistry with a dual-electrode detector. *Anal. Chem.* 55, 1458–1462 (1983).

M1. MacFadden, D. K., Hyland, R. H., Inouye, T., Edelson, J. D., Rodriguez, C. H., and Rebuck, A. S., Corticosteroids as adjunctive therapy in treatment of *Pneumocystis carinii* pneumonia in patients with acquired immunodeficiency syndrome. *Lancet* 1, 1477–1479 (1987).

M2. Maerker-Alzer, G., Diemer, O., Strümper, R., and Rohe, M., Neopterin production in inflamed knee joints: High levels in synovial fluids. *Rheumatol. Int.* 6, 151–154 (1986).

M3. Malchow, D., and Gerisch, G., Short-term binding and hydrolysis of cyclic $3':5'$-adenosine monophosphate by aggregating *Dictyostelium discoideum*. *Proc. Natl. Acad. Sci. U.S.A.* 71, 2423–2427 (1974).

M4. Manna, R., Gambassi, G., Papa, G., Greco, A. V., Fasciani, P., Liang, L., Manuppelli, C., Tafuri, A., Ghirlanda, G., Pozzilli, P., Sambo, A., Cittadini, A., Flamini, G., and Galeotti, T., Urinary neopterin levels of insulin dependent diabetes (IDDM) at onset. In "Biochemical and Clinical Aspects of Pteridines" (W. Pfleiderer, H. Wachter, and J. A. Blair, eds.), pp. 353–357. de Gruyter, Berlin and New York, 1987.

M5. Margreiter, R., Fuchs, D., Hausen, A., Huber, C., Reibnegger, G., Spielberger, M., and Wachter, H., Neopterin as a new biochemical marker for diagnosis of allograft rejection. *Transplantation* 36, 650–653 (1983).

M6. McRobbie, S. J., and Newell, P. C., Changes in actin associated with the cytoskeleton following chemotactic stimulation of *Dictyostelium discoideum*. *Biochem. Biophys. Res. Commun.* 115, 351–359 (1983).

M7. McRobbie, S. J., and Newell, P. C., Chemoattractant-mediated changes in cytoskeletal actin of cellular slime moulds. *J. Cell Sci.* 68, 139–151 (1984).

M8. Menz, G., Schmitz-Schumann, M., Binder, M., Virchow, C., Wachter, H., Fuchs, D., Reibnegger, G., and Grob, P., Neopterin serum and urinary levels in bronchial asthma are an indicator of the cause of exacerbation. In "Biochemical and Clinical Aspects of Pteridines" (W. Pfleiderer, H. Wachter, and J. A. Blair, eds.), pp. 321–329. de Gruyter, Berlin and New York, 1987.

M9. Meuer, S. C., Hussey, R. E., Penta, A. C., Fitzgerald, K. A., Stadler, B. M., Schlossman, S. F., and Reinherz, E. L., Cellular origin of interleukin 2 (IL 2) in man: Evidence for stimulus-restricted IL 2 production by $T4^+$ and $T8^+$ T lymphocytes. *J. Immunol.* 129, 1076–1079 (1982).

M10. Munzarova, M., and Kovarik, J., Is cancer a macrophage-mediated autoaggressive disease? *Lancet* 1, 952–954 (1987).

M11. Mura, P., Piriou, A., Tallineau, C., and Reiss, D. La néoptérine urinaire: Intérêt dans l'exploration de certaines néoplasies. *Ann. Biol. Clin. (Paris)* 44, 505–510 (1986).

N1. Nagatsu, T., Sawada, M., Yamaguchi, T., Sugimoto, T., Matsuura, S., Akino, M., Nakazawa, N., and Ogawa, H., Radioimmunoassay for neopterin in body fluids and tissues. *Anal. Biochem.* 141, 472–480 (1984).

N2. Nagatsu, T., Yamaguchi, T., Kato, T., Sugimoto, T., Matsuura, S., Akino, M., Tsushima, S., Nakazawa, N., and Ogawa, H., Radioimmunoassay for biopterin in body fluids and tissues. *Anal. Biochem.* **110**, 182–189 (1981).

N3. Nair, P. M., and Vaidyanathan, C. S., Anthranilic acid hydroxylase from *Tecoma stans*. *Biochim. Biophys. Acta* **110**, 521–531 (1965).

N4. Nathan, C. F., Peroxide and pteridine: A hypothesis of the regulation of macrophage antimicrobial activity by interferon-gamma. *In* "Interferon 7" (J. Gresser, ed.), pp. 125–143. Academic Press, London, 1986.

N5. Nathan, C. F., Murray, H. W., Wiebe, M. E., and Rubin, B. Y., Identification of interferon-gamma as the lymphokine that activates human macrophage oxidative metabolism and antimicrobial activity. *J. Exp. Med.* **158**, 670–689 (1983).

N6. Niederwieser, A., Staudemann, W., and Wetzel, E., High-performance liquid chromatography with column switching for the analysis of biogenic amine metabolites and pterins. *J. Chromatogr.* **290**, 237–246 (1984).

N7. Niederwieser, A., Blau, N., Wang, M., Joller, P., Atarés, M., and Cardesa-Garcia, J., GTP cyclohydrolase I deficiency, a new enzyme defect causing hyperphenylalaninemia with neopterin, biopterin, dopamine, and serotonin deficiencies and muscular hypotonia. *Eur. J. Pediatr.* **141**, 208–214 (1984).

N8. Niederwieser, A., Leimbacher, W., Curtius, H. C., Ponzone, A., Rey, F., and Leupold, D., Atypical phenylketonuria with "dihydrobiopterin synthetase" deficiency: Absence of phosphate-eliminating enzyme activity demonstrated in liver. *Eur. J. Pediatr.* **144**, 13–16 (1985).

N9. Niederwieser, D., Fuchs, D., Hausen, A., Judmaier, G., Reibnegger, G., Wachter, H., and Huber, C., Neopterin as a new biochemical marker in the clinical assessment of ulcerative colitis. *Immunobiology* **170**, 320–326 (1985).

N10. Niederwieser, D., Huber, C., Gratwohl, A., Bannert, P., Fuchs, D., Hausen, A., Reibnegger, G., Speck, B., and Wachter, H., Neopterin as a new biochemical marker in the clinical monitoring of bone marrow transplant recipients. *Transplantation* **38**, 497–500 (1984).

N11. Niederwieser, A., Joller, P., Seger, R., Blau, N., Prader, A., Bettex, J. D., Lüthy, R., Hirschel, B., Schaedelin, J., and Vetter, U., Neopterin in AIDS, other immunodeficiencies, and bacterial and viral infections. *Klin. Wochenschr.* **64**, 333–337 (1986).

N12. North, R. J., The murine antitumor immune response and its therapeutic manipulation. *Adv. Immunol.* **35**, 89–153 (1984).

O1. Ohkawa, H., Ohishi, N., and Yagi, K., New metabolites of riboflavin appear in human urine. *J. Biol. Chem.* **258**, 5623–5628 (1983).

P1. Pabst, W., and Rembold, H., Über das Verhalten des Biopterins im Säugetierorganismus. II. Einfluss von Vitaminmangel und eines Antagonisten auf die Biopterinausscheidung und das Wachstum in der Ratte. *Hoppe-Seyler's Z. Physiol. Chem.* **344**, 107–112 (1966).

P2. Pan, P., and Wurster, B., Inactivation of the chemoattractant folic acid by cellular slime molds and identification of the reaction product. *J. Bacteriol.* **136**, 955–959 (1978).

P3. Pan, P., Hall, E. M., and Bonner, J. T., Determination of the active portion of the folic acid molecule in cellular slime mold chemotaxis. *J. Bacteriol.* **122**, 185–191 (1975).

P4. Patterson, P. Y., Molecular and cellular determinants of neuroimmunologic inflammatory disease. *Fed. Proc., Fed. Am. Soc. Exp. Biol.* **41**, 2569–2576 (1982).

P5. Paterson, E. L., Saltza, M. H., and Stokstad, E. L. R., The isolation and characterization of a pteridine required for the growth of *Crithidia fasciculata*. *J. Am. Chem. Soc.* **78**, 5871–5873 (1956).

P6. Patterson, E. L., Broquist, H. P., Albrecht, A. M., von Saltza, M. H., and Stokstad,

E. L. R., A new pteridine in urine required for the growth of the protozoan *Crithidia fasciculata*. *J. Am. Chem. Soc.* **77**, 3167–3168 (1955).

P7. Pfefferkorn, E. R., Interferon-gamma blocks the growth of *Toxoplasma gondii* in human fibroblasts by inducing the host cells to degrade tryptophan. *Proc. Natl. Acad. Sci. U.S.A.* **81**, 908–912 (1984).

P8. Pfleiderer, W., Recent developments in the chemistry of pteridines. *Angew. Chem. Int. Ed. Engl.* **3**, 114–132 (1964).

P9. Pfleiderer, W., Chemistry of naturally occurring pterins. *In* "Chemistry and Biochemistry of Pterins" (S. J. Benkovic and R. L. Blakely, eds.), pp. 43–114. Wiley, New York, 1985.

P10. Pfleiderer, W., Bauer, R., Bartke, M., and Lutz, H., Reactivity of pteridines. *In* "Chemistry and Biology of Pteridines" (J. A. Blair, ed.), pp. 93–108. de Gruyter, Berlin and New York, 1983.

P11. Pheasant, A., Diurnal variation in urinary pterin excretion in man. *In* "Chemistry and Biology of Pteridines" (B. A. Cooper and V. M. Whitehead, eds.), pp. 268–270. de Gruyter, Berlin and New York, 1986.

P12. Plata, F., Enhancement of tumor growth correlates with suppression of the tumor-specific cytolytic T lymphocyte response in mice chronically infected by *Trypanosoma cruzi*. *J. Immunol.* **134**, 1312–1319 (1985).

P13. Polverini, P. J., Cotran, R. S., Gimbrone, M. A., and Unanue, E. R., Activated macrophages induce vascular proliferation. *Nature (London)* **269**, 804–806 (1977).

P14. Popovic, M., Sarngadharan, M. G., Read, E., and Gallo, R. C., Detection, isolation, and continuous production of cytopathic retroviruses (HTLV-III) from patients with AIDS and pre-AIDS. *Science* **224**, 497–500 (1984).

P15. Preble, O. T., Black, R. J., Friedman, R. M., Klippel, J. H., and Vilcek, J., Systemic lupus erythematosus: Presence in human serum of an unusual acid-labile leukocyte interferon. *Science* **216**, 429–431 (1982).

P16. Preble, O. T., Rothko, K., Klippel, J. H., Friedman, R. M., and Johnston, M. I., Interferon-induced 2' − 5' adenylate synthetase *in vivo* and interferon production *in vitro* by lymphocytes from systemic lupus erythematosus patients with and without circulating interferon. *J. Exp. Med.* **157**, 2140–2146 (1983).

P17. Prehn, R. T., and Prehn, L. M., The autoimmune nature of cancer. *Cancer Res.* **47**, 927–932 (1987).

P18. Prior, C., Frank, A., Fuchs, D., Hausen, A., Judmaier, G., Reibnegger, G., Werner, E. R., and Wachter, H., Immunity in sarcoidosis. *Lancet* **2**, 741 (1987).

P19. Prior, C., Fuchs, D., Hausen, A., Judmaier, G., Reibnegger G., Werner, E. R., Vogel, W., and Wachter, H., Potential of urinary neopterin excretion in differentiating chronic non-A, non-B hepatitis from fatty liver. *Lancet* **2**, 1235–1237 (1987).

P20. Prior, C., Bollbach, R., Fuchs, D., Hausen, A., Judmaier, G., Niederwieser, D., Reibnegger, G., Rotthauwe, H. W., Werner, E. R., and Wachter, H., Urinary neopterin, a marker of clinical activity in patients with Crohn's disease. *Clin. Chim. Acta* **155**, 11–22 (1986).

P21. Purrmann, R., Über die Flügelpigmente der Schmetterlinge. VII. Synthese des Leukopterins und Natur des Guanopterins. *Justus Liebigs Ann. Chem.* **544**, 182–190 (1940).

P22. Purrmann, R., Die Synthese des Xanthopterins. *Justus Liebigs Ann. Chem.* **546**, 98–102 (1940).

P23. Purrmann, R., Konstitution und Synthese des sogenannten Anhydroleukopterins. *Justus Liebigs Ann. Chem.* **548**, 284–292 (1941).

R1. Raedler, A., Studtmann, A., Emskoetter, T., Schulz, K. H., Greten, H., Fuchs, D., and Wachter, H., Correlation of urinary neopterin and activated peripheral lymphocytes

in inflammatory bowel disease, immunovasculitis and myasthenia gravis. *In* "Biochemical and Clinical Aspects of Pteridines" (W. Pfleiderer, H. Wachter, and J. A. Blair, eds.), pp. 203–212. de Gruyter, Berlin and New York, 1987.
R2. Rao, K. N., Trehan, S., and Noronha, J. M., Elevated urinary levels of 6-hydroxymethylpterin during malignancy and liver regeneration. A simple, noninvasive test for cancer detection. *Cancer (Philadelphia)* **48**, 1656–1663 (1981).
R3. Rao, K. N., Trehan, S., Shetty, P. A., and Noronha, J. M., Urinary 6-hydroxymethylpterin levels accurately monitor response to chemotherapy in acute lymphoblastic leukemia. *Cancer (Philadelphia)* **51**, 1425–1427 (1983).
R4. Reibnegger, G., Fuchs, D., Grubauer, G., Hausen, A., and Wachter, H. Neopterin excretion during incubation period, clinical manifestation and reconvalescence of viral infection. *In* "Biochemical and Clinical Aspects of Pteridines" (W. Pfleiderer, H. Wachter, and H. C. Curtius, eds.), pp. 433–447. de Gruyter, Berlin and New York, 1984.
R5. Reibnegger, G., Fuchs, D., Hausen, A., Kostron-Krainz, C., and Wachter, H., Urinary neopterin in malignant diseases of childhood. A marker for activity of the cell-mediated immunity. *Tumor Diagn. Ther.* **5**, 234–237 (1984).
R6. Reibnegger, G., Fuchs, D., Hausen, A., Werner, E. R., and Wachter, H., Activated macrophages and cancer. *Lancet* **1**, 1439 (1987).
R7. Reibnegger, G., Aulitzky, W., Huber, C., Margreiter, R., Riccabona, G., and Wachter, H., Neopterin in urine and serum of renal allograft recipients. *J. Clin. Chem. Clin. Biochem.* **24**, 770 (1986).
R8. Reibnegger, G., Boonpucknavig, V., Fuchs, D., Hausen, A., Schmutzhard, E., and Wachter, H., Urinary neopterin is elevated in patients with malaria. *Trans. R. Soc. Trop. Med. Hyg.* **78**, 545–546 (1984).
R9. Reibnegger, G., Fuchs, D., Hausen, A., Schmutzhard, E., Werner, E. R., and Wachter, H., The dependence of cell-mediated immune activation in malaria on age and endemicity. *Trans. R. Soc. Trop. Med. Hyg.* **81**, 729–733 (1987).
R10. Reibnegger, G., Fuchs, D., Hausen, A., Wachter, H., Bichler, E., and Böheim, K., Urinary neopterin in patients with head and neck cancer. *In* "Biochemical and Clinical Aspects of Pteridines" (H. Wachter, H. C. Curtius, and W. Pfleiderer, eds.), pp. 207–215. de Gruyter, Berlin and New York, 1982.
R11. Reibnegger, G., Egg, D., Fuchs, D., Günther, R., Hausen, A., Werner, E. R., and Wachter, H., Urinary neopterin reflects clinical activity in patients with rheumatoid arthritis. *Arthritis Rheum.* **29**, 1063–1070 (1986).
R12. Reibnegger, G., Hetzel, H., Fuchs, D., Fuith, L. C., Hausen, A., Werner, E. R., and Wachter, H., Clinical significance of neopterin for prognosis and follow-up in ovarian cancer. *Cancer Res.* **47**, 4977–4981 (1987).
R13. Reibnegger, G., Auhuber, I., Fuchs, D., Hausen, A., Judmaier, G., Prior, C., Werner, E. R., and Wachter, H., Urinary neopterin levels in acute viral hepatitis. *Hepatology* **8**, 771–774 (1988).
R14. Reibnegger, G., Bollbach, R., Fuchs, D., Hausen, A., Judmaier, G., Prior, C., Rottthauwe, H. W., Werner, E. R., and Wachter, H., A simple index relating clinical activity in Crohn's disease with T cell activation: Hematocrit, frequency of liquid stools and urinary neopterin as parameters. *Immunobiology* **173**, 1–11 (1986).
R15. Reibnegger, G. J., Bichler, A. H., Dapunt, O., Fuchs, D. N., Fuith, L. C., Hausen, A., Hetzel, H., Lutz, H., Werner, E. R., and Wachter, H., Neopterin as a prognostic indicator in patients with carcinoma of the uterine cervix. *Cancer Res.* **46**, 950–955 (1986).
R16. Rembold, H., Untersuchungen über den Stoffwechsel des Biopterins und über die polarographische Charakterisierung von Pteridinen. *In* "Pteridine Chemistry" (W. Pfleiderer, and E. C. Taylor, eds.), pp. 465–484. Pergamon, Oxford, 1964.

R17. Rembold, H., Catabolism of unconjugated pteridines, *In* "Chemistry and Biology of Pteridines" (K. Iwai, M. Akino, M. Goto, and Y. Iwanami, eds.), pp. 163–178. International Academic Printing, Tokyo, 1970.

R18. Rembold, H., and Buschmann, L., Trennung von 2-Amino-4-hydroxy-pteridinen durch Ionenaustauscherchromatographie. *Hoppe-Seyler's Z. Physiol. Chem.* **330**, 132–139 (1962).

R19. Rembold, H., and Buschmann, L., Struktur und Synthese des Neopterins. *Chem. Ber.* **96**, 1406–1410 (1963).

R20. Rembold, H., and Buschmann, L., Untersuchungen über die Pteridine der Bienenpuppe (Apis mellifica). *Justus Liebigs Ann. Chem.* **662**, 72–82 (1963).

R21. Rembold, H., and Gyure, W. L., Biochemistry of pteridines. *Angew. Chem. Int. Ed. Engl.* **11**, 1061–1072 (1972).

R22. Reynolds, J. J., and Brown, G. M., The biosynthesis of folic acid. *J. Biol. Chem.* **239**, 317–325 (1964).

R23. Rhodes-Feuillette, A., Druilhe, P., Canivet, M., Gentiline, M., and Peries, J., Presence d'interferon circulant dans le serum de malades infectes par *Plasmodium falciparum*. *C. R. Hebd. Seances Acad. Sci.* **293**, 635–637 (1981).

R24. Rifkin, J. L., and Wali, A. W., Effects of pteridines on the filopodia of *Dictyostelium discoideum* vegetative amoebae. *Cell Motil. Cytoskel.* **6**, 479–484 (1986).

R25. Robinson, B. W. S, McLemore, T. L., and Crystal, R. G., Gamma interferon is spontaneously released by alveolar macrophages and lung T lymphocytes in patients with pulmonary sarcoidosis. *J. Clin. Invest.* **75**, 1488–1495 (1985).

R26. Rokos, H., and Rokos, K., A radioimmunoassay for determination of D-*erythro*-neopterin. *In* "Chemistry and Biology of Pteridines" (J. A. Blair, ed.), pp. 815–819. de Gruyter, Berlin and New York, 1983.

R27. Rokos, H., Frisius, H., and Kunze, R., Neopterin and dihydroneopterin in serum of controls and patients with various diseases. Circadian rhythm of neopterin. Influence of cortisol? *In* "Chemistry and Biology of Pteridines" (J. A. Blair, and V. M. Whitehead, eds.), pp. 411–414. de Gruyter, Berlin and New York, 1986.

R28. Rokos, H., Rokos, K., Frisius, H., and Kirstaedter, H. J., Altered urinary excretion of pteridines in neoplastic disease. Determination of biopterin, neopterin, xanthopterin, and pterin. *Clin. Chim. Acta* **105**, 275–286 (1980).

R29. Rokos, K., Kunze, R. O. F., Koch, M. A., Rokos, H., and Nilsson, K., Kinetics of production and release of neopterin. Comparison of human pbmc with the permanent monocytic cell line U937 and its subclones. *In* "Unconjugated Pterins and Related Biogenic Amines" (H. C. Curtius, N. Blau, and R. A. Levine, eds.), pp. 177–184. de Gruyter, Berlin and New York, 1987.

R30. Röthler, F., and Karobath, M., Quantitative determination of unconjugated pterins in urine by gas chromatography/mass fragmentography. *Clin. Chim. Acta* **69**, 457–462 (1976).

S1. Sakurai, A., and Goto, M., Neopterin: Isolation from human urine. *J. Biochem. (Toyko)* **61**, 142–145 (1967).

S2. Sambo, A., Giannattasio, B., Flamini, G., Nicoletti, G., Magalini, S., Luciani, G., Leone, G., and Cittadini, A., Preliminary observations on urinary neopterin as a marker of neoplastic proliferation and lymphocyte activation. *Acta Med. Rom.* **23**, 533–545 (1985).

S3. Santelli, G., Marfella, A., Abate, G., Comella, P., Nitsch, F., and Perna, M., Urinary neopterin levels in hematologic malignancies. *Tumori* **72**, 139–143 (1986).

S4. Schäfer, A. J., Dreikorn, K., and Opelz, G., Comparison of neopterin and beta-2-microglobulin monitoring in renal transplantation. *Transplant. Proc.* **18**, 1060–1062 (1986).

S5. Schäfer, A. J., Daniel, V., Dreikorn, K., and Opelz, G., Assessment of plasma neopterin in clinical kidney transplantation. *Transplantation* **41**, 454–459 (1986).
S6. Schmutzhard, E., Fuchs, D., Hausen, A., Reibnegger, G., and Wachter, H., Is neopterin—a marker of cell-mediated immune response—helpful in classifying leprosy? *East Afr. Med. J.* 577–580 (1986).
S7. Schoedon, G., Troppmair, J., Adolf, G., Huber, C., and Niederwieser, A., Interferon-gamma enhances biosynthesis of pterins in peripheral blood mononuclear cells by induction of GTP-cyclohydrolase I activity. *J. Interferon Res.* **6**, 697–703 (1986).
S8. Schoedon, G., Troppmair, J., Fontana, A., Huber, C., Curtius, H. C., and Niederwieser, A., Biosynthesis and metabolism of pterins in peripheral blood mononuclear cells and leukemia lines of man and mouse. *Eur. J. Biochem.* **166**, 303–310 (1987).
S9. Schönitzer, D., Hönlinger, M., Fuchs, D., Hausen, A., Reibnegger, G., Werner, E. R., and Wachter, H., Experiences with serum neopterin for routine screening of blood donors. *In* "Biochemical and Clinical Aspects of Pteridines" (W. Pfleiderer, H. Wachter, and J. A. Blair, eds.), pp. 331–343. de Gruyter, Berlin and New York, 1987.
S10. Schöpf, C., and Becker, E., Über neue Pterine. *Justus Liebigs Ann. Chem.* **524**, 49–144 (1936).
S11. Schwedes, U., Teuber, J., Schmidt, R., and Usadel, K. H., Neopterin as a marker for the activity of autoimmune thyroid diseases. *Acta Endocrinol. (Copenhagen)* **111**, 51–52 (1986).
S12. Shimomura, O., Suthers, H. L. B., and Bonner, J. T., Chemical identity of the acrasin of the cellular slime mold *Polyspondylium violaceum*. *Proc. Natl. Acad. Sci. U.S.A.* **79**, 7376–7379 (1982).
S13. Shintaku, H., Isshiki, G., Hase, Y., Tsuruhara, T., and Oura, T., Normal pterin values in urine and serum in neonates and its age-related change throughout life. *J. Inherited Metab. Dis.* **5**, 241–242 (1982).
S14. Shiota, T., and Palumbo, M. P., Enzymatic synthesis of the pteridine moiety of dihydrofolate from guanine nucleotides. *J. Biol. Chem.* **240**, 4449–4453 (1965).
S15. Snyderman, R., Meadows, L., Holder, W., and Wells, S., Abnormal monocyte chemotaxis in patients with breast cancer—Evidence for a tumor-mediated effect. *J. Natl. Cancer Inst. (U.S.)* **60**, 737–740 (1978).
S16. Stastny, P., Association of the B-cell alloantigen DRw4 with rheumatoid arthritis. *N. Engl. J. Med.* **298**, 869–871 (1978).
S17. Stea, B., Abnormalities of folic acid and related pteridines; metabolism in cancer cells; application to the diagnosis and treatment of malignant diseases. Ph.D. dissertation, University of California, Los Angeles, 1979.
S18. Stea, B., Halpern, R. M., Halpern, B. C., and Smith, R. A., Quantitative determination of pterins in biological fluids by high-performance liquid chromatography. *J. Chromatogr.* **188**, 363–375 (1980).
S19. Stea, B., Halpern, R. M., Halpern, B. C., and Smith, R. A., Urinary excretion levels of unconjugated pterins in cancer patients and normal individuals. *Clin. Chim. Acta* **113**, 231–242 (1981).
S20. Stea, B., Backlund, P. S., Berkey, P. B., Cho, A. K., Halpern, B. C., Halpern, R. M., and Smith, R. A., Folate and pterin metabolism by cancer cells in culture. *Cancer Res.* **38**, 2378–2384 (1978).
S21. Stokstad, E. L. R., and Koch, J., Folic acid metabolism. *Physiol. Rev.* **47**, 83–116 (1967).
S22. Strober, W., An immunological theory of gluten-sensitive enteropathy. *In* "Perspectives in Coeliac Disease" (B. McNicholl, C. F. McCarthy, and P. F. Fottrell, eds.), pp. 169–182. MTP press, Lancaster, 1978.

S23. Strohmaier, W., Redl, H., Schlag, G., and Inthorn, D., D-*erythro*-Neopterin plasma levels in intensive care patients with and without septic complications. *CRC Crit. Care. Med.* **15,** 757–760 (1987).
S24. Sugimoto, T., Matsuura, S., Yamaguchi, T., Sawada, M., and Nagatsu, T., Determination of biopterin and neopterin in fluids and tissues by radioimmunoassay. *In* "Biochemical and Clinical Aspects of Pteridines" (H. C. Curtius, W. Pfleiderer, and H. Wachter, eds.), pp. 51–63. de Gruyter, Berlin and New York, 1983.
T1. Tashiro, M., Ciborowski, P., Klenk, H. D., Pulverer, G., and Rott, R., Role of *Staphylococcus* protease in the development of influenza pneumonia. *Nature (London)* **325,** 536–537 (1987).
T2. Taylor, E. C., Synthetic methods in pteridine chemistry: Some applications to pteridine natural products. *In* "Chemistry and Biology of Pteridines" J. A. Blair, ed.), pp. 23–49. de Gruyter, Berlin and New York, 1983.
T3. Thomas, P. D., and Hunninghake, G. W., Current concepts of the pathogenesis of sarcoidosis. *Am. Rev. Respir. Dis.* **135,** 747–760 (1987).
T4. Tietz, A., Lindberg, M., and Kennedy, E. P., A new pteridine-requiring enzyme system for the oxidation of glyceryl ethers. *J. Biol. Chem.* **239,** 4081–4090 (1964).
T5. Tilg, H., Margreiter, R., Scriba, M., Marth, C., Niederwieser, D., Aulitzky, W., Spielberger, M., Wachter, H., and Huber, C., Clinical presentation of CMV infection in solid organ transplant recipients and its impact on graft rejection and neopterin excretion. *Clin. Transplant.* **1,** 37–44 (1987).
T6. Tillinghast, H. S., and Newell, P. C., Chemotaxis towards pteridines during development of *Dictyostelium*. *J. Cell Sci.* **87,** 45–53 (1987).
T7. Todd, I., Pujol-Borrell, R., Hammond, L. J., Bottazzo, G. F., and Feldman, M., Interferon-gamma induces HLA-DR expression by thyroid epithelium. *Clin. Exp. Immunol.* **61,** 265–273 (1985).
T8. Trehan, S., Rao, K. N., Shetty, P. A., and Noronha, J. M., Urinary 6-hydroxymethylpterin levels accurately monitor response to chemotherapy in acute myeloblastic leukemia. *Cancer (Philadelphia)* **50,** 114–117 (1982).
U1. Unterweger, B., Fuchs, D., Fleischhacker, W. W., Hausen, A., Hengster, P., Reibnegger, G., Werner, E. R., Hinterhuber, H., Dierich, M. P., and Wachter, H., Neopterin as a possible predictor for the course of HIV infection in parenteral opiate dependence. *J. Cancer Commun.* **2,** 27–34 (1988).
V1. Vaith, P., Maas, D., Feigl, D., Hauke, G., Lang, B., Oepke, G., Stierle, H. E., Bross, K. J., Andreesen, R., Gross, G., Monner, D. A., Grote, W., and Mühlradt, P., *In-vitro-* und *in-vivo* Studien mit Interleukin-2 (IL-2) und verschiedenen Immunstimulanzien bei einer Patientin mit AIDS. *Immun. Infekt.* **13,** 51–63 (1985).
V2. VanBeelen, P., Stassen, A. P. M., Bosch, J. W. G., Vogels, G. D., Guijt, W., and Haasnoot, C. A. G., Elucidation of the structure of methanopterin, a coenzyme from *Methanobacterium thermoautotrophicum*, using two-dimensional nuclear-magnetic-resonance techniques. *Eur. J. Biochem.* **138,** 563–571 (1984).
V3. VanHaastert, P. J. M., DeWit., R. J. W., Grijpma, Y., and Konijn, T. M., Identification of a pterin as the acrasin of the cellular slime mold *Dictyostelium lacteum*. *Proc. Natl. Acad. Sci. U.S.A.* **79,** 6270–6274 (1982).
V4. Venet, A., Hance, A. J., Saltini, C., Robinson, B. W. S., and Crystal, R. G., Enhanced alveolar macrophage-mediated antigen-induced T lymphocyte proliferation in sarcoidosis. *J. Clin. Invest.* **75,** 293–301 (1985).
V5. Viscontini, M., Provenzale, R., Ohlgart, S., and Mallevialle, J., Synthese des natürlichen *D*-Neopterins und *L*-Monapterins. *Helv. Chim. Acta* **53,** 1202–1207 (1970).
V6. Voelter, W., Ebner, B., Bauer, H., Röhrer, H., Porcher, H., and Stiefel, T., Neopterin

levels in different human tumors under treatment with antitumoral peptides (factor AF2), course, response rate, prognosis. In "Biochemical and Clinical Aspects of Pteridines" (H. Wachter, H. C. Curtius, and W. Pfleiderer, eds.), pp. 546–558. de Gruyter, Berlin and New York, 1985.

V7. Volberding, P. A., Abrams, D., Beardslee, D., Gee, G., Moody, D., Stites, D., and Wofsy, C., Recombinant interleukin-2 therapy of acquired immune deficiency syndrome. *Abstr., Int. Conf. AIDS, 2nd, 1986,* 169 (1986).

V8. Volin, L., Jansson, S. E., Turpeinen, U., Pomoell, U. M., and Ruutu, T., Urinary neopterin in bone marrow recipients. *Transplant. Proc.* **19,** 2651–2654 (1987).

W1. Wachter, H., Grassmayr, K., and Hausen, A., Enhanced urinary excretion of 7,8-dihydro-6-hydroxylumazine by mice bearing the Ehrlich ascites tumour. *Cancer Lett.* **6,** 61–66 (1979).

W2. Wachter, H., Hausen, A., and Grassmayr, K., Erhöhte Ausscheidung von Neopterin im Harn von Patienten mit maligen Tumoren und mit Viruserkrankungen. *Hoppe-Seyler's Z. Physiol. Chem.* **360,** 1957–1960 (1979).

W3. Wachter, H., Fuchs, D., Hausen, A., and Reibnegger, G., Importance of pteridines for regulation of growth. In "Biochemical and Clinical Aspects of Pteridines" (H. Wachter, H. C. Curtius, and W. Pfleiderer, eds.), pp. 277–281. de Gruyter, Berlin and New York, 1982.

W4. Wachter, H., Hausen, A., Reider, E., and Schweiger, M., Pteridine excretion from cells as indicator of cell proliferation. *Naturwissenschaften* **67,** 610–611 (1980).

W5. Wachter, H., Grassmayr, K., Gütter, W., Hausen, A., and Sallaberger, G., *In-situ-*Fluorometrie niedermolekularer organischer Harnbestandteile nach Elektrophorese. *Wien. Klin. Wochenschr.* **84,** 586–590 (1972).

W6. Wachter, H., Fuchs, D., Hausen, A., Reibnegger, G., Werner, E. R., and Dierich, M. P., Who will get AIDS? *Lancet* **2,** 1216–1217 (1986).

W7. Wachter, H., Grassmayr, K., Gütter, W., Hausen, A., Sallaberger, G., and Gabl, F., Urinary low molecular weight products in neoplasia. *Br. Med. J.* **2,** 322–323 (1971).

W8. Wachter, H., Fuchs, D., Hausen, A., Huber, C., Knosp, O., Reibnegger, G., and Spira, T. J., Elevated urinary neopterin levels in patients with acquired immunodeficiency syndrome (AIDS). *Hoppe Seyler's Z. Physiol. Chem.* **364,** 1345–1346 (1983).

W9. Wachter, H., Fuchs, D., Hausen, A., Hengster, P., Reibnegger, G., Reissigl, H., Schoenitzer, D., Schulz, T., Werner, E. R., and Dierich, M. P., Are conditions linked with T-cell stimulation necessary for progressive HTLV-III infection? *Lancet* **1,** 97 (1986).

W10. Watson, B. M., Schlesinger, P., and Cotton, R. G. H., Dihydroxanthopterinuria in phenylketonuria and lethal hyperphenylalaninemia patients. *Clin. Chim. Acta* **78,** 417–423 (1977).

W11. Werner, E. R., Fuchs, D., Hausen, A., Reibnegger, G., and Wachter, H., Simultaneous determination of neopterin and creatinine in serum with solid-phase extraction and on-line elution liquid chromatography. *Clin. Chem. (Winston-Salem, N. C.)* **33,** 2028–2033 (1987).

W12. Werner, E. R., Hirsch-Kauffmann, M., Fuchs, D., Hausen, A., Reibnegger, G., Schweiger, M., and Wachter, H., Interferon-gamma induced degradation of tryptophan by human cells *in vitro*. *Biol. Chem. Hoppe-Seyler* **368,** 1407–1412 (1987).

W13. Werner, E. R., Bitterlich, G., Fuchs, D., Hausen, A., Reibnegger, G, Szabo, G., Dierich, M. P., and Wachter, H., Human macrophages degrade tryptophan upon induction by interferon-gamma. *Life Sci.* **41,** 273–280 (1987).

W14. Werner, E. R., Bichler, A., Daxenbichler, G., Fuchs, D., Fuith, L. C., Hausen, A., Hetzel, H., Reibnegger, G., and Wachter, H., Determination of neopterin in serum and urine. *Clin. Chem. (Winston-Salem, N. C.)* **33,** 62–66 (1987).

W15. Werner, E. R., Lutz, H., Fuchs, D., Hausen, A., Huber, C., Niederwieser, D., Pfleiderer, W., Reibnegger, G., Troppmair, J., and Wachter, H., Identification of 3-hydroxyanthranilic acid in mixed lymphocyte cultures. *Biol. Chem. Hoppe-Seyler* **366**, 99–102 (1985).
W16. Weygand, F., Simon, H., Dahms, G., Waldschmidt, M., Schliep, H. J., and Wacker, H., Über die Biogenese des Leucopterins, *Angew. Chem.* **73**, 402–407 (1961).
W17. Wiegele, J., Margreiter, R., Huber, C., Dworzak, E., Fuchs, D., Hausen, A., Reibnegger, G., and Wachter, H., Urinary neopterin excretion in breast cancer patients. *In* "Biochemical and Clinical Aspects of Pteridines" W. Pfleiderer, H. Wachter, and H. C. Curtius, eds.), pp. 417–424. de Gruyter, Berlin and New York, 1984.
W18. Wieland, H., and Schöpf, C., Über den gelben Flügelfarbstoff des Zitronenfalters *(Gonepteryx rhamni). Ber. Dtsch. Chem. Ges.* **58**, 2178–2183 (1925).
W19. Wolf, J., Musch, E., Neuss, H., and Klehr, U., Neopterin im Serum und Urin zur Differentialdiagnose von Nierenfunktionsstörungen nach Nierentransplantation. *Klin. Wochenschr.* **65**, 225–231 (1987).
W20. Woloszczuk, W., Köhn, H. D., Pohl, W., and Klech, H., Neopterin and interferon gamma serum levels in sarcoidosis. *In* "Biochemical and Clinical Aspects of Pteridines" (W. Pfleiderer, H. Wachter, and J. A. Blair, eds.), pp 279–287. de Gruyter, Berlin and New York, 1987.
W21. Woloszczuk, W., Schwarz, M., Havel, M., Laczkovics, A., and Müller, M. M., Neopterin and interferon-gamma serum levels in patients with heart and kidney transplants. *J. Clin. Chem. Clin. Biochem.* **24**, 729–734 (1986).
W22. Woloszczuk, W., Troppmair, J., Leiter, E., Flener, R., Schwarz, M., Kovarik, J., Pohanka, E., Margreiter, R., and Huber, C., Relationship of interferon-gamma and neopterin levels during stimulation with alloantigens *in vivo* and *in vitro*. *Transplantation* **41**, 716–719 (1986).
Z1. Ziegler, I., and Harmsen, R., The biology of pteridines in insects. *Adv. Insect Physiol.* **6**, 139–203 (1969).
Z2. Zitko, M., Andrysek, O., Cernovska, I., and Vasickova, M., Renal excretion of neopterin and biopterin in patients with malignant melanoma and Hodgkin's disease. *Neoplasma* **33**, 387–391 (1986).

BIOCHEMICAL DETECTION OF HEPATITIS B VIRUS CONSTITUENTS

Hsiang Ju Lin

Clinical Biochemistry Unit,
University of Hong Kong,
Hong Kong

1. Introduction... 143
   1.1 Epidemiology of HBV Infection 144
   1.2 Biology of the Virus.. 146
   1.3 Molecular Biology of the Virus 147
   1.4 Markers of HBV Infection 150
2. HBV DNA... 151
   2.1 Principles of DNA–DNA Hybridization 151
   2.2 Preparation of Specimens 153
   2.3 Probes... 154
   2.4 Dot Hybridization.. 158
   2.5 Analysis of HBV DNA Forms.. 159
   2.6 Significance of HBV DNA in HBV-Related Liver Disease............... 163
   2.7 HBV DNA in Extrahepatic Sites 170
3. HBV Enzymes.. 174
   3.1 HBV DNA Polymerase Assays....................................... 174
   3.2 HBV DNA Polymerase Activity in HBV-Related Disease 177
   3.3 HBV Protein Kinase 178
4. HBV Polypeptides... 179
   4.1 Pre-S Region Polypeptides... 179
   4.2 Region X Polypeptides 183
   References... 183

1. Introduction

Hepatitis B was recognized as a disease for many years prior to the identification of any of its viral constituents (Z4). In 1967, Blumberg described an antigen (now known as hepatitis B surface antigen, HBsAg) in human serum that occurred frequently in hepatitis, Down's syndrome, and leukemia (B11). Because it was first found in the serum of an Australian aborigine, it was called Australia antigen (B8). Identification of the hepatitis B virus (HBV) presented problems, because the serum of hepatitis patients contained three distinct virus-related forms: tubules, and spheres with diameters of about 22 or 42 nm (B10). All three forms reacted with antiserum directed against HBsAg. Because of their more complex structure, as seen under the electron microscope, Dane identified the 42-nm particles as virus, and correctly concluded that the other two forms were excess viral surface antigen (D1). Blumberg's discovery of HBsAg provided the first tool for the detection of hepatitis B infection. The identification of the HBV (Dane particle) led to the study of its biochemistry and the development of further specific tests for the virus. Serological and biochemical methods are now both used extensively in the study of hepatitis B infection. This article deals with biochemical methods for the detection of HBV constituents.

1.1. Epidemiology of HBV Infection

The factors that are implicated in HBV infection are gender, vertical transmission, and horizontal transmission. A person whose blood is positive for HBsAg for over 6 months is termed an HBV carrier (A7). Male adults have a higher HBsAg carriage rate than females (L15, Y3). Groups at high risk are babies born of HBV-infected mothers, patients receiving frequent blood transfusions, healthcare workers, homosexuals, drug addicts, renal patients, and persons in close contact with HBsAg carriers. The relative importance of any single factor in the spread of HBV infection in a given population varies. For example, perinatal and postnatal transmission from mother to child (vertical transmission) is extremely important in some Asian populations (N8, N9, S24, W7), but is less so in West Africa (P8) or India (N4). HBV infection among different groups living in one

TABLE 1
HBsAg Carrier Rate among Different Groups in Australia[a]

Category	HBV carriers (%)
Aboriginal/Torres Strait islanders	26.0
Southeast Asian origin	15.0
Institutionalized mentally retarded	14.9
Intravenous drug abusers	8.9
Male homosexuals	5.5
Mediterranean origin	2.5

[a]Adapted from Gust et al. (G11, G12).

TABLE 2
FREQUENCY OF HBsAg POSITIVITY IN VARIOUS AREAS

Area	No. tested	Subjects	HBsAg positive (%)	Reference
Taiwan	22,707	Men	15.2	B2
Hong Kong	16,334	Cross section	9.6	Y3
Senegal	2212	Children	9.0	B1
China	277,186	Cross section	8.8	Z2
Saudi Arabia	7020	Hospital patients	7.3	L15
India	8575	Pregnant women	3.7	N4
Greece	6606	Blood donors	2.9	E1
Japan	342,407	Blood donors	1.9	N9
United States	67,092	Blood donors	0.2	S33
Australia	12,585,167	Blood donors	0.1	G11

country would, accordingly, reflect such influences. Table 1 shows the carrier rate among different groups living in Australia.

HBV infection is endemic in Asia, central and southern Africa, and Oceania (M9). Table 2 shows the frequency of HBsAg positivity in large-scale screening studies. With due allowance for the effects of age, gender, and differences in the method of detecting HBsAg, the observed differences vary by one or two orders of magnitude.

It is estimated that about 200 million people worldwide are carriers (S33, W8). Figure 1 shows a general scheme for the possible outcomes and various sequelae of HBV infection. Approximately 5–10% of patients with acute hepatitis become chronic carriers (B15, K9). Because the carrier state may lead to chronic hepatitis, liver cirrhosis, and primary liver cell cancer, it has serious health implications (B9, B12, S32).

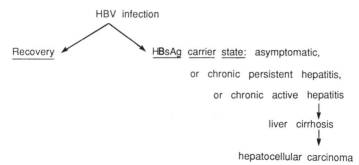

FIG. 1. Possible outcomes of HBV infection. Liver cirrhosis and hepatocellular carcinoma are frequent but not inevitable sequelae of chronic active hepatitis.

1.2. BIOLOGY OF THE VIRUS

Hepatitis B virus is the smallest of the DNA viruses that infect man. Only one other animal DNA virus, which infects pigs, is smaller (T8). The four viruses, HBV, woodchuck hepatitis virus, ground squirrel hepatitis virus, and duck hepatitis virus, form a distinct group of reverse-transcribing viruses that show strong similarities in virion size, in the length of the genome, and in other properties (G13, S26). Because these DNA viruses are hepatotropic, they are called the hepadna viruses (R4). Although these viruses utilize reverse transcription, they differ from retroviruses in their smaller size, in the fact that their genomes consist of DNA instead of RNA, and in their mode of replication (S6, S28, W5). The common tree squirrel also harbors a hepatitis B type of virus, but its exact relationship to the hepadna viruses remains to be defined (F7). It is possible that hepadna viruses and retroviruses are derived from a common ancestor, because there are similarities among some of their nucleotide sequences (M16).

Dane particles isolated from the plasma of HBV carriers consist of a double-shelled spheres that either may be "empty," or may contain circular double-stranded DNA and HBV DNA polymerase (K2, T3). Empty particles may be a defective form of the virus (G2). The particles containing DNA and polymerase are considered to be the complete, fully infective particle. The outer shell is composed of HBsAg. The inner shell, called the nucleocapsid, is composed of core protein (HBcAg) that is antigenically distinct from one of its components, e protein (HBeAg) (M4, O1). The viral DNA is a circle plaited from two linear strands of unequal length. There are two enzymes associated with the virion, DNA polymerase (K3, R3) and protein kinase (A5, G4).

There is no established cell culture system for growing HBV. Particles resembling HBV have been grown in hepatoblastoma cultures and in a hepatoma cell line transfected with cloned HBV DNA (S7, S31). Similar particles were detected in low yield from lymphoblastoid cell cultures that were grown from bone marrow aspirates taken from a hepatitis patient (R7). HBV in primary liver cell cultures is short-lived (S15).

Four subtypes of HBV are recognized, based on the HBsAg subdeterminants d/y and w/r. The commonest subtypes are adw, adr, and ayw (H18, L1). The distribution of the subtypes is along geographical lines, as might be expected from an infectious disease (B2, M8, N10). Several of the antigenic determinants exist in variant forms that are distinguishable by serological tests (C10). It has been deduced from DNA sequence analysis that individuals may be coinfected by more than one HBV (F14, O5, P1, Z3). Earlier studies, which were based on subtype analysis of HBsAg and anti-HBs, led to the same conclusion (C10).

HBV infection is a necessary condition for multiplication of the delta virus (HDV), an incomplete virus that contains an RNA with 1750 nucleotides with no sequence homology to HBV DNA (W1). Delta virus and HBV may coinfect man, aggravating liver damage. However, the prevalence of HBV and HDV infection among various groups differs. HDV infection is more frequent among homosexuals and drug addicts than in other groups that are at high risk for HBV infection (H17). Coinfection is frequent in Europe and in North and South America, but extremely rare among Asian Chinese, even among drug addicts (C6, L3, P10). Concurrent infection is more than coincidental, because the delta virus utilizes HBsAg as its coat (H21). Thus, HBV can multiply in the absence of HDV but the converse is not true (P10, R2).

1.3. Molecular Biology of the Virus

The structure of HBV DNA in its most common form, the circular double-stranded DNA with a single-stranded gap, is remarkable in many respects. The two linear strands, one long (L) and one short (S), have cohesive 5' ends, enabling a circle to be formed. Although the S strand is of variable length, its 5' end is always fixed at the same position on the L strand. Therefore, the end of single-stranded gap is also fixed. Figure 2 depicts the HBV genome. The L strand has about 3200 nucleotides and is attached to a protein at its 5' end (G3). The S strand is of variable length, ranging from about 1700 to 2800 nucleotides (R4). Examination of Dane particle preparations from one individual revealed that most HBV DNA molecules have a gap length of 600–700 nucleotides (D7). The 5' fixed end of the S strand is located over 200 nucleotides from the 5' fixed end of the L strand. There are four open reading frames, represented by L-strand transcripts that begin with AUG and end with UAA or UGA. HBcAg is encoded by gene C. HBsAg is encoded within an open reading frame that is longer than necessary to specify the 226 amino acids making up the antigen. This open reading frame is divided into two regions, pre-S and S. The pre-S region is further subdivided into the pre-S2 region (adjacent to gene S) encoding 55 amino acids, and the pre-S1 region further upstream, encoding another 100–110 amino acids. There is also a relatively short open reading frame, region X, that encodes polypeptides which have been detected in some human tissue specimens, but the function of which is unknown (F6). The fourth open reading frame, region P, covers over three-quarters of the genome; it probably encodes the viral DNA polymerase. The coding region for HBV protein kinase has not been identified.

HBV DNA was first sequenced in entirety in 1979 (G1). Since the L strand contained about twice as many stop codons as the S strand, it was

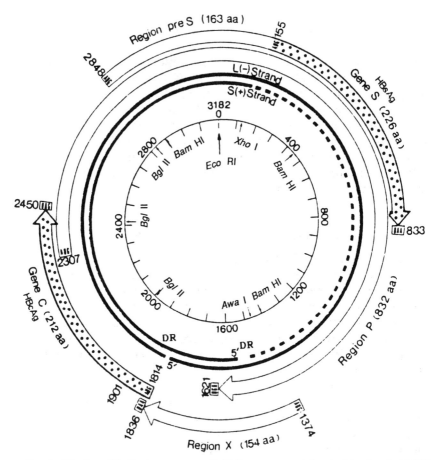

FIG. 2 The hepatitis B virus genome. Bold lines represent the double-stranded DNA with single-stranded region (dashes). Restriction enzyme sites are indicated in the innermost circle. The four arrows indicate the locations of the open reading frames, encoding the core antigen, surface antigen, polymerase, and region X polypeptides; DR, direct repeat sequence. Adapted from Tiollais and Wain-Hobson (T6).

determined that the S strand was the (+) strand, which is analogous to mRNA. The DNAs of all four subtypes and that of a complex subtype adyw are known. Table 3 summarizes the differences found within and between the subtypes. Nucleotide sequence differences within a subtype were less than 3%. Differences between subtypes were usually larger, about 9–12%, but those between ayr and adr were as small as those within the same subtype. As for the nucleotide sequence of the complex subtype adyw, it resembled that of ayw more than any of the other three subtypes.

TABLE 3
NUCLEOTIDE SEQUENCE COMPARISON OF CLONED HBV DNAs

Subtype		Length (base pairs)	Difference (%)		Reference
			Within subtype	Between subtype	
I	adr	3215	II (2.3)	—	K5
II	adr	3214	III (2.8)	—	F14
III	adr	3188	I (2.7)	IV (9.8)	O5
IV	adw	3200	V (1.6)	VII (10.0)	O5
V	adw	3221	—	VII (10.8)	V1
VI	ayw	3182	VII (2.6)	—	B4
VII	ayw	3182	—	III (11.2)	G1
VIII	ayr	3215	—	I–III (2.1–2.3) IV–V (8.4–9.0) VI–VII (10.4)	O2
IX	adyw	3182	—	III (11.1) V (9.4) VII (3.1) VIII (9.2)	P9
X	adyw	2743^a	IX (0.5)	V (11.6) VII (3.9)	P1

^{a}About 13% of the genome was missing from the cloned sequence.

The interpretation of these data is open, since it cannot be determined from serological subtyping whether a complex subtype is a mixture of two different subtype populations (say, adw and ayw), or that it is a hybrid form (C10, H18). The data would be consistent with the hypothesis that the DNA of a subtype ayw virus was selected in the course of cloning, but it is also possible that the clone was truly representative of a hybrid subtype.

There is evidence for subtype-independent heterogeneity in some constituents of HBV in different carriers. Examination of DNA lengths in preparations of Dane particles from HbsAg carriers revealed size heterogeneity (T3). DNA polymorphism was also demonstrated by restriction fragment analysis (B21, C5, P1, S18, S20) and molecular hybridization (L13). Microheterogeneity of the polypeptides found in small spherical HBsAg obtained from different HBsAg carriers was demonstrated by electrophoresis (S25).

The replication of HBV has been deduced from studies carried out on HBV-infected ducks (S27) and on chimpanzees infected with human HBV (C2, F9, W5). These observations have been confirmed by studies in man (B6, M15, M17–M19, S4). Although HBV constituents have been found in several extrahepatic tissues, the principal target is the liver cell. Once in the liver cell nucleus, elongation of the S strand is completed. The resulting fully double-stranded DNA is covalently closed and assumes a supercoiled form. The subsequent events take place in the cytoplasm. A RNA copy of the L strand is made with RNA polymerase and is extended with further sequences to form the pregenome. The RNA transcript of the L strand is then reverse transcribed to produce L strands. Most of the pregenome is then degraded and the viral DNA polymerase proceeds to synthesize the S strand on the L-strand template. This process is usually incomplete and the virion is excreted from the liver cell with one full-length strand and one shorter strand of DNA.

1.4. Markers of HBV Infection

The first tests that were useful in the diagnosis of hepatitis were serum transaminase (AST and ALT) and bilirubin assays (Z4). They continue to be used but their lack of specificity for this particular disease remains an obvious drawback.

The most frequently used serological tests in HBV infection are those for the detection of the viral antigens HBsAg and HBeAg, and the antibodies to HBc, HBe, and HBs (Table 4). The markers appearing earliest in acute hepatitis B infection are HBsAg, the IgM class of anti-HBc (L5, L14, T10), and HBeAg. Anti-HBc (IgG, or non-class-specific) may appear weeks, months, or years later. In many patients this event will be followed by the appearance of anti-HBe and finally anti-HBs, but in other patients

TABLE 4
Serological Markers of HBV Infection

Early markers	
HBsAg	Appears during acute infection; disappears during recovery; persists in chronic infection
Anti-HBc (IgM)	Test for recent acute infection; absent or weakly positive in chronic infection
HBeAg	Appears during acute infection; disappears during recovery; may be absent or present in chronic infection
Late markers	
Anti-HBc (IgG)	Absent during acute infection; appears during recovery and in chronic infection
Anti-HBe	Absent during acute infection; appears during recovery; may be present or absent during chronic infection
Anti-HBs	Absent during acute infection; appears during recovery; fails to appear in chronic infection

with chronic hepatitis seroconversion to the latter two antibodies may fail to occur. There are two excellent papers on the interpretation of standard serological tests for HBV infection (D3, M25).

The most important biochemical markers are HBV DNA and HBV DNA polymerase. Both of the analytes are contained within the complete, infective virus. In contrast, serological markers lack this specificity for the virion. The 22-nm spheres and tubules of HBsAg may outnumber the virion by a factor of 1000 (S23). Further, the HBeAg that is routinely assayed in serum is not the antigen present within HBV because the test is carried out without disruption of the viral coat. The e antigen occurs in serum in the unbound state or bound to IgG (besides being present within the virion as a component of the core) (T1). Prior to the development of methods for the detection of HBV DNA, the DNA polymerase assay was considered to be the most specific test for the presence of the virion. Currently, DNA hybridization is the most specific and sensitive test for HBV. Polyalbumin binding assays have received much attention, but their clinical usefulness has yet to be established.

Since the presence of HBV is a prerequisite for delta virus replication, the detection of delta virus markers is also meaningful in the context of HBV infection. Markers for delta virus include delta RNA, delta antigen, and anti-delta (IgM and IgG types) (R2, S19).

2. HBV DNA

Two biochemical techniques are applicable to the detection of HBV DNA: dot hybridization, which provides a semiquantitative test for its presence or absence, and Southern blot analysis, by means of which HBV DNA sequences are located in nucleic acid fragments that have been separated according to electrophoretic mobility. [*In situ* hybridization, which relies in large part on histological techniques, is not discussed herein, but the reader may consult some useful references (B7, G8, H1, N5, R1).]

2.1. Principles of DNA–DNA Hybridization

The technique of nucleic acid hybridization was developed in 1965 by Gillespie and Spiegelman (G6). The fundamental process in hybridization is the reversible formation of hydrogen bonds between bases on separate strands via pairing of guanine–cytosine and adenine–thymine (or adenine–uracil). The stability of a duplex DNA formed between two single strands of nucleic acid is determined principally by the degree of complementarity between the two strands, and is dependent on many other factors, including salt concentration, temperature, and guanine–cytosine (GC) content of the DNA. In naturally occurring DNAs, strand separation is half complete at a melting temperature (T_m) that is dependent on the GC content and on the sodium ion concentration of the solvent in moles/liter (M).

$$T_m = 81.5°C + 16.6(\log M) + 0.41(\% \text{ GC})$$

The GC content of HBV DNA is 48–49% (R6). Its melting temperature in 0.15 mol/liter NaCl–0.015 mol/liter Na citrate (pH 7.5) would be 88.4°C. Increasing the salt concentration to 1 mol/liter would raise the T_m to 101°C. Since the optimal temperature for formation of DNA–DNA hybrids is 20–25°C below the melting temperature (W4), the temperatures necessary for hybridization would in either case be above 65°C. Hybridization is often carried at lower temperatures in order to minimize heat-induced strand breakage. The addition of formamide to the hybridization medium enables lower temperatures to be used. The T_m decreases by 0.72°C for every 1% formamide (M10).

$$T_m = 81.5°C + 16.6(\log M) + 0.41(\%\text{GC}) - 0.72(\% \text{ formamide})$$

The effective T_m of HBV DNA would be 65°C in a solvent containing 50% formamide and 1 mol/liter sodium ion.

The actual temperature of hybridization is chosen according to the degree of stringency required. Both high and low stringency conditions may be used in studying the relatedness of different DNAs (H20). Highly stringent conditions should be used in tests for the presence or absence of a specific DNA. By way of illustration, serum from non-A, non-B hepatitis patients gave negative results with an HBV DNA probe when high-stringency conditions were used and positive results under low-stringency conditions (F11). Very high-stringency conditions would obtain at $T_m - 20°C$; high stringency, at $T_m - 25°C$; very low stringency, at $T_m - 50°C$.

The degree of mismatch between base pairs formed at the temperature of hybridization is 1.4% for every degree below the T_m. Under very high-stringency conditions ($T_m - 20°C$) duplexes formed may contain up to 14% mismatches. Perfect pairing of all the bases in such duplexes is seldom attained. It could be predicted that nucleotide sequences between HBV DNA of different subtypes, that do not exceed 12% (Table 3), would not be detectable by molecular hybridization.

The rate of hybridization is maximal in the temperature range $T_m - 18°C$ to $T_m - 32°C$ (W4). The time (t) required to achieve hybridization of half the strands is given by the equation

$$t = \frac{(N)(\ln 2)}{(3.5 \times 10^5)(L^{0.5})(C)}$$

where N is the complexity of the DNA to be probed, L is the number of nucleotides in the probe, and C is the concentration of probe in moles of nucleotide per liter (M11). N is approximately 3200 for HBV DNA. If a nick-translated probe ($L = 400$ nucleotides) was used at a concentration of 8 nmol/liter, t would be about 11 h.

2.2. Preparation of Specimens

2.2.1. Serum and Other Biological Fluids

The preparation of serum, plasma, or other fluids for dot or Southern blot hybridization involves lysing the viral envelope and nucleocapsid. Deproteinization is optional for dot hybridization samples but essential for samples that are used for Southern blot analysis. Direct spotting of serum on the membrane without deproteinization has the advantages of speed and convenience. The presence of alkali alone is sufficient to release the DNA from the virion and to denature it (S2). The alkaline serum is passed through a membrane filter, which binds DNA and proteins. The addition of Nonidet P-40, a nonionic detergent, to the serum mixture has two effects, lysing the viral envelope and improving the flow of fluids through the membrane filter (H8). 2-Mercaptoethanol added to the medium aids in lysis of the HBsAg coat by reducing its disulfide bonds. Addition of alkali is necessary to separate the DNA strands and for their binding to membranes. The presence of a high salt concentration is important in binding of DNA strands to nitrocellulose membranes, but much less so with nylon membranes.

Several methods for deproteinization have been reported. There are no clear advantages to the use of deproteinized extracts for dot hybridization, unless large samples of serum are to be tested. The disadvantages are the cost of the proteases and the extra time and labor involved in deproteinization. The sample is first incubated with a protease; enzymatic breakdown of all of the proteins must be aimed for to ensure release of the viral DNA. The digest is extracted with phenol and the resulting aqueous solution is made alkaline before application to the membrane (B3, W2). Two modifications of this procedure have been described. Serum may be directly spotted on the membrane, which is then treated with a protease (L7). The enzymatic digest can also be passed through the membrane filter without phenol extraction (M23). The procedures for applying urine, saliva, and semen specimens to membranes for dot hybridization are similar to those for serum, except that the specimens are usually first concentrated (Table 5).

Samples for Southern blot analysis must be deproteinized, using the protease–phenol method outlined above. After concentration of the HBV

TABLE 5
PREPARATION OF URINE, SALIVA, AND SEMEN FOR DNA–DNA HYBRIDIZATION

Specimen	Concentration step (factor)	DNA extraction	Reference
Urine	Ultracentrifugation (100-fold)	Phenol	D9
Urine	Dialysis vs. gel (100-fold)	Protease-phenol	K4
Urine	None	Alkali	D9
Saliva	Dialysis vs. gel	Protease–phenol	K4
Semen	Dialysis vs. gel	Protease–phenol	K4
Spermatozoa	Low-speed centrifugation	Protease–phenol	H3
Seminal fluid	None	Alkali	H3

DNA, the samples are subjected to electrophoresis and the separated bands are transferred to a membrane (S22).

2.2.2. Cells and Solid Tissues

The preparation of samples for dot or blot hybridization from blood cells (S13), cell cultures (E2), and solid tissues (L8, M15, R8) requires protease treatment and phenol extraction. If the objective is solely to detect supercoiled DNA, protease treatment is omitted and hot phenol–chloroform treatment is employed (L18, R8, Y7). Unlike other forms of HBV DNA, supercoiled forms are not attached to protein.

If the aqueous extracts are to be analyzed by Southern blotting they are also treated with RNase.

2.3. PROBES

Four kinds of probes have been developed for the detection of HBV DNA: probes consisting of HBV DNA isolated from plasma, probes obtained by recombinant DNA techniques for both HBV DNA strands or for either strand, and oligonucleotide probes.

2.3.1. Plasma HBV DNA as Probes

The concentration of HBV DNA in infected plasma is between 1 and 10 ng/ml (C12). HBV DNA isolated directly from plasma can be labeled and used directly as probes (L9, S18, S30). Figure 3 shows the flow diagram for the preparation of HBV DNA. Because the starting material is usually blood donor plasma, human DNA must be eliminated. It is degraded by DNase before the viral envelope and nucleocapsid are lysed. The DNase-treated preparation is layered over a deep cushion of 30% sucrose, and

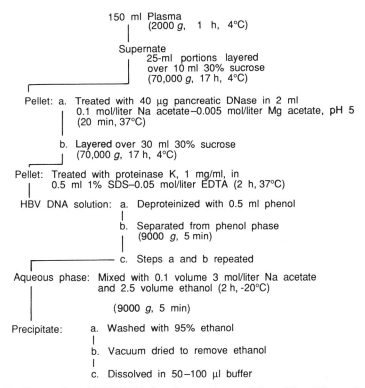

FIG. 3. Preparation of HBV DNA from human plasma. Adapted from Lin *et al.* (L9).

centrifugation effectively separates the HBV from the enzyme. Lysis of the viral envelope and nucleocapsid and deproteinization are carried out by means of proteinase K and phenol treatment.

The HBV DNA present in different carriers may vary considerably in the extent of hybridization with a given sample (L13). Such differences may affect the results of dot hybridization quantitatively but not qualitatively.

2.3.2. *Cloned HBV DNA as Probes*

Among the four kinds of cloning vectors that use *Escherichia coli* as host, bacteriophage λ and plasmids have been used for the production of double-stranded probes, and M13 has been used for single-stranded probes. HBV DNA cloned into λ strains were subsequently subcloned into plasmids (C5, C12, F14). The many plasmid strains employed as cloning vehicles for double-stranded HBV DNA include the commonly used pBR322 (B21).

Double-stranded probes representing various regions of the HBV genome have been devised using restriction enzyme fragments of cloned HBV DNA. Southern blot analysis carried out with specific probes for regions pre-S, S, C, and X revealed different hybridization patterns (M20).

Strand-specific probes are obtained with the M13 strains developed by Messing (H23, M12). In this system, single-stranded recombinant DNA would consist of the (+) strand of plasmid DNA and a sequence from either strand of the inserted DNA. The usefulness of the system was increased by the development of strains containing a series of unique restriction enzyme sites incorporated into the plasmid genome. Different strains carrying the restriction enzyme sites arranged in opposite orientations enable both strands of the insert DNA to be produced in single-stranded form.

As the first step in cloning, HBV may be purified by means of sucrose gradient centrifugation (S29) or isopycnic CsCl centrifugation (F13). Alternatively, the scheme illustrated in Fig. 3 may be used. Repair of the single-stranded gap is carried out prior to restriction enzyme cleavage by means of the "endogenous" DNA polymerase associated with the virion (C5, K3), *E. coli* DNA polymerase I (S29) or T4 DNA polymerase (C12). Inserts of HBV DNA are produced by ligating viral and vector DNAs that have been cut by the same restriction enzyme(s). The HBV subtype usually determines which restriction enzymes to employ (L16).

2.3.3. *Oligonucleotide Probes*

An oligonucleotide probe with the sequence 5'-d(CTTCGCTTCA CCTCTGCACGT) has been developed for detecting HBV DNA in serum (L10). This particular sequence was chosen because it occurs in the DNA of all subtypes cloned so far. It corresponds to the portion of the S strand immediately preceding its 5' fixed end. Since this sequence is located at the end of the single-stranded gap, there would be no competition from viral DNA strands. The sensitivity of the oligonucleotide probe is comparable to that of HBV DNA probes (Fig. 4). Moreover, the time required for hybridization to the oligonucleotide could be reduced to a few hours, as compared to the 16–24 h usually employed for HBV DNA probes.

2.3.4. *Labeling Techniques for Probes*

Table 6 lists the various techniques that have been used to label HBV DNA probes with radioisotopes and the specific activities of the products. Cloned double-stranded HBV DNAs are used as probes with or without removal of the vector DNA sequences. The latter may promote hybridization by the formation of networks (B3). However, vector DNA sequences may occasionally produce false-positive results in dot hybridi-

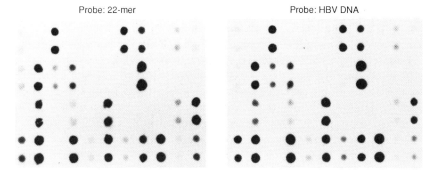

FIG. 4. Autoradiograms obtained with oligonucleotide and HBV DNA probes. Forty-eight serum specimens were applied in duplicate to each membrane. From Lin et al. (L10).

zation tests (D10), or would result in misinterpretation of Southern blots (G10). Current techniques for radiolabeling single-stranded HBV DNA probes depend on the presence of the vector DNA. The HBV DNA insert remains single-stranded and unlabeled, covalently linked to a vector DNA strand that is hydrogen-bonded to a complementary, radiolabeled sequence. Such probes must of course be employed without denaturation of the DNA. In the oligonucleotide-primed synthesis method, the radioisotope is incorporated into sequences of vector DNA by extension of an annealed oligonucleotide (M18). In the hybridization procedure the single-stranded M13 DNA bearing the HBV insert is labeled by annealing with

TABLE 6
METHODS FOR RADIOLABELING HBV DNA PROBES

Probe	Method	Radio-isotope	Specific activity (cpm/μg)	Reference
Probes for both strands				
Plasmid–HBV DNA	6-mer primed synthesis	^{32}P	2×10^9	P6, F4
Cloned HBV DNA	Nick translation	^{32}P	8×10^8	B6
Plasmid–HBV DNA	T4 DNA polymerase–exonuclease	^{3}H	5×10^6	F5
Strand-specific probes				
Plasmid–HBV DNA	Hybridization with radiolabeled plasmid DNA	^{32}P	4×10^8	F9
Plasmid–HBV DNA	17-mer primed synthesis	^{32}P	1×10^8	M18

nick-translated double-stranded M13 DNA (F9). In practice, hybridization tests with any type of probe containing vector DNA sequences are checked by reprobing the samples with vector DNA without the HBV DNA insert.

Biotin can be used to label HBV DNA probes. Nick translation in the presence of biotinylated dUTP results in the substitution of a large proportion of thymidine residues with biotin-labeled uridine (N5). Bound probe is detected by means of avidin complexed to an enzyme such as β-galactosidase and is quantitated by reaction with a suitable substrate (N1, Y8). The sensitivities of probes labeled with biotin and radiophosphorus (using exposure times of 1–3 days for the latter) are approximately comparable (M22, Y8). However, radiolabeled probes have the advantage that signals can be boosted by very long exposure times, whereas biotin labels do not have this flexibility. Furthermore, if nylon membranes are used with radiolabeled probes, they can be reused after the probe is removed by treatment with alkali (L9). Enzymatic staining of the membranes prevents reuse of membranes subjected to biotin-labeled probes (B13).

The limit of detection with radiophosphorus-labeled probes is approximately 0.1–0.5 pg (H2, L7, M7, S2), while that of biotin-labeled probes is about 1 pg (M22, Y8). The DNA content of 10^5 HBV particles is about 0.3 pg.

2.4. Dot Hybridization

Dot or (spot) hybridization is a semiquantitative technique suitable for handling large numbers of specimens. The amount of serum that can be directly applied to membranes is in the range of 5–100 µl (F2, S2). Larger samples can be tested if they are first concentrated by means of ultracentrifugation (F2) or aqueous phase separation with polyethylene glycol (S14). Dot hybridization is suitable for detection of picogram quantities of HBV DNA. A modification of the technique, slot hybridization, enables microgram quantities of purified DNA to be applied (M19).

Quantitation of HBV DNA based on the density of the autoradiogram presents some difficulties. Known quantities of HBV DNA may be applied to the membranes, but it must first be mixed with serum, because purified HBV DNA does not bind quantitatively to nitrocellulose membranes (S2). The relative densities of the sample and standard dots may be measured by densitometer. The linearity of such measurements depends on the optical density of the sample, with faint signals being overestimated and strong signals underestimated. The range over which densitometric measurements are linear is relatively short, 1–30 pg (K1).

Liquid scintillation counting has been employed. Since the amount of ^{32}P necessary to produce an image on X-ray film is less than 50 dpm, counting of such dots would lack precision unless large samples or unacceptably long counting times were used. Cerenkov counting of dots probed with ^{32}P-labeled DNA has been applied to samples representing 500 µl serum (F2). Liquid scintillation counting is also used to measure the samples (representing 1 ml serum) hybridized to an ^{3}H-labeled probe. The sensitivity of the tritium probe is about 10 pg, about two orders of magnitude lower than that obtained with ^{32}P-labeled probes (F5).

2.5. Analysis of HBV DNA Forms

The principal tool for studying the molecular forms of HBV DNA is the Southern blot technique. DNA forms are separated according to size and secondary and tertiary structure by means of gel electrophoresis and are transferred from the gel to a membrane either by blotting (S22) or by electrophoresis. Detection of HBV DNA sequences by probing the blots gives information on the state of the DNA, whether it is integrated or unintegrated ("free"), and on its molecular forms, which may signify an active or suspended state of replication.

Southern blots of ^{32}P-labeled DNA incorporated by HBV DNA polymerase have also been used as a detection method for HBV DNA (I3).

2.5.1. Molecular Forms of HBV DNA

HBV DNA can be found in a variety of forms that differ in the number and kinds of nucleic acid strands present, their lengths, and the secondary and tertiary structures of these molecules. Table 7 lists the various forms in order of increasing electrophoretic mobility. The large number of forms that the viral nucleic acid can assume may make the interpretation of Southern blots difficult. Other properties of the HBV forms, also given in Table 7, may be useful in their characterization.

All of the HBV forms have been detected in HBV-infected human liver (M19), and all of them have been found in the serum of human carriers, with one possible exception. The presence of supercoiled HBV DNA in human serum has been suggested by the transient appearance of the 3.6-kb form resulting from heat denaturation (S3) or digestion with a nuclease capable of introducing a nick in the supercoil (R8). Its occurrence in human serum remains to be firmly established (M17).

2.5.2. Interpretation of Southern Blots

The amount of information on HBV DNA that can be obtained by Southern blot analysis is significantly increased when the effects of treatment with certain restriction enzymes can be observed. Different samples (a, b, and c) of the preparation are (a) left untreated, (b) treated with an

TABLE 7
MOLECULAR FORMS OF HBV DNA[a]

Form	Electrophoretic mobility (kb)[b]	Sensitivity			Buoyant density (g/cm^3)
		Restriction enzymes	Nuclease S1	Heat denaturation	
Relaxed, circular, fully double stranded	3.6	+	+	+	1.55[c]
Circular, partially double stranded	3.5–1.8	+	+	+	
Linear, fully double stranded	3.2	+	−	+	1.42[d]
Supercoiled	2.0	+	+	±[e]	1.59[c]
Full length, single stranded	1.9	−	+	−	
Short, single stranded	<1.9	−	+	−	
RNA–DNA hybrid	3.6–2.3			+	1.45–1.60[d]

[a]Compiled from Miller et al. (M15, M17, M18), Ruiz-Opazo et al. (R8), and Yokosuka et al. (Y5).
[b]Expressed as the length of linear fully double-stranded DNA with the same electrophoretic mobility.
[c]Cesium chloride.
[d]Cesium sulfate.
[e]Relatively more resistant as compared to other double-stranded forms.

enzyme that does not cut HBV DNA, and (c) treated with an enzyme that cuts HBV DNA at a single site (S9). Although *Hin*dIII and *Eco*RI are often used in steps (b) and (c) respectively, two points should be noted. *Hin*dIII is usually employed as the restriction enzyme that does not cut HBV DNAs of any subtype. However, a few HBV DNAs that have been cloned or integrated into cellular DNA are sensitive to this enzyme (F8, K7, P5). Second, *Eco*RI cannot be used for cutting HBV DNA of subtype adr, since the latter does not possess the required site (L16, W10).

If integrated HBV DNA sequences are present, sample (a) will appear as a smear in the high-molecular-weight region (Fig. 5). The patterns of samples (b) and (c) will vary according to the sites of integration into the host genome and the length and arrangement of the HBV DNA insert. In the sample (b) pattern shown on the left side of Fig. 5, the HBV DNA sequences were found mainly in three *Hin*dIII restriction fragments, all longer than 3.2 kb. In another sample of the same specimen that was digested with *Eco*RI, there was a strong band at 3.2 kb, which could be explained by insertion into the host genome of two or more whole viral genomes, joined head to tail in a tandem arrangement. Sample (b) (on right, Fig. 5) showed a different pattern of hybridization. All five of the *Hin*dIII restriction fragments were also longer than 3.2 kb. Sample (c)

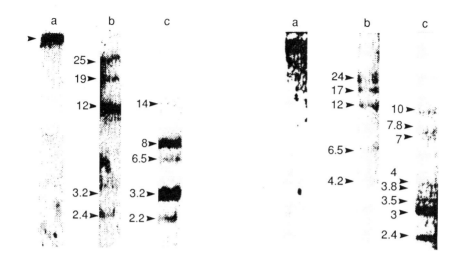

FIG. 5. Demonstration of integrated HBV DNA by means of restriction enzyme digestion and Southern blot analysis. Total cellular DNA was divided into three samples (a, b, and c) that were (a) not treated with any restriction enzyme, (b) treated with HindIII (which does not cut HBV DNA), and (c) cut with EcoRI. Arrows indicate the positions of the fragments hybridized to the HBV DNA probe. Fragment lengths are expressed as kilobases of linear double-stranded DNA. (Left) Pattern compatible with integration of the whole viral genome into host DNA. (Right) Pattern compatible with integration of shorter HBV DNA sequences. Adapted from Brechot et al. (B18).

contained eight bands, the most intense of which were 3.0 and 2.4 kb in length. This pattern was consistent with integration of HBV DNA sequences shorter than 3.2 kb. In other specimens (not shown), integration of short viral sequences could be inferred from the appearance of HindIII fragments shorter than 3.2 kb.

Figure 6 shows Southern blots of unintegrated HBV DNA. Sample (a) (left-hand side) is a typical blot of HBV DNA extracted from serum, a mixture of circular partially double-stranded molecules forming a smear from about 3.5 to 1.8 kb. Conversion to linear partially double-stranded molecules by means of EcoRI digestion [sample (c), left-hand side] did not markedly change the position of the smear. Sample (a) (right-hand side) prepared from infected hepatocyte nuclei, contained supercoiled and relaxed fully double-stranded circular forms at the 2.0 and 3.6 positions, respectively. EcoRI digestion converted both forms to linear double-stranded molecules of 3.2 kb. It follows that supercoiled DNA, if present in a serum sample, would be undetectable by Southern blot analysis, since

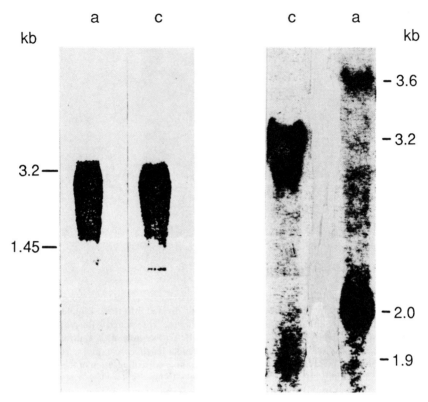

FIG. 6. Southern blots of unintegrated HBV DNA. Labeling of the lanes as given in the legend to Fig. 5; size markers given on both sides. (Left-hand side) HBV DNA extracted from serum: (a) a typical smear produced by undigested HBV DNA, with the pattern (c) only slightly altered by EcoRI digestion. Adapted from Scotto et al. (S3). (Right-hand side) Sample (a) prepared from liver nuclei contained HBV DNA in relaxed circular form (3.6 kb) and in supercoiled form (2.0 kb). EcoRI (c) converted both forms to linear fully double-stranded DNA of approximately 3.2 kb. Adapted from Miller and Robinson (M15).

the smear of partially double-stranded DNA forms that predominates in serum would obscure the presence of the supercoil and its derived forms.

Other patterns, suggesting the presence of HBV DNA in linear monomeric or oligomeric forms, have been observed. Figure 7 (left) shows the simplest pattern. HBV DNA is present as a discrete 3.2-kb band in all three samples. This pattern is consistent with the presence of unintegrated HBV DNA as a linear fully double-stranded monomer. The center panel shows the pattern encountered in Fig. 5 (left), in which (a) the undigested preparation appears as a smear in the high-molecular-weight region, (b) HindIII digestion resulted in the appearance of discrete bands

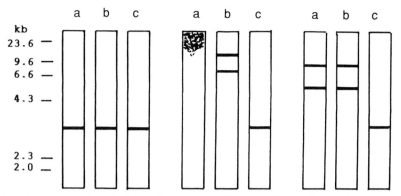

FIG. 7. Monomeric and oligomeric forms of HBV DNA. The labeling of the samples is given in the legend to Fig. 5. (Left) Unintegrated linear monomer. (Center) Integrated oligomeric forms. (Right) Unintegrated oligomeric forms.

that were longer than 3.2 kb, and (c) treatment with an enzyme making a single cut in HBV DNA produced one discrete band at 3.2 kb. As with the example shown in Fig. 5, it would be necessary to postulate the existence of oligomers formed by head-to-tail joining of whole HBV DNA genomes. The third pattern (right panel) consists of discrete high-molecular-weight fragments seen in (a) untreated and (b) *Hin*dIII-digested samples. These forms were converted to a single 3.2-kb form by *Eco*RI digestion (c). This pattern would be explained by the presence of unintegrated HBV DNA oligomers present as linked circles (concatenated form), or genome length sequences joined head to tail (L8).

RNA–DNA hybrids are detected by a combination of pancreatic RNase treatment, Southern blotting, and strand-specific probing (Fig. 8). The electrophoretic pattern of these hybrids is altered by digestion with RNase, with the appearance of additional shorter DNA bands detectable by probes for both strands. Strand-specific probing might furnish confirmatory proof: the DNA in RNA–DNA hybrids would be located at positions <1.9 kb and it would hybridize exclusively with probes specific for the L strand (M19).

2.6. Significance of HBV DNA in HBV-Related Liver Disease

HBV DNA has replaced HBeAg as the "gold" standard for the detection of active HBV replication (B13). In general, serum HBV DNA accurately reflects hepatic production of the virus. Paired serum and liver DNA specimens showed similar Southern blot patterns (Y7).

The various molecular forms of HBV DNA have come to be associated

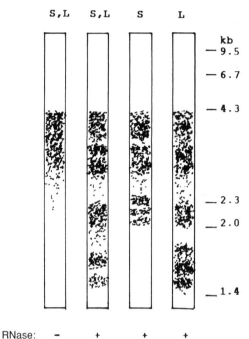

FIG. 8. Detection of RNA–DNA hybrids. The two lanes on the left were probed for both strands of HBV DNA. Probes specific for the S and L strands were used on the third and fourth lanes. RNA–DNA hybrids are sensitive to RNase digestion, and their DNA consists entirely of L strands. Adapted from Miller et al. (M19).

with different states of replication activity. The mixture of circular partially double-stranded DNA, RNA–DNA hybrids, and unpaired L strands which form the smear seen in Southern blots (Fig. 6, left-hand side) are replicative intermediates. Although supercoiled DNA is also a replicative intermediate, the finding of the supercoiled form in the absence of the other HBV DNA forms may signify a dormant state or "inactive replication." Yokusuka and co-workers showed by Southern blot analysis that free hepatic HBV DNA in chronic hepatitis patients could consist of supercoiled forms appearing either in the absence or presence of other replicative intermediates (Y5). Most patients with serum HBeAg showed the latter pattern, while supercoiled DNA was the dominant form in the majority of HBeAg-negative cases. Furthermore, interferon treatment in some patients resulted in the disappearance of all but the supercoiled form (Y7). In another study on hepatic HBV DNA in HBsAg carriers with minimal histological changes in the liver, integrated and 3.2-kb forms were found in some patients (K1). The analysis of hepatic HBV DNA forms appears to provide closer correlation with liver disease activity than could be provided by serological markers (S8).

2.6.1. Correlation of Serum HBV DNA with Serological Markers

The serial changes in HBV DNA, AST, and serological markers in a case of acute hepatitis are illustrated in Fig. 9. Serum HBsAg and HBeAg appeared first, followed shortly by HBV DNA, and then by rises in AST and IgM anti-HBc. Recovery was indicated by the disappearance of HBV DNA and eventual seroconversion to anti-HBe and anti-HBs. About 46% of the HBeAg-positive patients admitted to hospital with acute hepatitis B were found to be HBV DNA positive (K9). The relatively low frequency in this group of patients could be explained by the fact that the majority of them recovered from HBV infection. Serum HBV DNA was cleared before HBeAg, as shown in Fig. 9.

Persistence of serum HBV DNA for more than 8 weeks following the appearance of symptoms may be a sign of chronic infection (K9). In the chronic carrier state, there is a close association between serum HBV DNA and HBeAg in relatively healthy HBsAg carriers, such as blood donors (Table 8). However, this association does not always obtain in patients with liver disease. A proportion of HBeAg-positive patients may have no detectable serum HBV DNA. In the large series of patients studied by Matsuyama *et al.* (M7), this type of discordance was found to be absent in patients with no specific changes in liver histology, but increased progressively with CPH, CAH, and HCC. There is a second type of discordance, namely the HBV DNA positivity of patients who seroconvert to anti-HBe. As shown in Table 8, this anomaly was infrequent among blood donors in the United Kingdom (H8), but was found in one out of six donors in Taiwan (C8) and in a high proportion of patients with HBV-related conditions, chronic hepatitis, cirrhosis, and HCC. Approximately 15–55% of the patient serum specimens with anti-HBe contained HBV DNA. The HBe/anti-HBe system is not as accurate as serum HBV DNA as a marker for HBV replication (T5).

Given the positive (if loose) association of HBeAg with HBV DNA, the degree of correlation of anti-HBc and anti-HBs with the latter should only be examined in HBeAg-negative cases. Among 138 such anti-HBc-positive cases with acute hepatitis or chronic liver disease, only a small percentage was positive for serum HBV DNA (L18, M4, V2). The presence or absence of anti-HBs in the same serum had no significant effect. In seven HBeAg-negative, anti-HBc-positive HCC patients, however, all had serum HBV DNA (C7). This observation may be a further example of

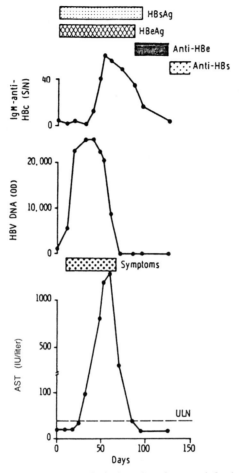

FIG. 9. HBV DNA, AST, and serological markers in acute infection. From Thomas et al. (T5). S/N, Ratio of cpm, specimen to negative control; OD, optical density.

the discordance of serum HBV DNA with serological markers in advanced liver disease.

2.6.2. Hepatic HBV DNA in HBV-Related Disease

Hepatic HBV DNA may be found free or integrated into cellular DNA in the liver. Both types of HBV DNA sequences may be transcribed into mRNAs and expressed. The significance of free hepatic HBV DNA was brought out in a study of liver biopsy specimens obtained from patients with chronic liver diseases (Y6). Those in whom hepatic free HBV DNA and mRNAs for both HBcAg and HBsAg were detected had active liver

TABLE 8
CORRELATION OF SERUM HBV DNA WITH HBeAg AND ANTI-HBe[a]

No. of specimens	HBe	Anti-HBe	HBV DNA positive (%)	Subjects	Reference
39	+		95	HBsAg-positive blood	H8
161		+	2	donors	
12	−	−	8		
56	+		89	HBsAg-positive blood	C8
24		+	17	donors	
9	−	−	11		
12	+		100	HBsAg carriers, 77% with	B14
17		+	54	CPH/CAH/cirrhosis	
1	−	−	0		
28	+		100	HBsAg carriers, many	L7
32		+	50	with CPH/CAH/ cirrhosis	
9	+		100	HBsAg carriers, 82% with	H4
26		+	45	CPH/CAH/cirrhosis/ HCC	
22	+		91	Patients with HBV	V2
46		+	13	markers and liver diseases	
446	+		82	HBsAg carriers, 40% with	M7
551		+	24	CPH/CAH/cirrhosis	
66	−	−	14		
109	+		77	Patients with HBV	W9
78		+	18	markers, 65% with	
17	−	−	47	acute HBV infection, CPH/CAH/cirrhosis/ HCC	
117	+		68	Patients with HBV	S2
26		+	27	markers or liver disease	
19	−	−	0		
9	+		67	HBsAg-positive patients	T11
151	−		6	28% with acute HBV infection, CAH/ cirrhosis/HCC	

[a]Abbreviations: CAH, chronic acute hepatitis; CPH, chronic persistent hepatitis; HCC, hepatocellular carcinoma.

disease. They were all seropositive for HBeAg. In contrast, patients who had no free hepatic HBV DNA and had only hepatic mRNA for HBsAg showed fewer histological signs of liver disease. Most members of the second group were seropositive for anti-HBe and a few had integrated hepatic HBV DNA.

The association of HBeAg positivity with the presence of free hepatic HBV DNA has been shown in other studies (Table 9). A large majority of HBeAg-positive patients had free hepatic HBV DNA, which was also found in HBeAg-negative cases but with significantly lower frequency. Integrated HBV DNA, on the other hand, was not associated with the HBeAg status. It was found in some patients but with no clearcut bias with respect to HBeAg seropositivity or -negativity, nor for any disease category.

In line with the observations on hepatic HBV DNA in HCC patients as shown in Table 9, Imazeki *et al.* (I2) found HBV DNA integrated into HCC tissue DNA in only 12 out of 34 cases. None of the tumors contained free HBV DNA. Clearly, the presence of free or integrated HBV DNA is unnecessary for the maintenance of the tumourous condition (F12).

Integration of HBV DNA into liver DNA can occur at an early stage in the progression of hepatitis to liver cirrhosis and HCC. Lugassy *et al.* found integrated HBV in 2 of 22 cases of acute benign hepatitis and in 7 of 26 patients with fulminant hepatitis (L18). Among the 48 cases, free hepatic HBV DNA was found in 9 patients who had either form of hepatitis. It is possible that the duration of infection with HBV would be positively related to integration of its DNA (S10).

2.6.3. *Characterization of Integrated HBV DNA in Liver*

Characterization of integrated HBV DNA has revealed some interesting features but it has not shed much light on the possible connection between HBV and oncogenesis. The role of integrated HBV DNA sequences in the development of HCC is not clear. Some inserted HBV DNA sequences are expressed. HBV-specific mRNAs have been detected in cell culture systems (C4, E2, P7) and in hepatoma tissues (Y2). Transformation of mouse cells with cloned HBV DNA resulted in integration of the latter into cellular DNA, and production of HBsAg particles (D11) and HBsAg and HBeAg (G7). The production of HBsAg in some human carriers with no free hepatic HBV DNA might, then, be explained by the integration in liver of fragments containing gene S.

There are a multiplicity of sites on human chromosomes at which HBV DNA sequences have been inserted. In the PLC/PRF/5 cell line, derived from a human hepatoma, inserts were found on chromosomes 11, 15, and 18 (B16) and chromosomes 1, 2, 16, and Y in another study (S11).

The host DNA regions flanking the viral DNA insert vary. They may be satellite sequences (S12), or coding regions that are expressed in a fusion protein (K7). Although much work has been done searching for oncogenes in tumors with integrated HBV DNA, no positive evidence was found (K6, S16, V3). However, in one specimen of hepatocellular

TABLE 9
LIVER HBV DNA IN HBV-RELATED STATES

Group	HBV DNA[a] Free	HBV DNA[a] Integrated	Reference
HBeAg-positive cases			
Asymptomatic HBsAg carriers	4/4	1/4	K1
CPH	4/4	1/4	F10
	3/3	0/3	B19
	2/2	—	B20
	10/10	3/10	S4
	15/18[b]	1/18[b]	F12
CAH	11/11	0/11	F10
	11/11	1/11	B19
	2/2	1/2	S4
HCC	1/1	1/1	B20
	2/2	2/2	B19
	3/4	4/4	H2
	0/1	1/1	S9
	1/1	1/1	C7
	4/9	2/9	F12
HBeAg-negative cases			
Asymptomatic HBsAg carriers	4/10	6/10	K1
CPH	0/1	0/1	F10
	0/3	—	B19
	0/1	0/1	S4
	8/36[b]	3/36[b]	F12
CAH	0/4	1/4	F10
	1/8	—	B19
HCC	0/2	0/2	F10
	0/6	0/6	B20
	2/10	3/10	H2
	0/11	6/11	S9
	7/7	4/7	C7
	2/19	9/19	F12

[a] Number positive/number tested.
[b] Chronic hepatitis cases.

carcinoma, integrated HBV DNA was flanked by sequences analogous to the v-*erb*-A oncogene and steroid receptor genes (D4). It is possible that some HBV DNA inserts are oncogenic while others are not.

Integration takes place at both single- and double-stranded regions of the HBV genome (Table 10). The sequences around viral–human DNA junctions in the various clones indicated that both single- and double-stranded regions of the viral genome have been inserted with comparable

TABLE 10
LOCATION ON THE HBV GENOME OF SEQUENCES AT THE JUNCTION SITES

Cells/tissues	Clone	S-strand insert		Reference
		5' end	3' end	
PLC/PRF/5	8	ds[a]	ds	Z3
	13	ss	ss	
	14	ds	vr	
	15	—	vr	
	23	ds	ds	
	26	ds	ss (DR)	
	27	ds	ds	
PLC/PRF/5	A-10.7	—	ss (DR)	K6–K8
	A-10.5	—	ss (DR)	
	A-6.0	ss	ds	
	A-4.0	—	ss	
huSP	hu-489	ds	ds	M21
HCC	DT1	ss (DR)	ds	D5
	1A22	ds (DR)	ds	
huH2-2	—	ds	ds	Y1

[a] Abbreviations: ss, single stranded; ds, double stranded; vr, variable (single or double stranded) region; DR, direct repeat sequence 5'-TTCACCTCTGC (D5) at terminus.

frequency. It is notable that the direct repeat (DR) sequence 5'-TTCACCTCTGC has been found in five clones, in part or in entirety, both at the 5' and 3' ends of the inserts. The DR sequence, strategically located on the S strand near the 5' fixed ends of the two strands (Fig. 2), may be favored as initiation points for integration (D5, T7).

The inserted sequences often contained deletions in the pre-S region (Z3). Deletions occur frequently in the inserted sequences. Rearrangements of the viral sequences may take the form of translocations, inverted duplications, and, in some clones, foldback structures (K6, M21, S12, Z3).

2.7. HBV DNA IN EXTRAHEPATIC SITES

The finding of HBV DNA in extrahepatic sites might be explained by deposition of immune complexes or by contamination with blood containing HBV DNA. Formation of immune complexes in HBV infection can result in accumulation of HBsAg in extrahepatic sites, particularly lymph nodes, spleen, blood vessel intima, and renal glomeruli (N13). Presumably, the presence of HBV DNA in some tissue specimens could be explained by formation of immune complexes with Dane particles. Positive evidence for the occurrence of extrahepatic HBV infection was demonstrated in two ways. Southern blot analysis has revealed integration of HBV DNA into cellular DNA in several tissues other than liver. Second, HBV replication or synthesis of HBV constituents has been demonstrated in some extrahepatic human and animal tissues. By these criteria, human spermatozoa, leukocytes, bone marrow, skin, kidney, pancreas, and placenta have been shown to be occasional sites of extrahepatic HBV infection.

2.7.1. Saliva, Seminal Fluid, and Urine

The detection of HBV DNA in saliva is complicated by the fact that specimens may be contaminated with blood. Some saliva specimens apparently free of blood were HBV DNA positive (F1, K4). The molecular form of HBV DNA in one saliva specimen was examined by Southern blot technique and unintegrated oligomeric HBV DNA forms were observed (F1). These were similar to some of the forms found in leukocyte HBV DNA (Section 2.7.2).

HBV DNA has also been detected in seminal fluid (K4) and in spermatozoa, in which the viral DNA was in integrated form (H3). Although the levels of HBV DNA in saliva and semen were well below those in serum, there were implications in regard to the mode of HBV infection in man, since these fluids have been shown to transmit HBV in primates (A6). HBV DNA has occasionally been detected in urine but it is not considered a major source of infection (D9).

2.7.2. Leukocytes

The occurrence of HBV DNA in leukocytes is well documented. There are several unusual features about leukocyte HBV DNA. Its occurrence did not correlate closely with the presence of HBV DNA in serum, as shown by the frequent finding of HBV DNA in leukocytes from blood specimens with no serum HBV DNA (Table 11).

Second, among the standard serological markers of HBV infection, anti-HBc and HBsAg were most closely associated with the occurrence of leukocyte HBV DNA (Table 12). Nearly all the leukocytes that contained HBV DNA were obtained in anti-HBc-positive blood specimens. In contrast, only half or less of the leukocyte preparations were obtained from blood positive for HBeAg, anti-HBe, or anti-HBs.

TABLE 11
FREQUENCY OF HBV DNA IN LEUKOCYTES AS COMPARED TO SERUM

Subjects	HBV DNA[a] Leukocytes	Serum	Reference
HCC patients	2/11	—	L8
Controls	0/21	—	
HBV-infected patients	8/16	2/16	P5
Controls	0/21	0/21	
HBsAg carriers	2/5	0/4	H16
Controls	0/6	0/6	
Hepatitis patients	14/50	22/49	G10
Controls	0/13	0/13	
HBsAg carriers	8/14	6/14	Y4
Controls	0/12	0/12	
HBsAg carriers; hepatitis patients	7/13	0/12	P6
Controls	2/9	1/9	
HBsAg carriers; patients with hepatitis, HCC, cirrhosis	24/70	19/70	S13
Controls	0/23	1/23	
HBsAg-positive and -negative cases, with hepatitis, HCC	42/72	14/76	P2

[a] Number positive/number tested.

TABLE 12
CORRELATION OF LEUKOCYTE HBV DNA WITH SEROLOGICAL MARKERS

HBV DNA-positive leukocytes (no. specimens)	HBV DNA[a]					
	Anti-HBc	HBsAg	HBeAg	Anti-HBe	Anti-HBs	Reference
2	2/2	2/2	NT[b]	NT	2/2	L8
8	8/8	8/8	4/8	5/8	0/8	P5
2	NT	2/2	0/2	2/2	NT	H16
14	ID[c]	13/13	11/13	ID	NT	G10
8	NT	8/8	5/8	NT	NT	Y4
7	6/6	6/7	1/7	5/7	3/6	P6
24	22/22	18/24	12/24	9/24	8/24	S13
42	6/7	30/42	ID	3/7	2/7	P2
Totals (%)	44/45 (98)	94/106 (89)	33/62 (53)	24/48 (50)	15/47 (32)	

[a] Number positive/number tested.
[b] Not tested.
[c] Insufficient data.

Third, the molecular forms of HBV DNA found in many leukocyte preparations were different from those usually seen in serum and liver. The existence of linear genome-length HBV DNA in the absence of partially single-stranded forms has been recorded (G10, H16, P5, P6). Leukocytes may also contain unintegrated oligomers of HBV DNA, possibly in concatenated form or joined head to tail (L8, S13, Y4). The latter may also be present in some preparations as integrated DNA (P5). In addition to these forms, smears in the 3.2-kb or shorter length range resembling the smears seen with serum or liver HBV DNA have been observed in some leukocyte preparations (G10, Y4).

It has been proposed that HBV DNA that is found in leukocytes is the result of phagocytosis of the virus or of HBV-infected cells. Polymorphonuclear cells distinctly contain more HBV DNA than do monocytes (H16, P6). Within a given blood specimen, B cell lymphocytes appear to contain more HBV DNA as compared to T cells (P2, Y4). The finding of HBV DNA in lymphoid cells suggested a possible connection with the immunological disorders in HBV infection (P2, P5). It was postulated that HBV infection of leukocytes and replication therein may interfere with their function, but as yet there has been no direct demonstration of such an effect.

2.7.3 *Bone Marrow*

Bone marrow may be a possible site for HBV replication. *In vitro* infection of bone marrow cells with HBV has been demonstrated (Z1). A culture of bone marrow obtained from an HBV-infected patient contained a mixture of supercoiled and relaxed circular fully double-stranded forms of HBV DNA, with no detectable integrated sequences (E3); intracellular particles resembling Dane particles together with HBcAg and HBsAg were detected in some cells (R7).

2.7.4. *HBV DNA in Other Cells and Tissues*

Table 13 summarizes other reports of HBV DNA in extrahepatic sites. Despite the small number of patients studied, some tentative conclusions may be drawn. In the study carried out by Dejean *et al.* (D6), HBV DNA was found in some specimens of skin, kidney, and pancreas of a carrier. The carrier's serum contained no HBV DNA, eliminating the possibility of serum contamination. The HBV DNA was integrated into the cellular DNA present in all three tissues, an observation that could not be explained by deposition of immune complexes. Experiments with another member of the hepadna viruses, duck hepatitis B virus, have demonstrated replicating forms of HBV DNA in the kidney and pancreas of infected ducks (H5). The relevance of these findings to clinical conditions is that HBV

TABLE 13
HBV DNA IN EXTRAHEPATIC CELLS AND TISSUES

Cells or tissue	HBV DNA form	No. positive/ no. specimens	Reference
Blood vessels			
Endothelial cells	—	3/3	B5
Smooth muscle cells	—	3/3	B5
Bile duct epithelium	—	3/3	B5
Heart	Free	1/2	D6
Brain	Free	1/2	D6
Lung	Free	1/2	D6
Intestine	Free	1/1	D6
Skin	Free/integrated	2/2	D6
Kaposi sarcoma (of skin)	Relaxed circle	1/1	S17
Pancreas	Integrated	1/1	D6
Kidney	Free/integrated	2/2	D6
	Free/integrated	1/1	N3
Placenta	Free/integrated	4/4	N3

infection of kidney has been implicated in membranous glomerular nephritis. While some cases had glomerular deposits of immune complexes containing HBeAg (H12) or HBsAg (H22), no such changes could be demonstrated in other HBsAg-positive patients.

The occurrence of integrated HBV DNA in placenta suggests transplacental infection as a possible source of vertical transmission, in addition to the oral route during delivery (L2). It should be noted that for all practical purposes serum HBV DNA assays are of greater value in the investigation of perinatal transmission from mother to child (D8). High levels of maternal serum HBV DNA were associated with babies infected despite immunization, and infants with detectable HBV DNA were almost all infected, as shown by the presence of HBsAg (L4).

3. HBV Enzymes

3.1. HBV DNA POLYMERASE ASSAYS

The term HBV DNA polymerase refers to the enzyme contained within Dane particles. A soluble DNA polymerase in the serum of HBsAg carriers, with properties different from Dane particle DNA polymerase, has also been characterized (M6). Further studies are required to establish the significance of the soluble enzyme.

All assays of HBV DNA polymerase call for the four deoxyribonucleoside triphosphates, at least one of which is radiolabeled with ^{3}H or

^{32}P. Enzyme activity is decreased with 5'-^{125}I-labeled dCTP (L12). Incorporation proceeds at a slow rate, even with suitable substrates. The amount of radioactivity incorporated is usually 0.1–5% of input, depending on the concentrations of all four nucleoside triphosphates used.

It is necessary to demonstrate the specificity of the assayed DNA polymerase activity because serum and other biological specimens may contain enzymes of cellular or bacterial origin, with similar activities. Two approaches have been used successfully. Serological tests have been used to show the association of DNA polymerase activity with HBsAg. Second, the biochemical properties of the viral enzyme that distinguish it from mammalian DNA polymerases have been exploited.

3.1.1. *Assays with Serological Tests of Specificity*

The assay of HBV DNA polymerase was first described by Kaplan and his co-workers (K3). Dane particles are concentrated by ultracentrifugation; typically, this is performed for 4 h at 66,000 g (H19). HBV may also be concentrated in an air-driven centrifuge for a shorter period of time (M13). The resulting pellet is suspended in 0.05–0.1 of the original sample volume and is incubated with Nonidet P-40 and radiolabeled deoxynucleoside triphosphates in a suitable buffer. After incubation the DNA is precipitated by means of trichloracetic acid and unincorporated precursors are removed by washing.

It was subsequently recommended that DNA polymerase activity be assayed before and after addition of appropriate antiserum, to test for the specificity of the assay for the viral enzyme (R3). Preparations containing HBV DNA polymerase would be expected to react positively in the serological test for HBsAg prior to lysis of the viral envelope, and to give a positive reaction with anti-HBc after lysis (R3, R5). Although many studies are carried out without such tests, their use is recommended, because pelleted fractions containing Dane particles may show DNA polymerase activity even before the step of nucleocapsid lysis, suggesting contamination with non-HBV enzymes (B17, H15, K3, L11, L17). Table 14 lists

TABLE 14
SEROLOGICAL TESTS USED IN THE HBV DNA POLYMERASE ASSAY

Precipitation test	Criterion for specificity	Reference
Anti-HBs	>30% enzymatic activity in particles precipitated	H19
Anti-HBs	>50% enzymatic activity in particles precipitated	F3, N2, C1
Anti-HBs	Enzymatic activity in particles precipitated before detergent treatment	R3
Anti-HBc	Enzyme precipitated after detergent treatment	R3

the serological procedures and criteria for specificity, which vary in their degree of stringency.

3.1.2. Differential Assays

These assays are based upon the differences in the properties of HBV DNA polymerase and the DNA polymerases present in human serum. Table 15 summarizes some of the properties of the viral enzyme. Comparison of its properties with those of mammalian DNA polymerases shows several differences. An unusual characteristic of HBV DNA polymerase is its inability to use a template other than its own circular DNA. The enzyme is unable to use linear HBV DNA molecules as template (R6). In contrast, the DNA polymerase activity present in human serum is stimulated by the addition of denatured DNA regardless of its source (H15). Human and HBV DNA polymerases differ also with their sensitivity to potassium ion. High concentrations have no effect on the viral enzyme, whereas human DNA polymerase activity is largely, but not completely,

TABLE 15
FACTORS AFFECTING HBV DNA POLYMERASE ACTIVITY

Factor(s)	Reference
pH	
Optimum: 7.5	K3
7.2–8.0	H7
Temperature	
Optimum at 37–41°C	K3
50% inactivation in 20 min at 60°C	N2
Template	
Absolute requirement for circular HBV DNA	R6
Cations	
Activated by 0.02–0.12 mol/liter Mg^{2+}; active in 0.4 mol/liter KCl	H7, H15
Inhibitors	
Phosphonoformate	H7, N12
Actinomycin D, daunomycin	K3
N-ethyl- or N-methylmaleimide	H7
Poly(dAT)	K3
Dideoxythymidine triphosphate	H7
Arabinosylnucleoside triphosphates	H6, H7
Ethidium bromide	H19
Detergents	
Active in 10% Nonidet P-40; inhibited by 0.1% sodium dodecyl sulfate	H15
Inactivators	
10 min in 70% ethanol; 1 min with 2500 ppm available chlorine	N2

inhibited. Third, the viral enzyme is more sensitive to phosphonoformate inhibition than is mammalian DNA polymerases (E4, L11).

Hirschman and co-workers developed a differential assay based on the resistance of HBV DNA polymerase to high KCl concentrations, its activation in 0.04 mol/liter $MgCl_2$, and the inhibitory effect of these conditions on other types of polymerases (H13–H15). Such tests are applicable to Dane particles prepared by ultracentrifugation from serum and saliva (M1).

It was later shown that high potassium and magnesium concentrations did not completely suppress the human DNA polymerases in all specimens. Therefore, a different assay was developed based on phosphonoformate inhibition of HBV DNA polymerase activity. DNA polymerase activity was measured in replicate samples with and without phosphonoformate. The difference in nucleotide incorporation in the absence of phosphonoformate and that in its presence represented HBV DNA polymerase activity. The phosphonoformate inhibition assay removed the need for performing serological tests of specificity. The original method required ultracentrifugation of 6-ml samples of plasma or serum (L11). It was replaced with a micromethod using 120 ul serum that, unlike other assays of HBV DNA polymerase, did not require concentration of Dane particles by ultracentrifugation (L12).

Phosphonoformate inhibits not only HBV DNA polymerase, but the DNA polymerases of woodchuck hepatitis B virus (H7), Epstein–Barr virus (D2), herpes simplex virus, and cytomegalovirus (E4, H10). The specificity of the assay in human serum for HBV DNA polymerase was in part attributable to the fact that infection of man with any of the latter three viruses did not result in their appearance in blood to any measurable extent. Phosphonoformate acts by binding to the DNA, blocking the further incorporation of nucleotides (H11), and inhibition is of the noncompetitive type (N11).

3.2. HBV DNA POLYMERASE ACTIVITY IN HBV-RELATED DISEASE

Serum HBV DNA polymerase activity appears to fluctuate spontaneously (H19). Peaks and troughs were observed in chronic hepatitis patients tested at short intervals over 2 days, with coefficients of variation ranging from 20 to 50% (A2). Polymerase activity followed in other patients over a period of weeks or months also revealed bursts of activity (S5). Spontaneous loss of both HBV DNA polymerase activity and HBeAg was recorded in 4 out of 10 CAH patients receiving placebo in a clinical trial (S1).

In acute infection HBV DNA polymerase peaks before the transaminases and, like the changes observed with HBV DNA and HBeAg, its activity falls with recovery. Persistence of DNA polymerase after the acute phase of HBV infection is prognostic of the chronic state (A1).

HBV DNA polymerase activity usually parallels serum HBV DNA and HBeAg. Correlation of DNA polymerase activity with HBeAg in a group of HBsAg carriers with and without symptoms of hepatitis was close: 21 of 22 with HBeAg were DNA polymerase positive, whereas none of the 12 HBeAg-negative specimens showed the enzymatic activity (N11). In another study, chronic hepatitis cases with neither HBV DNA polymerase nor HBeAg were compared to those with both markers (A8). The former group had more severe liver disease, as shown by clinical symptoms and biochemical tests, and the duration of their HBsAg carrier state was longer. HBV DNA polymerase activity was detected more often in CPH patients (14 of 15) than in CAH patients (4 of 8) (T9). Since both the enzyme and HBV DNA are markers for the complete virion, these findings were in accord with the reports mentioned above on the absence of serum HBV DNA in many patients with severe liver disease (Section 2.6.1).

Serum HBV DNA polymerase is detected in a small percentage of patients with HCC. It was found in only 2 of 72 (3%) Korean patients (C9) and in 12 of 106 (11%) South African black cases (S21).

The usefulness of the assay in monitoring the efficacy of several drugs was demonstrated by many reports. Rapid changes in HBV replication were seen in serum HBV DNA polymerase and HBV DNA (W3), but not by following HBeAg (C11). Figure 10 illustrates the decrease in DNA polymerase in a subject treated with three courses of interferon. Similarly, dramatic decreases in DNA polymerase activity have been reported in other trials (O4, S5). In contrast, serum HBsAg and HBcAg decreased slowly. Adenine arabinoside produced similarly rapid decreases in DNA polymerase activity with persistence of serological markers (C3, C11).

3.3. HBV Protein Kinase

Protein kinase activity copurifies with HBV core particles. Unlike HBV DNA polymerase, which is found principally in cores with higher density, protein kinase activities were found in both light and heavy cores, the lighter fraction having more activity (G4).

HBV protein kinase preferentially transfers the γ-phosphate residue of ATP to the HBV core protein. Only the serine residues in core protein are phosphorylated. Enzymatic activity is detected by incubating core protein with ^{32}P-labeled ATP in the presence of Nonidet P-40, followed by gel electrophoresis of the reaction mixture and autoradiography. The

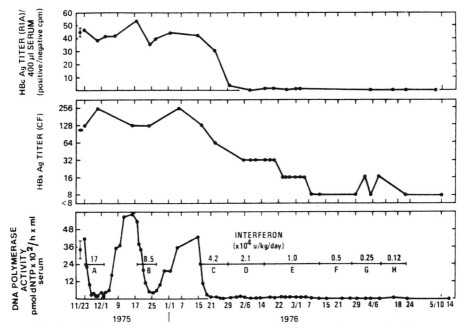

FIG. 10. DNA polymerase, HBsAg, and HBcAg levels in a patient treated with three courses of interferon. From Greenberg et al. (G9).

appearance of radioactivity at the 22-kDa position signifies phosphorylation of the core protein.

HBV protein kinase activities have been found in HBV core particles present in HBV-infected liver and in Dane particles isolated from human plasma. HBsAg particles present in plasma do not contain protein kinase activity (A5).

4. HBV Polypeptides

4.1. PRE-S REGION POLYPEPTIDES

Aside from HBV DNA polymerase, HBsAg, HBcAg, and HBeAg, there are several other polypeptides encoded by the viral genome. They are grouped as the HBsAg-related polypeptides and the region X polypeptide(s). Table 16 lists the HBsAg-related group, consisting of four polypeptides, including HBsAg, and their glycosylated derivatives. Their amino

TABLE 16
HBsAg-Related Polypeptides[a]

Polypeptide	Description (encoding regions)	Occurrence			Reference
		HBV	22-nm spheres	Tubules	
GP46	Minor constituent (pre-S, S)	+			W6
P39/GP42	Minor constituents (pre-S1, pre-S2, S)	+	−	+	H9, T2, T4
P31/GP35 (GP33/GP36)[b]	Minor constituents, pHSA receptor (pre-S2, S)	+[c]	+[c]	−	M2, T2
P24/GP27	HBsAg (S)	+	+	+	H9
P22[d]	No pHSA receptor activity (S)				M3
P8[d]	pHSA receptor (pre-S2)				M3

[a]Abbreviations: P, nonglycosylated polypeptides; GP, glycosylated polypeptides; pHSA, polymerized human serum albumin. Numbers denote molecular weight in kilodaltons. Symbols: /, related polypeptides; +, present; −, absent or scanty.
[b]Polypeptides designated GP33/GP36 (H9, S25) correspond to P31/P35 (O3).
[c]Present in HBeAg-positive serum, scanty in anti-HBe-positive serum.
[d]Derived from P31 by cyanogen bromide treatment.

acid sequences are encoded by the pre-S1, pre-S2, and S regions of the genome. Consequently, there is homology between them, demonstrable by cross-reactivity with antiserum (H9, W6). Polypeptides with pre-S2-encoded sequences possess polyalbumin receptor activity. All the polypeptides are found as constituents of Dane particles, and some are present in small spheres or tubules of HBsAg as well. It is interesting that P31/GP35 in Dane particles and small HBsAg spheres were detected only in HBeAg-positive serum (H9, M2, T2). These polypeptides correspond to GP33/GP36 described elsewhere, that were likewise increased in small HBsAg spheres isolated from HBeAg-positive serum, as compared to those present in anti-HBe serum (S25). The observed changes in polypeptide composition of some HBV constituents imply some kind of biological control or response in the different stages of HBV infection.

4.1.1. *Polyalbumin Binding/Receptor Assays*

The factors that affect polyalbumin binding assays are pH, temperature, ionic strength, and the polymerization method for albumin. The optimal pH and temperature for species-specific polyalbumin binding are 6.0–6.7

and 37°C, respectively. The species specificity of binding of human and chimpanzee polyalbumin to HBV forms was abolished at NaCl concentrations below 0.15 mol/liter; at concentrations above 0.5 mol/liter, binding to albumin from any species was abolished (M14, P3). All polyalbumin binding assays employ albumin that is prepared by cross-linking with glutaraldehyde. Polyalbumin that was cross-linked by treatment with carbodiimide, or cross-linked polyalbumin produced by aging of albumin solutions, did not bind as efficiently in the assay (Y9).

Assay of polyalbumin-binding sites is performed by hemagglutination or by immunoassay (Table 17). Test I was the first to be developed. It had the disadvantage that positive specimens had to be subjected to further assays to exclude the possibility of interference from antibodies to albumin (P4). Tests III and IV utilized anti-HBs, increasing the specificity of reaction for HBV forms, but like the first test depended on the interaction of pHSA and HBV forms. An antibody which precipitates HBV, but is not anti-pHSA, has been shown to react in this type of assay (A4). In contrast, tests II and V dispense with pHSA. They detect receptors directly by means of antibodies directed at sequences of 19 and 55 amino acids, respectively, which are encoded by the pre-S2 region.

4.1.2. *Polyalbumin Receptors in HBV Infection*

Polyalbumin assays may provide an additional marker for assessing the status of HBV infection or predicting its outcome, but their usefulness

TABLE 17
POLYALBUMIN BINDING/RECEPTOR ASSAYS

Assay (type)	Materials	Reference
Hemagglutination tests		
I	(Binding) sheep erythrocytes coated with pHSA[a]	I1, P4
II	(Receptor) sheep erythrocytes coated with monoclonal antibody directed against 19-amino acid sequence	O2
Immunoassays		
III	(Binding) pHSA-coated beads; detection with ^{125}I-labeled anti-HBs	M5, M14, N7
IV	(Binding) anti-HBs-coated wells; detection with pHSA linked to horseradish peroxidase	T2
V	(Receptor) wells coated with antibody directed against a 55-amino acid sequence	M3

[a]Human serum albumin polymerized by means of glutaraldehyde.

has not been proved. They appear to furnish little information that is missed by assays of HBV DNA and serological markers. Because the binding sites are located on the surfaces of HBV forms, the titers generally reflect their density per unit volume, without differentiating between the complete virion and the tubular or small spherical particles of HBsAg.

As a whole, HBeAg-positive serum specimens have higher pHSA-binding titers than do HBeAg-negative specimens or sera with anti-HBe (G5, I1, M5). However, there was considerable overlap with respect to the titers. About half of the HBeAg-positive serum specimens had titers within the range set by titers in the HBeAg-negative specimens (M5). In acute HBV infection, patients who recovered had initial pHSA-binding titers that were significantly lower than did patients who developed chronic infection (P4). The pHSA-binding titers were high during the HBeAg-positive period of acute infection and were significantly reduced at the time of recovery or seroconversion to anti-HBe (A3, P4). In chronic HBV infection, pHSA-binding titers in patients with CPH, CAH, or cirrhosis were in many cases within the range seen in asymptomatic carriers, indicating that this type of assay would not be useful in differentiating between the various disease states (P4).

4.1.3. Polyalbumin Receptors and Hepatotropism of HBV

Although there is little doubt that the polypeptide encoded by the pre-S2 region can specifically bind chimpanzee or human albumin polymers, the biological significance of this property is uncertain (N6). Polyalbumin binding was investigated in many laboratories because it provided a rationale for the hepatotropism of HBV. Lenkei reported in 1977 that liver cells had specific receptors for polyalbumin (L6). Two years later, Imai *et al.* reported that Dane particles could bind specifically to human and chimpanzee polyalbumins. It was postulated that since both HBV and the hepatocyte receptor could bind polyalbumin, the virus could gain entry into the cell by using polyalbumin as a bridge (I1).

There are two facts that make the hypothesis difficult to accept. Since the receptors do not have a great affinity for albumin that is polymerized by aging in aqueous solutions (such as plasma), it is difficult to see how HBV could effectively use them as a means to penetrate hepatocytes. Moreover, woodchuck hepatitis B virus, which, like HBV, produces hepatitis in the infected host, does not possess polyalbumin sites for polymerized woodchuck albumin (P3). It would appear that the hepatotropism of two hepadna viruses will possibly have to be explained by mechanisms other than polyalbumin binding.

4.1.4. *Pre-S Polypeptides*

Western blotting is used extensively in the detection of specific polypeptides. The specificity of the test depends almost entirely in the antibodies used to detect the amino acid sequences of interest. Recombinant DNA technology has been used to produce several of the antigens used to raise antibodies for the study of several HBV polypeptides. These techniques have resulted in the characterization of GP46 and GP42 (W6), P39/GP42, and P31/GP35 (H9, M2, T2).

A test for pre-S1 polypeptides has been developed based on use of antibodies directed against a fusion protein with 100 amino acid sequences encoded in the HBV genome (T4). The test is relatively new and further developments to make it more suitable for routine work may appear. The number of patient specimens studied in the initial trial was relatively small, but it appeared that the presence of the pre-S1 polypeptides correlated closely with HBeAg-positivity and with HBV DNA.

4.2. REGION X POLYPEPTIDES

Synthetic peptide sequences with 11–17 amino acid residues that were predicted from the nucleotide sequence of region X were used to raise antiserum and to detect naturally occurring antibodies to region X polypeptides. This strategy has enabled region X polypeptides to be detected in HBV-infected cells, HBV-infected liver, hepatomas, and in core particles prepared from liver (F6, M24). The HBV-infected tissues and hepatomas contained a 28-kDa protein reacting with the antiserum to region X peptides (M24).

In the same study, serum was tested for antibodies to region X peptides. Specimens from 254 cases with acute hepatitis B, asymptomatic HBV carriage, chronic hepatitis, liver cirrhosis, and HCC were screened. Antibodies to region X polypeptides were detected in 13, 16, and 73%, respectively, of the subjects in the last three disease groups. These observations suggest that region X may furnish further markers for HBV infection.

REFERENCES

A1. Alberti, A., Diana, S., Eddleston, A. L. W. F., and Williams, R., Changes in hepatitis B virus DNA polymerase in relation to the outcome of acute hepatitis type B. *Gut* **20**, 190–195 (1979).

A2. Alberti, A., Pontisso, P., and Realdi, G., Changes in hepatitis B virus DNA polymerase activity in patients with chronic infection. *J. Med. Virol.* **8**, 223–229 (1981).

A3. Alberti, A., Pontisso, P., Chemello, L., Schiavon, E., and Realdi, G., Virus receptors for polymerized serum albumin and anti-receptor antibody in hepatitis B virus infection. In "Viral Hepatitis and Delta Infection" (G. Verme, F. Bonino, and M. Rizetto, eds.), pp. 327–336. Alan R. Liss, New York, 1983.

A4. Alberti, A., Pontisso, P., Schiavon, E., and Realdi, G., An antibody which precipitates Dane particles in acute hepatitis type B: Relation to receptor sites which bind polymerized human serum albumin on virus particles. *Hepatology* **4**, 220–226 (1984).

A5. Albin, C., and Robinson, W. S., Protein kinase activity in hepatitis B virus. *J. Virol.* **34**, 297–302 (1980).

A6. Alter, H. J., Purcell, R. H., Gerin, J. L., London, W. T., Kaplan, P. M., McAuliffe, V. J., Wagner, J., and Holland, P. V., Transmission of hepatitis B to chimpanzees by hepatitis B surface antigen-positive saliva and semen. *Infect. Immun.* **16**, 928–933 (1977).

A7. Anderson, M. G., and Murray-Lyon, I. M., Natural history of the HBsAg carrier. *Gut* **26**, 848–860 (1985).

A8. Andres, L. L., Sawhney, V. K., Scullard, G. H., Smith, J. L., Merigan, T. C., Robinson, W. S., and Gregory, P. B., Dane particle DNA polymerase and HBeAg: Impact on clinical, laboratory, and histologic findings in hepatitis B-associated chronic liver disease. *Hepatology* **1**, 583–585 (1981).

B1. Barin, F., Perrin, J., Chotard, J., Denis, F., N'Doye, R., Mar, I. D., Chiron, J. P., Coursaget, P., Goudeau, A., and Maupas, P., Cross-sectional and longitudinal epidemiology of hepatitis B in Senegal. *Prog. Med. Virol.* **27**, 148–162 (1981).

B2. Beasley, R. P., Lin, C. C., Chien, C. S., Chen, C. J., and Hwang, L. Y., Geographic distribution of HBsAg carriers in China. *Hepatology* **2**, 553–556 (1982).

B3. Berninger, M., Hammer, M., Hoyer, B., and Gerin, J. L., An assay for the detection of the DNA genome of hepatitis B virus in serum. *J. Med. Virol.* **9**, 57–68 (1982).

B4. Bichko, V., Pushko, P., Dreilina, D., Pumpen, P., and Gren, E., Subtype ayw variant of hepatitis B virus. *FEBS Lett.* **185**, 208–212 (1985).

B5. Blum, H. E., Stowring, L., Figus, A., Montgomery, C. K., Haase, A. T., and Vyas, G. N., Detection of hepatitis B virus in hepatocytes, bile duct epithelium, and vascular elements by *in situ* hybridization. *Proc. Natl. Acad. Sci. U.S.A.* **80**, 6685–6688 (1983).

B6. Blum, H. E., Haase, A. T., Harris, J. D., Walker, D., and Vyas, G. N., Asymmetric replication of hepatitis B virus DNA in human liver: Demonstration of cytoplasmic minus-strand DNA by blot analysis and *in situ* hybridization. *Virology* **139**, 87–96 (1984).

B7. Blum, H. E., Figus, A., Haase, A. T., and Vyas, G. N., Laboratory diagnosis of hepatitis B virus infection by nucleic acid hybridization analysis and immunohistologic detection of gene products. *Dev. Biol. Stand.* **59**, 125–139 (1985).

B8. Blumberg, B. S., Polymorphisms of serum proteins and the development of isoprecipitins in transfused patients. *Bull. N.Y. Acad. Med.* [2] **40**, 377–386 (1964).

B9. Blumberg, B. S., and London, W. T., Hepatitis B virus and the prevention of primary hepatocellular carcinoma. *N. Engl. J. Med.* **304**, 782–784 (1981).

B10. Blumberg, B. S., Sutnick, A. I., London, W. T., and Millman, I., Electron microscopy. "Australia Antigen and Hepatitis," pp. 26–30. Butterworth, London, 1971.

B11. Blumberg, B. S., Gerstley, B. J. S., Hungerford, D. A., London, W. T., and Sutnick, A. I., A serum antigen (Australia antigen) in Down's syndrome, leukemia, and hepatitis. *Ann. Intern. Med.* **66**, 924–931 (1967).

B12. Blumberg, B. S., Larouze, B., London, W. T., Warner, B., Hesser, J. E., Millman, I., Saimot, G., and Payet, M., The relation of infection with the hepatitis B agent to primary hepatocellular carcinoma. *Am. J. Pathol.* **81**, 669–682, (1975).

B13. Bonino, F., The importance of hepatitis B viral DNA in serum and liver. *J. Hepatol.* **3**, 136–141 (1986).
B14. Bonino, F., Hoyer, B., Nelson, J., Engle, R., Verme, G., and Gerin, J., Hepatitis B virus DNA in the sera of HBsAg carriers: A marker of active hepatitis B virus replication in the liver. *Hepatology* **1**, 386–391 (1981).
B15. Bortolotti, F., Bertaggia, A., Crivellaro, C., Armigliato, M., Alberti, A., Pontisso, P., Chemello, C., and Realdi, G., Chronic evolution of acute hepatitis type B: Prevalence and predictive markers. *Infection* **14**, 64–67 (1986).
B16. Bowcock, A. M., Pinto, M. R., Bey, E., Kuyl, J. M., Dusheiko, G. M., and Bernstein, R., The PLC/PRF/5 human hepatoma cell line. II. Chromosomal assignment of hepatitis B virus integration sites. *Cancer Genet. Cytogenet.* **18**, 19–26 (1985).
B17. Bradley, D. W., Murphy, B. L., Smith, J. L., and Maynard, J. E., Naturally occurring antibody against unusually high serum DNA polymerase activity. *Nature (London)* **259**, 594–596 (1976).
B18. Brechot, C., Pourcel, C., Louise, A., Rain, B., and Tiollais, P., Presence of integrated hepatitis B virus DNA sequences in cellular DNA of human hepatocellular carcinoma. *Nature (London)* **286**, 533–535 (1980).
B19. Brechot, C., Hadchouel, M., Scotto, J., Degos, F., Charnay, P., Trepo, C., and Tiollais, P., Detection of hepatitis B virus DNA in liver and serum: A direct appraisal of the chronic carrier state. *Lancet* **2**, 765–768 (1981).
B20. Brechot, C., Hadchouel, M., Scotto, J., Fonck, M., Potet, F., Vyas, G. N., and Tiollais, P., State of hepatitis B virus DNA in hepatocytes of patients with hepatitis B surface antigen-positive and -negative liver diseases. *Proc. Natl. Acad. Sci. U.S.A.* **78**, 3906–3910 (1981).
B21. Burrell, C. J., Mackay, P., Greeenaway, P. J., Hoschneider, P. H., and Murray, K., Expression in *Escherichia coli* of hepatitis B virus DNA sequences cloned in plasmid pBR322. *Nature (London)* **279**, 43–47 (1979).
C1. Cappel, R., De Cuyper, F., and Van Beers, D., Diagnosis of hepatitis B by Dane particle associated DNA polymerase assay. *J. Med. Virol.* **3**, 77–80 (1978).
C2. Cattaneo, R., Will, H., and Schaller, H., Hepatitis B virus transcription in the infected liver. *EMBO J.* **3**, 2191–2196 (1984).
C3. Chadwick, R. G., Bassendine, M. F., Crawford, E. M., Thomas, H. C., and Sherlock, S., HBsAg-positive chronic liver disease: Inhibition of DNA polymerase activity by vidarabine. *Br. Med. J.* **2**, 531–533 (1978).
C4. Chakraborty, P. R., Ruiz-Opazo, N., Shouval, D., and Shafritz, D. A., Identification of integrated hepatitis B virus DNA and expression of viral RNA in an HBsAg-producing human hepatocellular carcinoma cell line. *Nature (London)* **286**, 531–534 (1980).
C5. Charnay, P., Pourcel, C., Louise, A., Fritsch A., and Tiollais, P., Cloning in *Escherichia coli* and physical structure of hepatitis B virion DNA. *Proc. Natl. Acad. Sci. U.S.A.* **76**, 2222–2226 (1979).
C6. Chen, D. S., Lai, M. Y., and Sung, J. L., Delta agent infection in patients with chronic liver disease and hepatocellular carcinoma—An infrequent finding in Taiwan. *Hepatology* **4**, 502–503 (1984).
C7. Chen, D. S., Hoyer, B. H., Nelson, J., Purcell, R. H., and Gerin, J. L., Detection and properties of hepatitis B viral DNA in liver tissue from patients with hepatocellular carcinoma. *Hepatology* **2**, 42S–46S (1982).
C8. Chen, D. S., Lai, M. Y., Lee, S. C., Yang, P. M., Sheu, J. C., and Sung, J. L., Serum HBsAg, HBeAg, anti-HBe and hepatitis B viral DNA in asymptomatic carriers in Taiwan. *J. Med. Virol.* **19**, 87–94 (1986).

C9. Chung, W. K., Sun, H. S., Park, D. H., Minuk, G. Y., and Hoofnagle, J. H., Primary hepatocellular carcinoma and hepatitis B virus infection in Korea. *J. Med. Virol.* **11**, 99–104 (1983).
C10. Courouce-Pauty, A.-M., Drouet, J., and Kleinknecht, D., Simultaneous occurrence in the same serum of hepatitis B surface antigen and antibody to hepatitis B surface antigen of different subtypes. *J. Infect. Dis.* **140**, 975–978 (1979).
C11. Craxi, A., Weller, I. V. D., Bassendine, M. F., Fowler, M. J. F., Monjardino, J., Thomas, H. C., and Sherlock, S., Relationship between HBV-specific DNA polymerase and HBe antigen/antibody system in chronic HBV infection: Factors determining selection of patients and outcome of antiviral therapy. *Gut* **24**, 143–147 (1983).
C12. Cummings, I. W., Browne, J. K., Salser, W. A., Tyler, G. V., Snyder, R. L., Smolec, J. M., and Summers, J., Isolation, characterization and comparison of recombinant DNAs derived from genomes of human hepatitis B virus and woodchuck hepatitis virus. *Proc. Natl. Acad. Sci. U.S.A.* **77**, 1842–1846 (1980).
D1. Dane, D. S., Cameron, C. H., and Briggs, M. Virus-like particles in serum of patients with Australia-antigen associated hepatitis. *Lancet* **1**, 695–698 (1970).
D2. Datta, A. K., and Hood, R. E., Mechanism of inhibitions of Epstein–Barr virus replication by phosphonoformic acid. *Virology* **114**, 52–59 (1981).
D3. Deinhardt, F., Predictive value of markers of hepatitis virus infection. *J. Infect. Dis.* **141**, 299–305 (1980).
D4. Dejean, A., Bougueleret, L., Grzeschik, K. H., and Tiollais, P., Hepatitis B virus DNA integration in a sequence homologous to v-*erb*-A and steroid receptor genes in a hepatocellular carcinoma. *Nature (London)* **322**, 70–72 (1986).
D5. Dejean, A., Sonigo, P., Wain-Hobson, S., and Tiollais, P., Specific hepatitis B virus integration in hepatocellular carcinoma DNA through a viral 11-base-pair direct repeat. *Proc. Natl. Acad. Sci. U.S.A.* **81**, 5350–5354 (1984).
D6. Dejean, A., Lugassy, C., Zafrani, S., Tiollais, P., and Brechot, C., Detection of hepatitis B virus DNA in pancreas, kidney and skin of two human carriers of the virus. *J. Gen. Virol.* **65**, 651–655 (1984).
D7. Delius, H., Gough, N. M., Cameron, C. H., and Murray, K., Structure of the hepatitis B virus genome. *J. Virol.* **47**, 337–343 (1983).
D8. De Virgiliis, S., Frau, F., Sanna, G., Turco, M. P., Figus, A. L., Cornacchia, G., and Cao, A., Perinatal hepatitis B virus detection by hepatitis B virus DNA analysis. *Arch. Dis. Child.* **60**, 56–58 (1985).
D9. Di Bisceglie, A. M., Dusheiko, G. M., and Kew, M. C., Detection of markers of hepatitis B virus infection in urine of chronic carriers. *J. Med. Virol.* **16**, 337–341 (1985).
D10. Diegutis, P., Keirnan, E., Burnett, L., Nightingale, B. N., and Cossart, Y. E., False-positive results with hepatitis B virus DNA dot-hybridization in hepatitis B surface antigen-negative specimens. *J. Clin. Microbiol.* **23**, 797–799 (1986).
D11. Dubois, M. F., Pourcel, C., Rousset, S., Chany, C., and Tiollais, P., Excretion of hepatitis B surface antigen particles from mouse cells transformed with cloned viral DNA. *Proc. Natl. Acad. Sci. U.S.A.* **77**, 4549–4553 (1980).
E1. Economidou, I., Hadziyannis, S., Paraskevas, E., Binopoulou, A., Hesser, J. E., Lustbader, E., and Blumberg, B. S., Australia antigen (HBsAg) carriers in a Greek community. Studies of transaminase (SGPT) levels. *Res. Commun. Chem. Pathol. Pharmacol.* **10**, 703–713 (1975).
E2. Edman, J. C., Gray, P., Valenzuela, P., Rall, L. B., and Rutter, W. J., Integration of hepatitis B virus sequences and their expression in a human hepatoma cell. *Nature (London)* **286**, 535–538 (1980).

E3. Elfassi, E., Romet-Lemonne, J. L., Essex, M., McLane, M. F., and Haseltine, W. A., Evidence of extrachromosomal forms of hepatitis B viral DNA in a bone marrow culture obtained from a patient recently infected with hepatitis B virus. *Proc. Natl. Acad. Sci. U.S.A.* **81,** 3526–3528 (1984).

E4. Eriksson, B., Oberg, B., and Wahren, B., Pyrophosphate analogs as inhibitors of DNA polymerases of cytomegalovirus, herpes simplex virus and cellular origin. *Biochim. Biophys. Acta* **696,** 115–123 (1982).

F1. Fagan, E. A., Alexander, G. J. M., Davison, F., and Williams, R., Persistence of free HBV DNA in body secretions and liver despite loss of serum HBV DNA after interferon-induced seroconversion. *J. Med. Virol.* **20,** 183–188 (1986).

F2. Fagan, E. A., Guarner, P., Perera, S. D. K., Trowbridge, R., Rolando, N., Davison, F., and Williams, R., Quantitation of hepatitis B virus DNA (HBV DNA) in serum using the spot hybridization technique and scintillation counting. *J. Virol. Methods* **12,** 251–262 (1985).

F3. Fang, C. T., Nath, N., Pielech, M., and Dodd, R. Y., A modified method for the detection of hepatitis B virus-specific DNA polymerase. *J. Virol. Methods* **2,** 349–356 (1981).

F4. Feinberg, A. P., and Vogelstein, B., A technique for radiolabelling DNA restriction endonuclease fragments to high specific activity. *Anal. Biochem.* **132,** 6–13 (1983).

F5. Feinman, S. V., Berris, B., Guha, A., Sooknanan, R., Bradley, D. W., Bond, W. W., and Maynard, J. E., DNA:DNA hybridization method for the diagnosis of hepatitis B infection. *J. Virol. Methods* **8,** 199–206 (1984).

F6. Feitelson, M. A., Products of the "X" gene in hepatitis B and related viruses. *Hepatology* **6,** 191–198 (1986).

F7. Feitelson, M. A., Millman, I., Halbherr, T., Simmons, H., and Blumberg, B. S., A newly identified hepatitis B type virus in tree squirrels. *Proc. Natl. Acad. Sci. U.S.A.* **83,** 2233–2237 (1986).

F8. Feng, Z., Zhou, Y., Wu, X., Gan, R., and Li, Z., A clone of hepatitis B virus (subtype adr) DNA with a new *Hin*dIII site. *Sci. Sin. (Engl. Ed.)* **28,** 827–834 (1985).

F9. Fowler, M. J. F., Monjardino, J., Tsiquaye, K. N., Zuckerman, A. J., and Thomas, H. C., The mechanism of replication of hepatitis B virus: Evidence of asymmetric replication of the two strands. *J. Med. Virol.* **13,** 83–91 (1984).

F10. Fowler, M. J. F., Monjardino, J., Weller, I. V. D., Lok, A. S. F., and Thomas, H. C., Analysis of the molecular state of HBV-DNA in the liver and serum of patients with chronic hepatitis or primary liver cell carcinoma and the effect of therapy with adenine arabinoside. *Gut* **25,** 611–618 (1984).

F11. Fowler, M.J.F., Monjardino, J., Weller, I. V., Bamber, M., Karayiannis, P., Zuckerman, A. J., and Thomas, H. C., Failure to detect nucleic acid homology between some non-A, non-B viruses and hepatitis B virus DNA. *J. Med. Virol.* **12,** 205–213 (1983).

F12. Fowler, M. J. F., Greenfield, C., Chu, C. M., Karayiannis, P., Dunk, A., Lok, A. S. F., Lai, C. L., Yeoh, E. K., Monjardino, J. P., Wankya, B. M., and Thomas, H. C., Integration of HBV-DNA may not be a prerequisite for maintenance of the state of malignant transformation. *J. Hepatol.* **2,** 218–229 (1986).

F13. Fritsch, A., Pourcel, C., Charnay, P., and Tiollais, P., Clonage de genome du virus de l'hépatite B dans *Escherichia coli. C.R. Hebd. Seances Acad. Sci.* **287,** 1453–1454 (1978).

F14. Fujiyama, A., Miyanohara, A., Nozaki, C., Yoneyama, T., Ohtomo, N., and Matsubara, K., Cloning and structural analyses of hepatitis B virus DNAs, subtype *adr. Nucleic Acids Res.* **11,** 4601–4610 (1983).

G1. Galibert, F., Mandart, E., Fitoussi, F., Tiollais, P., and Charnay, P., Nucleotide sequence of the hepatitis B virus genome (subtype ayw) cloned in *E. coli. Nature (London)* **281**, 646–650 (1979).
G2. Gerin, J. L., Ford, E. C., and Purcell, R. H., Biochemical characterization of Australia antigen: Evidence for defective particles of hepatitis B virus. *Am. J. Pathol.* **81**, 651–668 (1974).
G3. Gerlich, W. H., and Robinson, W. S., Hepatitis B virus contains protein attached to the 5' terminus of its complete strand. *Cell (Cambridge, Mass.)* **21**, 801–809 (1980).
G4. Gerlich, W. H., Goldmann, U., Muller, R., Stibbe, W., and Wolff, W., Specificity and localization of the hepatitis B virus associated protein kinase. *J. Virol.* **42**, 761–766 (1982).
G5. Gilja, B. K., Keh, W. C., Kasambalides, E. J., Thung, S. N., and Gerber, M. A., Correlation of polyalbumin receptors with hepatitis B surface antigen polypeptides in human sera. *Arch. Pathol. Lab. Med.* **110**, 1021–1024 (1986).
G6. Gillespie, D., and Spiegelman, S., A quantitative assay for DNA–RNA hybrids with DNA immobilized on a membrane. *J. Mol. Biol.* **12**, 829–842 (1965).
G7. Gough, N. M., and Murray, K., Expression of the hepatitis B virus surface, core and e antigen genes by stable rat and mouse cell lines. *J. Mol. Biol.* **162**, 43–67 (1982).
G8. Gowans, E. J., Burrell, C. J., Jilbert, A. R., and Marmion, B. P., Detection of hepatitis B virus DNA sequences in infected hepatocytes by *in situ* cytohybridisation. *J. Med. Virol.* **8**, 67–78 (1981).
G9. Greenberg, H. B., Pollard, R. B., Lutwick, L. I., Gregory, P. B., Robinson, W. S., and Merigan, T. C., Effect of human leukocyte interferon on hepatitis B virus infection in patients with chronic active hepatitis. *N. Engl. J. Med.* **295**, 517–522 (1976).
G10. Gu, J. R., Chen, Y. C., Jiang, H. Q., Zhang, Y. L., Wu, S. M., Jiang, W. L., and Jian, J., State of hepatitis B virus DNA in leucocytes of hepatitis B patients. *J. Med. Virol.* **17**, 73–81 (1985).
G11. Gust, I. D., Epidemiology of hepatitis B in Australia. *In* "Viral Hepatitis B Infection in the Western Pacific Region: Vaccine and Control" (S. K. Lam, C. L. Lai, and E. K. Yeoh, eds.), pp. 19–22. World Scientific Publ. Co., Singapore, 1984.
G12. Gust, I. D., Rankin, D. W., and Stephens, T., Hepatitis B in Victoria. *Med. J. Aust.* **2**, 264–265 (1982).
G13. Gust, I. D., Burrell, C. J., Coulepis, A. G., Robinson, W. S., and Zuckerman, A. J., Taxonomic classification of human hepatitis B virus. *Intervirology* **25**, 14–29 (1986).
H1. Haase, A. T., Gantz, D., Blum, H., Stowring, L., Ventura, P., Geballe, A., Moyer, B., and Brahic, M., Combined macroscopic and microscopic detection of viral genomes in tissues. *Virology* **140**, 201–206 (1985).
H2. Hada, H., Arima, T., Togawa, K., Okada, Y., Morichika, S., and Nagashima, H., State of hepatitis B viral DNA in the liver of patients with hepatocellular carcinoma and chronic liver disease. *Liver* **6**, 189–198 (1986).
H3. Hadchouel, M., Scotto, J., Huret, J. L., Molinie, C., Villa, E., Degos, F., and Brechot, C., Presence of HBV DNA in spermatozoa: A possible vertical transmission of HBV via the germ line. *J. Med. Virol.* **16**, 61–66 (1985).
H4. Hadziyannis, S. J., Lieberman, H. M., Karvountzis, G. G., and Shafritz, D. M., Analysis of liver disease, nuclear HBcAg, viral replication, and hepatitis B virus DNA in liver and serum of HBeAg vs. anti-HBe positive carriers of hepatitis B virus. *Hepatology* **3**, 656–662 (1983).
H5. Halpern, M. S., England, J. M., Deery, D. T., Petcu, D. J., Mason, W. S., and Molnar-Kimber, K. L., Viral nucleic acid synthesis and antigen accumulation in pancreas and kidney of Pekin ducks infected with duck hepatitis virus. *Proc. Natl. Acad. Sci. U.S.A.* **80**, 4865–4869 (1983).

H6. Hantz, O., Allaudeen, H. S., Ooka, T., De Clercq, E., and Trepo, C., Inhibition of human and woodchuck hepatitis virus DNA polymerase by the triphosphates of acyclovir, 1-(2'-deoxy-2'-fluoro-β-D-arabinofuranosyl)-5-iodocytosine and E-5-(2-bromovinyl)-2'-deoxyuridine. *Antiviral Res.* **4**, 187–199 (1984).

H7. Hantz, O., Ooka, T., Vitvitski, L., Pichoud, C., and Trepo, C., Comparison of properties of woodchuck hepatitis virus and human hepatitis B virus endogenous DNA polymerases. *Antimicrob. Agents Chemother.* **25**, 242–246 (1984).

H8. Harrison, T. J., Bal, V., Wheeler, E. G., Meacock, T. J., Harrison, J. F., and Zuckerman, A. J., Hepatitis B virus DNA and e antigen in serum from blood donors in the United Kingdom positive for hepatitis B surface antigen. *Br. Med. J.* **290**, 663–664 (1985).

H9. Heermann, K. H., Goldmann, U., Schwartz, W., Seyffarth, T., Baumgarten, H., and Gerlich, W. H., Large surface proteins of hepatitis B virus containing the pre-*s* sequence. *J. Virol.* **52**, 396–402 (1984).

H10. Helgstrand, E., Eriksson, B., Johansson, N. G., Lannero, B., Larsson, A., Misiorny, A., Noren, J. O., Sjöberg, B., Steinberg, K., Stening, G., Stridh, S., Oberg, B., Alenious, S., and Philipson, L., Trisodium phosphonoformate, a new antiviral compound. *Science* **201**, 819–821 (1978).

H11. Hess, G., Arnold, W., and Meyer zum Buschenfelde, K. H., Inhibition of hepatitis B virus DNA polymerase by phosphonoformate: Studies on its mode of action. *J. Med. Virol.* **5**, 309–316 (1980).

H12. Hirose, H., Udo, K., Kojima, M., Takahashi, Y., Miyakawa, Y., Miyamoto, K., Yoshizawa, H., and Mayumi, M., Deposition of hepatitis B e antigen in membranous glomerulonephritis: Identification by $F(ab')_2$ fragments of monoclonal antibody. *Kidney Int.* **26**, 338–341 (1984).

H13. Hirschman, S. Z., and Garfinkel, E., Differential assay for hepatitis B DNA polymerase. *Clin. Res.* **25**, 377A (1977).

H14. Hirschman, S. Z., and Garfinkel, E., Ionic requirements of the DNA polymerase associated with serum hepatitis B antigen. *J. Infect. Dis.* **135**, 897–910 (1977).

H15. Hirschman, S. Z., Gerber, M., and Garfinkel, E., Differential activation of hepatitis B DNA polymerase by detergent and salt. *J. Med. Virol.* **2**, 61–76 (1978).

H16. Hoar, D. I., Bowen, T., Matheson, D., and Poon, M. C., Hepatitis B virus DNA is enriched in polymorphonuclear leukocytes. *Blood* **66**, 1251–1253 (1985).

H17. Hoofnagle, J. H., and Alter, H. J., Chronic viral hepatitis. In "Viral Hepatitis and Liver Disease" (G. N. Vyas, J. L. Dienstag, and J. H. Hoofnagle, eds.), pp. 97–113. Grune & Stratton, Orlando, Florida, 1984.

H18. Hoofnagle, J. H., Gerety, R. J., Smallwood, L. A., and Barker, L. F., Subtyping of hepatitis B surface antigen and antibody by radioimmunoassay. *Gastroenterology* **72**, 290–296 (1977).

H19. Howard, C. R., The detection of DNA polymerase in the diagnosis of HBV infection. *J. Med. Virol.* **3**, 81–86 (1978).

H20. Howley, P. M., Israel, M. A., Law, M. F., and Martin, M. A., A rapid method for detecting and mapping homology between heterologous DNAs *J. Biol. Chem.* **254**, 4876–4883 (1979).

H21. Hoyer, B., Bonino, F., Ponzetto, A., Denniston, K., Nelson, J., Purcell, R., and Gerin, J. L., Properties of delta associated ribonucleic acid. In "Viral Hepatitis and Delta Infection" (G. Verme, F. Bonino, and M. Rizetto, eds.), pp. 91–97. Alan R. Liss, New York, 1983.

H22. Hsu, H. C., Lin, G. H., Chang, M. H., and Chen, C. H., Association of hepatitis B surface (HBs) antigenemia and membranous nephropathy in children in Taiwan. *Clin. Nephrol.* **20**, 121–129 (1983).

H23. Hu, N. T., and Messing, J., The making of strand-specific M13 probes. *Gene* **17**, 271–277 (1982).
I1. Imai, M., Yanase, Y., Nojiri, T., Miyakawa, Y., and Mayumi, M., A receptor for polymerized albumins on hepatitis B virus particles co-occurring with HBeAg. *Gastroenterology* **76**, 242–247 (1979).
I2. Imazeki, F., Omata, M., Yokosuka, O., and Okuda, K., Integration of hepatitis B virus DNA in hepatocellular carcinoma. *Cancer (Philadelphia)* **58**, 1055–1060 (1986).
I3. Imazeki, F., Omata, M., Yokosuka, O., Matsuyama, Y., Ito, Y., and Okuda, K., Analysis of DNA polymerase reaction products for detecting hepatitis B virus in serum—Comparison with spot hybridization technique. *Hepatology* **5**, 783–788 (1985).
K1. Kam, W., Rall, L. B., Smuckler, E. A., Schmid, R., and Rutter, W. J., Hepatitis B viral DNA in liver and serum of asymptomatic carriers. *Proc. Natl. Acad. Sci. U.S.A.* **79**, 7522–7526 (1982).
K2. Kaplan, P. M., Ford, E. C., Purcell, R. H., and Gerin, J. L., Demonstration of subpopulations of Dane particles. *J. Virol.* **17**, 885–893 (1976).
K3. Kaplan, P. M., Greenman, R. L., Gerin, J. L., Purcell, R. H., and Robinson, W. S., DNA polymerase associated with human hepatitis B antigen. *J. Virol.* **12**, 995–1005 (1973).
K4. Karayiannis, P., Novick, D. M., Lok, A. S. F., Fowler, M. J. F., Monjardino, J., and Thomas, H. C., Hepatitis B virus DNA in saliva, urine and seminal fluid of carriers of hepatitis B e antigen. *Br. Med. J.* **290**, 1853–1855 (1985).
K5. Kobayashi, K., and Koike, K., Complete nucleotide sequence of hepatitis B virus DNA of subtype adr and its conserved gene organization. *Gene* **30**, 227–232 (1984).
K6. Koch, S., Freytag von Loringhoven, A., Hofschneider, P. H., and Koshy, R., Amplification and rearrangement in hepatoma cell DNA associated with integrated hepatitis B virus DNA. *EMBO J.* **3**, 2185–2189 (1984).
K7. Koch, S., Freytag van Loringhoven, A., Kahman, R., Hofschneider, P. H., and Koshy, R., The genetic organization of integrated hepatitis B virus DNA in the human hepatoma cell line PRC/PRF/5. *Nucleic Acids Res.* **12**, 6781–6886 (1984).
K8. Koshy, R., Koch, S., Freytag von Loringhoven, A., Kahmann, R., Murray, K., and Hofschneider, P. H., Integration of hepatitis B virus: Evidence for integration in the single-stranded gap. *Cell (Cambridge, Mass.)* **34**, 215–223 (1983).
K9. Krogsgaard, K., Kryger, P., Aldershville, J., Andersson, P., Brechot, C., and the Copenhagen Hepatitis Acuta Programme, Hepatitis B virus DNA in serum from patients with acute hepatitis B. *Hepatology* **5**, 10–13 (1985).
L1. Le Bouvier, G. L., Bancroft, W. H., and Holland, P. V., On the rare ayr phenotype of hepatitis B antigen. *Intervirology* **1**, 405–408 (1973).
L2. Lee, A. K. Y., Ip. H. M. H., and Wong, V. C. W., Mechanisms of maternal–fetal transmission of hepatitis B virus. *J. Infect. Dis.* **138**, 668–676 (1978).
L3. Lee, S. D., Wang, J. Y., Wu. J. C., Chiang, Y. T., Tsai, Y. T., and Lo, K. J., Hepatitis B and D virus infection among drug abusers in Taiwan. *J. Med. Virol.* **20**, 247–252 (1986).
L4. Lee, S. D., Lo, K. J., Wu, J. C., Tsai, Y. T., Wang, J. Y., Ting, L. P., and Tong, M. J., Prevention of maternal–infant hepatitis B virus transmission by immunization: The role of serum hepatitis B virus DNA. *Hepatology* **6**, 369–373 (1986).
L5. Lemon, S. M., Gates, N. L., Simms, T. E., and Bancroft, W. H., IgM antibody to hepatitis B core antigen as a diagnostic parameter of acute infection with hepatitis B virus. *J. Infect. Dis.* **143**, 803–809 (1981).
L6. Lenkei, R., Onica, D., and Ghetie, V., Receptors for polymerized albumin on liver cells. *Experientia* **33**, 1046–1047 (1977).

L7. Lieberman, H. M., LaBrecque, D. R., Kew, M. C., Hadziyannis, S. J., and Shafritz, D. A., Detection of hepatitis B virus DNA directly in human serum by a simplified molecular hybridization test: Comparison to HBeAg/anti-HBe status in HBsAg carriers. *Hepatology* **3**, 285–291 (1983).
L8. Lie-Injo, L. E., Balesegaram, M., Lopez, C. G., and Herrera, A. R., Hepatitis B virus DNA in liver and white blood cells of patients with hepatoma. *DNA* **2**, 301–308 (1983).
L9. Lin, H. J., Lai, C. L., and Wu, P. C., Serum hepatitis B viral DNA in HBsAg-positive hepatocellular carcinoma treated with interferon or adriamycin. *Br. J. Cancer* **54**, 67–73 (1986).
L10. Lin, H. J., Wu, P. C., and Lai, C. L., An oligonucleotide probe for the detection of hepatitis B virus DNA in serum. *J. Virol. Methods* **15**, 139–149 (1987).
L11. Lin, H. J., Kwan, J. P. W., Wu, P. C., and Chak, W., Phosphonoformic acid-inhibitable nucleotide incorporation as a measure of hepatitis B viral DNA polymerase activity. *J. Med. Virol.* **12**, 61–70 (1983).
L12. Lin, H. J., Wu, P. C., Lai, C. L., and Chak, W., Micromethod for phosphonoformate inhibition assay of hepatitis B viral DNA polymerase. *Clin. Chem. (Winston-Salem, N.C.)* **30**, 549–552 (1984).
L13. Lin, H. J., Wu, P. C., Lai, C. L., and Leong, S., Molecular hybridization study of plasma hepatitis B virus DNA from different carriers. *J. Infect. Dis.* **154**, 983–989 (1986).
L14. Lindsay, K. L., Nizze, J. A., Koretz, R., and Gitnick, G., Diagnostic usefulness of testing for anti-HBc IgM in acute hepatitis B. *Hepatology* **6**, 1325–1328 (1986).
L15. Little, P. J., Some observations and implications of hepatitis B markers in Saudi Arabia. *J. Infect.* **7**, Suppl. I, 79 (1983).
L16. Lo, S. J., Lee, Y. H. W., Chiou, J. S., Ting, L. P., Liu, W. T., and Choo, K. B., Characterization of restriction endonuclease maps of hepatitis B viral DNAs. *Biochem. Biophys. Res. Commun.* **129**, 797–803 (1985).
L17. Loeb, L. A., Williams, R. O., Sutnick, A. I., O'Connell, A., and Millman, I., DNA polymerase activity in human serum: Studies with Australia antigen. *Proc. Soc. Exp. Biol. Med.* **143**, 519–525 (1973).
L18. Lugassy, C., Bernuau, J., Thiers, V., Krosgaard, K., Degott, C., Wantzin, P., Schalm, S. W., Rueff, B., Banhamou, J. P., Tiollais, P., and Brechot, C., Sequences of hepatitis B virus DNA in the serum and liver of patients with acute benign and fulminant hepatitis. *J. Infect. Dis.* **155**, 64–71 (1987).
M1. Macaya, G., Visona, K. A., and Villarejos, V. M., Dane particles and associated DNA polymerase activity in saliva of chronic hepatitis B carriers. *J. Med. Virol.* **4**, 291–301 (1979).
M2. Machida, A., Kishimoto, S., Ohnuma, H., Miyamoto, H., Baba, K., Oda, K., Nakamura, T., Miyakawa, Y., and Mayumi, M., A hepatitis B surface antigen polypeptide (P31) with the receptor for polymerized human as well as chimpanzee albumins. *Gastroenterology* **85**, 268–274 (1983).
M3. Machida, A., Kishimoto, S., Ohnuma, H., Baba, K., Ito, Y., Miyamoto, H., Funatsu, G., Oda, K., Usuda, S., Togami, S., Nakamura, T., Miyakawa, Y., and Mayumi, M., A polypeptide containing 55 amino acid residues coded by the pre-S region of hepatitis B virus deoxyribonucleic acid bears the receptor for polymerized human as well as chimpanzee albumins. *Gastroenterology* **86**, 910–918 (1984).
M4. Mackay, P., Lees, J., and Murray, K., The conversion of hepatitis B core antigen synthesized in *E. coli* into e antigen. *J. Med. Virol.* **8**, 237–243 (1981).

M5. Magrin, S., Craxi, A., Vinci, M., Greco, J., Maggiore, G., Scotto, J., and Pagliario, L., Assessment of HBV replicative status by receptors for polymerized human albumin. *Hepatogastroenterology* **33**, 6–8 (1986).
M6. Mao, J. C. H., Otis, E. R., Mushawar, I. K., and Overby L. R., Properties of soluble DNA polymerase from sera of hepatitis B virus carriers. *J. Med. Virol.* **6**, 285–299 (1980).
M7. Matsuyama, Y., Omata, M., Yokosuka, O., Imazeki, F., Ito, Y., and Okuda, K., Discordance of hepatitis B e antigen/antibody and hepatitis B virus deoxyribonucleic acid in serum. *Gastroenterology* **89**, 1104–1108 (1985).
M8. Mazzur, S., Burgert, S., and Blumberg, B. S., Geographical distribution of Australia antigen determinants d, y and w. *Nature (London)* **247**, 38–40 (1974).
M9. Mazzur, S., Falker, D., and Blumberg, B. S., Geographical variation of the "w" subtype of Australia antigen. *Nature (London)* **243**, 44–47 (1973).
M10. McConaughy, B. L., Laird, C. D., and McCarthy, B. J., Nucleic acid reassociation in formamide. *Biochemistry* **8**, 3289–3295 (1969).
M11. Meinkoth, J., and Wahl, G., Hybridization of nucleic acids immobilized on solid supports. *Anal. Biochem.* **138**, 267–284 (1984).
M12. Messing, J., and Vieira, A new pair of M13 vectors for selecting either strand of double-digest restriction fragments. *Gene* **19**, 269–276 (1982).
M13. Milborrow, H. M., Burnett, L., Lowe, S. B., Nightingale, B. N., and Cossart, Y. E., Rapid method for the concentration of hepatitis B virus particles using Beckman airfuge. *J. Clin. Pathol.* **39**, 808–809 (1986).
M14. Milich, D. R., Gottfried, T. D., and Vyas, G. N., Characterization of the interaction between polymerized human albumin and hepatitis B surface antigen. *Gastroenterology* **81**, 218–225 (1981).
M15. Miller, R. H., and Robinson, W. S., Hepatitis B virus DNA forms in nuclear and cytoplasmic fractions of infected human liver. *Virology* **137**, 390–399 (1984).
M16. Miller, R. H., and Robinson, W. S., Common evolutionary origin of hepatitis B virus and retroviruses. *Proc. Natl. Acad. Sci. U.S.A.* **83**, 2531–2535 (1986).
M17. Miller, R. H., Marion, P. L., and Robinson, W. S., Hepatitis B viral DNA-RNA hybrid molecules in particles from infected liver are converted to viral DNA molecules during an endogenous DNA polymerase reaction. *Virology* **139**, 64–72 (1984).
M18. Miller, R. H., Tran, C. T., and Robinson W. S., Hepatitis B virus particles of plasma and liver contain viral DNA–RNA hybrid molecules. *Virology* **139**, 53–63 (1984).
M19. Miller, R. H., Lee, S. C., Liaw, Y. F., and Robinson, W. S., Hepatitis B viral DNA in infected human liver and in hepatocellular carcinoma. *J. Infect. Dis.* **151**, 1081–1092 (1985).
M20. Miyaki, M., Sato, C., Gotanda, T., Matsui, T., Mishiro, S., Imai, M., and Mayumi, M., Integration of region X of hepatitis B virus genome in human primary hepatocellular carcinomas propagated in nude mice. *J. Gen. Virol.* **67**, 1449–1454 (1986).
M21. Mizusawa, H., Taira, M., Yaginuma, K., Kobayashi, M., Yoshida, E., and Koike, K., Inversely repeating integrated hepatitis B virus DNA and cellular flanking sequences in the human hepatoma-derived cell line huSP. *Proc. Natl. Acad. Sci. U.S.A.* **82**, 208–212 (1985).
M22. Molden, D. P., Nakamura, R. M., Suzuki, H., Greer, S., Pergolizzi, R. G., and Brakel, C. L., Comparison of radiolabeled DNA probe with a nonisotopic probe for assay of serum hepatitis B virus DNA. *Clin. Physiol. Biochem.* **3**, 174–183 (1985).
M23. Morace, G., von der Helm, K., Jilg, W., and Deinhardt, F., Detection of hepatitis B virus DNA in serum by a rapid filtration-hybridization assay. *J. Virol. Methods* **12**, 235–242 (1985).

M24. Moriarty, A. M., Alexander, H., Lerner, R. A., and Thornton, G. B., Antibodies to peptides detect new hepatitis B antigen: Serological correlation with hepatocellular carcinoma. *Science* **227**, 429–433 (1985).

M25. Mushawar, I. K., Dienstag, J. L., Polesky, H. F., McGrath, L. C., Decker, R. H., and Overby, L. R., Interpretation of various serological profiles of hepatitis B virus infection. *Am. J. Clin. Pathol.* **76**, 773–777 (1981).

N1. Nagata, Y., Yokota, H., Kosuda, O., Yokoo, K., Takemura, K., and Kikuchi, T., Quantification of picogram levels of specific DNA immobilized in microtiter wells. *FEBS Lett.* **183**, 379–382 (1985).

N2. Nath, N., Fang, C. T., and Dodd, R. Y., Inactivation of DNA-polymerase associated with hepatitis B virus. *J. Med. Virol.* **10**, 131–140 (1982).

N3. Naumova, A. K., Favorov, M. O., Keteladze, E. S., Nosikov, V. V., and Kisselev, L. L., Nucleotide sequences in human chromosomal DNA from nonhepatic tissues homologous to the hepatitis B virus genome. *Gene* **35**, 19–25 (1985).

N4. Nayak, N. C., Panda, S. K., Zuckerman, A. J., Bhan, M. K., and Guha, D. K., Dynamics and impact of perinatal transmission of hepatitis B virus in north India. *J. Med. Virol.* **21**, 137–145 (1987).

N5. Negro, F., Berninger, M., Chiaberge, E., Gugliotta, P., Bussolati, G., Actis, G. C., Rizetto, M., and Bonino, F., Detection of HBV-DNA by *in situ* hybridization using a biotin-labelled probe. *J. Med. Virol.* **15**, 372–382 (1985).

N6. Neurath, A. R., Interaction of albumin with hepatitis B virus envelope components—A phenomenon in search of a biologic function. *In* "Viral Hepatitis and Liver Disease" (G. N. Vyas, J. L. Dienstag, and J. H. Hoofnagle, eds.), pp. 191–200. Grune & Stratton, Orlando, Florida, 1984.

N7. Neurath, A. R., and Strick, N., Radioimmunoassay for albumin-binding sites associated with HBsAg: Correlation of results with the presence of e-antigen in serum. *Intervirology* **11**, 128–132 (1979).

N8. Nishioka, K., Predominant mode of transmission of hepatitis B virus: Perinatal transmission in Asia. *In* "Viral Hepatitis and Liver Disease" (G. N. Vyas, J. L. Dienstag, and J. H. Hoofnagle, eds.), pp. 423–432. Grune & Stratton, Orlando, Florida, 1984.

N9. Nishioka, K., Epidemiology of hepatitis B virus infection in Japan. *In* "Viral Hepatitis B Infection in the Western Pacific Region: Vaccine and Control" (S. K. Lam, C. L. Lai, and E. K. Yeoh, eds.), pp. 43–50. World Scientific Publ. Co., Singapore, 1984.

N10. Nishioka, K., Levin, A. G., and Simons, M. J., Hepatitis B antigen, antigen subtypes, and hepatitis B antibody in normal subjects and patients with liver disease. *Bull. W.H.O.* **52**, 293–300 (1975).

N11. Nordenfelt, E., and Andren-Sandberg, M., Dane particle-associated DNA polymerase and e antigen: Relation to chronic hepatitis among carriers of hepatitis B surface antigen. *J. Infect. Dis.* **134**, 85–89 (1976).

N12. Nordenfelt, E., Oberg, B. Helgstrand, E., and Miller, E., Inhibition of hepatitis B Dane particle DNA polymerase activity by pyrophosphate analogs. *Acta Pathol. Microbiol. Scand., Sect. B* **88B**, 169–175 (1980).

N13. Nowoslawski, A., Hepatitis B virus-induced immune complex disease. *Prog. Liver Dis.* **6**, 393–406 (1979).

O1. Ohori, H., Onodera, S., and Ishida, N., Demonstration of hepatitis B e antigen (HBeAg) in association with intact Dane particles. *J. Gen. Virol.* **43**, 423–427 (1979).

O2. Okamoto, H., Imai, M., Shimozaki, S., Hoshi, Y., Iizuka, H., Gotanda, T., Tsuda, F., Miyakawa, Y., and Mayumi, M., Nucleotide sequence of a cloned hepatitis B virus genome, subtype *ayr:* Comparison with genomes of the other three subtypes. *J. Gen. Virol.* **67**, 2305–2314 (1986).

O3. Okamoto, H., Imai, H., Usuda, S., Tanaka, E., Tachibana, K., Mishiro, S., Machida, A., Nakamura, T., Miyakawa, Y., and Mayumi, M., Hemagglutination assay of polypeptide coded by the pre-S region of hepatitis B virus DNA with monoclonal antibody: Correlation of pre-S polypeptide with the receptor for polymerized human serum albumin in serums containing hepatitis B antigens. *J. Immunol.* **134**, 1212–1216 (1985).

O4. Omata, M., Imazeki, F., Yokosuka, O., Ito., Y., Uchiumi, K., Moori, J., and Okuda, K., Recombinant leukocyte A interferon treatment in patients with chronic hepatitis B virus infection. *Gastroenterology* **88**, 870–880 (1985).

O5. Ono, Y., Onda H., Sasada, R., Igarishi, K., Sugino, Y., and Nishioka, K., The complete nucleotide sequences of the cloned hepatitis B virus DNA, subtypes adr and adw. *Nucleic Acids Res.* **11**, 1747–1757 (1983).

P1. Pasek, M., Goto, T., Gilbert, W., Zink, B., Schaller, H., Mackay, P., Leadbetter, G., and Murray, K., Hepatitis B virus genes and their expression in *E. coli*. *Nature (London)* **282**, 575–9 (1979).

P2. Pasquinelli, C., Laure, F., Chatenoud, L., Beaurin, G., Gazengel, C., Bismuth, H., Degos, F., Tiollais, P., Bach, J. F., and Brechot, C., Hepatitis B virus DNA in mononuclear blood cells. *J. Hepatol.* **3**, 95–103 (1986).

P3. Pohl, C. J., Cote, P. J., Purcell, R. H., and Gerin, J. L., Failure to detect polyalbumin-binding sites on the woodchuck virus surface antigen: Implications for the pathogenesis of hepatitis B virus in humans. *J. Virol.* **60**, 943–949 (1986).

P4. Pontisso, P., Alberti, A., Bortolotti, F., and Realdi, G., Virus-associated receptors for polymerized human serum albumin in acute and in chronic hepatitis B virus infection. *Gastroenterology* **84**, 220–226 (1983).

P5. Pontisso, P., Poon, M. C., Tiollais, P., and Brechot, C., Detection of hepatitis B virus in mononuclear blood cells. *Br. Med. J.* **288**, 1563–1566 (1984).

P6. Poon, M. C., Bowen, T., Cassol, S., and Hoar, D. I., DNA probing assay in the detection of hepatitis B virus genome in human peripheral blood cells. *Prog. Clin. Biol. Res.* **211**, 317–331 (1986).

P7. Pourcel, C., Louise, A., Gervais, M., Chenciner, N., Dubois, M. F., and Tiollais, P. Transcription of the hepatitis B surface antigen gene in mouse cells transformed with cloned viral DNA. *J. Virol.* **42**, 100–105 (1982).

P8. Prince, A. M., White, T., Pollock, N., Riddle, J., Brotman, B., and Richardson, L., Epidemiology of hepatitis B infection in Liberian infants. *Infect. Immun.* **32**, 675–680 (1981).

P9. Pugh, J. C., Weber, C., Houston, H., and Murray, K., Expression of the X gene of hepatitis B virus. *J. Med. Virol.* **20**, 229–246 (1986).

P10. Purcell, R. H., and Gerin, J. L., Epidemiology of the delta agent: An introduction. *In* "Viral Hepatitis and Delta Infection" (G. Verme, F. Bonino, and M. Rizetto, eds.), pp. 113–119. Alan R. Liss, New York, 1983.

R1. Rijntjes, P. J. M., Van Ditzhuijsen, T. J. M., Van Loon, A. M., Van Haelst, U. J. G. M., Bronkhurst, F. B., and Yap, S. H., Hepatitis B virus DNA detected in formalin-fixed liver specimens and its relation to serologic markers and histopathologic features in chronic liver disease. *Am. J. Pathol.* **120**, 411–418 (1985).

R2. Rizzetto, M., The delta agent. *Hepatology* **3**, 729–737 (1983).

R3. Robinson, W. S., DNA and DNA polymerase in the core of the Dane particle of hepatitis B. *Am. J. Med. Sci.* **270**, 151–159 (1975).

R4. Robinson, W. S., Genetic variation among hepatitis B and related viruses. *Ann. N. Y. Acad. Sci.* **354**, 371–378 (1980).

R5. Robinson, W. S., and Greenman, R. L., DNA polymerase in the core of the human hepatitis B virus candidate. *J. Virol.* **13**, 1231–1236 (1974).

R6. Robinson, W. S., Clayton, D. A., and Greenman, R. L., DNA of a human hepatitis B virus candidate. *J. Virol.* **14**, 384–391 (1974).
R7. Romet-Lemonne, J. L., McLane, M. F., Elfassi, E., Haseltine, W. A., Azocar, J., and Essex, M., Hepatitis B virus infection in cultured human lymphoblastoid cells. *Science* **221**, 667–669 (1983).
R8. Ruiz-Opazo, N., Chakraborty, P. R., and Shafritz, D. A., Evidence for supercoiled hepatitis B virus DNA in chimpanzee liver and serum Dane particles: Possible implications in persistent HBV infection. *Cell (Cambridge, Mass.)* **29**, 129–138 (1982).
S1. Schalm, S. W., and Heijtink, R. A., Spontaneous disappearance of viral replication and liver cell inflammation in HBsAg-positive chronic active hepatitis: Results of placebo vs. interferon trial. *Hepatology* **2**, 791–794 (1982).
S2. Scotto, J., Hadchouel, M., Hery, C., Yvart, J., Tiollais, P., and Brechot, C., Detection of hepatitis B virus DNA in serum by a simple spot hybridization technique: Comparison with results for other viral markers. *Hepatology* **3**, 279–284 (1983).
S3. Scotto, J., Hadchouel, M., Wain-Hobson, S., Sonigo, P., Courouce, A. M., Tiollais, P., and Brechot, C., Hepatitis B virus DNA in Dane particles: Evidence for the presence of replicative intermediates. *J. Infect. Dis.* **151**, 610–617 (1985).
S4. Scotto, J., Hadchouel, M., Hery, C., Alvarez, F., Yvart, J., Tiollais, P., Bernard, O., and Brechot, C., Hepatitis B virus DNA in children's liver diseases: Detection by blot hybridisation in liver and serum. *Gut* **24**, 618–624 (1983).
S5. Scullard, G. H., Greenberg, H. B., Smith, J. L., Gregory, P. B., Merigan, T. C., and Robinson, W. S., Antiviral treatment of chronic hepatitis B virus infection: Infectious virus cannot be detected in patient serum after permanent responses to treatment. *Hepatology* **2**, 39–49 (1982).
S6. Seeger, C., Ganem, D., and Varmus, H. E., Biochemical and genetic evidence for the hepatitis B virus replication strategy. *Science* **232**, 477–484 (1986).
S7. Sells, M. A., Chen, M. L., and Acs, G., Production of hepatitis B virus particles in Hep G2 cells transfected with cloned hepatitis B virus DNA. *Proc. Natl. Acad. Sci. U.S.A.* **84**, 1005–1009 (1987).
S8. Shafritz, D. A., Presence of hepatitis B virus-deoxyribonucleic acid in human tissues under unexpected circumstances. *Gastroenterology* **89**, 687–690 (1985).
S9. Shafritz, D. A., and Kew, M. C., Identification of integrated hepatitis B virus DNA sequences in human hepatocellular carcinomas. *Hepatology* **1**, 1–8 (1981).
S10. Shafritz. D. A., Shouval, D., Sherman, H. I., Hadziyannis, S. J., and Kew, M. C., Integration of hepatitis B virus DNA into the genome of liver cells in chronic liver disease and hepatocellular carcinoma. *N. Engl. J. Med.* **305**, 1067–1073 (1981).
S11. Shaul, Y., Garcia, P. D., Schonberg, S., and Rutter, W. J., Integration of hepatitis B virus DNA in chromosome-specific satellite sequences. *J. Virol.* **59**, 731–734 (1986).
S12. Shaul, Y., Ziemer, M., Garcia, P. D., Crawford, R., Hsu, H., Velenzuela, P., and Rutter, W. J., Cloning and analysis of integrated hepatitis virus sequences from a human hepatoma cell line, *J. Virol.* **51**, 776–787 (1984).
S13. Shen, H. D., Choo, K. B., Lee, S. D., Tsai, Y. T., and Han, S. H., Hepatitis B virus DNA in leukocytes of patients with hepatitis B virus-associated liver diseases. *J. Med. Virol.* **18**, 201–211 (1986).
S14. Shimizu, Y., Ida, S., Matsukara, T., and Yuasa, T., Determination of hepatitis B virus DNA in serum by molecular hybridization. *Microbiol. Immunol.* **28**, 1117–1123 (1984).
S15. Shimizu, Y., Nambu, S., Kojima, T., and Sasaki, H., Replication of hepatitis B virus in culture systems with adult human hepatocytes. *J. Med. Virol.* **20**, 313–327 (1986).
S16. Shish, C., Burke, K. J., Zeldis, J. B., Chou, M. J., Lee, C, S, Yang, C. S., Wands,

J. R., Isselbacher, K. J., and Goodman, H. M., Molecular characterization of HBV integration sites in human hepatoma tissues. *In* "Viral Hepatitis and Liver Disease" (G. N. Vyas, J. L. Dienstag, and J. H. Hoofnagle, eds), Abstr., p. 631. Grune & Stratton, Orlando, Florida, 1984.

S17. Siddiqui, A., Hepatitis B virus DNA in Kaposi sarcoma. *Proc. Natl. Acad. Sci. U.S.A.* **80,** 4861–4864 (1983).

S18. Siddiqui, A., Satler, F., and Robinson, W. S., Restriction endonuclease cleavage map and location of unique features of the DNA of hepatitis B, subtype adw2. *Proc. Natl. Acad. Sci. U.S.A.* **76,** 4664–4668 (1979).

S19. Smedile, A., Rizzetto, M., Denniston, K., Bonino, F., Wells, F., Verme, G., Consolo, F., Hoyer, B., Purcell, R. H., and Gerin, J. L., Type D hepatitis: The clinical significance of hepatitis D virus RNA in serum as detected by a hybridization-based assay. *Hepatology* **6,** 1297–1302 (1986).

S20. Sninsky, J. J., Siddiqui, A., Robinson, W. S., and Cohen, S. N., Cloning and endonuclease mapping of the heptatis B viral genome. *Nature (London)* **279,** 346–348 (1979).

S21. Song, E., Dusheiko, G. M., Bowyer, S., and Kew, M. C., Hepatitis B virus replication in southern African blacks with HBsAg-positive hepatocellular carcinoma. *Hepatology* **4,** 608–610 (1984).

S22. Southern, E. M., Detection of specific sequences among DNA fragments separated by gel electrophoresis. *J. Mol. Biol.* **98,** 503–517 (1975).

S23. Standring, D. N., and Rutter, W. J., The molecular analysis of hepatitis B virus. *Prog. Liver Dis.* **8,** 311–333 (1986).

S24. Stevens, C. E., Beasley, R. P., Tsui, J., and Lee, W. C., Vertical transmission of hepatitis B antigen in Taiwan. *N. Engl. J. Med.* **292,** 771–774 (1975).

S25. Stibbe, W., and Gerlich, W. H., Variable protein composition of hepatitis B surface antigen from different donors. *Virology* **123,** 436–442 (1982).

S26. Summers, J., Three recently described animal virus models for human hepatitis B virus. *Hepatology* **1,** 179–183 (1981).

S27. Summers, J., and Mason, W. S., Properties of the hepatitis B-like viruses related to their taxonomic classification. *Hepatology* **2,** 61S–66S (1982).

S28. Summers, J., and Mason, W. S., Replication of the genome of a hepatitis B-like virus by reverse transcription of an RNA intermediate. *Cell (Cambridge, Mass.)* **29,** 403–415 (1982).

S29. Summers, J., O'Connell, A., and Millman, I., Genome of hepatitis B virus: Restriction enzyme cleavage and structure of DNA extracted from Dane particles. *Proc. Natl. Acad. Sci. U.S.A.* **72,** 4597–4601 (1975).

S30. Summers, J., O'Connell, A., Maupus, P., Goudeau, A., Coursaget, P., and Drucker, J., Hepatitis B virus DNA in primary hepatocellular carcinoma tissue. *J. Med. Virol.* **2,** 207–214 (1978).

S31. Sureau, C., Romet-Lemonne, J. L., Mullins, J. I., and Essex, M., Production of hepatitis B virus by a differentiated human hepatoma cell line after transfection with cloned circular HBV DNA. *Cell (Cambridge, Mass.)* **47,** 37–47 (1986).

S32. Szmuness, W., Hepatocellular carcinoma and the hepatitis B virus: Evidence for a causal association. *Prog. Med. Virol.* **24,** 40–69 (1978).

S33. Szmuness, W., Hirsch, R. L., Prince, A. M., Levine, R. W., Harley, E. J., and Ikram, H., Hepatitis B surface antigen in blood donors: Further observations. *J. Infect. Dis.* **131,** 111–118 (1975).

T1. Takahashi, K., Miyakawa, Y., Gotanda, T., Mishiro, S., Imai, M., and Mayumi, M., Shift from free "small" hepatitis B e antigen to IgG-bound "large" form in the cir-

culation of human beings and a chimpanzee acutely infected with hepatitis B virus. *Gastroenterology* **77**, 1193–1199 (1979).
T2. Takahashi, K., Kishimoto, S., Ohnuma, H., Machida, A., Takai, E., Tsuda, F., Miyamoto, H., Tanaka, T., Matsushita, K., Oda, K., Miyakawa, Y., and Mayumi, M., Polypeptides coded for by the region pre-S and gene S of hepatitis B viral DNA with the receptor for polymerized human serum albumin: Expression in hepatitis B particles produced on the HBeAg or anti-HBe phase of hepatitis B virus infection. *J. Immunol.* **136**, 3467–3472 (1986).
T3. Takahashi, T., Nakagawa, S., Hashimoto, T., Takahashi, K., Imai, M., Miyakawa, Y., and Mayumi, M., Large-scale isolation of Dane particles from plasma containing hepatitis B antigen and demonstration of a circular double-stranded DNA molecule extruding directly from their cores. *J. Immunol.* **117**, 1392–1397 (1976).
T4. Theilman, L., Klinkert, M. Q., Gmelin, K., Salfeld, J., Schaller, H., and Pfaff, E., Detection of pre-S1 proteins in serum and liver of HBsAg-positive patients: A new marker for hepatitis B virus infection. *Hepatology* **6**, 186–190 (1986).
T5. Thomas, H. C., Karayiannis, P., Fowler, M. J. F., and Monjardino, J., Clinical uses of HBV-DNA assays. *J. Virol. Methods* **10**, 291–294 (1985).
T6. Tiollais, P., and Wain-Hobson, S., Structure and organization of hepatitis B virus DNA. *In* "Viral Hepatitis and Delta Infection" (G. Verme, F. Bonino, and M. Rizetto, eds.), pp. 11–22. Alan R. Liss, New York, 1983.
T7. Tiollais, P., Pourcel, C., and Dejean A., The hepatitis B virus. *Nature (London)* **317**, 489–495 (1985).
T8. Tischer, I., Gelderblom, H., Vettermann, W., and Koch, M. A., A very small porcine virus with circular single-stranded DNA. *Nature (London)* **295**, 64–66 (1982).
T9. Tong, M. J., Stevenson, D., and Gordon, I., Correlation of e antigen, DNA polymerase activity and Dane particles in chronic benign and chronic active type B hepatitis infections. *J. Infect. Dis.* **135**, 980–984 (1977).
T10. Tsuda, F., Naito, S., Takai, E., Akahane, Y., Furuta, S., Miyakawa, Y., and Mayumi, M., Low molecular weight (7s) immunoglobulin M antibody against hepatitis B core antigen in the serum for differentiating acute from persistent hepatitis B virus infection. *Gastroenterology* **87**, 159–164 (1984).
T11. Tur-Kaspa, R., Keshet, E., Eliakim, M., and Shouval, D., Detection and characterization of hepatitis B virus DNA in serum of HBe antigen-negative HBsAg carriers. *J. Med. Virol.* **14**, 17–26 (1984).
V1. Valenzuela, P., Quiroga, M., Zaldivar, J., Gray, P., and Rutter, W. J., The nucleotide sequence of the hepatitis B viral genome and the identification of the major viral genes. *In* "Animal Virus Genetics" (B. Fields, R. Jalnisch, and C. J. Fox, eds.), pp. 57–70. Academic Press, New York, 1981.
V2. Van Ditzhuijsen, T. J. M., Selten, G. C. M., van Loon, A. M., Wolters, G., Matthyssen, L., and Yap, S. H., Detection of hepatitis B virus DNA in serum and relation with the IgM class anti-HBc titers in hepatitis B virus infection. *J. Med. Virol.* **15**, 49–56 (1985).
V3. Varmus, H. E., Do hepatitis B viruses make a genetic contribution to primary hepatocellular carcinoma? *In* "Viral Hepatitis and Liver Disease" (G. N. Vyas, J. L. Dienstag, and J. H. Hoofnagle, eds.), pp. 411–413. Grune & Stratton, Orlando, Florida, 1984.
W1. Weiner, A. J., Wang, K. S., Choo, Q. L., Gerin, J. L., Bradley, D. W., and Houghton, M., Hepatitis delta (d) cDNA clones: Undetectable hybridization to nucleic acids from infectious non-A, non-B hepatitis materials and hepatitis B DNA. *J. Med. Virol.* **21**, 239–247 (1987).

W2. Weller, I. V. D., Fowler, M. J. F., Monjardino, J., and Thomas, H. C., The detection of HBV-DNA in serum by molecular hybridisation: A more sensitive method for the detection of complete HBV particles. *J. Med. Virol.* **9,** 273–280 (1982).

W3. Weller, I. V. D., Carreno, V., Fowler, M. J. F., Monjardino J., Makinen, D., Thomas, H. C., and Sherlock, S., Acyclovir inhibits hepatitis B virus replication in man. *Lancet* **1,** 273 (1982).

W4. Wetmur, J. G., and Davidson, N., Kinetics of renaturation of DNA. *J. Mol. Biol.* **31,** 349–370 (1968).

W5. Will, H., Reiser, W., Weimer, T., Pfaff, E., Buscher, M., Sprengel, R., Cattaneo, R., and Schaller, H., Replication strategy of human hepatitis B virus. *J. Virol.* **61,** 904–911 (1987).

W6. Wong, D. T., Nath, N., and Sninsky, J. J., Identification of hepatitis B virus polypeptides encoded by the entire pre-*s* open reading frame. *J. Virol.* **55,** 223–231 (1985).

W7. Wong, V. C. W., Lee, A. K. Y., and Ip, H. M. H., Transmission of hepatitis B antigens from symptom free carrier mothers to the fetus and infant. *Br. J. Obstet. Gynaecol.* **87,** 958–965 (1980).

W8. World Health Organization Scientific Group, Prevention of primary liver cancer. *Lancet* **1,** 463–465 (1983).

W9. Wu, J. C., Lee, S. D., Wang, J. Y., Ting, L. P., Tsai, Y. T., Lo, K. J., Chiang, B. N., and Tong, M. J., Analysis of the DNA of hepatitis B virus in the sera of Chinese patients infected with hepatitis B. *J. Infect. Dis.* **153,** 974–977 (1986).

W10. Wu, X., Zhou, Y., Feng, Z., Li, Z., and Xia, S., Cloning and restriction mapping of human HBV genome serotype adr. *Sci. Sin., Ser. B (Engl. Ed.)* **26,** 954–960 (1983).

Y1. Yaginuma, K., Kobayashi, M., Yoshida, E., and Koike, K., Hepatitis B virus integration in hepatocellular carcinoma DNA: Duplication of cellular flanking sequences at the integration site. *Proc. Natl. Acad. Sci. U.S.A.* **82,** 4458–4462 (1985).

Y2. Yaginuma, K., Kobayashi, H., Yoshida, E., Kobayashi, M., and Koike, K., Direct evidence for the expression of integrated hepatitis B virus DNA in human hepatoma tissue. *Gann* **75,** 743–746 (1984).

Y3. Yeoh, E. K., Chang, W. K., and Kwan, J. P. W., Epidemiology of viral hepatitis B infection in Hong Kong. *In* "Viral Hepatitis B Infection in the Western Pacific Region: Vaccine and Control" (S. K. Lam, C. L. Lai, and E. K. Yeoh, eds.), pp. 33–41. World Scientific Publ. Co., Singapore, 1984.

Y4. Yoffe, B., Noonan, C. A., Melnick, J. L., and Hollinger, F. B., Hepatitis B virus DNA in mononuclear cells and analysis of cell subsets for the presence of replicative intermediates of viral DNA. *J. Infect. Dis.* **153,** 471–477 (1986).

Y5. Yokosuka, O., Omata, M., Imazeki, F., and Okuda, K., Active and inactive replication of hepatitis B virus deoxyribonucleic acid in chronic liver disease. *Gastroenterology* **89,** 610–616 (1985).

Y6. Yokosuka, O., Omata, M., Imazeki, F., Ito, Y., and Okuda, K., Hepatitis B virus RNA transcripts and DNA in chronic liver disease. *N. Engl. J. Med.* **315,** 1187–1192 (1986).

Y7. Yokosuka, O., Omata, M., Imazeki, F., Okuda, K., and Summers, J., Changes of hepatitis B virus DNA in liver and serum caused by recombinant leukocyte interferon treatment: Analysis of intrahepatic replicative hepatitis B virus DNA. *Hepatology* **5,** 728–734 (1985).

Y8. Yokota, H., Yokoo, K., and Nagata, Y., A quantitative assay for the detection of hepatitis B virus DNA employing a biotin-labelled DNA probe and the avidin-β-galactosidase complex. *Biochim. Biophys. Acta* **868,** 45–50 (1986).

Y9. Yu, M. W., Finlayson, J. S., and Shih, J. W. K., Interaction between various polymerized human albumins and hepatitis B surface antigen. *J. Virol.* **55,** 736–743 (1985).
Z1. Zeldis, J. B., Mugishima, H., Steinberg, H. N., Nir, E., Gale, R. P., *In vitro* hepatitis B virus infection of human bone marrow cells. *J. Clin. Invest.* **78,** 411–417 (1986).
Z2. Zhao, K., Epidemiology of hepatitis B in China. *In* "Viral hepatitis B Infection in the Western Pacific Region: Vaccine and Control" (S. K. Lam, C. L. Lai, and E. K. Yeoh, eds.), pp. 23–28. World Scientific Publ. Co., Singapore, 1984.
Z3. Ziemer, M., Garcia, P., Shaul, J., and Rutter, W. J., Sequence of hepatitis B virus DNA incorporated into the genome of a human hepatoma cell line. *J. Virol.* **53,** 885–892 (1985).
Z4. Zuckerman, A. J., "Hepatitis-associated Antigen and Viruses," pp. 15–25. North-Holland Publ., Amsterdam, 1972.

MONITORING ACID–BASE AND ELECTROLYTE DISTURBANCES IN INTENSIVE CARE

A. Kazda, A. Jabor, M. Zámečník, and K. Mašek

Department of Clinical Biochemistry, Postgraduate Medical and Pharmaceutical Institute, Prague, Czechoslovakia

1. Introduction... 202
2. Osmolality.. 202
   2.1. Introductory Comments... 202
   2.2. Hyper- and Hypoosmolal States 204
3. Sodium... 206
   3.1. Introductory Comments... 206
   3.2. Interpretation of Natremia Findings 207
4. Potassium.. 215
   4.1. Introductory Comments... 215
   4.2. Calculations of Substitution and Correction 216
5. Renal Function from the Aspect of Water and Ion Balance................. 217
   5.1. Introductory Comments... 217
   5.2. Functional Renal Parameters 218
   5.3. Types of Diuresis 221
   5.4. Renal Function in ICU Patients.. 222
6. Acid–Base Balance 227
   6.1. Introductory Comments... 227
   6.2. Acid–Base Balance and Ions 232
   6.3. Metabolic Acidosis 235
   6.4. Respiratory Acidosis... 239
   6.5. Metabolic Alkalosis.. 240
   6.6. Respiratory Alkalosis.. 244
   6.7. Mixed Disorders 245
   6.8. Therapeutic Calculations 247
7. Computer Programs for Monitoring Water, Ion, and Acid–Base Metabolism.. 249
   7.1. Introductory Comments... 249
   7.2. Development and Contribution of Computer Programs 250
   7.3. Approach to biochemical monitoring of ICU patients................... 254
8. Conclusions.. 261
   References... 262

1. Introduction

Why plasma electrolytes, why acid–base control? Many publications available today address these questions. Current laboratory methods for investigation of disorders in the internal environmental balance are known and used. Yet, it is worthwhile to further study the medical aspects of electrolytes and problems of acid–base control. Interpretation of the laboratory findings requires up-to-date knowledge of the integrated function of the lungs, the kidneys, and the blood. Clinical chemists participate increasingly in the diagnosis of disorders and in the control of treatment of patients in intensive care units, in resuscitation wards, in specialized units for the treatment of burns or septic conditions.

Disorders in the internal environmental equilibrium and a disturbance of the internal regulatory mechanisms lead to a restricted supply of oxygen and nutrients (as well as drugs) and to an impaired excretion of catabolites. Hydrogen ion activity, osmolality, and ionic composition of body fluids are disrupted.

Interpretation of these problems is facilitated by the use of a computer. Problem-oriented laboratory findings are the result of the advances achieved in biochemical diagnosis and enable the clinician to take early and comprehensive measures. Computer control is useful not only for the establishment of the diagnosis of the disease and the follow-up of its further course, but also for the proposal of therapy and its monitoring. Computers are able to suggest treatment of disorders in water, ion, and acid–base metabolism as well as the composition of the defined nutrition. Accumulated data serve as a basis for the development of expert systems.

2. Osmolality

2.1. INTRODUCTORY COMMENTS

Reference intervals—serum, 275–295 mmol/kg; urine, 600–1200 mmol/day.

Data for the excretion of the daily osmotic load differ according to the situation. In healthy subjects without an elevated intake of salts and proteins and without an elevated physical activity, the values usually do not exceed 1300 mmol/day (M5, N4, S1). In the case of low intake of proteins and in hospitalized patients, the losses are under 600 mmol/day (W7). In critical states the excretion of the osmotic load is significantly increased. It is a characteristic feature of the catabolic situation and provides information about kidney performance. In 25 critically ill patients with multiorganic failure, the excretion of osmotically active substances in 300 whole-day collections was over 1500 mmol/day in 35% of the findings; a value over 2000 mmol/day was ascertained in not less than 15% of all findings (K4). Excretion of urea is chiefly responsible for the high values, but ions, frequently glycosuria in impaired glucose tolerance, and osmotherapeutic agents and/or low-molecular-weight substances excreted in renal intoxications also contribute.

In addition to the estimation of the excreted amount of osmotically active substances it is necessary to determine their concentration in the urine. This makes it possible to evaluate the existing concentration abilities of the kidneys and, in comparison with serum osmolality, it is one of the main parameters for establishing a differential diagnosis of prerenal or renal cause of the oligoanuric state (C7). The limits of osmotic urine concentration are 50–1200 mmol/kg; commonly occurring concentrations range from 300 to 900 mmol/kg. After a 12-h period of water deprivation, urine osmolality should rise to at least 850 mmol/kg (T2). When estimating the osmotic urine concentration, it is naturally necessary to consider also the decline in the kidney's concentration ability in relation to the patient's age (E2, S1).

Assessment of urine and plasma osmolality under conditions of dehydration with subsequent vasopressin administration serves for the establishment of differential diagnosis of the causes of hypotonic polyuria. This may be psychogenic, or it may occur in diabetes insipidus and in nephrogenic diabetes insipidus (S1, T2, W7).

Serum osmolality is maintained within relatively narrow limits. With its rise above 278 (ranging from 277 to 282) mmol/kg, the secretion of ADH begins to increase. The corresponding values of natremia are 137 (ranging from 136 to 140) mmol/liter (M16). The further increase in ADH secretion then continues until an osmolality of 296–298 mmol/liter is reached, i.e., when the maximum is attained. At the same time, osmotic urine concentration increases progressively. A further pathological rise in osmolality does not further increase the ADH secretion (E2). Another cause of stimulated ADH secretion is a decline in the circulating fluid volume of 10–20% or a decline in blood pressure of at least 5% (M16, W7). A similar effect is also produced by anxiety, trauma-induced pain, and certain drugs (opiates, barbiturates, etc.). Renal action of ADH is potentiated by some drugs (chlorpropamide, acetaminophen) and by oxytoxin. Certain tumors produce antidiuretic substances. Antidiuresis then also plays a part in normo- or hypoosomolal states. The effect of dopamine on ADH secretion has not been demonstrated (M16).

2.2. Hyper- and Hypoosmolal States

Causes of hyperosmolality include loss of water, acute catabolism, diabetic coma, hyperosmolal coma without acidosis, burns, nephritis, severe sepsis, acute intoxication with low-molecular-weight substances, diabetes insipidus, nephrogenic diabetes insipidus, and salt water ingestion (in swimming accidents).

The cause of hyperosmolality in acute catabolism, i.e., in shock, is the accumulation of metabolic intermediate products in cells. Because of energy disorders these can be neither metabolized into final products nor released from tissues. The result is hyperosmolality of intracellular fluids (ICFs) against extracellular fluids (ECFs), leading to a shift of water into cells. The reduction in the volume of ECFs often deteriorates the previously primarily affected circulatory system.

In intensive care, acute hyperosmolality may also develop iatrogenically. Among the causes, for example, are incorrect doses of parenteral or enteral nutrition, dialysis using hyperosmolal solution, high $NaHCO_3$ doses in cardiac resuscitation (M10), transcutaneous propylene glycol absorption in the treatment of burns, or osmotherapy with glycerol or other small-molecular-weight substances. In the latter situations, a large difference arises between the calculated and the measured osmolality, i.e., the osmolal gap, which can range between 100 and 150 mmol/kg (F3).

Causes of hypoosmolality include metabolic response to trauma, excess water intake, replenishment of lost isotonic fluid with water, chronic catabolism, fresh water ingestion (in swimming accidents), and excessive ADH excretion.

Following cerebrospinal injury, severe disorders may take place in the regulation of water and ion conservation. This regulation represents a complicated interconnection of diencephalo–hypothalamic functions with the functions of peripheral receptors and effectors (kidneys) as well as with the endocrine organs. Two types of disorders may develop (G1): (a) Diabetes insipidus may occur due to insufficient ADH secretion, with typical polyuria and serum osmolality at the upper limit of the reference values or elevated up to 330 mmol/kg. Hyperosmolality does not develop when fluids are suitably balanced. Urine osmolality is low, and vasopressin administration increases it significantly. (b) Unsuitable rise in ADH secretion may occur in brain injuries and tumors, leading to hyponatremia and hypoosmolality; the patients are at risk of cerebral edema. Urine osmolality is higher than serum osmolality.

Clinically, changes in osmolality are attended by the development of metabolic encephalopathy that resembles encephalopathies arising from other metabolic causes. Clinical symptoms appear in cases of acute sodium elevation (hypernatremia), with values above 150 mmol/liter, and with osmolality above 310 mmol/kg. In a chronic state symptoms develop only when Na^+ is above 160 mmol/liter and osmolality is above 330 mmol/kg. The decline in Na^+ and osmolality is manifest clinically in acute states when the values are below 127 mmol/liter and below 265 mmol/kg, respectively. In chronic states, the corresponding values are again more extreme, namely for Na^+ below 110–120 mmol/liter and for osmolality below 240–250 mmol/kg (M11).

Regardless of the causes, both hyper- and hypoosmolality produce symptoms of metabolic encephalopathy, the pathophysiological basis of which lies in diffuse neuronal functional disorders with a possible focal maximum. The states range from moderate neuropsychic disorders associated with unspecific motor symptoms, to delirium and, eventually, to coma. The development of a hyperosmolal state is accompanied with confusion and hallucinations that, in elderly persons, are sometimes erroneously considered to be manifestations of cerebral sclerosis. Typical symptoms are thirst and headache. In elderly persons certain factors controlling the water balance are obviously changed. The sensation of thirst is weaker and osmoregulatory secretion of vasopressin is increased but the renal response to it is paralyzed (E2). In hypernatremia hemorrhagic encephalopathy may develop. In cerebrospinal fluid, the number of cells is normal, proteins are over 0.5 g/liter, and the coloring is xanthochromic or hemorrhagic. An EEG demonstrates unspecific or focal changes in bleeding and computerized axial tomography (CAT) scans show a diminishing of the brain volume and/or the presence of hemorrhagia.

When acutely developed hyperosmolality persists, the concentration of active particles increases due to a compensatory mechanism of the brain. This enables a return of fluid levels to the original volume. The active particles are called idiogenic osmoles. They are probably metabolites of actual cell proteins, i.e., amino acids and ammonia (M11). In addition, other authors report an increase in the concentration of ions and glucose in hyperglycemia (M12, W7). Experimentally, this development takes place in hyperglycemia as well as in hypernatremia within the first few hours (W7); according to other authors it occurs in hypernatremia only in the course of a week (M12).

Hypoosmolal states are accompanied by weakness, nausea, apathy, and headache. Diffuse cerebral edema develops with a risk of brain stem herniation; protein content in cerebrospinal fluid is low, typically under 0.1 g/liter. ECG demonstrates unspecific or possibly epileptiform changes and CAT scans confirm diffuse cerebral edema.

Values of 230–490 mmol/kg are reported as limits of serum osmolality (T2). However, extreme values have also been described (H5, K5, L5):

hypoosmolality under 200 mmol/kg developed iatrogenically in a course of treatment requiring diuretics and a salt-free diet. The lowest serum osmolality measured by us was 213 mmol/kg in a patient with Addison's disease and diarrhea, who was treated at first with hypotonic solutions. After a long-term adequate treatment the patient survived. Hyperosmolality above 500 mmol/kg was ascertained in patients with combined ethyl and methyl alcohol intoxication.

Because no correlation has been proved to exist between mortality and the clinical and biochemical findings obtained at the time of admission of patients with hyperosmolal, hyperglycemic, or nonketonic syndrome, hyperosmolality cannot be regarded as being a prognostically important information in these states (R2).

Correction of osmolality disorders must proceed slowly, so that osmolality does not change more rapidly than by 2–4 mmol/kg/h, and/or natremia by 1–2 mmol/h, or that the change in osmolality within 24 h does not exceed 20–30 mmol/kg. Reasons for caution and for monitoring or biochemical findings are the above-mentioned alterations in the region of the brain, wherein—after a more rapid normalization, especially of hyperosmolality—the patient's clinical state may deteriorate due to intracranial expansion (N4). Correction of hyper- and hyponatremic states in relation to the patient's clinical situation is discussed in the Section 3. It should be recalled here that every patient should be approached individually and that only careful monitoring of both the biochemical and the clinical states is decisive for the estimation of whether treatment is adequate (W11).

3. Sodium

3.1. INTRODUCTORY COMMENTS

Reference intervals—serum, 132–142 mmol/liter; urine, 120–240 mmol/day.

Na^+ homeostasis is of constant concern in intensive care settings. Numerous outstanding studies of Na^+ homeostasis can be found in the literature (B4, E1, F1, G4, N3, R5, Z1); on the other hand, the interpretation of Na^+ levels remains nevertheless difficult, namely because of the different clinical situations accompanying disorders in Na^+ metabolism. The determination of optimum therapeutic doses is also not simple.

For the physician, these tasks can be facilitated by the computer—an example is given at the end of the paper. At this point an evaluation of the relationships between Na^+ levels, the Na^+ stores in the ECF, and the state of the patient's hydration will be discussed.

Hyponatremia is the expression of a relative water excess in relation to Na^+ supply. Hypernatremia is the expression of a relative water deficit in relation to Na^+ supply. Neither of these findings says anything about the real state of Na^+ supply in the ECF, which in both cases may be elevated, normal, or decreased, depending on the size of the distribution space, i.e., the ECF volume. Normonatremia may also be accompanied, depending on the size of the distribution space, with an elevated, normal, or decreased Na^+ supply.

The above-mentioned relationships may significantly influence the calculations used for the correction of Na^+ levels. Moreover, laboratory findings may be influenced by effective osmolality. Hyperglycemia, very frequently present in sepsis and other catabolic states, causes an osmotic shift of water from cells to the ECF. Per every 5 mmol/liter of its elevation, Na^+ levels decrease by 1.5 mmol/liter (F1). After the normalization of glycemia, Na^+ levels increase again. This should be taken into consideration in estimates of Na^+ concentration in severe hyperglycemic patients.

3.2. Interpretation of Natremia Findings

In the investigation of Na^+ metabolism, measurable Na^+ losses can be determined by commonly available methods, whereas unmeasurable losses (in sweat, in diarrhea, or into the ileum or peritoneal cavity) can be estimated only approximately. For the correct interpretation of Na^+ status, for an estimate of Na^+ stores in the ECF, and for a determination of the required therapeutic dose (i.e., the dose to compensate for the existing deficits and to meet current needs), two groups of data, laboratory and clinical, must be known. Laboratory data include Na^+ levels, glucose levels, osmolality, and measures of Na^+ losses (via urine, gastric tube, or catheter). Clinical data include state of hydration; a comparison is made of the normal and existing body weights, or at least the estimated body weight changes are taken as a basis. Further, it has to be decided whether the state is acute or chronic.

The following survey of the problems encountered in Na^+ homeostatis is based on a combination of clinical and laboratory findings, both of which should be considered inseparable in the interpretation of the problems.

3.2.1. *Normohydration with Hypo- or Hypernatremia*

3.2.1.1. *Normohydration with Hyponatremia.* This situation develops in both acute and chronic states. In acute states, the etiology and pathobiochemistry include losses of more or less isotonic liquids that are replaced sufficiently, but only by glucose or water. There is a decrease in Na^+ stores in the ECF, but the ECF volume remains unchanged. An osmotic gradient develops between the ECF and ICF, with some water shifted to the ICF.

Clinically, in acute states, hypoosmolality symptoms may appear in conjunction with a sudden decrease in Na^+ to less than 127 mmol/liter. In acute, asymptomatic states, isotonic salt solutions are usually sufficient to correct the problem. Only in symptomatic hyponatremia with an acute onset is Na^+ supplied in the form of hypertonic solutions; the dose is calculated according to the following equation:

$$\text{dose of } Na^+ \text{ mmol} = (\text{target } Na^+ - \text{measured } Na^+) \times F \times \text{kg} \quad (1)$$

The target Na^+ value is in the middle between measured hyponatremia and the reference value, i.e., 137 mmol/liter. F is a factor for total body fluid (for males, 0.6., for females, 0.55). The more acute the hyponatremia, the more careful its normalization must be. Target Na^+ levels should not be reached in less than 8 h; that means usually that Na^+ levels should not rise more rapidly than by 1–2 mmol/liter/h. The dose calculated using the factor F should be regarded as the upper limit. F is based on the change of the distribution space of Na^+ after its administration when part of water is transferred from the ICF to the ECF. In extreme hyponatremia, the doses thus determined are very high. Then, considering the patient's clinical status, even a lower value of the factor F can be selected; in extreme cases it may be as low as 0.2 (the calculation applies only to the ECF). Depending on the course of Na^+ homeostasis, the correction doses are then specified.

In chronic states, the etiology and pathobiochemistry include catabolic atrophy of cell structures and decreased tonicity in the ICF, with water migration from cells to the ECF and consequently dilution of the ECF. Hypoproteinemia is usually present and the osmostat is reset (A3). Patients have hyponatremia that does not return to normal with an increased Na^+ supply; the Na^+ excretion increases. Moreover, in chronic conditions the course described in acute situations may develop.

Clinically, in chronic states hyponatremia is symptomatic only when Na^+ levels decrease to 110–120 mmol/liter and lower (M11). Corrective therapy primarily involves management of the causes, control of hypoproteinemia, and monitoring and maintaining the patient's energy balance.

3.2.1.2. *Normohydration with Hypernatremia.* The etiology and pathobiochemistry include the iatrogenic condition following the administration of concentrated salt solutions. Na^+ supply is increased, the amount of water is physiological. The ECF becomes hypertonic as compared with the ICF, and water leaves the cells. Expansion of the ECF takes place to the detriment of the ICF.

Clinically, symptoms of hyperosmolality are present; in extreme cases there is a risk of pulmonary edema. For therapeutic management, corrective calculations are of value only for orientation to the patient's status and should be made immediately after the onset of the disorder; later they are influenced by renal excretion of water and ions.

$$Na^+ \text{ excess} = F \times kg \times (\text{measured } Na^+ - 137) \qquad (2)$$

$$H_2O \text{ deficit} = F \times kg \times \left(\frac{\text{measured } Na^+}{137} - 1\right) \qquad (3)$$

Diuretics that act in Henle's loop are administered. Because the diuretic-induced loss of water is higher than the loss of Na^+, hyperosmolality is brought under control by 5% glucose infusions. In case of renal failure, dialysis treatment is necessary.

3.2.2. Dehydration with Normo-, Hypo-, or Hypernatremia

For the assessment of the seriousness of dehydration it is helpful to know the difference between the normal and the existing body weight. Because in adult patients this information is generally not available, the deficit is only estimated. When the clinical symptoms include slight dehydration, a deficit of 2.5% of total body weight is adequate; with more serious dehydration, a deficit of 5% of total body weight is calculated (D1).

3.2.2.1. Dehydration with Normonatremia. The etiology and pathobiochemistry include loss of water and ions in the ratio in which they are physiologically present in the ECF. The losses may be external (gastrointestinal, renal) or into the third space (ileum, peritoneal cavity). Na^+ levels and osmolality are normal, no osmotic gradient develops between the ECF and ICF, and there is no movement of water from cells to the ECF. The whole loss thus takes place to the detriment of the ECF. Oliguria is present, Na^+ in the urine is very low, and osmolality is high owing to the concentration of urea. A gradual increase is found in serum urea and hemoglobin.

Clinically, there is a risk of circulatory collapse, metabolic acidosis, and oliguria; turgor of the skin is decreased and cramps may occur in the calves. Laboratory findings cannot be used for corrective calculations. Isotonic liquids are replaced according to the estimated fluid loss. Their composition depends on the biochemical findings, chiefly the acid–base status. With long-term pathological losses, the substituted amount of water and ions is often high.

3.2.2.2. *Dehydration with Hyponatremia.* The etiology and pathobiochemistry include extrarenal or renal losses of salts and water. They may be isotonic with body fluids but are replaced by an insufficient volume of liquids without ions, e.g., glucose. The decline in Na^+ supply is relatively higher than the loss of fluids. Some water shifts from the ECF to the ICF.

Extrarenal losses may be due to gastrointestinal injury, burns, or shift into the third space. Hypovolemia stimulates the regulatory renin–angiotensin system; aldosterone and ADH increase. Consequently, changes develop in renal hemodynamics. The amount of urine is small and the concentration of Na^+ and Cl^- is below 20 mmol/liter. An exception may be metabolic alkalosis when, together with HCO_3^-, more Na^+ is excreted.

Renal losses may be due to overdose of diuretics, primary adrenal insufficiency, nephropathy with a rise in Na^+ losses, or proximal renal tubular acidosis. Na^+ and Cl^- concentration is usually above 20 mmol/liter. An exception is found in states following diuretic overdose, when, after discontinuation of treatment, ion concentration in the urine is low.

Clinically, symptoms are chiefly the result of ECF deficits and the decrease in the volume of circulating fluids and/or hypoosmolality. For corrective therapy, complete substitution for the water and Na^+ deficit has two parts. The first involves replacing the lost isotonic fluids with salt solutions, possibly by as much as 5–10% of the total body weight. The second part involves normalization of measured hyponatremia by calculating the target Na^+ according to Eq. (1). The pathophysiological mechanisms, however, are primarily normalized by administration of isotonic salts. Often, colloid solutions also have to be administered (e.g., in hypoproteinemia).

3.2.2.3. *Dehydration and Hypernatremia.* In dehydration associated with hypernatermia the supply of Na^+ may be decreased, physiological, or increased.

3.2.2.3.1. *Dehydration with hypernatremia and physiological Na^+ supply* (Fig. 1, situation A). The etiology and pathobiochemistry develop as a consequence of losses of more or less pure water, e.g., in fever, during exposure to a dry and hot environment, in hyperventilation, because of insufficiently moisturized artificial ventilation, or in cases of losses of heavily diluted urine in diabetes insipidus (also after brain injuries).

Osmolality of the ECF increases, the produced osmotic gradient leads to the shift of water from the ICF to the ECF. The loss of water thus involves all of the body fluids, not only the ECF. Condensation of plasma proteins facilitates the maintenance of circulating fluids.

Clinically, symptoms of hyperosmolality predominate. Development of vascular collapse is late, and so is the increase in urea. Oliguria is present,

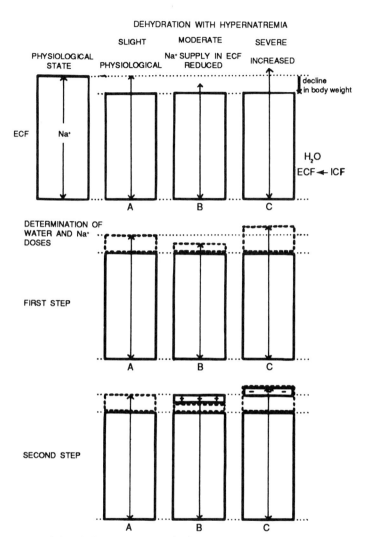

FIG. 1. Possible relations between Na^+ supply and water deficit in dehydration with hypernatremia. First step: relation between actual deficit of body weight and deficit of free water, calculated according to Eq. (4). Second step: relation between free water deficit and deficit of isotonic fluid in three types of dehydration with hypernatremia: (A) only deficit of free water is present; (B) it is necessary to administer both isotonic fluid and free water; (C) deficit of free water is present together with Na^+ excess.

urine is characterized by high osmolality and high urea levels, and concentration of Na^+ in urine is very low. Turgor of the skin does not decrease markedly. Corrective calculations for administration of liquids are based on Na^+ levels:

$$H_2O \text{ deficit} = kg \times F \times \left(1 - \frac{137}{\text{measured } Na^+}\right) \quad (4)$$

To prevent a sudden change in osmolality, glucose is administered in combination with physiological salt solutions, diluted to various degrees (to two-thirds, one-half, or even one-fourth of the usual concentration) according to the state of natremia. In a severe disorder the treatment is over 2–3 days, at the beginning using 9 g saline/liter.

3.2.2.3.2. *Dehydration with hypernatremia and decreased Na^+ supply* (Fig. 1, situation B). The etiology and pathobiochemistry include losses of hypotonic juices in gastroenteritis, osmotic diuresis (glycerol, manitol, urea), perspiration, and peritoneal dialysis using the hypertonic dialyzer. The loss of hypotonic fluid can be understood partially as a loss of pure water, partially as a loss of isotonic fluid. Na^+ levels do not serve as a reliable indicator for estimating the loss of fluids.

Clinically, symptoms of hyperosmolality are present. Disturbance of the ECF is more serious than in the previous situation because the isotonic proportion of the lost fluids was to the detriment of the ECF only. Therapeutic correction has two steps: (1) By calculation according to Eq. (4) the deficit of pure water is determined. This can once again be replaced by administering glucose with a suitably diluted saline. (2) For that part of the decrease in body weight that remains to be increased after water supplementation based on the preceding calculation, solutions of isotonic salts can be administered.

3.2.2.3.3. *Dehydration with hypernatremia and increased Na^+ supply* (Fig. 1, situation C). This situation is more or less hypothetical; such a state develop in the case of a combination of the previous states, leading to dehydration during increased (chiefly iatrogenic) Na^+ administration (salts of antibiotics, inappropriate concentrations of hypertonic salt solutions). Therapeutic correction according to the calculated water deficit due to natremia, i.e., according to Eq. (4), would supply more fluids than the patient actually needs. That is why it is necessary to decrease the calculated substitution to an amount by which the supply would exceed the patient's usual body weight. Even heavily diluted salt solutions used in combination with glucose for the prevention of a rapid decline in osmolality further increase Na^+ supply. For this reason it is necessary, after correction of the volume with persisting natremia, to administer diuretics and 5% glucose.

To summarize, in dehydration with hypernatremia the Na^+ supply is not typically known in practice. By means of relationships discussed above and given in Fig. 1, Na^+ supply is estimated. There exist only three possibilities derived from Eq. (4): the H_2O deficit (1) is equivalent to the decrease in body weight—Na^+ supply is physiological, only water is decreased; (2) is less than the decrease in body weight—Na^+ supply is decreased, water and isotonic fluids are decreased; and (3) is greater than the decrease in body weight—Na^+ supply is increased, water is decreased, and Na^+ is in excess.

3.2.3. Hyperhydration with Normo-, Hypo-, or Hypernatremia

For the estimation of the seriousness of hyperhydration, it is best to know the difference between the normal and the existing body weights. Because in adult patients this information is frequently unavailable, the degree of fluid excess is often the only parameter used.

3.2.3.1. *Hyperhydration with Normonatremia.* The etiology and pathobiochemistry involve states in which water and ions are gained in the ratio in which they are physiologically present in the ECF. This happens, for example, after excessive infusions of salt solutions in hemorrhage and in states of impaired excretion of water and ions. Water and ion supply in ECFs increase, osmolality and Na^+ levels do not change. No osmotic gradient develops between the ECF and ICF, and no fluid shifts are present. Hemoglobin and total proteins decrease.

Clinically, blood supply to the jugular veins and central venous pressure are elevated, body weight increases, and edema appears. Diuresis is often decreased because of the primary disease. Therapeutic correction involves control of intake of salt solutions, diuretics, and cardiotonics.

3.2.3.2. *Hyperhydration with Hyponatremia.* The situations may vary, with Na^+ supply being increased, physiological, or decreased. The increase in volume of the ECF is to some extent moderated by the shift of water to cells, the degree of which depends on the magnitude of the osmotic gradient between the ECF and ICF. Essentially, two states may develop, acute and chronic.

In acute states, the etiology and pathobiochemistry include water intoxication that occurs during rapid rehydration, e.g., in the treatment of acute hypovolemia with glucose infusions. The risks rises after administration of drugs with an antidiuretic effect (barbiturates, opiates), during labor (oxytocin acts as an antidiuretic hormone), in states of chronically impaired water excretion, when drinking nonionic solutions after sweating, and in schizophrenia (release of ADH); in addition, water intoxication in mental patients is not uncommon (J3).

Clinically, Na^+ levels that decrease more rapidly than 1–2 mmol/liter/h lead to a feeling of bloating and to anorexia, nausea, vomiting, muscle spasms, and possibly lethargy and disorientation. These complaints are obviously related to changes in the volume of tissue cells. Physical examination demonstrates signs of elevated intracranial pressure due to brain expansion. Mortality is over 50%; the frequency of irreversible cerebral effects in survivors is not known. Therapeutic correction involves calculation of the target Na^+ according to Eq. (1). The measures used are the same as those given for normohydration with hyponatremia. The more extreme the initial situation, the more important it is to provide treatment. Correction of the disorder in an interval shorter than 12–24 h is wrong; the risk is high chiefly in old persons and in cardiovascular patients.

In chronic states, the etiology and pathobiochemistry include edema due to heart failure, nephrotic syndrome, and cirrhosis with hypoproteinemia. An osmotic gradient develops between the ICF and ECF and fluids shift outside the vascular blood bed. The kidneys respond as in hyponatremia and hypovolemia. Diuresis decreases to less than 800 ml/liter and urinary Na^+ decreases to less than 10–20 mmol/liter.

Clinically, symptoms of the primary disease are present—edema and depletion of fluid volume to various degrees. Hyponatremia, as long as it is not severe, is usually asymptomatic. Symptoms are present when Na^+ levels are 110–120 mmol/liter or lower. Therapeutic correction involves consideration of the seriousness of the hyponatremia. With Na^+ values less than 130 mmol/liter, it is necessary to ensure that water balance is not positive; at 120–130 mmol/liter the intake of liquids should be restricted to less than 1000 ml/day. When symptoms of hypoosmolality appear, the supply of liquids from the proximal tubule to the distal portion of the nephron must be increased by means of plasma expanders, including blood derivatives, acetozolamide, or manitol. Then diuretics acting distally may be useful, such as furosemide, thiazides, and spironolactone. Improvement of the state of the body energy balance is likewise an important part of the treatment.

Pathobiochemically, a similar situation may also arise in acute renal failure due to water and Na^+ overloading. In case of such symptoms of acute hypoosmolality, therapy consists of the infusion of a hypertonic salt solution with simultaneous dialysis.

3.2.3.3. *Hyperhydration with Hypernatremia.* The etiology and pathobiochemistry include iatrogenic administration of concentrated salt solutions. The ECF is becomes hypertonic in relation to the ICF, and water leaves the cells. Expansion of the ECF takes place to the detriment of the ICF.

Clinically, there is risk of pulmonary edema. With normal renal function, a prompt diuresis develops following intake of diuretics. Therapeutic correction involves evaluation of the seriousness of the state by calculating the Na$^+$ excess according to the Eq. (2). In fact, the calculation takes into consideration the part of the Na$^+$ increase that is not covered by the simultaneous increase in water supply. The calculation of the water deficit according to Eq. (3) determines the amount of water that would bring the measured hypernatremia to 137 mmol/liter.

The actual treatment consists of administering diuretics that act in Henle's loop. With this treatment the loss of water will be relatively higher than the loss of Na$^+$. For this reason, it is essential to administer simultaneously infusions of 5% glucose so that osmolality will not increase.

4. Potassium

4.1. Introductory Comments

Reference intervals—serum, 3.8–5.4 mmol/liter; urine, 40–90 mmol/day.

Surveys of problems concerning K$^+$ metabolism have been published repeatedly (B4, C3, C8, E4, N4, T2). The focus herein will therefore be limited only to the evaluation of potassium status and the methods of determining therapeutic doses for patients receiving intensive care. The estimation of K$^+$ status and K$^+$ balance is based on their links with energy metabolism and acid–base status (ABS). Although it is not possible in practice to examine intracellular conditions, the pH and K$^+$ status are their reflection. With a pH decline of 0.1, an elevation of K$^+$ levels takes place, on an average by 0.4 mmol/liter (A1, D3, E4), with a variability from 0.1 to 0.9 mmol/liter. In acidosis due to an accumulation of organic acids, this effect is almost negligible (O2); the influence of inorganic acids, on the other hand, is more marked (D3). The ABS effect on K$^+$ status is modified, in a general way, by further factors: $-\Delta K^+/\Delta pH$ is more pronounced in acidemia than in alkalemia and in metabolic disorders than in respiratory ones, and it increases in intensity with the duration of acidosis (B11).

The evaluation of the relationship of pH and K$^+$ status in 1409 findings of hospitalized persons (K5) revealed the dependence given in Fig. 2. The graph shows the logical implications of the hazard of K$^+$ doses related to the actual metabolic situation of the patient. If, for example, in simultaneous severe acidemia and sufficient diuresis, normo- or hypokalemia is present, the risk of overdosage is negligible. If alkalemia is accompanied with normo- or hyperkalemia, the risk of overdosage rises with the pH normalization.

When considering the K$^+$ dose, it is also necessary to estimate other

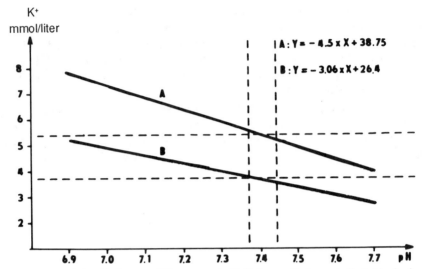

FIG. 2. Relationship between K^+ status and pH. Reference limits are indicated by broken lines.

relationships, such as K^+ excretion in the urine and in other secretions, diuresis, and nitrogen balance.

4.2. Calculations of Substitution and Correction

The substitution dose is calculated to include renal losses and/or other secretions, corrected to nitrogen balance. Per every gram of a positive nitrogen balance, 3 mmol of K^+ are added and vice versa. The correction dose for the modification of K^+ to 4.4 mmol/liter is

$$K1 = [\text{body weight} \times 0.2 \,(4.4 - \text{measured } K^+)] \times F$$

The correction dose for the modification of K^+ to 4.4 mmol/liter with the assumed change of pH to 7.40 is

$$K2 = \{\text{body weight} \times 0.2\,[(33.05 - 3.87 \times \text{pH}) - \text{measured } K^+]\} \times F$$

F is the factor for the modification of the correction dose with regard to diuresis:

$F = 1$ with diuresis under 300 ml/day

$F = \left(\dfrac{\text{diuresis in ml/day} - 300}{450}\right) + 1$ with diuresis 300–1200 ml/day

$F = 3$ with diuresis over 1200 ml/day

5. Renal Function from the Aspect of Water and Ion Balance

5.1. INTRODUCTORY COMMENTS

In view of the central role played by kidneys in the control of water and ion balance (B4, S1, T2), it is necessary to test regularly renal function in patients in critical states. Acute renal failure may be produced by a number of factors. Besides pathophysiological mechanisms related to the underlying disease, iatrogenic influences are also of importance. The possible causes can be classified as follows (B1, B2, C7, K5):

1. Prerenal: reduction of effective intravascular volume, decrease of cardiac output
2. Renal: ischemia of the kidneys, nephrotoxins (chlorohydrocarbons, ethylene glycol, heavy metals such as *cis*-platinum, *Amanita phaloides*, antibiotics, aminoglycosides, radiopaque solutions, etc.), allergic or drug-induced interstitial nephritis
3. Postrenal: ureteral, cystic, urethral obstructions

Pathophysiological mechanisms influencing renal function in critical states include hypotension, in which a redistribution of intrarenal blood flow has been ascertained. The fraction flowing through the region of internal cortical nephrons and through the marrow increases. In case of a relatively large decline of perfusion in the renal cortex, an impediment occurs in the supply of urea and ions from tubules to the medullary interstitial tissue, where they otherwise continuously complete the medullary osmotic gradient. This leads to a gradual "washing out" of this gradient, and the kidney loses the ability to form concentrated urine (B10, M14, S1). The medullary osmotic gradient is disturbed also by an increased flow of the tubular fluid that is due to a decline in the fractionary resorption of solutes and water, e.g., in diuretic treatment, often using high doses of drugs (S1).

There are practical implications: in catabolism, the patient must often excrete a high amount of osmotically active substances. Thus, for example, when he has to excrete 2000 mmol/day and his ability to concentrate urine is maximally only about 500 mmol/kg, he must produce 4000 ml of urine. If such a diuresis is not ensured, hyperosmolality develops. These aspects must be considered in the estimation of whether diuresis is adequate.

In critical states the glomerular filtration rate (GFR) may also be influenced significantly. In addition to the more common GFR decrease seen in prerenal or renal failure, an increase in the GFR may be present. This has been observed in patients suffering from burns and in postoperative states. The elevation is regarded as a physiological compensation following an adequate hydration. In uncomplicated postoperative states, the correlation between the cardiac index and the GFR was positive (B10, L6). It has also been reported that water and osmotic diuresis, including the action of furosemide and ethacrynic acid, increase the GFR. Manitol increases blood flow through the kidneys (B5, C1, S1). These circumstances also influence the administered doses of drugs (antibiotics, cardiotonics, etc.) excreted by the kidneys (K12, L4).

Most renal failures in patients in critical states have a nonoliguric course, namely thanks to intensive rehydration and diuresis-stimulating drugs (furosemide, manitol) as well as drugs improving blood supply to the kidneys (dopamine) (H1). The onset of oliguria and the elevation of creatinine are late symptoms of renal failure. Thus certain renal function parameters have been proposed as predictors of the onset of renal failure. The most important parameters are creatinine clearance (C_{cr}) and free water clearance (C_{H_2O}) (B10). These measures warn of the onset of renal failure 12–24 prior to the development of oliguria and the elevation of creatinine. Because the medullary osmotic gradient is markedly impaired, it is indispensable for the prediction and the monitoring of nonoliguric renal failure, in turn monitoring also other parameters characterizing tubular functions, i.e., fractional excretions (see below). Oligoanuria is the terminal stage, occurring in cases where it was impossible to prevent the progressive deterioration of renal functions. An early diagnosis and an intensive therapy enable reversion of this adverse course and contribute to a gradual normalization of renal functions.

5.2. Functional Renal Parameters

5.2.1. Clearances

Reference intervals—C_{cr}, 1.150–2.350 ml/s/1.73 m³; C_{osm}, 0.050 ml/s; C_{H_2O}, −0.027 to −0.007 ml/s.

Calculations of clearances:

$$C_{cr} = \frac{U_{cr} \times V}{P_{cr}}$$

$$C_{osm} = \frac{U_{osm} \times V}{P_{osm}}$$

$$C_{H_2O} = V - C_{osm}$$

where U_{cr} and P_{cr} are the creatinine concentrations in the urine and the plasma in mmol/liter, U_{osm} and P_{osm} are the osmolality of the urine and plasma in mmol/kg, V is the urine volume in ml/s, and C_{cr}, C_{osm}, and C_{H_2O} are the creatinine, osmolal, and free water clearances.

In the evaluation of clearances, the interpretation of C_{cr} is common and is mentioned only briefly here. Attention was drawn earlier to the possible causes of C_{cr} elevation in critical states. In its evaluation it should be taken into consideration that an age-related decline in the values takes place (B5, P1, S1). For the calculation of the mean age-related value, the following equation can be used:

$$Y = -0.00946 \times \text{age in years} + 2.118$$

The following evaluation based on this calculation can be used also in computer processing:

Actual C_{cr}	Estimation of filtration
$1.3 \times Y$ or more	Increased
$0.71 \times Y$ to $1.30 \times Y$	Reference interval
$0.51 \times Y$ to $0.70 \times Y$	Decreased
$0.26 \times Y$ to $0.50 \times Y$	Considerably decreased
$0.26 \times Y$ or less	Critically decreased

Despite all reservations, the C_{cr} remains the only method available in routine practice for the estimation of the GFR.

Osmolal clearance (C_{osm}) provides information on the amount of plasma cleared from osmotically active substances during the flow through kidneys. In hypercatabolic states with a high production of osmotically active substances (urea, hyperglycemia) the value of C_{osm} is usually increased, unless C_{cr} is decreased. The condition is generally accompanied by polyuria and a high rate of excretion of osmotically active substances in the urine. This is termed overflow osmotic diuresis.

Free water clearance (C_{H_2O}) is the measure of urinary excretion of water that does not contain osmotically active solutes. Because the urine is usually osmotically more concentrated than the plasma, it would be necessary to add water to the urine to attain isoosmolality with the plasma. The reference interval is therefore negative. In hypotonic urine, pure water is present in the urine in excess and C_{H_2O} is positive and expresses the amount of water that has to be removed from the urine to make it isoosmolal with the plasma. In critical states, the medullary osmotic gradient is often impaired for the reasons mentioned above. Then C_{H_2O} rises, typically above the reference interval. This increase can be regarded as pathological only when plasma osmolality is increased simultaneously. The point is that C_{H_2O} also increases physiologically after a water load, reflecting an adequate reaction of the kidneys. In such cases plasma osmolality naturally does not rise!

C_{H_2O}, in addition to C_{cr}, has been recommended as a valuable parameter in abdominal sepsis, particularly when the progression of pathological values is considered to be one of the indications for laparotomy (S4).

5.2.2. Fractional Excretions

Fractional excretions supply information on the amount of substances excreted in the urine as opposed to the quantity originally filtered in the glomerules. FE_{H_2O}, FE_{osm}, FE_{Na}, and FE_K represent water, osmolal, sodium, and potassium fractional excretions.

FE_{H_2O} is calculated as the P_{cr}/U_{cr} ratio. Fractional excretions of the other substances equal the ratio between the clearance of the given substance and C_{cr}, and can therefore be calculated without knowledge of the volume of the urine. Thus, e.g., in the calculations of FE_{Na}

$$FE_{Na} = U_{Na}/U_{cr} \times P_{cr}/P_{Na}$$

where U_{Na} and P_{Na} are Na^+ concentrations in the urine and in the plasma, respectively.

	Reference interval	Maximum attainable value
FE_{H_2O}	0.010–0.020	0.350
FE_{osm}	≤0.035	0.350
FE_{Na}	0.004–0.012	0.300
FE_K	0.040–0.190	2.000

In the evaluation of fractional excretions, FE_{H_2O} increases in water, tubular osmotic, and combined diuresis (see below). With minimal intake of liquids, it may even fall to 0.004. FE_{osm} increases in tubular osmotic and combined diuresis, and FE_{Na} increases in tubular osmotic diuresis. FE_{Na} belongs to the parameters used for the differential diagnosis of oligoanuric states: if the cause is prerenal, FE_{Na} is <0.01; if the cause is renal, it is >0.02 (C7). In case of a previous treatment with furosemide or manitol, these criteria are not applicable.

In burns, an FE_{Na} elevation with two peaks was proved to exist in the first and the fourth weeks. The first peak is attributed to a hypoxia-induced affection of tubules and to escharectomy, the second peak is attributed to treatment with aminoglycosides (C2). When FE_{Na} values exceed 0.160, an increased administration of diuretics has no effect (S1). A decline in FE_{Na} accompanies extracellular losses and Na^+ depletions.

The maximum attainable values of FE_K, exceeding the amount filtered in the glomerules, are obtained through tubular secretion. Among the causes of FE_K elevation are diuretics, tubular compensation of a glomerular filtration rate decline, hypercatabolism, and hyperaldosteronism. An FE_K elevation with K^+ levels below 3.8 mmol/liter justifies the consideration whether treatment with aldosterone antagonists and other K^+-sparing diuretics might be suitable. An FE_K decline accompanies extrarenal K^+ losses, depletions, and anabolic states.

5.3. Types of Diuresis

Patients receiving intensive care may develop water diuresis, osmotic diuresis, and combined diuresis. The diagnostic parameters are given in Table 1. Water diuresis may be produced by the following pathological states:

1. Diabetes insipidus (insufficient production of ADH)—primary, with an inclination to familial incidence, a rare disorder; or secondary, which is much more frequent, due to cranial injuries, a consequence of neurosurgical treatment, meningitis, encephalitis, vascular aneurysma, so-called cerebral death due to brain hypoxia, brain tumors, tuberculosis, and sarcoidosis.

2. Nephrogenic diabetes insipidus (insufficient response of tubules to ADH)—congenital, a rare disorder; or secondary, which is much more frequent, due to K^+-deficit nephropathy, drug treatment (lithium salts, demeclocyline, methoxyflurane), acute tubular necrosis, removal of an obstruction of urinary pathways, and certain types of organic nephropathy.

3. Psychogenic polydipsia, a frequent cause of water intoxication (J3).

4. Certain drugs that decrease the threshold of thirst perception in the hypothalamus so that patients drink excessively (laxatives, diuretics).

5. High intake of liquids of iatrogenic origin.

TABLE 1
Differential Diagnostic Criteria for Evaluation of Type of Diuresis[a]

Parameter	Type of diuresis		
	Water	Tubular osmotic	Combined
Urine osmolality/serum osmolality	<1	≥1	<1
FE_{H_2O}	>2	>2	>2
FE_{osm}	≤3.5	>3.5	>3.5
C_{H_2O}	>0	≤0	>0

[a] Modified from Ref. S1.

Osmotic diuresis may be produced by the following pathological states:

1. Overflow osmotic diuresis occurs in an elevation of GFR × P_{osm} (a high level of urea, glycemia, or manitol). C_{osm} increases, polyuria is present, and the excretion of osmotically active substances in the urine is high. If a simultaneous increase in GFR is present (see above), an increased osmotic load is excreted in polyuria without the development of plasma hyperosmolality and without an elevation of FE_{H_2O} and FE_{osm} (K5).

2. Tubular osmotic diuresis; the resorptive capacity is decreased due to the action of diuretics or a pathological process (in chronic renal insufficiency; polyuria need not be present).

3. Combined water diuresis and osmotic diuresis develop when both of the above-described disorders are present.

5.4. Renal Function in ICU Patients

The renal function parameters given in the preceding text are examined as a rule in nephrological practice. Their comprehensive monitoring in critically ill patients has not yet become, however, common practice, and, for this reason, they are discussed below.

Renal functions were examined daily, a combined total of 961 times, in 174 ICU patients (A. Jabor and A. Kazda, unpublished), 71 females and 103 males. The group consisted of patients with a complicated course after gastrointestinal tract operations (25), gynecological and urological operations (30), and vascular operations (16), and patients with intracranial hemorrhage (9), with contusion of the brain (16), with multiple injuries (10), with intoxications (9), with cardiogenic shock and after CPR (14), with respiratory insufficiency (9), and with other diagnoses (36). The mean frequency of investigations in one patient was 2.65 (the geometric mean), with a minimum of 1 and a maximum of 45 investigations. Altogether, 129 patients (74.1%) survived and 45 (25.9%) died.

In view of the high number of the examinations performed and the wide range of the diagnoses established, it is possible to regard the reported findings as characteristic for hypercatabolic patients of ICU.

Table 2 gives a survey of the frequency of selected parameters in the reference interval as well as those above and under this interval. It is possible to see a relatively high frequency of hypernatremia. Also, serum osmolality is far more frequently pathologically increased than decreased. The reason is that, besides electrolyte changes, hyperglycemia, elevation of urea, and iatrogenic influences (manitol and possibly treatment with high doses of antibiotics) also play a part in it. The relatively high frequency of hypokalemia necessitates a more intensive monitoring of K^+ status.

TABLE 2
FREQUENCY OF SELECTED PARAMETERS[a]

Parameter/reference interval[b]	Number of findings		
	Under reference interval	In reference interval	Above reference interval
S-creatinine/≤110 μmol/liter		813	148
S-urea/4–9 mmol/liter	12	546	403
S-osmolality/275–295 mmol/kg	62	497	402
S-Na$^+$/132–142 mmol/liter	87	654	220
S-K$^+$/3.8–5.4 mmol/liter	164	718	79
DU-osmolality/600–1200 mmol/day	92	423	446
DU-Na$^+$/120–240 mmol/day	312	336	313
DU-K$^+$/50–100 mmol/day	226	543	192
DU-urea/330–420 mmol/day	404	171	386
C_{cr}/1.150–2.350 ml/s	153	429	379
C_{osm}/≤0.050 ml/s		582	379
C_{H_2O}/−0.027 − −0.007 ml/s	196	517	248
Diuresis/1000–1500 ml/day	47	125	789
FE_{osm}/≤0.035		781	180
FE_{H_2O}/0.01–0.02	235	470	256
FE_{Na}/0.004–0.012	222	521	218
FE_K/0.040–0.190	53	765	143

[a]Data From 961 examinations of renal functions in 174 ICU patients.
[b]U, Urine; S, serum; see text for discussion.

Evaluation requires, however, the actual pH to be taken into consideration. The daily recording of Na$^+$ losses demonstrated a high, equally frequent, incidence of increased and decreased excretion. The excretion of osmotically active substances differs from these findings. Pathologically increased values clearly predominate here with regard to the higher demands on the excretion of osmotically active substances in catabolic states. Excretion rates of over 2000 mmol/day were found 105 times (10.9% of findings) and over 2500 mmol/day, 43 times (4.5% of findings).

The frequency of pathologically elevated C_{cr} values predominates over decreased values (the possible causes were discussed in the preceding section). The frequent increase in C_{osm} is related to the excretion of high amount of osmotically active substances under conditions of normal or increased GFR, i.e., overflow osmotic diuresis. There is also a high percentage of C_{H_2O} in the pathological range. Markedly negative values are due to the excretion of a high amount of osmotically active substances in concentrated urine and are indicative of the fact that the medullary osmotic gradient is being maintained.

A pathological C_{H_2O} elevation suggests either diuresis with the presence of water (glucose) load or a disorder in the medullary osmotic gradient. In the former case, serum osmolality is in the reference interval or is lower; in the latter case, it is increased. Only a small number of findings related to diuresis were under the reference limit; values under 500 ml/day were found only 4 times (0.4% of findings). On the other hand, polyuria is very frequent, e.g., values over 2500 ml/day were found 512 times (53.2%) and values over 3500 ml/day were found 234 times (24.3%)

The frequent elevation of FE_{H_2O} and FE_{osm} corresponds to the trend of these parameters in the different types of diuresis. Water diuresis was found 21 times (2.2%), tubular osmotic diuresis was found 150 times (15.6%), and mixed diuresis was found 14 times (1.5%) (see also Table 1).

The frequent FE_{Na} elevation is likewise in keeping with the incidence of osmotic diuresis. With values over 0.160, more intensive diuretic treatment has no effect. This situation was found only four times (0.4%). Also, the frequency of pathologically decreased FE_{Na} values was high, and particularly, the relationship of decreased FE_{Na} to a simultaneous hypernatremia was of interest. A simultaneous FE_{Na} decrease to under 0.004 and a Na^+ level above 142 mmol/liter took place in 59 cases (6.1%). The index Na^+/K^+ in the urine, considered as the index of hyperaldosteronism, was under 1.0 in these cases 36 times. This was produced more frequently by a pathological FE_{Na} decrease than by an FE_K elevation. These patients were not in an oligoanuric state caused by dehydration. This is confirmed by diuresis (the mean was 2304 ml/day) as well as by C_{cr} (the mean was 2.204 ml/s). Findings of hypernatremia accompanied with a low FE_{Na} should therefore be examined from the aspect of hyperaldosteronism, and, in such cases, treatment with aldosterone antagonists should be considered.

Among the pathological FE_K findings, an elevation was more frequent than a decline. An elevation above 1.000, indicative of a tubular secretion, was found six times (0.6%). Special attention was given to the estimation of the incidence of hypokalemia associated with an inadequate renal regulation, i.e., with an increased FE_K. This finding was obtained 44 times (4.6%) and was accompanied with a decline in the Na^+/K^+ index below 1.000 8 times, and by a C_{cr} decline to under 0.833 ml/s 17 times. Treatment with aldosterone antagonists is indicated in hypokalemia with an FE_K elevation. Monitoring of K^+ status is particularly important during such a treatment when a simultaneous GFR decline is present. The point is that there is a risk that hyperkalemia may develop.

The investigation of relations existing between C_{cr}, diurnal excretion of osmotically active substances, serum osmolality, and diuresis revealed that patients with a C_{cr} decline under 1.15 ml/s were capable of excreting only a part of the daily osmotic load and remained predominantly hyperosmolal. This situation was observed among 153 findings with a C_{cr} decline 126 times (82.4%). The excretion of the osmotic load exceeding 1500 mmol/day in patients with a GFR decline always necessitated diuresis above 3000 ml/day and was found 23 times. It proves that the adequacy of diuresis in intensively treated patients must be evaluated not only from the aspect of the actual volume of the urine, but also in relation to GFR and osmolality of body fluids. At the same time, efforts should be made to decrease catabolism (a high rate of formation of urea, hyperglycemia), which represents, with the existence of a C_{cr} decline, a difficult problem from the point of view of osmolality of body fluids. Patients with a C_{cr} elevation above 2.35 ml/s are, on the contrary, usually polyuric and often excrete high osmotic load.

The ability of the kidneys to excrete a relatively or absolutely increased quantity of osmotically active substances is mediated through two mechanisms: through osmotic tubular diuresis or through overflow osmotic diuresis.

Diagnostic criteria of osmotic tubular diuresis are clearly defined and were described in the previous text and in Table 1. Overflow osmotic diuresis was also described in a general way in the foregoing text but no exact numerical criteria are reported in the literature (CR4, S1). The results of a study reporting 150 findings meeting the criteria of tubular osmotic diuresis is given in Table 3. Parameters determining this type of diuresis (i.e., the urine and serum osmolality ratio, FE_{H_2O}, FE_{osm}, and C_{H_2O}) and values facilitating a more detailed characterization of these findings are given. The latter include C_{cr}, C_{osm}, serum and urine osmolality, and daily excretion of osmotically active substances.

In contrast to this group of findings Table 3 also gives values of the same parameters seen in 379 findings with C_{osm} over 0.050 ml/s. As compared with the preceding group, marked differences exist here in all parameters. It is believed that in this second group there are included findings of overflow osmotic diuresis. It appears that even a considerable osmotic load can be eliminated, with a physiological or even increased GFR, only by filtration. For the most part tubular compensatory mechanisms do not play any part in this, as shown by findings of FE_{H_2O} and FE_{osm}. The situation is made clear by the third group of findings, where among the criteria of selection, besides the elevation of C_{osm}, an excretion rate of osmotically active substances over 2000 mmol/day was also included. Altogether, the findings met this condition. In this last group a significant rise was further observed in C_{cr}, C_{osm}, and the urine and serum osmolality, while FE_{H_2O} and FE_{osm} did not change significantly. These findings are regarded as

TABLE 3
Selected Renal Function Parameters[a]

Group	Criteria of selection	n	U_{osm}/S_{osm}	FE_{H_2O}	FE_{osm}	C_{H_2O} ml/s	C_{cr} ml/s	C_{osm} ml/s	S_{osm} mmol/kg	U_{osm} mmol/kg	DU_{osm} mmol/d
1	Tubular osmotic diuresis (see Table 2)	150	1.361 ± 0.263	0.046 ± 0.027	0.058 ± 0.027	−0.014 ± 0.012	1.067 ± 0.609	0.053 ± 0.024	317.4 ± 23.1	431.2 ± 82.6	1461 ± 667
2	$C_{osm} > 0.050$ ml/s	379	1.778 ± 0.516	0.020 ± 0.018	0.031 ± 0.018	−0.028 ± 1.667	2.727 ± 1.089	0.071 ± 0.018	297.4 ± 19.5	526.4 ± 147.8	1828 ± 484
3	$C_{osm} > 0.050$ ml/s $DU_{osm} > 2000$ mmol/d	105	1.938 ± 0.481	0.020 ± 0.014	0.034 ± 0.016	−0.043 ± 0.018	3.226 ± 1.181	0.095 ± 0.016	302.6 ± 23.2	581.6 ± 131.7	2480 ± 387
Statistical significance between groups 2 and 3			$p < 0.01$	n.s.[b]	n.s.	$p < 0.01$	$p < 0.01$	$p < 0.01$	$p < 0.05$	$p < 0.01$	$p < 0.01$

[a] $x \pm s$ in findings meeting the criteria of osmotic tubular diuresis (group 1), in findings with $C_{osm} > 0.050$ ml/s (group 2), and in findings with $C_{osm} > 0.050$ ml/s and daily excretion > 2000 mmol (group 3). Findings in groups 2 and 3 represent situations with overflow osmotic diuresis.
[b] n.s., No significance.

typical for overflow osmotic diuresis. Because of the relatively high frequency of these findings, they should also receive attention regarding the drugs excreted by the kidneys. On the basis of these results, we used as a diagnostic limit of overflow osmotic diuresis the higher value of 0.693 mmol/s. This value is the product of the upper reference interval C_{cr} and serum osmolality and a factor value of 0.001. It is equal to the amount of filtered mmol/s.

Further, the relationship between serum osmolality and C_{H_2O} was estimated, depending on the different types of diuresis. When hyperosmolality and C_{H_2O} elevation are ascertained and when changes in other parameters indicating water diuresis are present, diabetes insipidus should by suspected. Such findings were noted 6 times in 961 examinations. On the other hand, in the case of serum hypoosmolality and a distinctly negative C_{H_2O} value, the syndrome of unsatisfactory secretion of antidiuretic hormone (SIADH) should be suspected. Such findings were noted 13 times.

After completing the aforementioned research, we observed the types of diuresis in another group of ICU patients. In 82 subjects, treated largely on the basis of neurological and internal diseases and injuries, renal functions were examined 330 times. Overflow osmotic diuresis appeared in 14% of the findings. Its occurrence was largely related to osmotherapy. Renal function findings inclusive as to the types of diuresis will vary between different groups of intensively treated patients, depending not only on the diagnosis and rate of catabolism but on the therapy as well.

To conclude, monitoring renal functions provides information about water and electrolyte balance and qualitatively higher aspects of the management of patients.

6. Acid–Base Balance

6.1. Introductory Comments

Reference intervals—pH, 7.36–7.44; HCO_3^-, 22–26 mmol/liter; pCO_2, 4.8–5.9 kPa; *BE*, −2.5–+2.5 mmol/liter.

The values of the standard and the actual bicarbonate do not differ at pCO_2 5.3; in hypercapnia the actual bicarbonate value is higher than the standard value, and in hypocapnia the situation is reversed.

In the evaluation of buffer bases, the base excess *(BE)* is used more frequently in Europe, and the plasma bicarbonate is used in the United States (A4, M13, O1, R9).

The relationship between the main parameters of the acid–base balance is expressed in terms of the simplified Henderson–Hasselbalch equation:

pH ~ HCO_3^-/pCO_2. The numerator of the fraction indicates the metabolic component of the acid–base equilibrium; the denominator expresses the respiratory component, that is, one of the buffer acids. The acid–base balance is determined by following three parameters: by H^+ concentration, expressed as its negative logarithm, i.e., pH; by buffer bases, expressed as HCO_3^- or BE; and by acids, practically measured only as pCO_2. Each of these components may undergo a primary change so that there are six possible main causes of acid–base status (ABS) disorders. Two of them, an increase of H^+ and a loss of HCO_3^-, lead to metabolic acidosis (MAC); two others, a loss of H^+ or a gain of HCO_3^-, lead to metabolic alkalosis (MAL). A primary CO_2 retention and thus a pCO_2 increase leads to respiratory acidosis (RAC); an increased CO_2 output and thus a pCO_2 decline leads to respiratory alkalosis (RAL).

The development of acid–base disturbances passes through two stages: the buffer and the compensatory response. The buffer response takes place at the very onset of the primary disorder and is in proportion to its seriousness. In its course, the pH and one component of the buffer pair, base or the acid, change. In metabolic disorders the level of the bases changes: in MAC it falls, in MAL it increases. In respiratory disorders pCO_2 changes in this reaction: in RAC it increases, in RAL it falls. Then the organ-mediated compensatory response follows. In this response the other component of the buffer pair changes. The compensation of metabolic disorders is mediated by the lungs. In MAC, hypocapnia develops; in RAL, hypercapnia develops. The maximum rate of this compensation is reached in 12–24 h. The compensation of respiratory disorders is mediated by the kidneys. In RAC, the bases are retained and the urine is acidic; in RAL, the bases are excreted and the urine is alkaline. The compensatory reaction starts during the first day and reaches a maximum within 2–4 days. After the cessation of the primary disorder the compensatory efforts persist, which should be considered in assessing treatment. The buffer reaction has prevented extreme pH changes and the compensatory reaction shifts pH in the direction of the reference interval.

The acid–base findings can be interpreted by calculations or graphically. Either method can be of diagnostic importance. The choice depends on the customary practice of the institute as well as on the approach of the clinician.

Calculations facilitate the determination as to whether the findings are typical for simple acid–base disturbances and whether maximum compensation has been reached. It is possible to judge by means of the findings the presence of certain mixed disorders. Yet the calculations alone do not enable diagnosis of all mixed disorders nor a number of other relations and ratios. In this respect the calculations are necessarily limited, but so are the interpretations obtained from a graph. These problems are dealt with later on. Here the most frequently used equations are presented (B3, D2, D3, P3, W3).

Metabolic alkalosis—calculation of a compensatory rise in pCO_2:

$$pCO_2 = 0.093 \times \Delta HCO_3^- \pm 0.2 \quad \text{(upper limit } pCO_2 = 7.33 \quad \text{kPa)}$$

In MAC (acidemia plus ↓ HCO_3^-) lasting 12–24 h, when this pCO_2 calculation is lower than the measured one, then the maximum compensation has not been reached and a respiratory insufficiency is present. On the other hand, when the calculation is higher than the measured pCO_2, a mixed MAC + RAL is present.

Metabolic alkalosis—calculation of a compensatory rise in pCO_2:

$$pCO_2 = 0.093 \times \Delta HCO_3^- \pm 0.2 \quad \text{(upper limit } pCO_2 = 7.33 \quad \text{kPa)}$$

In MAL (alkalemia plus ↑ HCO_3^-) lasting 12–24 h, when the value of the calculation of pCO_2 is higher than the measured pCO_2, then the maximum compensation has not been reached. on the other hand, when the calculation is lower than the measurement, a mixed MAL + RAC is present.

Respiratory acidosis—in an acute disorder as a result of the buffer reaction there is an elevation of HCO_3^- by 2–4 mmol/liter; the upper limit is 30 mmol/liter. The next calculation pertains to a HCO_3^- change that correlates with the development of the maximum compensation within 2–4 days:

$$HCO_3^- = 2.63 \times \Delta pCO_2 \pm 0.2 \quad \text{(upper limit } HCO_3^- = 45 \quad \text{mmol/liter)}$$

In RAC (acidemia plus ↑ pCO_2), when the calculation of HCO_3^- is higher than the measured value, the patient is not compensated maximally either because of a renal disorder or MAC of another etiology is present. When the calculation is lower than the measurement, a mixed RAC + MAL is present.

Respiratory alkalosis—in an acute disorder as a result of the buffer reaction there is a decline in HCO_3^- by 2–4 mmol/liter. The next calculation pertains to a HCO_3^- change that correlates with the development of the maximum compensation within 2–4 days:

$$HCO_3^- = 3.76 \times \Delta pCO_2 \pm 2 \quad \text{(lowest limit } HCO_3^- = 12\text{–}15 \quad \text{mmol/liter)}$$

In RAL (alkalemia plus ↓ pCO_2), when a greater decrease in HCO_3^- is calculated than is measured in the patient, then the disorder is not maximally compensated. If, on the other hand, the measured decrease in

HCO_3^- is more severe than the calculated decrease, a mixed RAL + MAC is present.

A finding of simultaneous acidemia, ↑ pCO_2, and ↓ HCO_3^- is indicative of a mixed MAC + RAC. A finding of simultaneous alkalemia, ↓ pCO_2, and ↑ HCO_3^- suggests the presence of MAL + RAL.

Even when we decide to use the above calculations, relationships will become more understandable when they are compared with graphed findings of acid–base regulations. The graphic representations of relationships from the Henderson–Hasselbalch equation are often used for the interpretation of ABS findings. There are three variables, two of them always being pH and pCO_2 and the third one either HCO_3^- (C5, K3, L2, V2) or BE (A2, C6, L1, W10). Sometimes just pH and pCO_2 are used (M4, O1).

We use a graph with three variables, pH, pCO_2, and BE (see Fig. 3). This graph is also a part of our programmed data processing. The physiological range is in the middle of the graph. The fields ruled-off by thick

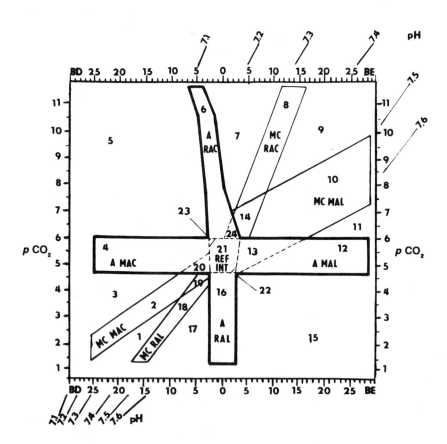

lines demarcate areas to which, with a 95% probability, the finding will move in the course of the buffer reaction. The fields ruled-off by thin lines demarcate areas of maximally compensated disorders, to which the finding will move, with 95% probability, during the compensatory reaction. At the time when the compensation is only developing, the findings lie in the transition areas between the acute and the maximally compensated disorder. If the ABS findings are in the fields that on the graph are the areas between two of the four main disorders, mixed disturbances can be assumed. The graph is instructive, providing information on the relationship between the fundamental ABS parameters and on changes occurring during the development of the disorders. From the ABS finding alone it is, however, possible to evaluate only the diagnostic field in which the finding is situated. For the diagnosis of the acid–base status it is essential to know also the medical history, clinical data, and various laboratory values. The ABS findings can be allocated to the appropriate field of the graph with

---

FIG. 3. Evaluation of findings according to their classification in the particular areas of the graph of acid–base regulation. Modified from Ref. C6. In unventilated patients: (1) mixed MAC and RAL or MAC in alkalization treatment; (2) maximally compensated MAC; (3) less than maximally compensated MAC; (4) acute MAC; (5) mixed MAC and RAC; (6) acute RAC; (7) less than maximally compensated RAC; (8) maximally compensated RAC; (9) mixed MAL and RAC; (10) maximally compensated MAL; (11) less than maximally compensated MAL; (12) acute MAL; (13) acute to maximally compensated MAL; (14) maximally compensated RAC or MAL or mixed MAL and RAC; (15) MAL and RAL; (16) acute RAL; (17) less than maximally compensated RAL; (18) maximally compensated RAL; (19) moderate, compensated MAC or RAL; (20) moderate MAC, acute to maximally compensated; (21) ABS values within reference limits; (22) ABS values still within reference limits (pH is slightly increased); (23) ABS values still within reference limits (pH is slightly decreased); (24) moderate RAC, acute to maximally compensated.

In ventilated patients: (1, 2, and 3) hyperventilation-induced hypocapnia—acidemia and *BD* are indicative of MAC; (4) ventilation is proportional, the patient is in MAC; (5) hypoventilation—this RAC is accompanied with MAC; (6) unsatisfactory ventilation, the patient is in RAC; (7) unsatisfactory ventilation, RAC already compensated less than maximally; (8) unsatisfactory ventilation, RAC maximally compensated; (9) unsatisfactory ventilation, RAC accompanied with MAL; (10) hypoventilation-induced hypercapnia—warnings: patient is in MAL; (11) moderate hypercapnia, severe MAL; (12 and 13) normal ventilation, MAL (sudden improvement of ventilation in chronic hypercapnia?); (14) hypoventilation with complete metabolic compensation and normal pH; (15) hyperventilation-induced RAL, MAL is also present; (16) hyperventilation-induced acute RAL, so far without metabolic compensation; (17) hyperventilation-induced RAL that is less than maximally compensated; (18) hyperventilation-induced hypocapnia that is maximally compensated; (19) moderate hypocapnia, maximally compensated; (20) proportional ventilation, the patient is in moderate MAC; (21) ABS values within reference limits; (22) ABS values still within reference limits (pH slightly increased); (23) ABS values still within references limits (pH slightly decreased); (24) hypoventilation, moderate RAC, acute to maximally compensated.

regard to the mode of ventilation, as discussed in the legend to Fig. 3. In any case, even the best possible graph cannot provide the following information:

1. How long the disorder lasts and whether the compensation proceeds in the corresponding way—let us give an example: In case of an ABS finding in the field of a less than maximally compensated MAC, three different situations may be present: (a) an acute disorder that is being just compensated; (b) a disorder lasting several days when, owing to respiratory insufficiency, compensation could not have developed fully; and (c) the MAC was already fully compensated but, because of respiratory insufficiency, it has become decompensated.

2. It is not possible to recognize mixed disorders when their parameters are in mutual opposition, i.e., a metabolic acidosis and a metabolic alkalosis.

3. Multiple causes of one disorder cannot be identified, e.g., MAC due to starvation and hypoxia; recognition of multiple causes is of principal importance for a comprehensive therapy.

4. In critical conditions a finding in the area of acute MAC is rarely observed. The typical response to a trauma is hyperventilation and therefore on initial examination hypocapnia is present with a base deficit.

5. A finding in the area of respiratory alkalosis is often accompanied with unrecognized lactic, hypoxic acidosis.

6. In MAL the corresponding hypercapnia is encountered only exceptionally, even in the case of a several-day-long course, although this is exceptional. The reason is that, with the development of compensatory hypercapnia, hypoxemia develops at the same time. A hypoxia-induced irritation of the breathing center leads to the increase of ventilation, and compensatory hypercapnia is thus limited.

7. A particular situation develops in chronic hypoperfusion states, when alkalizing processes predominate in the arterial bed but when acidosis persists in the tissues.

6.2. Acid–Base Balance and Ions

For the comprehension of this relationship, calculations pertaining to anions are of importance—serum buffer bases (BB_s) and the anion gap (AG). Interrelationships of serum Na^+ and Cl^- are always evaluated (N1):

$$BB_s = (Na^+ + K^+) - Cl^- = 42 \quad \text{mmol/liter}$$

When K^+ is left out, the value is 4–5 mmol/liter lower. In all bases, when the value of BB_s increases, that part of the anion column (see Fig.

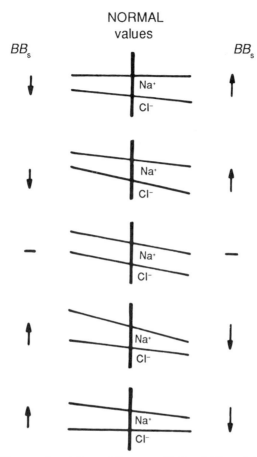

FIG. 4. Possible mutual variations of the serum Na^+ and Cl^- and their relation to the serum buffer base value (BB_s).

4) increases that corresponds to the sum total of HCO_3^- and of the negative charge of plasma proteins. To maintain electroneutrality, the amount of HCO_3^- increases and the patient develops MAL. In all cases when, on the contrary, the relationship changes to favor Cl^-, the space for BB_s decreases and hyperchloremic MAC develops. It follows from Fig. 4 that not every hyperchloremic state signals MAC, and that not every hypochloremic state is indicative of MAL. Thus, e.g., when a higher Na^+ level and a higher Cl^- level are found in the serum, but when the increase of Na^+ is relatively higher, hyperchloremia is not a symptom of MAC and, on the contrary, MAL develops.

The anion gap represents a sum of negative charges of plasma proteins, chiefly of albumin (L3, M12, N1, 02):

$$AG = (Na^+ + K^+) - (Cl^- - HCO_3^-) = 15.2 \pm 1.6 \quad \text{mmol/liter}$$

When K^+ is left out of the calculation, then the value is 4–5 mmol/liter lower. The value increases with an increased concentration of anions of organic and inorganic acids, which occurs in certain MAC cases. The consequence of the increasing charge of these acids is a decrease of HCO_3^-, which functions to preserve electroneutrality. The causes of AG elevation in MAC will be dealt with below in the discussion of this disorder. MAC is far from being the only cause of AG changes. The value of AG may also increase in alkalemia (G3, G6); the causes of this are as follows:

1. Alkalemia may develop also in a situation when the production of organic acids is increased and an acid–base finding is obtained that indicates MAL or RAL (that is, when a mixed disorder with a preponderance of the alkalizing component is present).

2. Alkalemia may increase the production of organic acids. In RAL the lactate increases by 2–3 mmol/liter, as the activity of anaerobic glycolysis becomes more intensive.

3. Certain exogenously administered anions (sodium lactate or sodium acetate) may, after their partial metabolization, lead to alkalemia, but, at the same time, they persist to some degree in the circulation and increase the AG.

4. In alkalemia the negative charge of proteins increases.

5. An increase of AG in MAL is related also to the contraction of the volume and to the condensation of total proteins (TP). The magnitude of this increase can be determined by a calculation (G6); let us call it AG_{MAL}: $AG_{MAL} = 0.11 \times \Delta pH \times TP$ (g/liter) $+ 0.24 \times \Delta TP$ (g/liter), where ΔpH = the difference between the actual pH and pH 7.40, and ΔTP = the difference between the value of total protein in dehydrated state and its value after rehydration. Only when the increase of AG is higher than would correspond to the increase of AG_{MAL} can a mixed disorder with a concomitant MAC be suspected.

Under certain circumstances, however, a decrease may also occur in AG (G3, M12, 02); the causes of this are as follows:

1. Increase of unmeasured cations—(a) cationic IgG paraproteins: have the isoelectric point above pH 7.40 and therefore have a positive charge with physiological pH; some authors doubt the importance of this influence (B10); (b) increase of Ca^{2+}, Mg^{2+}, and K^+: less probable, all would have

to increase at the same time; (c) acute lithium intoxication; (d) administration of polymixin B.

2. Unsystematic analytical error—(a) hyponatremia: certain types of photometers indicate falsely low Na^+ when levels are above 170 mmol/liter; (b) serum hyperviscosity: falsely low Na^+ because of aspiration of a smaller amount of specimen using a peristaltic pump during the dilution prior to examination (possible error up to 50 mmol/liter) (O2); (c) intoxication with bromides and iodine-containing drugs: halogens react with agents used for the determination of Cl^- by autoanalyzers; (d) hyperlipemia: falsely increases colorimetric determination of Cl^-.

3. Systematic error in the determination of Na^+, Cl^-, and HCO_3^-: the calculation of AG was, moreover, proposed as a quality control (B8).

4. Decrease of unmeasured anions—(a) dilution in hypoosmolal states, mainly with inappropriate secretion of the antidiuretic hormone; (b) hypoalbuminemia: this is probably the most frequent cause of an AG decrease in nephrotic syndrome; investigation of AG in a set of 82 patients with albuminemia under 30 g/liter demonstrated, regardless of the diagnosis, a highly significant decrease of AG as compared with the normal value. No correlation existed between individual hypoalbuminemia and AG (N1).

A very interesting recent study has proved that hypoproteinemia and the AG decline resulting from it may be the cause of MAL (M12). Proteins participate as a buffer in the ABS. Plasma proteins therefore have to be measured and assessed also from the point of view of their effect on the ABS. In severe hypoproteinemia, primary hypoproteinemic MAL develops. In the study by McAuliffe et al., it was reported that a decrease in albumin by 10 g/liter was accompanied by a mean increase in HCO_3^- by 3.4 mmol/liter and of BE by 3.7 mmol/liter; AG falls by about 3 mmol/liter. It follows from the relations existing among ions that BB_s does not change and that the decline in AG is compensated by a rise in HCO_3^-.

The knowledge of the existence of these relations is a fundamental contribution to the interpretation of ABS findings in patients with marked hypoproteinemia. They help in the explanation of findings obtained in MAL patients with normal BB_s values and with a low AG. Patients with a simultaneous hyperchloremic MAC will have low BB_s and low AG.

6.3. METABOLIC ACIDOSIS

Metabolic acidosis is represented by two primary findings: (1) an increase in the anion gap and (2) an increase in chloride (but with a normal anion gap). The causes of these findings are as follows (B3, B4, D2, G3, K9):

Increase in the anion gap
  Fast-developing diabetic ketoacidosis with disorders in the excretion of ketoacids and possibly of other organic acids
  Starvation with ketoacidosis
  Lactic acidosis with an increase of *l*-lactate (M1) (rarely of *d*-lactate) (M9) to 5 mmol/liter and more
  Renal failure: retention of anions of organic as well as inorganic acids is relatively late, with a decline in glomerular filtration rate under 0.33 ml/s or with a rise in creatininemia above 350 μmol/liter (W2) or, according to other authors, above 600 μmol/liter (O2)
  Intoxication
    With ethyl alcohol (keto-acids, mainly β-hydroxybutyric acid, lactic acid)
    With methanol (formic acid, lactic acid)
    With ethylene glycol (lactic acid, glycolic acid, oxalic acid)
    With salicylates (several organic acids) and with other agents, their acidifying component not being always defined (toluene, papaverine, paraldehyde, nalidixic acid)
  Parenteral nutrition with protein hydrolysate (exceptional) (T3)
Increase in chlorides, normal anion gap
  With an inclination to hyperkalemia
    Subsiding diabetic ketoacidosis or ketoacidosis with a protracted course: HCO_3^- was consumed for neutralization of protons and replaced by Cl^-; a larger distribution space of HCO_3^- than of ketoanions may probably also play a role (O2)
    Early uremic acidosis: in medullary disease of the kidney, H^+ excretion may be disturbed before the glomerular filtration is affected. This is a decrease in HCO_3^- levels while inorganic strong anions are still excreted
    Obstructive uropathy in incipient stages; later, the *AG* also rises
    Renal tubular type 4 acidosis
    Hypoaldosteronism
    Infusion or ingestion of HCl, NH_4Cl, lysine–HCl, and arginine–HCl
    Potassium-sparing diuretics in patients with parenteral nutrition: due to inhibition of renal H^+ excretion and to an elevated intake of Cl^- (K14)
  With an inclination to hypokalemia:
  Renal tubular type 1 (classical distal) acidosis and type 2 (proximal) acidosis
  Acetazolamide, by the inhibition of carboanhydrase, leads to a decrease in H^+ excretion and to an increase in HCO_3^- losses
  Acute diarrhea: losses of HCO_3^- and of K^+
  Ureterosigmoideostomy: resorption of H_2O and Cl^-, losses of HCO_3^- and of K^+
  Obstruction of artificial ileal bladder
  Dilution acidosis in cases of rapid hydration with physiological saline solution; as compared with body fluids, 24 mmol of HCO_3^- are missing per every liter and, from the aspect of the blood, a further 24 mmol of strong conjugated bases of protein anions are missing, i.e., of hemoglobin and proteins (N4); with a frequent disorder of tubular functions, hyperchloremic MAC easily develops in intensively treated patients

Clinical features of MAC include hyperventilation, usually accompanied by an enlargement of the pneumovascular system. Acidotic Kussmaul's respiration is tolerated by patients better than are cardiac dyspnea or the breathing complications in pulmonary edema. The myocardium is influenced directly by intracellular acidosis, causing decreased contractility, and indirectly by the release of catecholamines. Effects in the circulation are seen: the arterial blood bed is dilated, the venous blood bed is constricted, and the cerebral blood flow is increased. In cases of sudden renal deterioration in lactic or diabetic acidosis, a coma may develop. In renal insufficiency there is sometimes pseudouremic–acidotic coma, and a differentiation from coma caused by renal deterioration is essential. Renal-induced coma is preceded by mental changes, headache, and vision disturbances, and later by loss of orientation, somnolence, coma, and possibly spasms (R5, R9, Z1). The changes depend on the pH decline in the cerebrospinal fluid and are even more marked in RAC (R5, R9, Z1). Gastrointestinal symptoms include nausea, vomiting, and diarrhea in uremia and diabetic ketoacidosis.

Further laboratory findings include a decrease in pH accompanied by an increase in K^+ levels. With a shift of the hemoglobin dissociation curve to the right, an improved liberation of oxygen occurs. Protein-bound calcium is reduced and Ca^{2+} rises.

A frequent cause of MAC in critical states is hypoxia. For the establishment of its diagnosis one of the following information items is sufficient: a low value of PO_2 in arterial or central venous blood, low hemoglobin or evidence of its disturbance, a slightly reduced blood volume, reduced oxygen consumption, or an increased lactate level (N4). Some of these parameters are not measured in routine practice. Lactate levels, which are routinely measured, are thus indicated in suspected hypoxia.

Caution should be taken regarding findings of P_aO_2; when they are within reference limits or are even increased, tissue hypoxia cannot be excluded. This follows from the fact that P_aO_2 is a relative indicator, which informs us about the relationship between alveolar ventilation and perfusion. The relationship between hyperlactatemia and P_aO_2 findings in 57 patients from ICU is given in Fig. 5 (K6). A recent excellent report on lactic acidosis problems states, in agreement with our own experiences, that it is impossible to take the pH decline as an indicator of lactic acidosis. At the same time, alkalizing factors may play a role, and even an extreme hyperlactatemia need not be accompanied with acidemia (M1).

Therapy includes the following measures:

1. Elimination of the cause of MAC.
2. Alkalizing agents are to be used for a prolonged period of time only when the basic cause cannot be determined and resolved.
3. At pH values greater than 7.15–7.20 in acute states, alkalization must be done carefully.

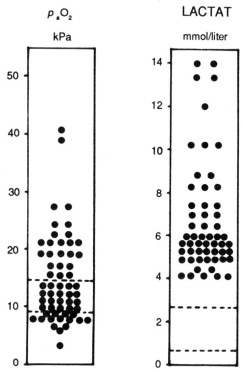

FIG. 5. The relationship between hyperlactatemia and p_aO_2 in 57 patients from ICU (W2). Reference limits of both parameters are denoted by broken lines.

4. In acute MAC only $NaHCO_3$ is administered. Sodium lactate and sodium citrate require time for the metabolization of the anion. THAM is not fully utilized for buffering immediately after infusion because it penetrates slowly into cells (R9).

The choice of suitable alkalization in cardiopulmonary resuscitation (CPR) is difficult. Recently it was proved that the percentage of survivors decreases rapidly when pH 7.55 is exceeded from 10 min to 1 h after alkalizing CPR treatment. The prognosis is in indirect proportion to degree of lactatemia and depends on the maintenance of reference interval K^+ levels (W5). Bicarbonate cannot penetrate quickly into cells; this leads to an elevation of H_2CO_3 in the ECF. The CO_2 liberated from buffer reactions enters the cells easily; counters intracellular acidemia and hence myocardial efficiency. No intracellular increase of pH was described for at least 15 min following $NaHCO_3$ administration (G7, W4). An experimental investigation of arteriovenous differences in blood gases confirms these findings (W6). There are formulas for therapeutic calculations (see below).

6.4. Respiratory Acidosis

Respiratory acidosis has numerous causes, which are listed below (K1, N2, R9).

1. Depression of the breathing center (narcotics, barbiturates).
2. Injury to CNS (tumors and degenerative diseases).
3. Cerebrovascular complications.
4. Airway obstruction.
5. Injury to the thorax and diaphragm.
6. Shock (lung disease in advanced stages).
7. Chronic obstructive lung disease.
8. Asthmatic complications (severe).
9. Pulmonary infection (severe).
10. Pulmonary emphysema and embolism (large areas of lungs ventilated but not perfused).
11. Neuromuscular disorders.
12. Unsuitably managed artificial ventilation.

RAC is rare in critically ill patients with the exception of those with a previous chronic, mainly obstructive, lung disease. Chronically ill patients show a more frequent susceptibility to hyperventilation.

In the evaluation of 500 findings of ABS in randomly selected unventilated surgery patients the following frequencies were found (K5): MAC, 13%; RAC, 4%; RAL, 32%; and MAL, 14%; 9% represented mixed disorders and 28% were in the reference interval. Clinical features indicated that acute, severe RAC is usually accompanied with neurological symptomatology. The risk of the brain acidemia is higher in RAC than in MAC. CO_2 penetrates well into lipid structures and thus also through the hematoencephalic barrier and markedly decreases the pH of the brain. HCO_3^-, which is water soluble, penetrates more slowly. Cerebral vasodilatation is typical. Intracranial hypertension develops and headache is present as well as motor disorders, tremor, myoclonia, and susceptibility to spasms. Transient psychoses may alternate with euphoria and states of delirium. Coma occurs at pCO_2 9–14 kPa (in chronic RAC at over 14 kpa) (R9). Hemorrhage into the vitreous body and edema of papillae may ensue. The stimulation of the sympathicus may produce peripheral vasodilatation, warm, erythematous skin, and elevated blood pressure. With severe hypoxia the cardiac output decreases. On the other hand, if hypoxia is absent, the effect of RAC is more modest.

Further laboratory findings indicate that renal tubular compensation stimulates the exchange of Na^+ for H^+, and Cl^- and H^+ losses increase. Kidneys with a frequently disturbed tubular function are not functional. With a sudden restoration of compensated hypercapnia, alkalemia threatens. A decrease occurs in Ca^{2+} and K^+; treatment with digitalis is accompanied with the risk of intoxication. The hemoglobin dissociation curve shifts to the left and liberation of oxygen in tissues is difficult.

Treatment in acute RAC includes the following measures:

1. Elimination of the cause as far as possible.
2. The biochemical indication for ventilation therapy is pCO_2 8 kPa and more. Normalization of hypercapnia takes place while pH is being checked to prevent the development of alkalemia. Alkalemia does not threaten unless compensatory hyperbasemia has developed already.
3. In hypoxemia, oxygen therapy should be introduced. If this does not succeed in changing p_aO_2 to at least a value over 8 kPa, ventilation therapy is indicated from the biochemical point of view.
4. Possible alkalization in marked acidemia (see therapeutic calculations).

Treatment in chronic RAC is more conservative; patients are adapted to hypercapnia and are compensated so that acidemia generally is not extreme. The following guidelines may be considered:

1. Conservative treatment has priority over ventilation, which often cannot be interrupted.
2. When oxygen therapy is indicated, it should be introduced only slowly; the ABS should be monitored to reduce the risk of hypercapnia.
3. In cases of ventilation, pCO_2 should not decline more rapidly than by 0.3–0.7 kPa/h with simultaneous acidification—hyperbasemia is usually present.

6.5. Metabolic Alkalosis

Metabolic alkalosis has numerous causes, which are listed below (B4, C4, M12, R10).

1. Increased intake of bases of bicarbonate (treatment of acidosis), carbonate (antacid, milk–alkali syndrome), acetate (hyperalimentation), or citrate (transfusion).
2. Gastrointestinal losses of acids (gastric juice).
3. Escape of acids in the urine: increase of tubular transportation of Na^+ (diuretic therapy); increase in mineralocorticoid activity, including states with hyperreninemia (contraction of the ECF, deficiency of Mg^{2+}, and Bartter's syndrome) and primary hypermineralocorticoid states (primary hyperaldosteronism, Cushing's syndrome, and adrenogenital syndromes); increase in negativity of tubular lumen from nonreabsorbable anions (phosphates and penicillin); posthypercapnic states; and hyperparathyroidism and hypercalcemic states.

4. Severe hypoproteinemia leading to a decline in *AG*.
5. Contraction alkalosis with the loss of Na^+ as well as Cl^- and with preservation of HCO_3^- (cystic fibrosis and villous adenoma).

It is important to know not only the causes of MAL but also the factors that maintain MAL. Otherwise, MAL may persist after the elimination of causes. Kidneys are not capable of excreting the excess of HCO_3^-. The cause lies in Cl^- deficiency accompanied with a decrease in the volume of ECF, K^+ deficiency, and hypermineralocorticism. The pathophysiology of MAL persistence has not been fully clarified (C4, J2).

Clinical features of MAL are often influenced by volume depletion (weakness, spasms, postural vertigo) or by hypokalemia (polyuria, polydipsia, muscular weakness, intestinal obstructions, cardiac arrhythmias). Neurological symptoms accompany only severe hyperbasemia, the reason being probably that HCO_3^- penetrates the hematoencephalic barrier only slowly and that this reduces changes in the pH value of the CSF. But in severe hyperbasemia, cerebral dysfunction takes place: muscle twitching, disorientation, lethargy and even grand mal convulsions, leading to a dramatic, even though short-term, increase in lactate (exceeding 12 mmol/liter), and thus leading to a temporary decrease in HCO_3^- (G6). The symptoms are obviously also due to intoxication with ammonium, the concentration of which increases with increasing pH (R7).

In surgical patients, mortality increases progressively with the pH above 7.5 (R10, W1). MAL usually does not occur immediately after a trauma but begins to appear only from the second day, and the frequency increases in a complicated course within 7–10 days.

Causes include persisting renal correction of MAC or compensation of RAC, alkalizing infusions, transfusions, saluretics, potassium losses, hyperaldosteronism, glucocorticoids, and gastric juice losses.

According to calculations and graphic evaluations, compensatory hypercapnia ought to be more or less in proportion to the elevation of bases. The reality is often different. Hypercapnia leads to hypoxia; in severe MAL p_aO_2 is usually between 6.7 and 8.0 kPa. Hypoxia stimulates the breathing center and hypercapnia disappears. The whole situation may repeat itself cyclically, as can be seen from the example of our patient given in Fig. 6.

Frequent dehydration contributes to the resorption of almost all of the Na^+ as well as of the HCO_3^- in the tubules. Because only a little Cl^- is available, the rest of the Na^+ is resorbed in exchange for K^+ and H^+. In the urine K^+ losses increase and, paradoxically, the urine may be acid. In addition to K^+, Ca^{2+} declines in MAL, and a slight increase in lactate is generally seen.

Treatment of metabolic alkalosis is as follows:

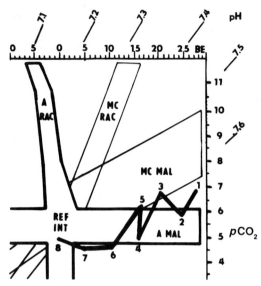

FIG. 6. Typical finding of unsatisfactory compensatory hypercapnia in MAL, which alternates with normocapnia. The numbers correspond to the chronology of investigations in the course of days 1 to 8. See text for details.

1. Elimination of factors that produced alkalosis (modification of diuretic therapy, management of vomiting, and ventilation treatment of hypercapnia).

2. Restoration of Cl^- balance: MAL usually responds to Cl^- treatment (Cl^- values in the urine are typically under 10 mmol/liter). Irrespective of the origin of the losses of Cl^- and the ECF, renal mechanisms preserving Na^+ and HCO_3^- are activated. Rarely, Cl^- values in the urine may be higher in this type of MAL in cases of simultaneous diuretic therapy and increasing of hypercapnia. In all of these MAL cases, treatment with saline solution rapidly influences the renal factors responsible for the persistence of MAL. The glomerular filtration rate goes up and renin and renin-maintained hyperaldosteronism decline. Kidneys start to excrete HCO_3^- and the pH of the blood returns to normal. The supply of K^+ partially improves these types of MAL. The saline solution alone is normalizing, regardless of the persisting K^+ deficiency. Nevertheless, it is recommended to supply both Cl^- and K^+.

3. Rarely MAL is resistant to treatment with saline solution. The urine contains Cl^- above 20 mmol/liter. In these conditions hypermineralocorticism is present without the depletion of the ECF. The causes are a high activity of the adrenal glands or external influences (licorice contains a substance with mineralocorticoid activity). Plasma renin is low; any is

due to stimuli other than depletion of the ECF (Mg^{2+} deficiency, stenosis of renal artery, or Bartter's syndrome). Volume expansion and hypertension are typical and are absent only in severe K^+ and/or Mg^{2+} deficiency and in Bartter's syndrome. The treatment must abolish the cause of mineralocorticoid activity and normalize ion deficiencies (K^+ and Mg^{2+}). Mineralocorticoid influences are blocked by spironolactone or amiloride in Bartter's syndrome. In symptomatic MAL, HCO_3^- excretion may be increased by the administration of acetazolamide (attention should be given to the effect this treatment has on the volume of the ECF and K^+).

Replenishment of Cl^- supply may be accomplished with the following solutions:

1. Saline solution: 1 liter contains in 155 mmol each of Na^+ and Cl^-. As compared with physiological conditions, there is a surplus of acidifying Cl^- (50 mmol). More severe MAL cases, with hypernatremia, cannot, of course, be returned to normal with saline solution.

2. Ammonium chloride, lysine hydrochloride, and arginine hydrochloride: As a rule, molar solutions are used and are added to glucose or other infusions. Traditional indications: NH_4Cl, 5.3%—indicated in severe MAL, particularly with hypernatremia; contraindication in hepatic and severe renal insufficiency. Arginine HCl, 21%—indicated in severe MAL, particularly with hypernatremia including hepatic insufficiency; contraindication in severe renal insufficiency. Recently, certain objections have been expressed to the use of these solutions. It has been found that they acidify the ECF but that intracellular HCO_3^- does not decrease and that pH goes up (R10). Experimentally, an increase in intracellular pH was proved after NH_4Cl administration due to ammonia diffusion (R8). The same trend was observed after administration of arginine HCl, thus tissues have either a high buffering activity or cells are penetrated by arginine, which has strong basic properties.

3. Hydrochloric acid: Indications are MAL that is resistant to treatment using the other salt solutions, MAL with hepatic insufficiency, and edema treated with diuretics (W9). Intracellular decline in pH is detectable (R6). HCl cannot be administered by the peripheral venous route or in situations where clinical and laboratory monitoring is impossible. The infusion speed is 0.25 mmol per kilogram of body weight per hour; as a rule 0.1 mol is used, less frequently a 0.2 mol solution is used (R6, W1). Hydrochloric acid is sometimes added to solutions of amino acids administered parenterally. The advantages are a higher pH and a "gentler" solution than HCl alone (F2, R6). Some clinicians also have good results introducing this infusion into a peripheral vein (K11). The question is whether the same drawbacks will be encountered as with arginine HCl (R6).

6.6. Respiratory Alkalosis

Causes of respiratory alkalosis are as follows (K1, N2, R10, W2).

1. Central disorders and brain stem hypoxia due to (a) anxious, hysterical hyperventilation, (b) metabolic encephalopathy in hepatic failure, (c) CNS infections (meningitis, encephalitis), (d) cerebrovascular complications, (e) gram-negative sepsis, (f) salicylate poisoning, and (g) pregnancy in the third trimester.
2. Systemic hypoxia due to (a) pulmonary shunts, (b) imbalance in the ventilation–perfusion ratio, (c) defects in pulmonary diffusion, and (d) hypotension.
3. Physical stimuli due to (a) irritation in respiratory pathways (tumors, inflammations, and spasms), (b) early interstitial pulmonary diseases (fibrosis), and (c) limited mobility of thorax and diaphragm (not extremely limited).
4. Artificial ventilation.

The hemostatic response to a stress situation is increased ventilation. With p_aCO_2 higher than 5.3 kPa in sepsis, shock, or trauma, the patient's ventilation ability is doubtful.

Clinically, symptoms are most marked in an acute disorder with distinct alkalemia. Patients have circumoral and digital paresthesia, more intensive tendinous reflexes, carpopedal spasms, and muscle contractions. CNS symptoms may be present and are evidently due to hypoxia induced by cerebral vasoconstriction. They include photophobia, nausea, and vomiting. Concerning the cardiovascular system, in the alert patient tachycardia develops that prevents a decline in cardiac output and blood pressure. In the anesthetized patient cardiac output and blood pressure decrease because tachycardia does not develop. Moreover, in artificial ventilation the venous blood return is decreased (K1, R10).

The fact that RAL leads to cerebral vasoconstriction is utilized sometimes by anesthesiologists as a means of prevention or treatment of cerebral edema. In such cases they decrease p_aCO_2 to 3.3–4.0 kPa. Naturally, hypoxemia must not be present. In combination with osmotherapy the effect of this treatment is immediate and lasts at most 5 days. Major hypocapnia is harmful; venous blood from the brain contains a greater amount of lactate than is found in arterial blood. Further laboratory findings involve development of hypobicarbonatemia, hyperchloremia, hypokalemia, and possibly hypophosphatemia (N2). A moderate elevation of AG by 4–7 mmol/liter, due to lactate accumulation, may be present with accelerated glycolysis and an increase in the charge of plasma proteins in alkalemia. Hyperlactatemia is demonstrated only in some of the patients with RAL.

Of 48 unventilated patients with acute RAL, 22 had increased lactate (46%). Among 28 other patients with either less than maximally compensated or maximally compensated RAL, 19 had increased lactate (68%); in one-third of the patients the increase exceeded 4 mmol/liter (K7). Treatment of respiratory alkalosis is as follows:

1. At pH less than 7.55, provoking causes should first be eliminated.
2. In symptomatic patients the p_aCO_2 can be increased by having them breathe into a paper bag or mask.
3. In severe alkalemias with cardiac arrhythmia and mental changes, decrease HCO_3^- with acetazolamide. This measure will be effective only when hypobicarbonatemia has not yet developed. When this treatment is not successful, HCl is administered.
4. In extreme situations mechanical ventilation is used, chiefly when extreme alkalemia is present in mixed RAL + MAL.

6.7. Mixed Disorders

A mixed disorder is defined as a state in which two to three underlying acid–base disorders are present simultaneously (A3, G3, M1, 02, W3). In a broader sense this type of disorder also includes states with several underlying causes of one and the same disorder (R10).

The diagnosis of mixed acid–base disorders must be approached in a comprehensive way. Medical history data, clinical state, and laboratory data have to be considered. It is necessary to estimate the influences that may have led to the simultaneous occurrence of two or more acid–base disorders (e.g., the drugs used, pathological losses of fluids, and organic functional disorders). Laboratory investigations should be performed that confirm or refute the clinical suspicion of a simultaneous development of several underlying disorders or of several causes of one disorder.

Of crucial importance is the evaluation of relationships between ions and ABS, chiefly the relationship between Na^+ and Cl^-, i.e., BB_s and the calculation of AG.

To recognize a mixed disorder is not a theoretical problem, but it is important from the diagnostic and therapeutic aspects. Thus, in a febrile septic patient with RAL, a decline in HCO_3^- with an increase in AG may be the first sign of the development of lactic acidosis and it signals septic shock. Patients with chronic RAC are usually treated with diuretics for congestive heart failure, and also have MAL. When the latter disorder is not diagnosed and treated, during artificial ventilation, hyperbasemia persisting after the treatment of hypercapnia may lead to respiratory depression, making termination of ventilation impossible.

A survey of mixed disorders and examples of typical causes follows (for a better orientation, see the graph in Fig. 3):

1. MAC + RAC, leading to a marked decline in pH: Causes include (a) cardiac arrest, (b) chronic obstructive lung disease plus septic shock, (c) severe renal and respiratory failure, (d) severe pulmonary edema, and (e) treatment with carboanhydrase inhibitors plus β-adrenergic blockers (B7). Treatment involves artificial ventilation and alkalization to prevent a pH decrease to below 7.15.

2. MAL + RAL, leading to a marked increase in pH: Causes include (a) chronic liver disease plus gastric juice losses, (b) congestive heart failure plus diuretic therapy, (c) breathing center disorders plus diuretic therapy or gastric juice losses, (d) pregnancy plus MAL of different etiology, (e) intentional hyperventilation in MAL, and (f) successful treatment of compensated MAC plus intensive alkalization. Treatment involves acidification with the use of HCl and artificial ventilation.

3. MAL + RAC, with almost normal pH: Causes include chronic obstructive lung disease plus diuretic therapy or vomiting. It is necessary to determine whether hypercapnia exists per se or whether it is secondary in MAL. Differentiation can be made by two methods: (a) if during several minutes of inhalation of 100% O_2 the pCO_2 further increases, then RAC is primary; (b) if the alveoloarterial O_2 gradient is above 2 kPa, then a serious pulmonary disease is present and RAC is primary. Treatment involves artificial ventilation to eliminate RAC; the persisting increase in HCO_3^- would lead again to an increase in pCO_2 and a decline in pO_2. Thus a simultaneous acidification with chloride is essential.

4. MAC + RAL, pH decreased but usually only slightly: Causes include (a) septic shock, (b) hepatorenal syndrome, (c) renal failure plus sepsis, (d) salicylate poisoning, and (e) congestive heart failure and renal failure. Treatment is according to the underlying cause with monitoring and modification of ABS according to need.

5. MAC + MAL, pH is shifted to the side of the predominating disorder, but may be also normal: Causes include (a) intoxication with ethyl alcohol plus vomiting, (b) uremia plus vomiting or diuretic therapy, (c) ketoacidosis plus vomiting or diuretic therapy, (d) lactate acidosis plus vomiting, and (e) serious hypoproteinemia plus various causes of MAC. Diagnosis will be facilitated by calculations of BB_s that are increased in MAL and AG. This value rises only slightly in MAL, more markedly in MAC. (A rough determination of the participation of both disorders in the AG increase has already been discussed.) Both disorders must be treated simultaneously.

6. MAC + MAL + RAL, pH is usually only slightly alkaline: Causes in acute states often involve hypoxia (systemic or local in the area of the

breathing center), which leads to hyperventilation (G6) (see also causes a–d listed for MAC + MAL, step 5 above).

7. MAC + MAL + RAC, pH is generally normal to slightly decreased: Causes are as given in step 5, a–d (MAC + MAL); moreover, see causes of RAC (Section 6,4).

8. A special situation is presented by chronic RAC in which a sudden change takes place in pCO_2. Two situations may arise: (a) A further increase in pCO_2 due to heart failure, pneumonia, excessive supply of O_2, or treatment with tranquilizers. There is risk of stupor and even coma and hypoxia. Treatment involves artificial ventilation. (b) A sudden decrease in pCO_2 due to excessive artificial ventilation. Serious alkalemia develops due to persisting compensatory hyperbasemia. Treatment involves acidification and normalization of ventilation.

9. Hyperchloremic MAC + MAC with a high *AG*. Causes include (a) diarrhea plus lactic acidosis or ketoacidosis, (b) renal tubular acidosis plus uremic acidosis, and (c) a possible further combination of causes of MAC of both types.

6.8. Therapeutic Calculations

The quantity of alkalizing or acidifying solution to administer is calculated according to the ABS: correction to $BE = 0$ or to the target pH.

The calculation for the correction to $BE = 0$ is formulated in the same way as for the treatment of acidosis and alkalosis, the only difference being that in the former case the base deficiency *(BD)* is eliminated, and in the latter case the base excess *(BE)* is removed.

Calculation of HCO_3^-:

$$\text{mmol } HCO_3^- = BD \times 0.3 \times kg \quad (1)$$

Calculation of Cl^-:

$$\text{mmol } Cl^- = BE \times 0.3 \times kg \quad (2)$$

In acidosis, when it is useful to adjust the pH only to the target value, then the following corrected formula is used (N4):

$$\text{mmol } HCO_3^- = BD \times 0.3 \times kg \times \frac{\text{target pH} - \text{actual pH}}{\text{pH } X - \text{actual pH}} \quad (3)$$

The target pH is selected according to the clinical situation. In acute states (lactic acidosis, ketoacidosis) this value is 7.1 to 7.2 The value pH X can be looked up in the table according to the patient's pCO_2 value. It is a theoretical value to which the pH of the patient's blood would be adjusted after the correction of *BD* with the actual p_aCO_2 being retained (Values are in kPa):

pCO_2	pH X
2.0	7.670
2.5	7.610
3.0	7.567
3.5˙	7.517
4.0	7.487
4.5	7.457
5.0	7.427
5.5	7.397
6.0	7.377
6.5	7.350
7.0	7.332
7.5	7.314
8.0	7.298
8.5	7.280
9.0	7.263
9.5	7.248

Equation (3) can also be used (with a small correction) for the correction of MAL if hypercapnia is present. Then a correction will be made to the target pH (generally 7.400) and not to $BE = 0$, because we would shift the patient from alkalemia to acidemia. In the calculation

$$\text{mmol } Cl^- = BE \times 0.3 \times kg \times \frac{\text{actual pH } - \text{ target pH}}{\text{actual pH } - \text{ pH } X} \quad (4)$$

The pH X is subtracted depending on pCO_2.

General information about our approach to the calculation of the necessary amount of acidifying or alkalizing infusion can be seen in Fig. 7. Whether and to what extent a disorder in ABS is going to be treated with infusions depends always on the decision of the attending physician, who takes into consideration not only laboratory findings and calculations but also the patient's medical history and clinical status. The corrective treatment of the acid–base disorder as far of the treatment of its underlying cause is then under control.

Correction of chronic RAC + MAC by means of infusions should be avoided (despite the discussion of calculations) as far as possible, unless pH is critically low. Oxygen treatment alone involves the risk of a sudden increase in pH after the elimination of the hypoxic component of the disorder. In acute disorders the primary cause should first be eliminated. As far as metabolic acidoses are concerned, greatest care should be given to hypoxic MAC and to ketoacidosis, which subside rapidly and spontaneously with effective treatment of the cause. Alkalization is recommended only when pH is less than 7.100 (maximally, 7.200). In cases of renal insufficiency, acidosis must likewise be corrected very carefully. Evidently a chronic disorder is usually associated with anemia and hypocalcemia, and patients tolerate a moderate acidemia much better than alkalemia. We must be more resolute in those MAC cases that would persist if uncorrected. These cases cover the frequent hyperchloremic dilution acidoses or acidoses accompanying diarrhea. On the other hand, in MAL, as a rule, the whole calculated dose of the corrective solutions will have to be administered.

7. Computer Programs for Monitoring Water, Ion, and Acid–Base Metabolism

7.1. Introductory Comments

Data processing with the use of computers has made it possible to improve the control of vital functions of critically ill patients (B9, B12, M6, V2). With the rapid advances achieved in the management of these problems, it is necessary to overcome objective difficulties as well as traditional approaches in the thinking of healthcare personnel. This is evident from the following data:

1. Technical development of computer systems is very progressive. Compromises are repeatedly made between what is already feasible and what is actually done in practice. Conceptions are subject to new modifications (M6).
2. In combination with monitors, supervising the patients with computers enables complex systems. Computer use requires a change in the thinking of the attending staff and also a nontraditional approach to the monitoring of patients.
3. New information items, estimations of relationships, and cumulative and graphic evaluations enables a more profound view to be taken of the patient's situation. Thus both practical and theoretical knowledge is enhanced. At the same time, there are increased demands for continuing education and postgraduate studies for medical personnel (M15).

Many accounts have been given of the critical-care approach of clinicians with respect to computer interpretation. Yet, comparative studies show that the reaction of the clinical staff to critical findings of ABS and p_aO_2 is slow and unsatisfactory (B9). It has been found that most clinicians overestimate their ability to interpret data, and only a minority interpret the findings correctly and propose adequate treatment (H4). At the neonatal intensive care unit, computer ABS data processing was found to lead to an earlier normalization of greatly abnormal results, and serious pathological trends developed less frequently. On the other hand, the treatment used for the correction of ABS was sometimes too aggressive and, in extreme cases, gave rise to the development of pneumothorax. The duration of oxygen therapy, tracheal intubation, and mortality remained unchanged (H3).

7.2. Development and Contribution of Computer Programs

The first program interpreting acid–base findings was published toward the end of the 1960s (C6). The graph on acid–base regulation (see Fig. 3) was divided into 28 areas. The location of the finding in any area was commented on with due regard to the mode of ventilation.

At the beginning of the 1970s extensive interactive programs for interpretation appeared (B6, G5). The first of them used information on the basic parameters of ABS, serum ion concentrations, ketonemia, creatin-

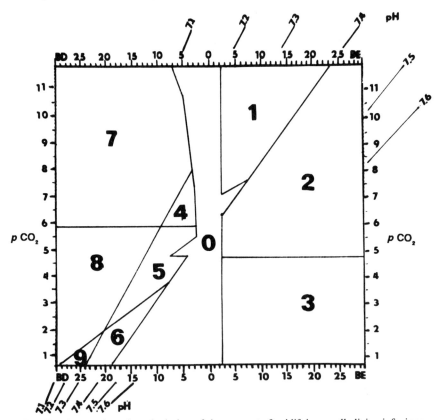

FIG. 7. Approach to the calculation of the amount of acidifying or alkalizing infusions.

FIG. 7
EXPLANATION

Number of zone	Equation used	Corrected to	Note
1	2	$BE = 0$	Under the condition of a simultaneous elimination of hypercapnia
2	4	Target pH, 7.4	—
3	2	$BE = 0$	—
4	1	$BE = 0$	Be on your guard against inhibition of the breathing center! Often managed by oxygen therapy with monitoring of ABS; in acute disorders by elimination of the cause
5	3	Target pH, 7.4	As long as acidosis due to hypoxia or ketoacidosis is not present
6	3	Target pH higher by 0.1 than the actual pH (if it exceeded 7.4, pH 7.4 is regarded as the target)	As long as acidosis due to hypoxia or ketoacidosis is not present
7	3 /1/	Target pH higher by 0.1 than the actual pH (Exceptionally to $BE = 0$)	Be on your guard against inhibition of the breathing center! Often managed by oxygen therapy with monitoring of ABS; in acute disorders by elimination of the cause
8	3	Target pH, 7.1	In acidosis due to hypoxia, or in ketoacidosis
		Target pH higher by 0.1 than the actual pH	In other types of acidosis
9	3	Target pH higher by 0.1 than the actual pH	—

inemia, and glycemia. The mode of ventilation, types of medication, etc., were also among the data set. If the user was unable to answer, he could find the background information on the display. The computer report consisted of a numerical part and of a commentary. The latter described the diagnostic area in which the finding was situated and assured the printing of the most probable and less probable diagnoses. According to the user's wish, the least probable diagnoses and remarks on possible treatment were printed.

The other program also requires data on urinary excretions and also gives attention to water and ion metabolism. The dialogue with the user as well as the actual output are naturally more detailed. In the 1980s this program was improved (M15). It considers not only diagnostic possibilities but also explains the pathophysiology, recommends treatment, and draws attention to data inconsistencies and any lack of information.

In the second half of the 1970s a further interactive program was published that was intended to serve both the treatment of critically ill patients and the medical education community (S2). Biochemical investigations, together with data on the patient's clinical state, form a basis for calculations of water, blood, and ion deficiencies. Continual monitoring using the program also requires a daily supplementation of information on the vital signs and symptoms, on the intake and excretion of liquids, and on temperature, medication, further biochemical findings, and cardiovascular and respiratory functions. In addition to ABS, the program also evaluates pulmonary shunts and respiratory functions. When the cardiac output and blood gas findings (arterial and central venous blood) are to be estimated, the program calculates the respiratory quotient and determines the need for water, ions, and energy. Commentaries draw attention to problems of major importance, but the solution must be found by the physician.

In the meantime, a number of studies have been published on the problem of seeking optimum variants and of conveying and commenting on acid–base parameters and oxygen measurements (R1, S5, T1, V1, W8). Many of these programs also deal with patients treated with artificial ventilation (B9, D1, E3, K10, L7, M15, O1, S3, V2). The program estimates the adjustment of the ventilator and indicates treatment with the use of positive end-expiratory pressure (R3).

A further and qualitatively higher degree of utilization of biochemical information is represented by continual monitoring of p_aO_2 and of p_aCO_2 values used for on-line management of ventilation therapy. Electrodes are introduced into the artery (K13); also, the transcutaneous measurement of p_aO_2 and end-expiratory CO_2 as the degree of p_aCO_2 are estimated (H2). In both cases the continually scanned data are transferred to the microprocessor, which compares the desirable values, set in advance, with the actual ones. The data obtained are converted into impulses that adjust the ventilator and the valves, ensuring O_2 supply to the mixture of gases.

Many programs, even recent ones, are not of an interactive nature and therefore do not require a direct dialogue with the clinician. They work only with laboratory findings, supplemented maximally with a limited number of data, set in a standard way, on the mode of ventilation, on the patient's temperature, etc. (B9, D1, K2, K10, L7, M3, M4, O1, V2). With such a manner of processing, the physician's possible subjective error in the evaluation of the patient's clinical state is eliminated; the processing is simpler, it does not require a direct availability of hardware at every clinical department, and the program can thus be used to a wider extent. The time factor limiting the clinician in his dialogue with the computer does not come into consideration.

Interactive programs, based on a dialogue of the physician with the computer, require technical resources, but, in return, provide higher order information from the qualitative as well as the quantitative point of view (B6, G5, M15, S2, S3). Thus they surpass the framework of a mere interpretation of results and also serve as teaching programs in undergraduate and postgraduate education. The findings obtained in the course of their use serve as background material for the development of expert systems. The contribution of programs that process and evaluate biochemical findings for ICU patients are summarized as follows:

1. Warn of serious pathological findings and thus speed up the action of the healthcare team.

2. Draw attention to impossible and improbable data.

3. Complete calculations and evaluate a number of parameters not commonly utilized in evaluations of results.

4. Substitute for that part of the clinician's work for which there is limited time in routine practice; provide expertise sometimes not possessed by the clinician.

5. Provide an evaluation with a commentary and possibly also with a graphic representation, in addition to the usual numerical expression of the laboratory results.

6. Facilitate a cumulative evaluation of the results and monitor the prevailing trends.

7. Relate the evaluation of biochemical findings to functional parameters, chiefly cardiopulmonary ones.

8. Suggest therapeutic methods including ventilation therapy and its modifications.

9. Enable, with suitable communication media, an interpretation of the results at a considerable distance.

10. May be supplemented with new findings.
11. Serve both for undergraduate and postgraduate studies.
12. May be adjusted in accordance with the users' experiences and wishes.
13. Are opening up new possibilities for research work.
14. Are becoming the physician's helpers (their purpose not being, however, to replace the physician). The final decision rests in the physician's hands. The programs afford him, however, new and qualitatively higher evaluations of biochemical findings than those performed by himself. In this way they facilitate his diagnostic as well as therapeutic decision making.

7.3. Approach to Biochemical Monitoring of ICU Patients

7.3.1. *Acid–Base Status, Metabolic Balances, and Renal Functions*

The programs described below have been used since the beginning of the 1980s in the daily ICU practice for reporting of biochemical findings (K8, M7, M8). They have been further developed and gradually introduced into a number of hospitals. Findings of water and electrolyte metabolism, acid–base balance, and pO_2 are processed. The programs compute the accessible lung function values as well as renal functions, and they make it possible to estimate the quality and quantity of the parenteral nutrition and diet formulas. The outputs have numerical and graphical data and comments. They enable a cumulative evaluation of data.

(A) Input data: are both clinical and laboratory. Data are entered into the computer either using the terminal in the clinical ward or by the staff of the emergency laboratory using the order form accompanying the biological specimens. When all of the data are not available, the programs offer the maximum possible information with respect to the missing data.

Clinical data include (1) weight, height, and sex, (2) artificial ventilation (used or not; if used, then ventilation (liters) and FiO_2 data are demanded), (3) type of blood sample for ABS and pO_2 investigations, (4) body temperature at the time of specimen collection and its mean for the period under study, (5) urine volume and the time of the urine collection and/or the same parameters for the tube or drain contents, (6) and estimated amount of sweat. As far as metabolic balances are ascertained, all received infusions, transfusions, blood derivatives, and formula diets and intakes of selected antibiotics (Na^+ or K^+ salts) are put in by means of a simple dialogue. Depending on the diagnosis and clinical status, the clinician sets the degree of catabolism, used later for the calculation of the energy demand (P2).

Laboratory data include (1) ABS, i.e., pH, pCO_2, pO_2, hemoglobin, and possibly saturated hemoglobin (S_{O_2}), (2) serum (plasma) Na^+, K^+, Cl^-, osmolality, urea, creatinine, total protein, glucose, lactate, and possibly other electrolytes, (3) urine Na^+, K^+, Cl^-, osmolality, urea, creatinine, glycosuria, ketonuria, and possibly other electrolytes, and (4) Na^+, K^+, Cl^-, and urea contents of tube and drain. (B) Program processing proceeds after the clinical and laboratory data have been set. Its individual parts (ABS plus electrolytes, renal functions, and metabolic balances) are arranged as a brick-box so that they can be used also as separate units. (C) Output: In addition to the usual identification data, the chronology of findings is indicated in alphabetically order within each subgroup. The output has three parts (see Fig. 8), numerical, graphical, and comments.

The numerical part, wherein results are always expressed in SI units, comprises the first page, which is logically arranged into three blocks. The first block deals with ABS. The two left-hand columns contain information on the type of blood, ventilation, FiO_2, hemoglobin, and the measured temperature used for correction of blood gases. Another type of blood, withdrawn at the same time, is indicated separately here. The two center columns include the measured and the calculated ABS parameters. The two right-hand columns cover pO_2 findings and calculations derived from them, i.e., the volume of O_2 dissolved in 1 liter of blood (V_{O_2}) and S_{O_2}. Lung arteriovenous shunts are estimated from FiO_2 and p_aO_2, as well as the alveoloarterial difference for oxygen ($A - aD_{O_2}$). If central venous blood is analyzed, then the arteriovenous V_{O_2} difference is calculated and arteriovenous shunts are evaluated more precisely.

The second block is concerned with further biochemical parameters that are related to the parameters examined and to the evaluation of the balances. The parameters used are given in the left column; the substances measured and the calculations are in the upper line, and are arranged from left to right in the following way:

1. Serum: glucose and various other parameters are examined (lactate from plasma is also included).
2. Urine: timed volume and other parameters are examined; the conversion of excreted urea into grams of nitrogen is also given.
3. Loss: total losses of particular parameters for the time of the collection are summarized; the energy output is given in kilojoules (Harris–Benedikt's formula, corrected to body temperature and to estimation of the degree of catabolism). Alternately, this calculation can be replaced by data obtained by indirect calorimetry.
4. "Income": includes summary intake of the balanced items using constants from data bases. Included are intake levels of certain ions, the investigation of which in the emergency laboratory is not always feasible. The total intake is given.

```
2.07.87  * HOSPITAL           * ACUTE LABORATORY * PATIENTS NAME
7:30 "H" * BULOVKA PRAGUE *        REPORT        * ICU
********************************************************************

BLOOD TYPE   ARTERY   PH        7.184   PO2            6.59
VENTILATION  12.0     PCO2      5.52    VO2          120.5
FIO2          .60     BE       -12.4    SO2            .75
HB           120      ST  BIC   15.2    SHUNT ESTIM   .46
TEMP ACTUAL  38.0     ACT BIC   15.2    AA DO2        44.3
PEEP          5.0


BLOOD TYPE   CENTRAL  PH        7.164   PO2            4.99
             VEIN     PCO2      6.04    VO2           92.1
                      BE       -12.3    SO2            .57
                      ST  BIC   15.3    SHUNT          .64
                      ACT BIC   15.9    AV DVO2       28.2
..................................................................

             SERUM    URINE   LOSS    INCOME  BALANCE   JUICE   DRAIN

VOLUME                 450    2644     2750    106       800     350
  /TIME                 24      24       24     24        24      12
GLUCOSE      10.8      34.6    34.6
KETONURIE               1
LACTAT        3.2
OSMOLALITY  310        216
CREATININ     1.8
UREA         13.0       38      31                        9.6     2.8
NITROGEN                1.2      1.7     3.1    1.4        .3      .1
NA          127         11      57       56     -1        32      14
K             3.5       13      28       27     -1        10       5
CL           92         19      85       80     -5        44      22
TOT PROT     68
CA                                       6.7
MG                                       1.2
P                                        6.3
ZN                                        .15
ENERGY/TIME            108   14105    14113      8
..................................................................

BBS          38.5    HEIGHT       175    BLOOD         500
AG           23.3    WEIGHT        80    ERYMASS       200
DEF K/PH     91      SURF AREA    1.96   ALBUMIN       100
DEF K        43      PF          3600    PLASMA         50
DEF NA      160      ECF        16000    E I.V.         .71
DIUR/D      450      ICF        32000    E P.O.         .29
CLR CR        .135   TBW        48000    E GLYCIDS      .43
CLR OSM       .007   TEMP /D      37.5   E FAT          .36
CLR H2O      -.003   ENERG/D    14105    E PROTEIN      .21
FE  OSM       .053   MM OUT W/D  1044    THIAMIN       20.5
FE  H2O       .034   MM INC W/D   350    RIBOFLAVIN     8.4
FE  NA        .007   IDX STRESS  -2.0    PYRIDOXIN      6.1
FE  K         .293   IN E:N      4553    PANTOTENAT    10.0
U/S CR       29.3                        NIKOTIAMID    80.5
U/S OSM       1.5    FEA H2O      .149   C VIT       1108
U/S U         3.2    FEA NA       .007   A VIT      13085
U   NA/K       .8    FEA K        .298   K VIT        21.5
                                         E VIT         4.5
```

FIG. 8. Sample program output, used daily in the intensive care unit for biochemical and metabolic monitoring of the patients. All the values are SI units.

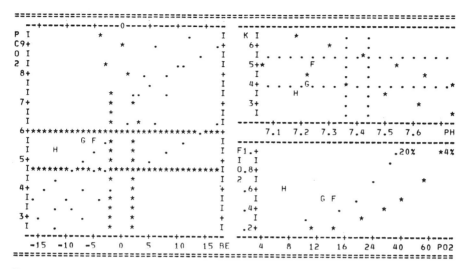

EVALUATION OF AB- STATUS AT VENTILATED PATIENT:

[PH,PCO2,BE,PO2:] AB FINDING DEMONSTRATES METABOLIC ACIDOSIS. FROM THE POINT
OF VIEW OF ITS COMPENSATION THE VENTILATION IS NOT OPTIMAL. PROGRESSION OF
FINDING FROM THE LAST EXAMINATION. FOR CORRECTION OF BASE EXCESS TO BE=0
IT HAS TO BE GIVEN 298 ML NAHCO3 8.4%, TO PH=7.284 GIVE 146 ML OF THIS SALT.
HYPOXEMIA, INDICATION TO CORRECTION OF THE OXYGENOTHERAPY.
[U,CR,KETO,LACT,BBS,AG:] MIXED DISTURBANCE WITH SHARE OF METABOLIC ALKALOSIS
IS NOT PROBABLE.
[K,PH:] HYPOKALEMIA IN DISCREPANCY WITH PH. AT CORRECTION OF PH FURTHER FALL
MAY BE EXPECTET.

EVALUATION OF KIDNEY FUNCTIONS (CORRECTED TO BODY SURFACE AREA):

[CLR-CR:] CREATININ CLEARANCE IS VERY HEAVY DEPRESSED!
[S-OSM,U-OSM/S-OSM,DIURESIS:] KIDNEYS CONCENTRATE URINE WITH DEPRESSED EFFORT,
SERUM OSMOLALITY IS ELEVATED. OLIGURIE! DANGER OF FURTHER RETENTION OF OSMOTIC
ACTIVE SUBSTANCES. CHECK THE SUPPLY OF GIVEN LIQUIDS!
[FE-H2O,FE-OSM,CLR-H2O,U-OSM/S-OSM:] EXCRETION OF WATER AND OSMOTIC ACTIVE SUB-
STANCES IS WITH REGARD TO GLOMERULAR FILTRATION ELEVATED, OSMOTIC TYPE OF
DIURESIS IS PROVED. CAUSE OF ELEVATED FREE WATER CLEARANCE MAY BE IMPAIRED
CONCENTRATING ABILITY. AT GIVEN DEVELOPMENT OF SERUM OSMOLALITY WOULD OTHERWAY
THE FREE WATER CLEARANCE BE NEGATIVE.
[FE-NA,FE-K,FEA-NA,FEA-K:] FENA IS IN PHYSIOLOGICAL RANGES, FEK IS ELEVATED -
-(TUBULAR COMPENSATION OF DECREASED GF?, DIURETICS?, HYPERALDOSTERONISM?).
COMPARISON OF FE ADEQUATES WITH FE ACTUALS DEMONSTRATES THAT KIDNEYS DO NOT
INFLUENCE BALANCE OF NA OR K NEITHER IN SENSE OF RETENTION NOR DEPLETION.
[SUMMARY:] DIFFERENTIAL DIAGNOSTIC PARAMETERS DO NOT ENABLE CLEAR DISTINCTION
OF PRERENAL OR RENAL FAILURE. CREATININ AND FREE WATER CLEARANCES DOCUMENT
MIXED DISTURBANCE OF GLOMERULAR AND TUBULAR FUNCTION !!

FIG. 8. *Continued.*

5. Balance: given as the difference between the intake and the output.

6. Tube ("juice") and drain: the columns contain data on losses during the time of collection. If the volume of sweat was estimated, then losses of electrolytes are calculated again with the use of constants from data bases.

The third block contains in the two left-hand columns values calculated from the column (Fig. 4) of anions (BB_s, AG) and calculations of Na^+ and K^+ deficits; the value for K^+ deficit is an alternative parameter to the measured pH. Then follow parameters of renal functions, i.e., clearances and fractional excretions. Also included are indices of urine/serum parameters, used for differential diagnosis of oligoanuric states. The two center columns contain basic anthropometric data and theoretical calculations of the body spaces. Then auxiliary calculations for metabolic balances follow, i.e., the unmeasurable output of water (NM OUT W/D) and the intake of metabolic water (NM INC W/D), which were considered above in the water balance. The calculation of the stress index (K5) is followed by the calculation of the ratio between the intake of nonprotein kilojoules and the received nitrogen (IN E:N) and calculations of adequate fractional excretions for water, Na^+, and K^+. Finally, the two right-hand columns contain data on intake for the period under study, where a total is given for the quantity of transfusions and blood derivatives, the proportion of energy received by the intravenous and oral routes (E I.V. and E P.O.), the evaluation of the proportion of glycides, fats, and proteins in the total intake of energy, and, finally, data on the amount of vitamins taken.

The graphical part of Fig. 8 contains three graphs. On the left is a graph of acid–base regulations in which the existing finding and two preceding ones are marked with letters in alphabetical order so that the course of the disorder may be observed. The same symbols are used in the other two graphs. The upper right graph shows the K^+:pH ratio; reference intervals of both parameters are delineated by points, and the relationship between the parameters is marked with asterisks. The bottom right graph illustrates pulmonary shunts and their development. The curve of asterisks corresponds to the upper physiological limit of 4% of the shunts; the curve of points indicates the clinically significant 20% of the shunts. The graphic interpretation is enabled by a commercially available line printer.

Comments pertain to the evaluation of the acid–base finding and to the estimation of renal functions. First, the location of the finding in the graph of the acid–base regulation is described and the mode of ventilation is taken into consideration. Then follows the evaluation of the course of the acid–base findings since the last investigation; the length of the interval since that investigation is taken into account. Next comes the calculation of the correction dose suitable either for the attainment of the target pH or BE according to the procedures described in Section 6.8. The presence of a mixed disorder is then evaluated by means of suitable biochemical parameters. The comments assessing K^+ status are presented in relationship to pH and estimate the development of this relationship. The last remark is devoted to biochemical indications for the modification of ventilation therapy and for oxygen therapy.

The comments dealing with renal functions evaluate chiefly C_{cr}, then diuresis in relation to serum osmolality and its trend, and to the osmolal index, i.e., to the urine and serum osmolality ratio. An estimation is further made of the type of diuresis and, in cases of C_{H_2O} elevation, the possible causes are considered. In the evaluation of FE_{Na} and FE_K the most probable and least probable causes of their changes are discussed. In extremely elevated FE_{Na} a further possible effect of diuretic treatment is taken into consideration; in a simultaneous incidence of a high FE_K and hypokalemia, attention is drawn to the possible treatment with aldosterone antagonists. In the estimation of relationships between fractional excretions of H_2O, Na^+, and K^+ and their adequate fractional excretion, it is considered whether the kidneys influence the balance via retention or depletion. Finally, in the summary, the parameters of prerenal and renal failure are evaluated in indicated cases, and attention is drawn to findings indicative of a combined disorder of the glomerular and tubular function. Wherever necessary, a note is added concerning any doubts about the validity of the input data. Attention is thus given to a methodological error made in the collection of the material or in its processing.

The discussed programs, written in the language FORTRAN IV and alternatively in TURBO PASCAL, are a part of a far wider data-processing scheme in the laboratory information system (LIS).

7.3.2. Proposal of Sodium and Potassium Doses

A qualitatively higher step in program processing is a proposal of the particular components of parenteral nutrition and formula diet. We present programs for calculations of Na^+ and K^+ doses used daily for ICU patients (J1).

Program processing of the sodium dose proposal requires input variables, including both laboratory data (sodium status, glycemia, Na^+ excretion in the urine, and extrarenal Na^+ losses) and clinical data (state of hydration, determination whether the state is acute or chronic, common and measured body weight or estimation of changes in body weight, and data on diuresis and the patient's age and sex).

Program processing with respect to the proposed dose has two purposes—substitution and correction. The substitution dose replaces losses and is based on the knowledge of renal and/or extrarenal Na^+ losses. Extreme losses need not be replaced fully. The correction dose addresses two concerns, requiring two calculations:

1. Calculation of a dose for attaining a target natremia: this is used chiefly in hyponatremia and, to a limited extent, in normonatremia.
2. Calculation of a dose to correct a Na^+ deficit or excess; this is based on the knowledge or estimate of changes in body weight during dehydration or hyperhydration and natremia.

Both types of calculations, depending on the circumstances, account for different theoretical distribution spaces, clinical situations, and the constellation of laboratory findings. With the varying proportion of free water and isotonic fluid losses, one proceeds according to Section 3 (see also Fig. 1). By a combination of different doses and their summation, three proposed values are obtained: the optimum proposed dose and the lower and upper limit of the proposed dose.

Together with the proposed doses a commentary is printed, which forms an inseparable part of the output. It substantiates the proposal of the optimum dose as well as the proposal of the limits, it estimates Na^+ supply with regard to the ECF:TBW ratio, it gives the factors used in estimates of body weight changes, it draws attention to the proposal of an incomplete replacement of loses in suitable cases, it recommends convenient concentrations of salts, and it contains comments on the suitability of diuretic treatment in indicated cases. For an example of the output see Fig. 9.

Program processing of the K^+ dose proposal requires input variables, including both laboratory data (potassium status, pH of arterial or arterialized blood, and nitrogen balance) and clinical data (diuresis and body weight or its estimate).

Program processing with respect to the proposed dose has two purposes—substitution and correction. Substitution is based on the knowledge of renal and/or extrarenal losses and accounts for the negativity or positivity of the N balance. Correction is based on the doses, which are different in the six areas given by the K^+:pH relations. The border of the areas is demarcated by a K^+ level of 4.4 mmol/liter, pH 7.4 and the ideal K^+:pH ratio, defined by the equation

$$K^+ \text{ mmol/liter} = 33.05 - 3.87 \times pH$$

The specific areas differ from each other by the risk of developing hyperkalemia, which may increase with the modification of pH. A standard dose has been proposed for the normalization of K^+ to 4.4 mmol/liter.

```
NAME: DEMONSTRATION              WARD: ICU
-----------------------------------------------------------------

LABORATORY:    NATREMIA              155  MMOL/L
               GLYCEMIA                5  MMOL/L
               NA LOSS URINE          30  MMOL/D
               NA LOSS DRAIN          70  MMOL/D

CLINICALY:     STATE OF HYDRATION    DEHYDRATION
               DEVELOPMENT           ACUTE
               CURRENT WEIGHT        75   KG
               ACTUAL  WEIGHT        NOT GIVEN
               DIURESIS              650 ML/D
               SEX                   MALE

-----------------------------------------------------------------

NA DOSE - PROPOSAL:  191  (144 - 289) MMOL/D
-----------------------------------------------------------------

THE CONDITION CANNOT BE TREATED WITH WATER ONLY - CHANGES
IN OSMOLALITY! THE PROPOSAL CONSIDERS THE USE OF DILUTED
NACL SOLUTIONS, EVEN THOUGHT THE STORE OF NA INCREASES.
THE LOSS OF FLUIDS IS ESTIMATED TO BE 1.9 - 3.8 LITERS.

THE PROPOSED BASIC DOSE CONSIDERS THE REPLENISHMENT OF 1.9
LITERS USING 0.68 % NACL, THE LOWER LIMIT WITH 1.9 LITERS
OF 0.45 % NACL, THE UPPER LIMIT WITH 3.8 LITERS OF 0.45 %
NACL.

IN CASE OF PERSISTENCE OF HYPERNATREMIA AFTER REHYDRATATION
THE TREATMENT WITH SALURETICS AND WATER SHOULD BE CONSIDERED.
```

FIG. 9. Program output for the proposal of sodium dose.

When the measured K^+ is higher than 4.4 mmol/liter, this dose becomes "negative" and is subtracted from the substitution dose. A change in the dose is always considered when the pH is outside the reference limits, provided it is going to be normalized.

The upper and the lower limits of the dose and the optimum dose are also suggested in the output. The commentary substantiates these proposals. In certain situations it stresses the risk of K^+ administration and estimates the necessity of changing dosed in connection with ABS changes. Attention is drawn here to the N balance. The output is arranged similarly to that for the Na^+ dose proposal.

8. Conclusions

The information obtained from biochemical investigations of patients in intensive care can be exploited in an optimum way only when certain conditions are met. Among these conditions are a simultaneous evaluation of the medical history, the clinical state, and the laboratory results. From the medical history it is necessary to estimate the effect of the previous course of the disease and of its effect on the water, electrolyte, and acid-base balance and on energy metabolism. Equally important is the evaluation of the actual clinical state. Of greatest importance are the assessments of the state of hydration and the cardiopulmonary findings. All of these items of information must be considered in the interpretation of the laboratory values. In view of the interrelationships existing between biochemical parameters, specimen collection for blood gases and other tests must be performed simultaneously. Computing methods make it possible to complete the calculation of a number of parameters and to evaluate them along with the measured ones. Of importance are estimates of interrelationships between ABS and electrolytes and oxygen parameters, including pulmonary shunts and renal functions. In addition to clearances, it is important also to assess parameters providing information about tubular functions. The monitoring of renal functions is of fundamental importance for the assessment of water and electrolyte balance as well as for the estimation of the adequacy of diuresis.

Computer data processing enables a qualitatively higher evaluation of the biochemical findings supplemented with anamnestic and clinical information. The increasing amount of information to put into a system and the new methods of data processing decisively indicate that "interpreting programs" will be replaced by expert systems.

REFERENCES

A1. Adrogue, H. J., and Madias, N. E., Changes in plasma potassium concentration during acute acid-base disturbances. *Am. J. Med.* **71**, 456–467 (1981).
A2. Anderssen, O. S., "The Acid Base Status of the Blood," pp. 92–144. Munksgaard, Copenhagen, 1976.
A3. Assadi, F. K., Agrawal, R., and Jocher, K., Hyponatremia secondary to reset osmotat. *J. Pediatr.* **108**, 262–264 (1986).
A4. Astrup, P., and Severinghaus, J. W., "The History of Blood Gases, Acids and Bases," pp. 272–276. Munksgaard, Copenhagen, 1986.
B1. Bennet, W. M., Antibiotic induced renal failure. *Semin. Nephrol.* **1**, 43–49 (1981).
B2. Bennet, W. M., Singer, I., and Golper, T., Guidelines for drug therapy in renal failure. *Ann. Intern. Med.* **86**, 754–784 (1977).
B3. Bia, M., and Thier, S. O., Mixed acid base disturbances, a clinical approach. *Med. Clin. North Am.* **65**, 341–361 (1981).
B4. Bidani, A., Electrolyte and acid-base disorders. *Med. Clin. North Am.* **70**, 1013–1036 (1986).
B5. Black, A. K., and Cameron, J. S., Renal function. In "Chemical Diagnosis of Diseases" (S. S. Brown, F. L. Mitchell, and D. S. Young, eds.), pp. 483–513. Elsevier, Amsterdam, 1977.
B6. Bleich, H. L., Computer based consultation. *Am. J. Med.* **53**, 285–291 (1972).
B7. Boada, J. E., Estopa, R., Izquierdo, J., Dorca, J., and Mauvesa, F., Severe mixed acidosis by combined therapy with acetazolamide and timolol eyedrops. *Eur. J. Respir. Dis.* **68**, 226–228 (1986).

B8. Bockelman, H. W., Cembrowski, G. S., and Daniel, F. I., Quality control of electrolyte analyzers. Evaluation of the anion gap average. *Am. J. Clin. Pathol.* **81**, 219–223 (1984).
B9. Broughton, J. O., and Kennedy, T. C., Interpretation of arterial blood gases by computer. **85**, 148–149 (1984).
B10. Brown, R., and Babcock, R., Renal function in critically ill postoperative patients. *CRC Crit. Care Med.* **8**, 68–72 (1980).
B11. Brown, R. S., Extrarenal potassium homeostasis. *Kidney Int.* **30**, 116–127 (1986).
B12. Burridge, P. W., Skakun, E. N., and King, E. G., Evaluation of a computer-based clinical reference library in an ICU. *CRC Crit. Care Med.* **13**, 763–766 (1985).
C1. Cantarovich, F., Locatelli, A., and Fernandez, J. C., Furosemid in high doses in the treatment of acute renal failure. *Postgrad. Med. J.* **47**, 13 (1971).
C2. Chen, Y.-L., Fang, Z.-Y., and Xu, F.-X., Clinical significance of changes in excreted fraction of filtered sodium in severe burn patients. *In* "Recent Advances in Burns and Plastic Surgery—The Chinese Experience" (T.-H. Chang, Y.-X. Shi, and Z.-J. Jang, eds.), pp. 205–210. MTP Press, Lancaster, 1985.
C3. Clark, J., and Walker, W. F., Acid-base problems in surgery. *World J. Surg.* **7**, 590–598 (1983).
C4. Cogan, M. G., Liu, F. Y., Berger, B. E., Sebastian, A., and Rector, F. C., Metabolic alkalosis. *Med. Clin. North Am.* **67**, 903–914 (1983).
C5. Cohen, J. J., and Kassirer, J. P., Acid-base metabolism. *In* "Clinical Disorders of Fluid and Electrolyte Metabolism" (H. M. Maxwell and CH. R. Kleeman, eds.), pp. 181–232. McGraw-Hill, New York, 1980.
C6. Cohen, M. L., A computer program for the interpretation of blood gas analysis. *Comput. Biomed. Res.* **2**, 549–557 (1969).
C7. Corwin, H. L., and Bonventer, J. V., Acute renal failure. *Med. Clin. North Am.* **70**, 1037–1053 (1986).
C8. Cox, M., Potassium homeostasis. *Med. Clin. North Am.* **65**, 363–384 (1981).
D1. Doyle, D. J., A simple computer program for obtaining gas exchange indices. *CRC Crit. Care Med.* **13**, 775–776 (1985).
D2. Dubose, T. D., Clinical approach to patients with acid-base disorders. *Med. Clin. North Am.* **67**, 799–813 (1983).
D3. Dzúrik, R., "Disturbances of Internal Medium" (in Slovak), pp. 84–121. Osveta, Martin, 1984.
E1. Eastham, R. D., "Biochemical Values in Clinical Medicine." 7th ed. Wright, Bristol, 1985.
E2. Editorial, Thirst and osmoregulation in the elderly. *Lancet* **3**, 1017–1018 (1984).
E3. Enrichens, F., Foco, A., Festa, T., Mao, P., Benedetto, G., Manconi, P., Aiello, D., and Rozzio, G., L'usso dell'elaboratore in terapia intensiva. *Minerva Med.* **75**, 39–50 (1984).
E4. Epstein, F. H., Hyperkalemia. *Med. Grand Rounds* **1**, 147–155 (1982).
F1. Feig, P. U., Hypernatremia and hypertonic syndromes. *Med. Clin. North Am.* **65**, 271–289 (1981).
F2. Finkle, D., and Dean, R. E., Buffered hydrochloric acid: A modern method of treating metabolic alkalosis. *Am. Surg.* **47**, 103–106 (1981).
F3. Fligner, C. L., Jack, R., Tuiggs, G. A., and Ralsys, A., Hyperosmolality induced by propylene glycol. *JAMA, J. Am. Med. Assoc.* **253**, 1606–1609 (1985).
G1. Gaab, M., Trost, H. A., Haubitz, I., and Pflughaupt, K. W., Osmolalität und Osmotherapie in Intensivterapie und Prognose nach Schädel-Hirn-Trauma. *In* "Zentraleuropäische Anaesthesiekongress" (B. Haid and G. Mitterschiffthaler, eds.), pp. 153–163. Springer-Verlag, Berlin, 1981.

G2. Gabel, R. A., and Hooper, A., Calculation of blood gas and acid-base variables with a versatile computer program. *Anesth. Analg. (Cleveland)* **60**, 889–896 (1981).
G3. Gabov, P. A., Disorders associated with an altered anion gap. *Kidney Int.* **27**, 472–483 (1985).
G4. Goldberg, M., Hyponatremia. *Med. Clin. North Am.* **65**, 251–269 (1981).
G5. Goldberg, M., Green, S. B., and Moss, M. L., Computer based instruction and diagnosis of acid-base disorders. *JAMA J. Am. Med. Assoc.* **223**, 269–275 (1973).
G6. Gradner, L. B., and Fournier, A. M., Extraordinary alkalemia and triple acid-base disturbance. *South. Med. J.* **77**, 255–268 (1984).
G7. Grundler, W., Weil, H. M., Rackow, E. C., Falk, J. L., Bisera, J., Miller, J. M., and Michaels, S., Selective acidosis in venous blood during human cardiopulmonary resuscitation: A preliminary report *CRC Crit. Care Med.* **13**, 886–887 (1985).
H1. Heinze, V., Nierenversagen. *In* "Aktuelle Probleme der Intensivbehandlung II" (P. Lavin and M. Wendt, eds.), pp. 68–78. Thieme, Stuttgart, 1981.
H2. Heller, K., and Völkel, V., Die messwertgesteuerte Beatmung-eine Möglichkeit zur Optimierung der Respirator Behandlung. *In* "Rechnergestützte Intensivpflege" (E. Epple and H. Junger, eds.), pp. 75–78. Thieme, Stuttgart, 1981.
H3. Hermansen, M. C., Perlstein, P. H., Atherton, H. D., and Edwards, N. K., A computer-generated blood gas display in a newborn intensive care unit. *J. Pediatr.* **103**, 825–828 (1983).
H4. Hingston, D. M., Irwin, R. S., Pratte, M. R., and Dalen, J. E., A computerized interpretation of arterial pH and blood gas data. Do physicians need it? *Respir. Care* **27**, 809–815 (1982).
H5. Holmes, J. H., Measurement of osmolality in serum, urine and other biologic fluids. *In* "Pre Workshop Manual of the Workshop of Urinanalysis and Renal Function Studies," pp. 1–21. Am. Soc. Clin. Pathol., Chicago, Illinois, 1962.
J1. Jabor, A., and Kazda, A., Program for the proposal of kalium dose in intensive care patients. *Biochem. Clin. Bohemoslov.* (in Czech, Engl. summ.). (In press).
J2. Jacobson, H. R., and Seldin, D. W., On the generation, maintenance and correction of metabolic alkalosis. *Am. J. Physiol.* **14**, F425–F432 (1983).
J3. Jos, C. J., Evenson, R. C., and Mallya, A. R., Self-induced water intoxication: A comparison of 34 cases with matched controls. *J. Clin. Psychiatry* **47**, 368–370 (1986).
K1. Kaehny, W. D., Respiratory acid-base disorders. *Med. Clin. North Am.* **67**, 915–928 (1983).
K2. Kaldor, G., and Rada, R., Computerized evaluation of acid-base disorders based on a nine-cell decision matrix. *Med. Biol. Eng. Comput.* **23**, 269–273 (1985).
K3. Kaufmann, W., and Griebenow, Z., Säure-Basen-Elektrolythaushalt. *In* "Pathophysiologie" (H. E. Bock, ed.), pp. 108–114. Thieme, Stuttgart, 1985.
K4. Kazda, A., Nutzbarkeit der biochemischen Überwachung von Kranken in kritischen Zuständen. *Z. Med. Laboratoriums diagn.* **26**, 146–150 (1985).
K5. Kazda, A., "Biochemical Monitoring of Patients in Intensive and Resuscitation Care" (in Czech). Avicenum, Prague, 1986.
K6. Kazda, A., and Zítko, K., Diagnostic and prognostic value of lactate examination. *Biochem. Clin. Bohemoslov.* **13**, 335–344 (1984) (in Czech, Engl. summ.).
K7. Kazda, A, Hendl, J., and Nejedlý, B., Importance of the determination of lactate for diagnosis of mixed disturbance of the internal medium. *Vnitr. Lek.* **25**, 685–692 (1979) (in Czech, Engl. summ.).
K8. Kazda, A., Jabor, A., and Zámečník, M., Evaluation of renal functions—A computer programme. *Int. J. Bio-Med. Comput.* (1989) (in press).
K9. Keller, V., Diabetic ketoacidosis: Current views on pathogenesis and treatment. *Diabetologia* **29**, 71–77 (1986).

K10. Kirkeby, O. J., and Risoe, C., Computer interpretation of acid-base status. A practical way of dealing with a difficult problem. *Br. J. Clin. Pract.* **39**, 377–378 (1985).
K11. Knutsen, O. H., Salzsäurebehandlung bei metabolisher Alkalose. *Anaesth., Intensivther. Notfallmed.* **20**, 220 (1985).
K12. Koller, F., and Nagel, G. A., "Internitische Notfallsituationen," pp. 232–238. Thieme, Stuttgart, 1981.
K13. Kunke, S., and Schulz, V., Ein System zur p_aCO_2 und p_aO_2 generelten Beatmung. *In* "Rechergestützte Intensivpflege" (E.Epple and H. Junger, eds.), pp. 71–74. Thieme, Stuttgart, 1981.
K14. Kushner, R. F., and Sitrin, M. D., Metabolic acidosis, development in two patients, receiving a potassium sparing diuretic and total parenteral nutrition. *Arch. Intern. Med.* **146**, 343–345 (1986).
L1. Lang, F., Deetjen, P., and Reissigl, H., "Handbuch der Infusions-therapie und klinischen Ernährung," vol. 1, pp. 95–116. Karger, Basel, 1984.
L2. Levinski, N. G., Acidosis and alkalosis. *In* "Harrison's Principles of Internal Medicine," pp. 279–286. McGraw-Hill, New York, 1983.
L3. Linter, S. P. K., and Ryan, D. W., The anion gap. *Br. J. Med.* **35**, 79–80 (1986).
L4. Lison, A. E., Nierenfunktion und Antibiotikabehandlung. *In* "Nierenfunktion während Anästhesie und Intensivbehandlung" (P. Laurin and R. Wisdorf, eds.), pp. 56–60. Thieme, Stuttgart, 1981.
L5. Loeb, J. N., The hyperosmolar state. *N. Engl. J. Med.* **290**, 1184–1187 (1974).
L6. Loirat, P., Increased glomerular filtration rate in patients with major burns. *N. Engl. J. Med.* **299**, 915–919 (1978).
L7. Lynn, L. A., and Sunderrajan, E. V., Bedside respiratory analysis by pocket computer. *CRC Crit. Care Med.* **14**, 62–64 (1986).
M1. Madias, N. E., Lactic acidosis. *Kidney Int.* **29**, 752–774 (1986).
M2. Madias, N. E., Ayus, J. G., and Adrogue, H. J., Increased anion gap in metabolic alkalosis: The role of plasma protein equivalency. *N. Engl. J. Med.* **300**, 1421 (1974).
M3. Magalini, S. I., Proietti, R., and Lorino, F., Diagnostic, therapeutic, didactic, data bank utilization of computerized monitoring in a critical care center. *Resuscitation* **11**, 193–205 (1985).
M4. Manconi, P., Enrichens, F., Festa, T., Mao, P., and Benedetto, G., Uso dellélaboratore per il controllo delléquilibrio acido-base nel paciente ricoverato in unità di terapia intensiva chirugica. *Minerva Med.* **75**, 2063–2066 (1984).
M5. Manz, F., Vecsei, P., and Wesch, H., Renale Säureausscheidung und renale Molenlast bei gesunden Kindern und Erwachsenen. *Monatsschr. Kinderheilkd.* **132**, 163–167 (1984).
M6. Martin, L., and Jeffreys, B., Use of a minicomputer for storing, reporting and interpreting arterial blood gases/pH and pleural fluid. *Respir. Care* **28**, 301–308 (1983).
M7. Mašek, K., and Kazda, A., New approach to biochemical monitoring in intensive care. *J. Clin. Chem. Clin. Biochem.* **19**, 765 (1981).
M8. Mašek, K., Zámečník, M., and Kazda, A., Problem oriented lab reports in clinical decision making. *In* "Medical Decision Making: Diagnostic Strategies and Expert Systems" (J. H. Benned, F. Grémy and J. Zvárová, eds.), pp. 282–284. Elsevier, Amsterdam, 1985.
M9. Mason, P. D., Metabolic acidosis due to d-lactate. *Br. Med. J.* **292**, 1105–1106 (1986).
M10. Mattar, J. A., Cardiac arrest in the critically ill. *Am. J. Med.* **56**, 162–168 (1974).
M11. Mattle, H., Neurologische Manifestationen der gestörten Osmolalität. *Schweiz. Med Wochenschr.* **115**, 882–889 (1985).
M12. McAuliffe, J. J., Lind, L. J., Leith, D. E., and Fencl, V., Hypoproteinemic alkalosis. *Am. J. Med.* **81**, 86–89 (1986).

M13. Medinger, A., Interpretation of arterial blood gases in the emergency patient. *In* "Handbook of Pulmonary Emergencies" (S. V. Spaguola and A. Medinger, eds.), pp. 17–31. Plenum, New York, 1986.
M14. Miller, P. D., and Krebs, R. A., Polyuric prerenal failure. *Arch. Intern. Med.* **140**, 907–909 (1980).
M15. Moore, M. J., and Bleich, H. L., Consulting the computer about acid-base disorders. *Respir. Care* **27**, 834–838 (1982).
M16. Morton, J. J., Connel, J. M. C., Hughes, M. J., Ingless, G. G., and Wallace, E. C., The role of plasma osmolality, angiotensin II and dopamine in vasopressin release in man. *Clin. Endocrinol. (Oxford)* **23**, 129–138 (1985).
N1. Nanji, A. A., Campbell, D. J, and Pudek, M. R., Decreased anion gap associated with hypoalbuminemia and polyclonal gammopathy. *JAMA, J. Am. Med. Assoc.* **246**, 859–860 (1981).
N2. Narins, R. G., and Gardner, L. B., Simple acid-base disturbances. *Med. Clin. North Am.* **65**, 321–346 (1981).
N3. Narins, R. G., Jones, E. R., Stone, M. C., Rudnick, M. R., and Bastl, C. P., Diagnostic strategies in disorders of fluid, electrolyte and acid-base homeostasis. *Am. J. Med.* **72**, 496–520 (1982).
N4. Nejedlý, B., "Internal Medium, Clinical Biochemistry and Practice" (in Czech), pp. 193–227. Avicenum, Prague, 1980.
O1. Osswald, P. M., Bernauer, J., Bender, H. J., and Hartung, H. J., Graphic presentation of blood gas data. *In* "Computing in Anesthesia and Intensive Care" (O. Prakash, S. H. Mey, and R. W. Patterson, eds.), pp. 342–351. Martinus Nijhoff Publ., Boston, Massachusetts, 1983.
O2. Overlack, A., and Krück, F., Klinische Bedeutung der Anionenlücke. *Dtsch. Med. Wochenschr.* **110**, 687–691 (1985).
P1. Payne, R. B., Creatinine clearance, a redundant clinical investigation. *Ann. Clin. Biochem.* **23**, 243–250 (1986).
P2. Phillip, C. P., "Parenteral and Enteral Nutrition," pp. 31–46. Flinders University of South Australia, 1982.
P3. Pruden, E. L., and Siggaard-Andersen, O., Blood gases and pH. *In* "Textbook of Clinical Chemistry" (N. W. Tietz, ed.), pp. 1191–1253. Saunders, Philadelphia, Pennsylvania, 1986.
R1. Richards, B., "Computer Assistance in the Treatment of Patients with Acid-Base and Electrolyte Disturbances," pp. 407–409. Medinfo, North Holland, 1977.
R2. Rimalho, A., Rion, B., Dader, E., Richard, C., and Anzépy, P., Prognostic factors in hyperglycemic hyperosmolar nonketotic syndrome. *CRC Crit. Care Med.* **14**, 552–554 (1986).
R3. Rinecker, H., Überwachung der Atmung und Regelung der Beatmung. *In* "Rechnergestützte Intensiv-pflege" (E. Epple and H. Junger, eds.), pp. 53–65. Thieme, Stuttgart, 1981.
R4. Rock, R. C., Walker, W. G., and Jennings, C. D., Nitrogen metabolites and renal function. *In* "Textbook of Clinical Chemistry" (N. W. Tietz, ed.), pp. 1254–1316. Saunders, Philadelphia, Pennsylvania, 1986.
R5. Rose, B. D., "Clinical Physiology of Acid-Base and Electrolyte Disorders," pp. 295–370. McGraw-Hill, New York, 1977.
R6. Rothe, K. F., Die Salzsäuretherapie bei der Schweren metabolischen Alkalose. *Anäesth. Intensivther. Notfallmed.* **20**, 221–223 (1985).
R7. Rothe, K. F., and Schimek, F., Paradoxe Reaktion des intrazellulären pH-Wertes bei der Therapie des Säuren-Basen-Haushaltes mit Argininhydrochlorid. *Anaesthesist* **32**, 532–537 (1983).

R8. Rothe, K. F., Schimek, F., and Harzmann, R., Experimental studies on therapy of metabolic alkalosis during the beginning of uremia. Influences of ammonium chloride on the intra- and extracellular acid-base status of the rat. *Eur. Urol.* **12,** 54–58 (1986).
R9. Rothe, K. F., and Schore, R., Der Säuren-Basen-Haushalt in der Anästhesiologie und operativen Intensivmedizin. Physiologie, Pathophysiologie und Klinik der Azidosen. *Anäesth. Intensivther. Notfallmed.* **20,** 69–75 (1985).
R10. Rothe, K. F., and Shorer, B., Der Säuren-Basen-Haushalt in der Anästhesiologie und operativen Intensivmedizin. Physiologie, Pathophysiologie und klinik der Alkalosen und gemischten Störungen. *Anäesth. Intensivther. Notfallmed.* **20,** 111–118 (1985).
S1. Schück, O., "Examination of Kidney Function," pp. 157–197. Martinus Nijhoff Publ., Boston, Massachusetts, 1984.
S2. Siegel, J. H., Computer based consultation in "care" of the critically ill patient. *Surgery (St. Louis)* **80,** 350–363 (1976).
S3. Skaredoff, M. N., A computerized system for rapid interpretation of acid-base disorders. *Int. J. Bio-Med. Comput.* **18,** 229–238 (1986).
S4. Sporn, P., Fürnschlief, E., and Kolocny, M., Die freie Wasser-clearance als renaler Verlaufsparameter in der Intensivtherapie des akuten Abdomens. *In* "Zentraleuropäischer Anaesthesiekongress" (B. Haid and G. Mitterschiffthaler, eds.), pp. 174–177. Springer-Verlag, Berlin, 1981.
S5. Swezey, C. B., Computer based diagnostic reporting for serum electrolytes. *Am. J. Clin. Pathol.* **74,** 812–819 (1980).
T1. Thomas, L. J., Algorithms of selected blood acid-base and blood gas calculations. *J. Appl. Physiol.* **66,** 154–158 (1972).
T2. Tietz, N. W., Pruden, E. L., and Siggard-Andersen, O., Electrolytes, blood gases and acid-base balance. *In* "Textbook of Clinical Chemistry" (N. W. Tietz, ed.), pp. 1172–1190. Saunders, Philadelphia, Pennsylvania, 1986.
T3. Tzamaloukas, A. H., Froelich, C. J., and Jeffery, W. H., High anion gap acidosis during intravenous infusion of protein hydrolysate. *South. Med. J.* **77,** 661–663 (1984).
V1. Vallbona, C., Computer analysis of blood gases and of acid-base status. *Comput. Biomed. Res.* **4,** 623–633 (1971).
V2. Voight, E., Teaching the interpretation of acid-base and blood gas parameters by computer application. *In* "Computing in Anesthesia and Intensive Care" (O. Prakash, S. H. Mey, and R. W. Patterson, eds.), pp. 354–365. Martinus Nijhoff Publ., Boston, Massachusetts, 1983.
W1. Wagner, C. W., Nesbit, R. R., and Mansberger, A. R., Treatment of metabolic alkalosis with intravenous hydrochloric acid. *South. Med. J.* **72,** 1241–1245 (1979).
W2. Walmsey, R. N., and White, G. H., Normal "anion gap" (hyperchloremic) acidosis. *Clin. Chem. (Winston-Salem, N.C.)* **31,** 309–313 (1985).
W3. Walmsey, R. N., and White, G. H., Mixed acid-base disorders. *Clin. Chem. (Winston-Salem, N.C.)* **31,** 321–325 (1985).
W4. Weil, M. H., Rackow, E. C., and Trevino, R., Difference in acid-base state between venous and arterial blood during cardiopulmonary resuscitation. *N. Engl. J. Med.* **315,** 153–156 (1986).
W5. Weil, M. H., Ruiz, C. E., Michaels, S., and Rackow, C., Acid-base determinants of survival after cardiopulmonary resuscitation. *CRC Crit. Care Med.* **13,** 888–892 (1985).
W6. Weil, M. H. Grundler, W., Yamagughi, M., Michaels, S., and Rackow, E. C., Arterial blood gases fail to reflect acid-base status during cardiopulmonary resuscitation: A preliminary report. *CRC Crit. Care Med.* **13,** 884–885 (1985).
W7. Weisberg, H. F., Unraveling the laboratory model of a syndrome: The osmolality model. *In* "Clinician and Chemist" (D. S. Young, H. Nippe, D. Uddin, and Y. Hicks, eds.), pp. 200–237. Am. Assoc. Clin. Chem., Washington, D. C., 1979.

W8. Wiener, F., Fayman, M., Teitelman, U., and Bursztein, S., Computerized medical reasoning in diagnosis and treatment of acid-base disorders. *CRC Crit. Care Med.* **11,** 470–475 (1983).
W9. Williams, D. B., and Lyons, J. H., Treatment of severe metabolic alkalosis with intravenous infusion of hydrochloric acid. *Surg., Gynecol. Obstet.* **150,** 315–321 (1980).
W10. Winters, R. W., Engel, K., and Dell, R. B., "Acid Base Physiology in Medicine." Radiometer, Copenhagen, 1969.
W11. Worthley, L. I. G., Hyperosmolar coma treated with intravenous sterile water. *Arch. Intern. Med.* **146,** 945–947 (1986).
Z1. Zumkley, H., "Klinik des Wasser-Elektrolyt-und Säure-Basen-Haushalts," pp. 163–226. Thieme, Stuttgart, 1977.

INFORMATION IN THE CLINICAL LABORATORY: COMPUTER-ASSISTED ORGANIZATION AND MANAGEMENT

Stanley J. Robboy and Ronald Trost

Department of Pathology
University of Medicine and Dentistry of New Jersey,
Newark, New Jersey

1. Introduction	270
2. Ease of Use	273
2.1. Menus	273
2.2. Formatted Screens	273
2.3. Branching	273
2.4. Help Functions (Written and On-Line)	274
3. Hidden Enhancements to Functionality	274
3.1. Access	274
3.2. Audit Trails	275
3.3. Reduction of Unproductive Efforts	275
4. Safeguards	275
4.1. Patient Security	275
4.2. Prevention of Access to Unauthorized Data	276
4.3. Check-Digit	276
5. Enhancements for the User	278
5.1. Electronic Mail	278
6. Common Core Functions to All Lab Modules	278
6.1. Patient Registration	278
6.2. Order Entry	279
6.3. Patient Inquiry	280
7. General Chemistry Functions	280
7.1. Specimen Accessioning	280
7.2. Label Generation	280
7.3. Specimen Collection	281
7.4. Worklist/Loadlist Generation	282
7.5. Result Entry	283
7.6. By-products	284
7.7. Quality Control	285
7.8. Workload Reporting	286
7.9. Instrument Interfaces	286
8. Data Presentation and Retrieval	289
8.1. Patient Charting	289
8.2. Emergency (STAT) Printing	291

  8.3. Quality of Chart Presentation.. 291
  8.4. Long-Term On-Line Storage: The Lab's Medical Abstract 292
9. Management 292
  9.1. Management Tools 293
  9.2. Quality Assurance... 294
  9.3. Human Resource Management....................................... 296
  9.4. Management Information Systems................................... 297
  9.5. Utilization Reporting 297
  9.6. Labor Productivity Reporting....................................... 298
  9.7. Cost Accounting 299
  9.8. Service Level Analysis.. 300
10. Conclusion... 301
  References.. 301

1. Introduction

We live in what has been called the "age of information." Yet, for all the promise envisioned for this new era, a paradox has emerged. Rather than obtaining pertinent data, reduced and organized to exactly that which is needed, i.e., information, we are overwhelmed by minute bits of data thrown into reams of reports. This data overload, in reality, obscures the real paucity of information.

The senior author of the present article has designed, and both authors have implemented, computer systems in teaching institutions. Herein, we describe our concepts of an integrated, computer-assisted healthcare architecture crucial to cope with the expanding role of laboratory medicine. The guiding tenet, well established for more than a decade, is that information, i.e., converted data, be compact, informative, timely, and easy to access.

The laboratory professional recognizes that every datum has its costs and intended benefits. Associated with that datum are several factors, which include (1) the degree of difficulty associated with its acquisition, (2) the ability to transmit it on a timely basis, (3) organization in a format amenable to fostering patient care, and (4) correlative organization with other data, thereby enhancing its inherent value. Collectively, the data must be condensed so that they tell a story in as efficient and clear a manner as possible.

In developing a philosophy pertinent to synthetic integration of information, the modern laboratory should be cognizant and exploit the range of benefits now feasible, which, under manual systems, were unattainable without Herculean effort. These benefits should ultimately allow healthcare institutions to achieve the following goals:

1. Improve the quality of patient care by providing timely and accurate information to the care team. This will be accomplished with methodology and delivery locations which best ensure that the information is promptly received, assimilated, and utilized.

2. Increase labor productivity by reducing the number of manual steps involved in clinical procedures and, therefore, the time required to complete those procedures.

3. Facilitate active management of departmental and institutional resources through systems measurement and reporting of production costs, employee productivity, resource utilization, and quality assessment.

4. Capture more complete administrative and financial data allowing managers to identify, track, and cost each event involved in producing and processing clinical information. Control of variable costs will also be forthcoming, thereby enhancing system use where marginal costs are favorable.

5. Accommodate technological developments in computer hardware, communications, operating systems, and software development in areas such as expert clinical decision systems.

6. Foster scientific research, through comprehensive integration of the data, making its access and manipulation easy for data analysis.

Application of this philosophy dictates that the integrated system allow for data collection in detail sufficient for providing the desired information. The system must define the specific activities of each procedure and track those activities to enable full exploitation of each of their attributes to include the "who, what, where, and when" of each event. This detail enables the pathologist to exercise rational, insightful control over laboratory operations.

From the parochial view of pathology, the schema in Fig. 1 depicts a working concept of a comprehensive healthcare architecture, aspects of which are currently being implemented at the New Jersey Medical School and University Hospital of the University of Medicine and Dentistry of New Jersey (UMDNJ). It is organized on three levels: (1) who can access the system and from where; (2) which clinical information systems are available or planned for a university system, along with their communications schemata, and the available specialized software packages which enhance the system; and (3) examples of management and advanced information systems which enhance the value of integrated data where the composite information far exceeds the collective value of the separate components.

The basis of the system is software. In this article, software functionality will be stressed from the vantage point of how laboratory personnel, daily

A COMPREHENSIVE, INTEGRATED HOSPITAL COMPUTER ARCHITECTURE
As viewed from Pathology
– A complete patient care, laboratory, scientific and managerial system –

```
        USERS          ----------------------→ |←---ACCESS points (with approp security)
Lab                                             |    Duty stations
   Technologist/staff                           |    Hosp floors
   Pathologists                                 |    Remote locations
   Machines                                     |    Any telephone anywhere, e.g., homes
   Remote users (e.g., off site labs)           |    Via commercial & education networks,
Hospital (HIS)                                  |      e.g., CompuServe, NJ Educa Network
   Hosp floors (direct report/inquiry),         |
      including operating room visuals          |←---ACCESS methods (with call playback)
   Hosp staff (access individualized)           |    Hardware via terminal
   Other institut clinical/research systems     |    Local area network
Administration                                  |    Telephone via modem: home, offices
   Key hosp chiefs, management                  |      Dumb terminal, PC linkage with
Educational                                     |      up/download abilities
   Medical School                               |    Touch tone telephone directly
   School of Allied Health                      |    Other information systems
                                                ↓
                        --------------------INFORMATION SYSTEMS/TOOLS--------------------
              |                   |                         |                        |
              ↓                   ↓                         ↓                        ↓
     LABORATORY INFORMATION SYSTEMS              UMDNJ COMPUTERS/INFO SYSTEMS      TOOLS
General laboratory system    Common clinical systems   In Place or Committed    Word processing
  Chemistry/toxicology         Infection control       HIS                      Data managers
  Hematology/coagulation       Quality control         MISS (material manage, purchasing,  Spread sheets
  Urinalysis                   Automated instruments     accounts payable, receiving)     Graphics
  Serology                     Management information systems  Path (LIS, adm system)    Electronic mail
  Immunology                     Cost accounting       Sammy Davis Jr. Nat'l Liver Inst  Bulletin boards
Microbiology system              Utilization reporting Lab computer science (general,    Financial tools
  Bacteriology                   Labor productivity reporting   Crit Care Bedside Network ICUs)  Languages
Mycology                         Service level analysis George Smith University Library  Statistics
  Virology                       DRG analysis          Systems planned                   Expert systems
  Parasitology                   Inventory control     Radiology, Pharmacy, operating room  Operating systems
  Mycobacteriology               Ext query, PC interface  Individual PC systems
Blood bank                     Commercial information system  Multiple sites - IBM/HP/Apple/VAX
  Transfusion system             Client accounting
  Donor system                   Marketing analysis    EXTERNAL SYSTEMS via NETWORK
Anatomical system                Profitability analysis Bit-Net
  Surgical Pathology             Operations control    Princeton Supercomputer (Cray III)
  Cytology                       Data communication management  Medline (Nat'l Lib Medicine)
  Autopsy                      Advanced information systems
  Histology                      Expert systems
                                 Resource management
                        ------------------------------------
                                  ↓
                        ENHANCEMENTS (Examples)
```

PATIENT MANAGEMENT
Graphic analysis/display individual lab data
Expert systems to enhance graphic and data display by time, possible disease entity, organ system evaluation
Computer initiated calls warn of panic values

CROSS SYSTEM ENHANCEMENTS
Resident from home via LIS through library system performs literature search to deal with hosp emergency
Pathology drives tumor registry for new cases
Pathology provides feedback data for radiologist about his diagnoses
Infection control early outbreak alert
Therapeutic drug monitoring

ORGANIZATIONAL PATIENT MANAGEMENT
Graphic analysis/display lab data by class disease, day, run, etc.
Permit MD to query computer for groups of patients by MD, floor, class
DRG intervention (permit pathologist to query use of test before performed)
Logistics (printing results on floor bypasses dispatch, allow better grouping)

BUSINESS MANAGEMENT
Organization (equip control; table organization & responsibilities; annual reports, 5 year plans, goals monthly/yearly)
Production orientation (work lists; reports of incomplete specimens)
Fiscal Productivity: cost analysis per test
  Cost/vol/profit analysis for new tests (do or refer decision)
  Vol/manpower analysis for shifting volumes
  Excess capacity analysis (delete or refer or interhosp share)

FIG. 1. A comprehensive, integrated hospital computer architecture, partially in place at UMDNJ, as viewed from the perspective of the pathology department.

operations, and end-users are affected; technical and hardware aspects hidden from the user will be generally ignored.

2. Ease of Use

No system is trivial to use. Several features, though, greatly enhance its usability.

2.1. MENUS

Menus are a formatted listing of programmatic applications available to the user. Menus, which contain multiple levels of submenus, are a great boon to both the novice and the experienced user. To the novice, the menus act as a guide to navigate through the many options of the system. Menus markedly shorten the training time for a beginning user (e.g., technician, secretary, or pathologist) to become proficient. For advanced users, menus can act as a major aid in performing difficult tasks, especially those performed before, but with such infrequency that they have been mostly forgotten. Menus reduce the need for consulting reference books. Menus, thus, are a tool to make available the computer programs invisible to the user. As such, they make no bearing on the intrinsic worth of the computer programs to be run.

2.2. FORMATTED SCREENS

Formatted screens, which combine graphics and text, enhance efficiency of entry data. Each screen should be designed to facilitate data entry by using fields which are clearly labeled and logically arranged on the screen. Efficiency and accuracy are key goals. Sensitivity to the clinical concept and the user's qualifications and intentions is important; parsimony of visual presentation and cognitive aspects of "ease-of-use" are closely related. As a by-product, screens can solicit entry of data that are desirable, but would otherwise likely be forgotten, and which might be difficult to obtain at other times.

2.3. BRANCHING

Branching is the direct access to other on-line functions without returning to a menu. Users should be able to "branch" from one transaction to any other, without menu intervention and without losing the original process in progress.

Most systems permit only one function to be accessed at any one time. With branching, the user is able to suspend operations in one function, move to a second or third function, and, after its completion, return to an earlier function and continue where the former was interrupted. For example, if a markedly abnormal lab value is being entered, the user can suspend the operation, call forth other data on the patient to determine its medical significance, and then return to the first operation at the point of interruption, rather than starting the entire first function anew. If a telephone inquiry is received, the user can suspend the current transaction, investigate, and respond to the inquiry and then return to the first transaction.

Branching affords the experienced user flexibility to break out of the normal controlled information flow and bypass menus without sacrificing work already performed on the transaction in progress, without sacrificing productivity by manually transcribing data, and without sacrificing security. Functionally, branching improves accuracy by electronically transferring the necessary information, eliminating the possibility of error that exists when the information is reentered by the user.

2.4. HELP FUNCTIONS (WRITTEN AND ON-LINE)

Vendors design systems and programmers write the software. All too often, neither document the system, reducing its utility and thus its power. Written documentation should be constantly updated in three regards: (1) an in-depth system manager's guide, (2) user's guide in English, and (3) a pocket book of useful commands.

Newer systems incorporate on-line help functions of several types. On-line help functions, at the most detailed level, should list all relevant responses to the question being asked on the screen. For example, help to "sex" might be "F" for female, "M" for male, and "U" for unknown. A second level of on-line help is more generic in scope and provides information about the screen as a whole, such as what is known as a Levy–Jennings plot in quality control. Further, on-line interactive training modules guide, and thereby teach, the novice user how to use the system in a step-by-step fashion.

3. Hidden Enhancements to Functionality

3.1. ACCESS

In older systems, certain functions were often accessible only through specific terminals, thus making the physical location of hardware a primary determinant of access. In newer systems, any terminal, regardless of location, is in principle capable of performing any function. The keys to functionality are the software and the levels of authority given to any particular user/administrator (see Section 4). Where control of capabilities by physical location is important, e.g., remote inquiry-only terminals, the software can so limit the terminal to such desired limits.

3.2. Audit Trails

Audit trails of all transactions are important to determine who added/altered/deleted data items and when. For example, the system should capture the identification of the employee who alters a result after it was initially entered, register the date and time of the correction, and identify the function(s) used to effect the change.

3.3. Reduction of Unproductive Efforts

Manual systems are inherently decentralized. Each work station operates in a more or less independent manner with little or no communication beyond itself. It must have detailed logs at any given time so as to be able to function. In a centralized system, the computer is likened to the hub of a wheel, with all work stations at the periphery of the spoke, but nonetheless always integrally in touch with each other. Many functions essential in a manual system become redundant and can be eliminated. Many unproductive (so called "scut") tasks can be abridged, freeing up time for more productive effects. Each person who works on the system must add value. Each area must add a unique bit of useful data. Never should the addition of data duplicate the work of others.

4. Safeguards

4.1. Patient Security

Security of data is the most difficult area in today's laboratory systems, both in terms of access to the system and the ability to allow only authorized persons who have a need to see data.

The most basic of security measures is the use of passwords. A password known only to the user is, and will likely remain, the first and main level of security. The user name associated with the password, which should be unique for each user, resides in a table that identifies that user's levels of authorization.

In the past, a second common form of security was the physical limitation of access to terminals hard-wired into the computer. With the presence today of inexpensive modems, with a 2400-baud rate of transmission being already common and 9600 baud being introduced, it is fully expected that any telephone line from any area in the country will represent a potential user. To thwart access by "hackers," which has been publicized in the media, some systems use a secondary password which changes weekly. As the password is specific to the system rather than the user, the authorized users must be given the new password weekly or must know how to calculate the new password.

Given the highly sensitive data in a medical data base, it is likely that extra degrees of security will be needed to ensure integrity. For example, once the user has successfully identified himself by both primary and secondary passwords, the system might ask the user to hang up, and the computer will then dial the user's telephone number (from a prestored approved list of numbers) to achieve a valid connection.

4.2. Prevention of Access to Unauthorized Data

All users are not equal. Even though an individual may be authorized to use the laboratory computer, he may not be authorized to have access to all lab modules. Conceptually, today's systems must be able to segregate patients into groups of general patients, hospital employees, VIP patients, and, in some cases, special contract groups (e.g., data for research groups of patients not connected with the hospital, residing in the lab computer). Employees of the healthcare system must be stratified in their level of security, to permit access only to the classes of patients to which they are authorized.

An example of a test result requiring strict confidentiality is the test which detects antibody to the AIDS virus. At a data field level, the system must have the capacity to distinguish who should have access. The laboratory technologist who runs the HIV test must have the ability to work with the data, even though she may have limited privileges in any other lab area. Conversely, other workers, regardless of their privileges in other areas, must have no access to the results of AIDS data (lack of security clearance). Security must be such that the inquiring physician must have access to the AIDS data on his own patient, but to no other patient. Similarly, the hospital nursing coordinator or resident rotating on the floor must have access to data at the appropriate time. This requires that systems should have sophisticated and flexible security systems, such that, e.g., the system knows each resident's service and on-call schedules.

Rules issued by the Department of Health for the State of New Jersey allow results of HIV tests to appear on the patient charts. The slips may contain the HIV results and the patient's name and unit number. Thus, the computer system should print cumulative summaries for all patients on a daily basis. Yet, at the same time, the computer security system should not present on a video screen the HIV results or even the fact that a HIV test had been ordered if the person inquiring has inappropriate authorization.

4.3. CHECK-DIGIT

An important, albeit infrequent, occurrence in computer systems is for data to be erroneously entered and summoned from the wrong patient's file. For example, if a patient's hospital unit number is accessed, but two digits are reversed, the wrong patient's file will have been entered. Although the person entering the data is expected to read the name of the patient on the terminal screen, this does not always happen.

To help prevent such accidents, the concept of a check-digit was devised. The check-digit is a single terminal digit added to the main number, e.g., hospital unit number, which is calculated from the main number. The terminal digit detects transpositions or substitutions of digits. Upon detection, the computer notifies the user than an invalid number was entered and inhibits further transactions until a valid number is provided. Table 1 details the calculation methodology and the practical import of such a computer-generated check-digit.

TABLE 1
CALCULATION OF COMPUTER-GENERATED CHECK-DIGIT

Purpose: Method to detect erroneous transposition or substitution of digits in a unit number or pathology accession number. In the case of a hospital unit number, at the time the number is first assigned, the computer assigns also the check-digit, e.g., 09832149, where 0983214 is the base number and 9 is the check-digit.

Multiply first digit by 1	$0 \times 1 = 0$
Multiply second digit by 2	$9 \times 2 = 18$
Multiply third digit by 1	$8 \times 1 = 8$
Multiply fourth digit by 2	$3 \times 2 = 6$
Multiply fifth digit by 1	$2 \times 1 = 2$
Multiply sixth digit by 2	$1 \times 2 = 2$
Multiply seventh digit by 1	$4 \times 1 = \underline{4}$
	40

Add results (40), divide result by 10 (quotient = 4, remainder = 0), subtract the remainder from 9, and the result n becomes the check-digit ($9 - 0 = 9$).

Note: Each time the unit number is entered into the computer, it calculates a check-digit which is constant for any given sequence of numbers. When, for example, the secretary errs and inquires about 09823149 instead of 09832149, the computer calculates the check-digit "0" for the former, notes the discrepancy with the check digit "9" entered by the secretary, and immediately rejects the entry as invalid.

5. Enhancements for the User

5.1. ELECTRONIC MAIL

Electronic mail is a feature enabling all clinical laboratory personnel to communicate with one another. Whether management or staff, an individual can send a message to another individual, a section, an entire department, or to selected individuals based on subject material. Since the message does not depend upon availability of transcriptionists or distribution of memos, the communication is both timely and direct. If the anticipated recipient is not using the machine at the time the message is sent, the message will be announced when the recipient next signs onto the system. Within the near future, it is anticipated that lab systems will use electronic mail increasingly as a mailbox function to notify physicians of abnormal lab results. Thus, electronic mail will become an alarm and early warning system. Hard-copy notification should follow.

6. Common Core Functions to All Lab Modules

6.1. PATIENT REGISTRATION

Patient registration (admission/discharge/transfer, ADT) introduces the patient to the lab system. Increasingly, direct interfaces are joining the hospital computer and lab systems, which allow real-time admissions, transfers, updates, and dismissals to be sent immediately from the hospital information system to the lab systems. Some interfaces transmit ADT information from the lab system to update the hospital information system (HIS). Full validation of information should be completed within seconds. If the HIS fails during a transaction, the lab system should recognize an incomplete record was sent. Once the HIS returns on-line, the lab system should be able to retransmit the record.

Demographic and insurance data should be comprehensive. In addition to expected basics (name, unit number, etc.), the system should record maiden name, patient type (inpatient, outpatient, etc.), birth date and conception date for newborns, height, weight, admitting and consulting physicians, location (room and bed), diagnosis, and admission date.

When an inexperienced person accessions a specimen, such as a resident on night call performing a frozen section, it is wise for the system to restrict registration to a minimum of data, e.g., patient name and unit number, and then have the system queue that record for more complete entry the next workday by an experienced worker.

In many institutions patients must be admitted to the hospital on an emergency basis before proper identity can be ascertained, such as in cases of auto accidents and stabbings. The system should allow registrations of "John Doe" patients, with temporary names and numbers, to be corrected at a later time. Once the correct information has been entered, the system should readjust reference values, e.g., normal and critical ranges appropriate to the patient's age, sex, etc.

6.2. Order Entry

Order entry is a key function in any system since it is one of the most effective methods to save cost and labor without sacrifice of quality. With HIS interfaces, requests can be entered through any HIS terminal, allowing doctors, nurses, and patient floor unit clerks to enter procedures from all hospital locations, and simultaneously verifying that the test ordered was in fact the test desired.

In some systems procedures can be ordered as *individual tests* (e.g., glucose, BUN, hemoglobin), as *group tests* (collections of detailed procedures, e.g., CBC, chemistry panel), as *super groups* (collection of group as well as individual tests, e.g., prenatal screen consisting of CBC, ABO typing, and rubella screen), as *interval tests* (multiple specimens drawn at specified intervals, e.g., glucose tolerance), and as *prompted data* (predefined additional information needed at order entry (e.g, total volume, duration of collection). It is useful if multiple ordering priorities are supported, such that terminolgy (STAT, routine, early AM, etc.) can be defined, each with unique processing criteria. The collection priority and reporting priority should be independent of each other and should, respectively, indicate how soon the specimen must be processed and the results reported. For example, the sample for LDH isoenzyme may need to be collected as a STAT procedure, so as to catch enzyme changes at a particular moment, but can be performed and reported in a routine manner. In such cases the system should indicate the LDH isoenzymes are available for STAT collection priorities.

Flexibility in procedure ordering enhances the utility of order entry. Orders should be validated as entered. The system should automatically include the procedure on the appropriate collection list, based upon parameters such as collection date and time, collection priority, and patient location, which were entered in advance. An important aspect of order entry is for the system to identify duplicate procedures ordered within a specified time interval defined for each procedure, e.g., 24 h, and display a warning message. Newer systems allow for automatic rejection of duplicate tests, unless overwritten by the ordering physician. Procedures should be able to be scheduled at least a month in advance. During a user-specified period, additional orders should be allowed as "add-ons" to existing orders. The additional procedures should be assigned the identical accession number of the original procedure request.

6.3. Patient Inquiry

On-line inquiry permits quick access to information. Most systems permit inquiry by medical record number or name. More advanced systems increasingly permit inquiry by parts of last name and first name, nursing station, room and bed, service (e.g., pediatrics, oncology, emergency), and doctor name. A useful function of inquiry is the capability to carry information (e.g., patient registration, order entry) accessed in inquiries over to other functions easily, by a stroke of a single key.

7. General Chemistry Functions

Given the intended audience of this article—persons interested in clinical chemistry—advances in medical computing will be limited to this general discipline. Where pertinent, selected advances in other laboratory modules will also be addressed.

7.1. Specimen Accessioning

The specimen accession number can play an important role in graphic display of data and in the use of expert systems. Coupling the accession number with the use of a calendar or Julian date not only speeds accession, but facilitates data presentation. Time of day, treatment protocols, and significant clinical events can be integrated into data analysis.

7.2. Label Generation

Labels, generally considered a mundane aspect of computing, can markedly reduce clerical error and effort, particularly when they have bar codes. Labels should be generated automatically as part of the order entry process, either immediately or held for a collection list for each priority. For highlighting special collections, such as STAT, designated printers can produce color-coded labels. Dependent upon the label size employed, much useful data, in addition to demographic data, can be printed, e.g., workcenter, priority, isolation code, collection date/time, collection tube type, volume, and doctor. Recent advances indicate bar coding will become much more frequent in the near future. By providing extra checks for patient identity at various steps throughout the processing system, errors should further decrease.

7.3. Specimen Collection

Specimen collection can be time consuming and inefficient. A major benefit of computers is assistance in organizing laboratory orders for the efficient dispatching and laboratory-defined routing of laboratory personnel for the collection of specimens. Each phlebotomist should be assigned unique collection lists, which can be modified at any time. In some cases, laboratory-defined parameters will determine when specimens, e.g., urine and sputum, are to be collected by nursing personnel. Specimen volumes and the number and types of containers required for each draw should be automatically calculated. The system should assess the parameters whenever a test is ordered and "net" specimens together based on collection class, specimen type, container type, volume required, and special handling instructions. Commonly, the system will detect, for example, that three tests, each requiring a 10-ml specimen when collected singly (30 ml total), might require only 15 ml when collected together. Special handling and/or preparation instructions can be defined at the test level so that special instructions appear on the collection list any time a test is ordered. Appropriate labels, and reports which accompany each label to indicate the required container type, infectious hazard, or other partinent information, should also be produced.

After a specimen has been received, verification of the receipt should be performed quickly and the receipt time automatically recorded. If a collection is entered as "missed," the system should require a reason why the collection was missed along with the technologist's name and identification number. Thus, an audit trail is established to ensure that a specimen is collected and delivered to the laboratory. Missed collections should be continuously tracked. Specimens that are missed should continue to appear on a Collection Pending Report and the Pending Test Report until results are entered or the test is cancelled. Cancelled tests should require a comment indicating reason for cancellation so that uncollected specimens cannot simply "disappear" from the system without qualification.

By manipulating various aspects of the audit trail data, the status of an order can be easily monitored in entirety throughout its natural transaction cycle. Statuses provided at each step of the process include the following examples:

Order	Collection	Location	Result
Ordered	Pending	STAT lab	Pending
Drawn	Dispatched	Remote	Performed
In-process	Drawn	In-transit	Verified
Complete	—	In-lab	New result
Cancelled	Missed	—	Cancelled

For those labs with physically remote collection or testing centers, specimens can be tracked from remote sites via packing lists and courier status monitoring screens.

With the recent introduction of reliable and economical bar-coding hardware in general industry, bar-coded specimen-tracking systems are becoming available. A bar-coded specimen number on each collection label tracks the times and appropriate identification at the time of draw and at each receiving location. This information is quickly captured and used to distribute the specimens to the appropriate laboratory section as well as to an aliquotting area. Recently, the Health Care Industry Bar-Code Council (HIBCC) has proposed a single bar-code symbology, code 39, as the standard.

7.4. Worklist Loadlist Generation

Worklist/loadlist generation typically is both cumbersome and a resource hog. New systems now have the capacity to automatically assign the appropriate workstation based upon priority, day of the week, time of day, patient age, and method of test request, i.e., ordered singularly or as part of a profile. Parameters set at the test level assign accessions to the workstation that is the desired location for testing. Such a rapid, efficient approach of automatically assigning tests to the appropriate testing site also helps build a virtually paperless environment, eliminating the need for manual worksheets.

A system that logically reflects a laboratory's policies and procedures applies numerous laboratory-defined system parameters, such as times of day and days of week each instrument is available, and what ordering priorities should or should not be performed on each instrument. Using such parameters, the system literally recreates the pathologists' thought processes each time a test is received by assigning the tests to workbenches, automated instruments, and reference laboratories. Systems become even more "paperless" when each workbench has on-line access to the worklists for which it is responsible, thus eliminating the need to generate hard-copy reports of worklists, which eliminates transcription errors or misplaced tests while assuring that laboratory personnel and instruments are fully utilized.

7.5. RESULT ENTRY

Entry of results can be dramatically shortened when each workstation is given a range of options as to the method most suitable for it. Common methods include batch test entry, single accession entry, single test entry, profile test entry (i.e., urine analysis, CSF analysis), interval test entry (the technologist uses the keyboard in place of the cell counter), and specialized instrument interface entry. Formatted screens with preformed rules speed result entry by a number of methods, such as automatically displaying the first test's accession number according to priority, filling in the associated demographic data, and inserting any required decimal points automatically based upon the institution's predefined format for each result.

In general, all results should require verification with identification before becoming available to inquiries external to the lab or for patient reports. Verification can be performed either at result entry or as a separate function. Results can be verified by single specimen or, if so desired, by entire worksheet/loadlists at one time. All verified results should be immediately released for reporting. Calculation, error checking, flagging of abnormal results, and comparison of abnormal against normal results should be provided. Systems should distinguish between when a test has been "performed" (results maintained in the system, but not released outside of the laboratory) and "verified" (results available outside the laboratory).

Automated result entry should be technologist controlled. Interface abilities should include monitoring of the automated instruments, processing the raw data, and performing on-line error and quality control checking. Operators should have the ability to start and stop data acquisition, restart and repeat, disable and enable channels, synchronize instrument and computer, add samples during instrument operation, correct or void results, capture calculated test results for diluted specimens, verify results and release them for reporting, handle results that exceed the instrument's linearity, accept diluted samples, and accept otherwise altered samples.

Delta checks, which are quantifications of percentage variance or absolute variance of two consecutive testing values, should be automatically performed on-line by the system on all tests as specified by the user. The test result with the drawn date/time immediately preceding the result currently entered is used to compute the delta value.

7.6. BY-PRODUCTS

A common shortcoming in many systems is the inability of the pathologist to obtain scientific and management reports without additional insertion of information into the computer's data base. A useful philosophy in system design and for the prospective purchaser evaluating systems is to find every datum element used in an integrated, generalized fashion for multiple subsequent purposes such that a new report or analysis requires only computer effort, not the pathologist's effort. Such a system greatly enhances the pathologist's productivity by reducing the effort to attain the product needed. Table 2 lists several types of reports/manage-

TABLE 2
MANAGEMENT REPORTS

Delivery of service
    Turnaround times
    Phlebotomy audit summary
    STAT phlebotomy response time

Quality assurance
    Test utilization monitors
    Policy compliance audit
    Section supervisor and pathologist review reports
    Result exemption

Productivity
    CAP workload recording productivity monitor
    Employee productivity reporting

Financial
    Test cost analysis
    Contribution reporting

Test volumes
    Laboratory volume
    Laboratory revenue/cost report
    Sendout analysis
    Testing site volume
    Instrument utilization

Equipment and supplies
    Preventive maintenance schedules
    Materials management

Employee
    Continuing education records
    Employee scheduling
    Employee time and attendance records

ment tools typically needed for most hospital labs. By-products are further discussed in Section 9.2.

7.7. QUALITY CONTROL

Quality control is the assurance of conformity of clinical laboratory activities to established specifications by systematic observation and regulation of starting materials, workmanship, environmental factors, and other influences. This activity has become paramount in the era of mushrooming technological growth. Increasingly, there is scrutiny from all sides—patients, courts, and government regulators. In contrast to the concept of quality assurance (which is less process oriented and more directed toward outcome, encompassing more global issues of appropriateness of test use), quality control refers to the periodic, systematic surveillance of people, instruments, methods, and reagents.

Each laboratory generally defines its own unique set of rules for quality control, which apply for each test and every testing site/instrument. This includes definitions of means and standard deviations of controls, rules for the percentage deviation, and parameters against which a result is validated. Manufacturers' values, current period, prior period, and inception-to-date values are included.

Quality control specimens should be processed in a manner similar to specimens from patients. Automatic scheduling of control tests should be available for both automated instruments and manual testing sites. Controls are usually run multiple times each day for each test. In many cases multiple specimens are included in each run or batch, producing an avalanche of data. Control results should be included on all pending reports, audit trails, worksheets, and loadlists. Some systems have statistical reports by testing site and include all test results, dates, and times, at user-definable intervals (shift, day, week, month, and year).

Corrective action must be taken when a control is out of range. The laboratory information system (LIS) systems should provide options which permit each institution, when an out-of-range control is entered at the CRT of via an interfaced instrument, to continue normal processing or to prevent patient result entry until controls are rerun or corrected.

Levy–Jennings plots, based on laboratory-defined criteria, indicate multiple data points with identical values. LIS systems should permit them to be printed on demand or during the course of daily operations, for all lab sections or selected lab sections, and for all tests or selected tests. The plots should include information on report period, laboratory section, instrument identification, all shifts, shift 1 period, test name, test number, level, corrective action, date and time, technologist and control identification, expected range, number of range, mean, standard deviation, and result. CRT terminal presentations of such plots are also valuable.

7.8. Workload Reporting

Workload reporting (usually by the College of American Pathologist methodology) is a proved management tool for pathology in that its norms are based on nationwide experience. The workload reporting system, which allows interlaboratory comparison, and to some degree year to year intralaboratory comparison, is continuously revised to incorporate both new tests and old tests using newer automated machinery. In a fully computerized system, workload is automatically captured for all procedures, including phlebotomies by codes that can be defined according to CAP or laboratory-defined guidelines. Today's computer systems should allow assignment of workload units by instrumentation or methodology, where each testing site can have different workload units assigned, based on instrumentation, methodology, etc. Many tests can be performed by multiple methods (i.e., automated on a multichannel instrument and for backup on a single-channel instrument), and at more than one location during the day. Today's systems should capture these differences. In addition, LIS systems should capture workload assigned to each testing site, and, based on the nature of each testing site, a third component, handling units, should be incorporated into either the venipuncture units or the testing units, or both.

Workload reports should be computer generated (Table 3). The timing of workload reports should be user specified, but these generally occur daily, weekly, monthly, or yearly. Common workload reports include workload by time period; workload by procedure; workload by shift; number of each procedure per shift; list of procedures by instrument; raw counts year to date; workload units year to date; workload by section, by month, and year to date; and over time. Other useful comparisons include year to date: raw counts with previous year, workload units with previous year, workload by section with previous year, and workload by technician with prior year.

7.9. Instrument Interfaces

Automated instruments are more common in clinical chemistry than in any other area of laboratory medicine. Direct interfaces to high-volume instruments achieve marked savings, especially substantial in the area of laboratory personnel. Instrument interface programs have the capacity to process raw data and to perform on-line error and quality control checking.

Results collected by the system should be displayed on a screen for verification. Additional functions in some systems include the ability to start and stop data acquisition, restart and repeat, disable and enable channels, synchronize instrument and computer, add samples while the instrument is operating, correct or void results, calculate test results for diluted specimens, verify results and release them for reporting, handle off-the-chart specimens, accept diluted samples, accept otherwise altered samples, and record and flag repeated tests for management and workload statistics while not adding additional charges to the patient's bill.

Some operating systems perform parity checking and the software performs check-sum routines. Further checking should also be done by the laboratory system against normal values and previous patient results. This has the clear advantage of quickly alerting the pathologist that the clinical status of the patient has begun a possible life-threatening change. Error messages are posted to the system log during periods of communication failure and a list is maintained of specimens for which results have already been processed.

8. Data Presentation and Retrieval

This thesis of this article, that a paucity of information often exists in the midst of a flood of data, is nowhere better exemplified than in reports of today's laboratory information systems. How often does the clinician miss the critical abnormal value in a sheet painted with hundreds of data items? How often are data inaccessible because they have been purged from the system after a short interval of time, or, if retained longer, have been buried with all other data collected during the same period? Finally, how often does the clinician, even with today's computer systems, still have to graph out serial data collected over a patient's hospital stay to make sense of trends, especially in relation to some triggering or therapeutic event? Various aspects of charting are discussed below.

8.1. PATIENT CHARTING

Charting is a complex procedure that has an important impact on patient care. Today, it is the major mechanism by which computers help organize data into useful information. Some systems have the ability to produce patient charts at multiple levels: encounter, order, or accession. The encounter level generates a report which shows all results accumulated on a patient. The order level facilitates outpatient result reporting. The accession level supports "case-type" reporting.

Since results (charting) are continually reported during a patient's hospital course, it is useful to determine data to be included. Types of data

TABLE 3
WORKLOAD REPORT

Report number:	LR-WLR-						Time:	19:36:41
Period:	07 Jan 86 to 07 Jan 86						Prepared:	08 Jan 86
Section:	300 Chemistry						Page number:	1
Workcenter:	00200 Special							

	Procedure			Raw counts			Workload units		
Site	Name	Units	CAP number	In-patient	Out-patient	Total	Period 1	Period 2	Total
220	ACA/Abbott/bili/manual	5.0	82250.000	5	0	5	25.0	0.0	25.0
	Bili neont	2.5*	82249.034	1	0	1	3.0	0.0	3.0
	Bili dirct	2.5*	83690.034	0	1	1	3.0	0.0	3.0
	Lipase	2.5*	84103.034	1	1	2	6.0	0.0	6.0
	Phosphorus	2.5*	84465.034	1	1	2	6.0	0.0	6.0
	SGRP/ALT	2.5*	84720.034	1	1	2	3.0	3.0	6.0
	PC estrase	2.5*	83735.034	1	0	1	3.0	0.0	3.0
	Magnesium	.0	0.000	6	0	6	0.0	0.0	.0
	CSF anal	5.5*	0.012	1	0	1	6.5	0.0	6.5
	CSF prot								
	Patient total			17	4	21	55.5	3.0	58.5
	Bili neont	5.0	82250.000	2	0	2	10.0	0.0	10.0
	Lipase	2.5*	83690.034	0	1	1	2.5	0.0	2.5
	Phosphorus	2.5*	84105.034	1	0	1	2.5	0.0	2.5
	SGPT/ALT	2.5*	84465.034	1	0	1	2.5	0.0	2.5
	PC estrase	2.5*	84720.034	0	1	1	2.5	0.0	2.5
	Magnesium	2.5*	83735.034	1	0	1	2.5	0.0	2.5
	CSF prot	5.5*	0.012	10	0	10	55.0	0.0	55.0
	Repeat total			15	2	17	77.5	0.0	77.5

Bili neont	5.0	82250.000	9	0	9	30.0	15.0	45.0
Bili dirct	2.5*	82249.034	1	0	1	2.5	0.0	2.5
Lipase	2.5*	83690.034	0	1	1	2.5	0.0	2.5
Phosphorus	2.5*	84105.034	1	0	1	2.5	0.0	2.5
SGPT/ALT	2.5*	84465.034	1	0	1	2.5	0.0	2.5
PC estrase	2.5*	84720.034	1	1	2	2.5	2.5	5.0
Magnesium	2.5*	83735.034	1	0	1	2.5	0.0	2.5
CSF prot	5.5*	0.012	3	0	3	16.5	0.0	16.5
QC control total			17	2	19	61.5	17.5	79.0
Bili neont	5.0	82250.000	2	0	2	10.0	0.0	10.0
CSF prot	5.5*	0.012	6	0	6	33.0	0.0	33.0
QC standard total			8	0	8	43.0	0.0	43.0
Testing site total			57	8	65	237.5	20.5	258.0
Workcenter total			57	8	65	237.5	20.5	258.0

include cumulative to date (all work during current admission), cumulative since a specified date, interim since last chart (only new results not previously reported), and interim since last cumulative (all results since last cumulative report.

Once a patient has left the hospital, several reports can be produced. A "discharge chart" is an ad hoc chart which can be produced if results are entered after the patient is discharged and before the final chart is produced. This report is intended to be temporary and for the convenience of the clinician. A "final chart" is typically a cumulative listing produced for physicians and the medical records department and is produced automatically a user-defined number of days following the patient's discharge. Not all lab orders are completed at the time the final report is produced, and therefore "addendum charts" are generated if results are entered or corrected after the final chart has been produced.

Chart distribution is an important aspect of information flow. Given governmental impacts such as DRG's, where costs associated with delays in information flow are borne entirely by the hospital, speed is essential and it becomes desirable to route the same report or tailored reports to multiple places simultaneously, e.g., to the patient floor and to the private physician. Given advances in equipment, such as high-speed lasers (printing 40 pages/min), it is also now feasible to eliminate forms with multiple carbons and instead print original copies in the appropriate distribution queues. Such an approach saves on the purchase of mechanical bursters and on secretaries who must collate individual copies into distribution routes, and permits documents with elaborate header information on the first sheet but minimal header information on the second and subsequent sheets. The following list is a typical chart distribution route for today's medical center:

Hospital patient
  Copy 1: Nursing station, further sorted by room and bed
          If discharged → by medical record number (send to Medical Records)
          Clinic, then patient name or medical record number
          For emergency room, final copy → Medical Records
  Copy 2: Physician (alphabetical)
          (Print only abnormal and pending values)
  Copy 3: Consultant physician and/or responsible resident (alphabetical)
          (Print only abnormal and pending values)
  Copy 4: Medical service type (for quality control officer)
  Copy 5: Special-interest group (pertinent data to Tumor Registry, Infection Control, etc.)

Referred lab work (by physician)
  Copy 1: To physician (alphabetical), then patient name (alphabetical)
  Copy 2: If patient has medical record number, then to Medical Records

8.2. Emergency (STAT) Printing

With today's capabilities of an extended computer network (LIS connected to the hospital floors directly or via HIS systems), it is becoming feasible to print all STAT reports and all critical values directly on patient floors/nursing stations. Such "expedited" reports, which can be automatically printed at a designated printer location or, ideally, at multiple locations, are sent based on hospital location (e.g., emergency room, operating room, intensive care units), ordering and/or processing priority (STAT), critical results (life threatening), or manual selection by the technologist, pathologist, or laboratory clerk.

8.3. Quality of Chart Presentation

The presentation of data on the chart is enhanced substantially by the layout of the data and by background features which help highlight abnormal data. For the ease of physician use, it is important that all data be presented in logical sections and that any given procedure be reported in as many sections as needed for logical interpretation ("chart sections"). Grouping of tests into logical system/organ orientations aids the user in determining the type of abnormality present. In some instances, the results may even be duplicative, e.g., portions of the CSF analysis consist of tests unrelated to the lab section where many of the multiple-component tests are performed. Such a grouping allows tests that are performed in multiple areas of the lab to be reported together to enhance interpretation by the physician. Broader examples of logical groupings which enhance the physician's ability to amalgamate scattered data into useful information (and thereby save unnecessary extra orders of new tests) include immunology, hematology, liver profiles, transfusion, and cross-match summary. Good grouping can also subliminally help orient the physician as to which test must next be ordered.

In our system, the data can be printed horizontally or vertically, with a theoretical maximum exceeding the number of lines and columns that can physically be comprehended per page. Other printing choices include inclusion or exclusion of day of week and day of stay, printing abnormal results in boldface type, use of underlining or color (if the device supports color), listing pending orders, determining the best location of result qualifiers (e.g., footnotes, interpretive data, interpretations, or reference lab explanation and whether they should be in the body of the report, after the results section, at end of page, at end of section, or at end of report), and beginning with newest versus oldest results and specification of the maximum number of results to be included in any given page.

Organization of the demographic features, which generally are user defined for each hospital, enhances the ability of the medical staff to use productively the lab results. For example, page headers and footers can be designed to highlight the patient's name and medical record number, nursing station and room/bed, admitting and consulting doctors, admitting diagnoses, height and weight, social security number and other demographic data, birthdate/time, admission date/time, chart date/time, discharge date/time, and print location. The importance of these features in strengthening the report should not be underestimated.

With the recent introduction of inexpensive laser printers, header and demographic materials, reference values, and organization lines can be produced in various tones and fonts so as to appropriately and insightfully highlight the lab data, aiding its assimilation. Lasers offer the ability for the system to highlight in different tones and size fonts values which are mildly abnormal, dynamically changing (beyond a delta threshold), or at such thresholds as to be considered critical or life threatening.

8.4. LONG-TERM ON-LINE STORAGE: THE LAB'S MEDICAL ABSTRACT

In some instances, the laboratory must retain, on-line, records of procedures performed for a patient for many years after completion. Examples wherein such information is particularly useful include the blood bank, anatomical pathology, and, increasingly, certain chemical tests where there is long-term significance, e.g., HIV test results and hepatitis screening. To accommodate the need, some systems have developed a selective archiving information system which provides an efficient solution to long-term on-line storage requirements. Laboratory medical abstract systems should not be confused with complete backup and archiving capabilities. The abstract system is economical on space utilization. If all test information were to be retained in active storage files, tremendous and wasteful disk capacity would otherwise be required.

9. Management

The spiraling costs of medicine, the substantial percentage (about 10%) of a hospital's costs attributable to the laboratory, and the increased burden of regulatory agencies to which laboratories must respond have forced all laboratory staff to devote increasing time to managerial activities. The effort basically falls into three categories: (1) the thought process to decide what should be managed and how—a productive process; (2) the effort required to collect the data and organize them into information—generally a time-consuming and unproductive process; and (3) interpretation of the information and action based upon the interpretations—a useful process. It is at step number 2 (data collection) and to some degree at step number 3 (organization of data for interpretation—query systems) that today's modern computer systems can reduce markedly "scut" time, returning that time to the manager for more productive thinking processes.

9.1. Management Tools

Marked advances in software packages for query of large data bases and for graphic display of such manipulations have become readily affordable and should be integrated in any modern laboratory system. Such query systems should, for example, in an automated fashion be able to extract thousands of pieces of data, organize them, and then present them in a simple pie chart, saving the pathologist the hours of time required to do the same by hand. All data bases in the system should be accessible for such ad hoc applications. The essential components of ad hoc reporting include data extraction, report writing, and personal computer (PC) interface.

Ad hoc applications have witnessed much growth over the past few years. For some time we have used Datatrieve, a powerful query language. More recently, we have used the Cerner Command Language (CCL), a fourth-generation structured query language (SQL), and a relational data base language. SQL, a popular language used for data management in commerce, science, and industry, is an English-like language that is easy to use and efficiently searches and joins data base files. CCL enables non-data-processing personnel total access to archived and active data bases. Users may define report headings; time and date formats; center, right, and left justification; and padding, sorting, and selection criteria. Customized, ad hoc reports can be produced on-line, in printed format, or extracted to a file. This extracted information can be input into a standard generic report writer, spreadsheet, or statistical software package.

Of conceptual import, some lab vendors have developed query systems which are too limiting and permit access only to specific fields with specific rules for application. A query system should afford total flexibility, allowing access to any field in the system, and should permit complete manipulation of the field with a wide range of reporting methods. For example, through combinations of Boolean attributes tying together various data fields, the user should be able in English or with simple rules to ask for the following search: find all males between the ages of 15 and 30 yr with malignant tumors where at least three serial determinations were performed on a battery of tests for disseminated intravascular coagulation (e.g., PT, PTT, fibrinogen) and where a value changed greater than one standard deviation, and report the results, giving values, calendar date, and Julian date organized alphabetically by patient name.

As described earlier, assist (help) modes which use menus and prompts should be available. They help guide the inexperienced user in choosing information to extract from the data base. As each option is selected, the actual commands are displayed on the screen. In this way, the assist mode acts as a tutorial (again saving effort). As the novice user gains more experience, commands may be executed without the assist mode. On-line help functions, user documentation, and editing are available to ease the transition. Eventually, the user may issue compound commands to save additional time in defining the collection of data. A wide range of powerful commands should be available to permit the pathologist to browse through displayed data, execute stored programs, and produce quick ad hoc queries with default headings and default print formats by referencing multiple tables at any one time. For purposes of security, the pathologist can assign special privileges to individuals or groups, and can define protection levels for data base tables and programs.

Among new advances in today's highly interconnected computer world, extracted files of information can be downloaded to a microcomputer for manipulation and/or analysis using third-party software products (e.g., with Lotus 1-2-3 or DBASE).

9.2. QUALITY ASSURANCE

Unlike quality control, which assesses whether test performance was technically correct and produced the correct result, quality assurance assesses whether the test itself is medically useful and has helped the patient. A major function of a good computer system is that information fosters better care. The systems we have devised have stressed information output tailored to the needs of the pathologist/healthcare worker, which can have impacts in any of multiple ways.

As an example, the following generic report formats were devised for surgical pathology, and simultaneously affect patient care, quality assurance, and research. A general module allows any pathologist to inform the computer at will of the type of cases he wishes to know of on a daily basis (R1). The listing of cases is individualized for each pathologist and is of cases encountered by the department during the previous day (Fig. 2). In the example given, the pathologist desired knowledge each day of all malignant tumors (SNOMED morphology (M) field, first digit "8" or "9," but fifth digit not "0") of bone (SNOMED topography (T) field, first digit "1"), and any tumor, whether benign or malignant, of synovium (SNOMED morphology field, M904**), osteoid, cartilage, giant cell, mis-

BONE TUMOR

KIRKHAM
==
DEMOGRAPHIC DATA:
(*** DENOTES EXCLUSIVE CONDITION EXISTS)
NO COMPARISONS

SNOMED DIAGNOSES DATA:
SNOMED CODE: ((T1*))
AND CODE: ((M8-9 AND NOT M8***0 AND NOT M9***0))
OR
SNOMED CODE: ((M904*) OR (M918-934))
==

FIG. 2. Example algorithm for a daily report listing cases of bone tumors of interest to the pathologist. Such a listing is useful for patient care, teaching, research, and quality assurance. [Reprinted with the permission of Robboy *et al.* (R1) and the *American Journal of Clinical Pathology.*]

cellaneous rare lesions, or odontogenic lesions (SNOMED morphology field, M918**–M934**). Not uncommonly, the pathologist, because of his special knowledge of the field, would review the case, find something of immediate import, and communicate the finding to the patient's doctor with a timeliness that affected patient care. Over time, such review of cases also helped to identify weaknesses in procedures, procedures that might be abandoned or revamped, or new procedures that needed to be added. Of utility, collection of cases identified daily often led in practice to endeavors in new areas of research.

Integrating computer systems has also had the effect of producing useful information that heretofore has been nearly impossible to achieve. Linking radiology and pathology systems resulted in follow-up lists for the radiologist, informing him of the final pathology diagnosis rendered on any X-ray study he performed with during the past 3 months (G1). Such listings become educational for the individual radiologist, because confirmation or denial of radiological diagnoses points to strengths or errors and hence aids learning. These listings become useful for teaching in general, since it becomes trivial to identify, for example, the spectrum of X-ray manifestations of lung tumors when using such prepared lists.

Finally, linkage of systems provides information about patient care. Linkage of pathology with an institution's tumor registry permits rapid identification of new cases and follow-up on patients already in the registry. The fact that a patient has a pathology specimen, whether positive or negative for cancer, nonetheless indicates the patient is alive, and hence provides valuable follow-up information for the tumor registry (see Fig. 3). In the example given, the pathology department notifies the tumor registry of every patient if the patient location is a clinic associated with cancer patients (COXASU or MEDONC), if the pathology specimen is a malignant tumor (SNOMED morphology of tumor, but not benign tumor), or, finally, if the patient is already known to the tumor registry, regardless of the current findings (Fig. 3).

9.3. HUMAN RESOURCE MANAGEMENT

Managing human resources is a necessary and important function, but one that is often ignored. To manage 150 employees and keep track of all their functions in multiple areas throughout the laboratory requires substantial effort. If done manually, the effort is often considered as unproductive administrative overhead. Maintenance of an employee profile data base helps to streamline the centralization of pertinent data. Examples of data items needed include employee name, address, job classification, title, office phone extension, office mail address, employee identification number, social security number, residence phone number, emergency contact (name, phone number, relationship), birthdate, date of employment, date of termination, employee status (part time, full time, etc.), section, shift assignment, salary, and security level.

Such profiles save substantial effort every year in preparing credentialing statements for state government, for reporting types of manpower to the

```
TUMOR REGISTRY LIST                    7:06 PM 04/04/80

UNIT NUMBER            NAME
                       LISTED FIRST:   PATHOLOGY FILE NAME
                       LISTED SECOND:  TUMOR REGISTRY FILE NAME
================================================================

DEMOGRAPHIC DATA:
(*** DENOTES EXCLUSIVE CONDITION EXISTS)
***ORDERING LOCATION:         (COXASU)
                                OR
                              (MEDONC)

SNOMED DIAGNOSES DATA:
SNOMED CODE: ((M8-9 AND NOT M8***0 AND NOT M9***0 AND NOT M809*))
================================================================
```

FIG. 3. Example algorithm for a daily report listing patients of interest to the hospital Tumor Registry. Potential cases are identified on the basis of hospital location, pathology diagnosis, or whether the patient is already known to have cancer. [Reprinted with the permission of Robboy *et al.* (R1) and the *American Journal of Clinical Pathology.*]

hospital, for real-time security for personnel access to computer functions, for contacting employees in times of crisis, and for job relocation within the lab. It should be obvious that the data, once updated in the data base, are available for all functions. Conversely, no datum item should ever be maintained in more than one file.

Given the increasing importance of human resource management, newer systems might maintain continuing education records, reprimand records, absentee journals, salary history, hours worked and hours paid, demographic profile reporting, and payroll maintenance and reporting.

9.4. MANAGEMENT INFORMATION SYSTEMS

Management information systems summarize information about laboratory activity. The information they produce supports laboratory management in review and decision-making processes. This same information also assists the pathologist in working with the hospital administrator in decision-making-processes.

Many management data bases are updated daily and include extractions and updates of laboratory activity files such as raw counts, revenue, and cost information for each ordered procedure. Monthly and yearly maintenence functions are also usually included which provide totals for the current fiscal period for each orderable procedure. To accommodate the variety of informational needs in management, several options should be available regarding the level of detail stored. Examples include assigning a time window to the activity (within a 24-h period), including/excluding detailed information at the testing site level, including/excluding details about the patient location, and including/excluding details about the client.

Such flexibility permits management to choose the requisite information required for prudent decisions. Grouping features provides for specification of detailed information for additional focus: clients, physicians, physician specialty services, patient types, collection priorities, patient locations, reporting priorities, client classes, etc.

9.5. UTILIZATION REPORTING

Utilization reporting (analysis of usage patterns and demand for specific procedures) provides an analysis of each lab section's activity by user-defined priorities. Utilization reports rank usage and demand by each client or group of clients, medical specialty service, ordering physician, or time window. Reports should be able to be generated in either ascending or descending order to include all procedures (including their reporting priority) ordered within a specific time window.

Utilization reporting, in conjunction with cost accounting data, should help identify contributions by margin, thus, fostering identification where additional use of equipment can be made for marginal costs. As a variation on the theme, physicians—individually or grouped to reflect group practices or medical speciality—clients, or even wards can be monitored as individual clients and reviewed as health maintenance organizations (HMOs). Time windows can help monitor and identify activity for specific shifts and/or peak periods. Some variables for which sorts by frequency help identify productivity include employee shifts, priority, physician, medical specialty, client, client class (e.g., HMOs, PPOs, corporations), time windows, volume, revenue, contribution margin, section and department, ordering location (e.g., emergency room), sales territory, and courier route. Such reviews often help identify optimal as well as wasteful practices, in addition to the actual experiences of the groups.

Utilization reporting is useful in the analysis of operations. Examples include identification of number of STAT orders for a specific collection time (i.e., 6 AM), physician, or specialty service; analysis of volume by time window (which can correspond to shifts), allowing managers to adequately staff peak operations; and, finally, detection of trends. The latter is especially important to assist pathologist intervention at an earlier time to avoid unnecessary costs and procedures, or, alternatively, to identify new and useful situations on which to capitalize.

9.6. Labor Productivity Reporting

Labor is the lab's single most expensive commodity, accounting for about 50% of all costs in many institutions. In manually operated organizations, especially those with lax central administration, these costs can quickly spiral out of control. With DRG fiscal policy, where laboratories have become cost rather than revenue centers and where reimbursements is fixed, regardless of real costs, active management is crucial.

Yet, in the area of productivity assessment, to gather data manually is near impossible and is, exceedingly wasteful of personnel time. To develop economic productivity reporting, computer systems are needed to capture all raw procedure counts and workload units for each procedure by testing site within each workcenter. Each time a procedure is performed, the data for that procedure should be updated. Raw counts should be maintained by patient type and workload should be maintained by user-defined time windows.

Given the extensive system of standard workload units developed by the College of American Pathologists, maintenance of counts by testing site within each workcenter should be relatively easy. Phlebotomy workload is captured at the phlebotomy section level by type (e.g., venous or fingerstick), and can be incorporated into the overall procedure.

As a consequence of detailed procedure counts, reports should be easily generated for any time period, such as shift, daily, weekly, monthly, or annually. Further, as workload units can be totaled at testing site, workcenter, lab section, and laboratory levels, and also against patient type, etc., this type of analysis becomes a potent tool to assess hospital/lab areas for (1) maldistribution of personnel; (2) true resources for two competing types of instrumentation, e.g., 24-h SMAC coverage versus 8-h SMAC and 16-h ACA combination; (3) types of resources required for a hospital program, e.g., 24-h trauma center; and (4) marginal cost for handling new business from an external source during the evening shift when the equipment is underutilized.

9.7. Cost Accounting

Cost accounting, which defines the costs involved with the performance of each procedure within the lab, is the obvious culmination of the aforementioned management systems to assess and report contribution dollars and/or contribution margin. Cost accounting provides the pathologist the tool necessary to assess and maintain costs of production. The standard cost report should examine all internal costs associated with performing a clinical procedure and permit comparison between costs if the procedures were to be sent elsewhere to be performed, thereby maximizing profitability. Such systems foster continuous standard costing information. Standard costs for labor (both technical and professional components), materials, depreciation, and testing site overhead at the testing site level can be easily maintained and updated.

As materials and quantities and their costs to perform each procedure are known, the cost per unit is easily calculated. Because the system also knows the specific testing site where the procedure was performed, it accurately captures costs based on the methodology used to perform the procedure.

Overhead costs are a real and high cost of doing business. At many institutions they are difficult to assess directly, causing predetermined formulas to be used, such as 2.3 times "direct cost." When procedure/testing site overhead costs can be precisely measured, the true cost can be applied on a recurring basic each time the procedure is ordered. These costs include set-up and shut-down costs as well as equipment depreciation costs and the cost of being in the facility. In the absence of precise data, overhead costs must be allocated in an imprecise manner at various levels of the organization (e.g., section, departmental, location, client, hospital, or institutional levels) based also to some degree on volume, revenue, and known direct expenses. A useful listing is "procedure costs by testing site." This report details the costs for each procedure: labor, materials, testing equipment depreciation, and testing site overhead.

Cost accounting, when combined with other types of management systems described above, further expands the pathologist's management capabilities. Section operating statements can be generated which report costs by client, client class, physician, specialty service, and time window and procedure rankings by contribution margin (percentage or dollar amounts).

In all instances, it is important that data are summarized in a format useful as information for monitoring budgets. We have had as a long-term goal to decentralize (delegate) authority and responsibility so that each division director (e.g., hematology) and section head (e.g., coagulation) become responsible for the costs of his own unit. We therefore have desired to develop monthly operating statements by section. In addition, a summary report for the entire organization can be generated and compared to the monthly budget. Each operating statement will show the revenue generated by patient type within the given section, direct labor, direct materials, test site burdens accumulated as a result of volume, direct margin (revenues less direct expenses), allocated overhead expenses, and the contribution margin (direct margin less indirect expenses).

9.8. SERVICE LEVEL ANALYSIS

A final aspect of management, and yet probably the most crucial of all, is the service the pathologist provides to clinicians. All too often clinicians claim that service is slow, e.g., an 8-h turnaround time, whereas the pathologist believes it to be but 1 h. The former measures the time elapsed between ordering of a test and receipt of a result. The latter measures the interval between when the specimen arrived in the lab to when the report was ready.

Analysis of turnaround times by procedure is based on collection priority, patient location, and performing location (testing site). In a system where timings are ascertained at several checkpoints (i.e., requested collection, specimen drawn, specimen received in the lab, procedure completion), bottlenecks become easily identified. The value of this information to pathologists and the hospital is high, because areas of inefficiency, once identified, can often be corrected.

10. Conclusion

LIS systems have progressed greatly during the past decade in how they help manage information and improve the quality of patient care. With the current expanding scopes of litigation and regulations, LIS systems have proved they can produce needed documents as by-products of data already collected. Such systems, which help the pathologist transform bits of data into organized information, hold even greater promise for the future. Through computer based laboratory expert systems, the pathologist will be better able to serve as an active laboratory manager. This will permit more time for the pathologist to utilize his personal knowledge in consultation for patient care.

REFERENCES

G1. Greenes, R. A., Bauman, R. A., Robboy, S. J., Wieder, J. F., Mercier, B. A., and Altshuler, B. S., Immediate pathologic confirmation of radiologic interpretation by computer feedback. *Radiology (Easton, Pa.)* **127**; 381–383 (1978).

R1. Robboy, S. J., Altshuler, B. A., and Chen, H. S., Retrieval in a computer-assisted pathology encoding and reporting system (CAPER). *Am. J. Clin. Pathol.* **75**; 654–661 (1981).

MONOCLONAL ANTIBODIES: PRODUCTION, PURIFICATION, AND TECHNOLOGY

David Vetterlein

Genentech,
Recovery Process Research and Development,
South San Francisco, California

1. Introduction.. 303
   1.1. Monoclonal Antibody Structure and Diversity 304
2. Monoclonal Antibody Production.. 306
   2.1. Mouse Monoclonal Antibody Production............................. 307
   2.2. Quadroma Generation 311
   2.3. Human Monoclonal Antibody Production and Difficulties.............. 313
   2.4. Hybridoma Cultivation... 315
3. Monoclonal Antibody Purification 317
   3.1. Common Purification Techniques 318
   3.2. Monoclonal Antibodies for Medical Applications..................... 322
4. Monoclonal Antibody Technology 329
   4.1. *In Vitro* Diagnostic Assays 330
   4.2. Immunoaffinity Purification.. 333
   4.3. Mouse Monoclonal Antibody Therapeutic Products 334
   4.4. Human Monoclonal Antibody Products 335
   4.5. Developing Areas of Monoclonal Antibody Technology................ 338
   References... 342

1. Introduction

Few biological achievements have so thoroughly captured the imagination of scientists, the medical community, and entrepreneurs as has the development of monoclonal antibody technology. Monoclonal antibodies recognizing hundreds of different antigens have been developed since Kohler and Milstein first reported the production of monoclonal antibodies to red blood cell antigens in 1975 (K9). Contributions by many investigators have led to today's unprecedented understanding of B cell immunity and the harnessing of B cell antibody production for man-made purposes (B16,

C13, H10, L13). B cell immortalization techniques have provided human and mouse monoclonal antibody reagents for the study of a large array of biological substances, including proteins, nucleic acids, carbohydrates, lipids, and drugs. Innovative applications abound. New monoclonal antibodies are being generated at an enormous rate to configure novel diagnostic tests, study protein function, serve as therapeutics, advance basic biological research, and establish new products in biotechnology.

Considering the size of the field, a comprehensive review of monoclonal antibody technology is not possible here. Therefore, this review will cover certain important aspects of monoclonal antibody production, purification, and technology. The methods employed for human and mouse monoclonal antibody production and generation are illustrated along with the generation of hybrid hybridomas (quadromas). Several key problems have been encountered in the production of human and mouse monoclonal antibodies and will be discussed with an emphasis on manufacturing considerations. In addition, various purification methods are surveyed and key regulatory and characterization issues are highlighted for monoclonal antibodies applied therapeutically. Monoclonal antibody products are becoming more common and several successful commercialized products are described. This review concludes with a brief discussion of future directions of the technology. Several novel and evolving areas of the technology are also briefly discussed, including immunosensors, recombinant antibodies, bispecific antibodies, and catalytic antibodies.

1.1. Monoclonal Antibody Structure and Diversity

The mouse B cell repertoire is capable of producing over 10 million different antibody molecules (S14) with dissociation constants for specific antigens ranging from 10^{-5} to 10^{-10} mol/liter. Figure 1 illustrates the basic structure of a mouse IgG molecule and how this diversity is generated at the DNA level. Most monoclonal antibodies currently employed in biotechnology are of murine origin—usually mouse IgG antibodies. The diverse binding specificity of an IgG molecule is a consequence of the unique structure of the IgG molecule and the amino acids contained within two identical binding pockets at the N-terminal end. An IgG molecule is symmetrical and consists of one pair of heavy (H) 55,000-Da polypeptide chains and one pair of light (L) 25,000-Da polypeptide chains linked by interchain disulfide bonds. The binding pocket is formed by orientation of variable or hypervariable stretches of the first 108 or so H and L chain amino acids (R3). The H and L chains are woven into a tight barrellike structure formed by repeating β-pleated sheets that juxtapose variable and hypervariable

Monoclonal Antibody Diversity

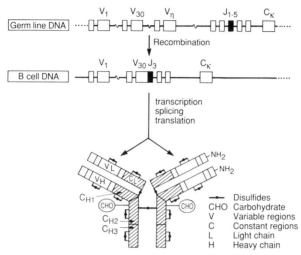

FIG. 1. Schematic illustrates the generation of light chain diversity at the genome level by recombination of V and J gene segments with the κ constant region gene segments during β cell maturation. Heavy chain diversity is generated in a similar way but includes an additional small D region gene segment (S1). The conventional structure for an IgG molecule is also illustrated. Each IgG molecule is a symmetrical disulfide-linked structure composed of identical pairs of heavy and light polypeptide chains forming two equivalent binding sites. The N-terminal, variable region amino acid sequences differ from antibody to antibody, forming binding sites specific for a given antigen. Constant (CH2 – CH3) regions of the antibody molecule confer specific effector functions.

stretches of amino acids and fold to form two identical antigen combining sites (A5).

The IgG binding pocket is generated at the DNA level by specific recombination events that consolidate H and L chain gene segments. Only one of the H and L chain germ line maternal or paternal alleles is properly rearranged (allelic exclusion). One pair of H or L chain alleles is excluded in the chromosome (E1). Thus, each B cell or hybridoma cell produces only one type of antibody with unique binding characteristics. In the mouse, kappa (κ) light chains are coded by gene segments on chromosome 6, lambda (λ) light chains on chromosome 16, and heavy chains on chromosome 12 (D5, G3). The variable and hypervariable regions of light and heavy polypeptide chains are generated in two basic ways, first, by the selection and recombination of spatially separated constant region gene segments with one of several hundred different germ line variable (V) region gene segments (V region gene segments encode about 96 N-terminal amino acids), and second, by junctional diversity arising from the consolidation of short DNA segments into the final recombined H and L chain genes. In the case of the light κ chain (Fig. 1), a specific joining segment referred to as a J region (13 residues long), present in about five distinct copies, is recombined at the 3' end of one out of several hundred variable region gene segments (C11). Each H and L chain contains three regions of amino acid hypervariability and the mechanism of recombination is somewhat inaccurate, leading to additional diversity around position 96—a critical site for antibody–antigen combination (P1). H chain gene recombination events are similar, but involve an additional DNA element referred to as a D region, forming a V/D/J heavy chain variable region (S1). Additional antibody diversity can arise through somatic mutation by a mechanism still not well understood (S21).

Further diversity results from variations in antibody constant regions that result in different classes and subclasses of antibody molecules. All antibody molecules contain either κ or λ light chains, which have no known effector functions. H chain constant regions (C2, C3; Fig. 1) confer functional differences, including complement fixation, type of antigen bound, opsonization activity, macrophage binding, mast cell mobilization, and activity in secretions (G10). There are five major classes of immunoglobulins known as A, D, E, G, and M. In the mouse, four different subclasses of IgG are known and referred to as 1, 2a, 2b, and 3. IgG subclasses have distinct structures (O4) and also differ with respect to biological activity.

2. Monoclonal Antibody Production

The production of monoclonal antibodies by B cell immortalization has several advantages over conventional immunization and polyclonal antibody production. Advantages include the following examples:

1. Each cloned B cell produces only one type of antibody of predefined specificity.
2. Impure antigens can be used to generate monoclonal antibodies.
3. Monoclonal antibodies can be generated against rare or weakly immunogenic antigens.
4. Monoclonal antibodies have a predefined isotype and thus variable effector functions.
5. Reproducible antibody can be made in unlimited amounts.
6. Specific human monoclonal antibodies can be generated without immunization.

7. Second-generation monoclonal antibodies (such as bispecific antibodies) can be designed which cannot be obtained by conventional immunization.

The principal areas to be covered in this section will be methods for generating mouse and human monoclonal antibodies and bispecific antibodies. Technical limitations are also considered, particularly for the production of human monoclonal antibodies. In addition, hybridoma propagation methods are discussed, along with the advantages of serum-free media.

2.1 Mouse Monoclonal Antibody Production

Today most monoclonal antibodies used in research, medicine, and biotechnology are of mouse origin. There are several reasons for the emphasis on mouse monoclonal antibodies. First, excellent myeloma fusion partners are available for mouse hybridoma generation. Second, reproducible standard procedures have been developed for mouse immunization and myeloma fusion to immunocompetent B cells. Third, selection and cultivation of antibody-producing hybridoma clones is relatively straightforward.

The procedures most commonly employed to generate mouse hybridoma producing monoclonal antibodies have been thoroughly described by others (E6, F2, G2, G9) and are outlined schematically in Fig. 2. For immunization, a primary injection of 1–10 μg of protein antigen emulsified in complete Freund's adjuvant is injected intraperitoneally. (Other injection sites, such as intramuscular hind legs and hind foot pads are also commonly employed.) Secondary injections are then given using incomplete Freund's adjuvant at 2- to 3-week intervals until an antibody response is detected by assay for serum antibodies. To ensure an optimum antigenic response, a final intravenous injection of the antigen in microgram amounts is often given 2 to 3 days before spleen or lymph cells are removed.

Although B cell immortalization has been accomplished by direct transfection using Sendai or other viruses (K12, S23), immortalization by fusion with myeloma cells is usually more successful. Reliable mouse myeloma (BALB/c) fusion partners are available for mouse hybridoma generation (K5, S12). Mouse myeloma fusion partners can be obtained that grow aggressively in cell culture as myelomas and as heterokaryons, produce no endogenous antibody or fragments, and are stable to fusion in high concentrations of polyethylene glycol (PEG). Mouse B cells are typically fused with myeloma cells using 30% polyethylene glycol. As many as several hundred antigen-specific hybridoma cells can be obtained from a single spleen cell fusion. Normally, one hybridoma cell is expected for every

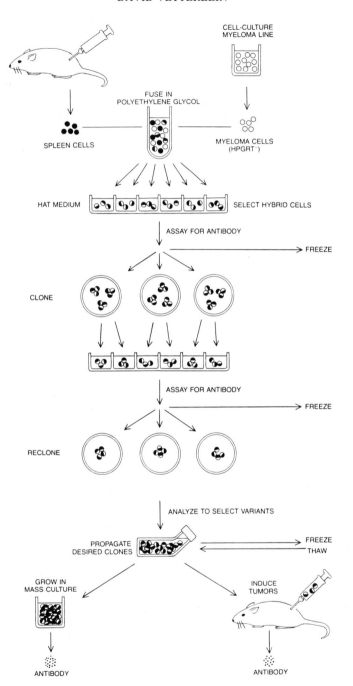

10^4 lymphocytes fused. B cell–myeloma fusion may also be carried out using inactivated Sendai virus (C13), or by means of an electric field (V5). The electrofusion method may offer some important advantages for the production of human monoclonal antibodies where high fusion frequencies are necessary to generate antibodies from a limited availability of immunocompetent spleen or lymph node-derived immune cells (G11). Electrofusion methods, however, are technically more demanding, require relatively expensive equipment, and have a greater risk of culture contamination.

Hypoxanthine/aminopterin/thymidine (HAT) selective media, first described by Littlefield (L13), are commonly employed for hybridoma generation. B cells will die in cell culture unless immortalized by fusion with myeloma cells. Mouse myeloma cells commonly employed for the purpose of cell fusion lack hypoxanthine guanine phosphoribosyltransferase (HGPRT$^-$) or thymidine kinase (TK$^-$). These are two essential DNA or RNA salvage pathway enzymes which utilize hypoxanthine (H) and thymidine (T) substrate intermediates. Aminopterin (A) inhibits the normal pathways for DNA and RNA synthesis. Thus, nonfused (TK$^-$) or (HGPRT$^-$) myeloma cells also die in the presence of aminopterin or HAT media. Since B cell salvage pathway enzymes provide for nucleic acid biosynthesis, only fused hybridoma cells survive and propagate in HAT media.

Once hybridoma cells begin propagating, antibody-producing colonies are usually detected by enzyme-linked immunosorbent assays (ELISA) (E5). A number of different ELISA methods may be utilized for screening monoclonal antibody harvest fluids. ELISA methods for hybridoma screening have been recently reviewed by Brennand *et al.*, who have suggested that the sandwich-type ELISA assay formats are the most sensitive for initial screening of harvest fluids (B12). Antigen agglutination, RIA, or biological activity measurements may also be employed (E6). Antibody-producing colonies are usually cloned by limiting dilution to establish growth from a single cell. Cloning prevents overgrowth by competing hybridomas and should be repeated several times to ensure monospecificity (D2). Other less common methods of cloning mouse hybridoma cells include selection of single-cell colonies in soft agar (G2), direct separation

---

FIG. 2. Classical procedure for generating mouse monoclonal antibodies. Generation proceeds in stages, mouse immunization, polyethylene glycol fusion with myeloma cells, selection in HAT medium, then cloning and detection of antibody-secreting cells by assay. Cloned hybridoma cells are either propagated as a mouse ascites tumor or in cell culture for production of large quantities of antibody. Adapted from Ref. M8, with the permission and courtesy of Cesar Milstein and Scientific American Publications.

by fluorescence-activated cell sorting (D1), or antigen-specific cell adherence assays (N1), where antibody-producing cells are cloned and selected on the basis of direct antibody-mediated recognition of solid-phase antigen.

After cloning, hybridoma cells can be expanded in cell culture or grown as an ascites tumor to produce desired quantities of the monoclonal antibody required for a particular application (see Section 2.4).

2.1.1. *Technical Difficulties*

Generating hybridoma cells for particular antigens can be a difficult and time-consuming process and can pose a number of problems:

1. Difficulties with media components or toxic reagents (PEG preparations are sometimes contaminated with toxin).
2. Contamination of cultures by fungus, bacteria, yeast, or mycoplasma.
3. Difficulties in generating an immune response.
4. Clonal instability leading to loss of antibody production by approximately 50% of the producing clones.
5. Absence of hybridoma clones that produce monoclonal antibodies with desired physical properties (e.g., affinity, specificity, pH stability, or biological properties such as complement fixation).
6. Difficulties in obtaining desired quantities of monoclonal antibody due to poor growth in cell culture or ascites.

Consequently, it is often necessary to perform multiple fusion experiments over a period of many months before hybridoma clones with the desired characteristics are obtained.

When a protein is weakly or poorly immunogenic, new immunization strategies are generally employed. Often simple changes in immunization protocols, such as employing high or low doses of immunogen, haptenization, varying injection routes, and changing adjuvants, give remarkable improvements in immune response. Recently, Nilsson *et al.* demonstrated that intrasplenic implantation of nitrocellulose- or Sepharose-coupled antigens greatly enhanced the immune response in mouse and rabbits for several antigens (N4). Conjugation of an antigen to an immunogenic protein or aggregration with cross-linkers may also be helpful in eliciting an immune response to conserved proteins or low-molecular-weight peptides (L15).

In vitro approaches have also been employed. Most procedures have involved *in vitro* stimulation of B cells by antigens and mitogens 3–4 days prior to fusion (F8). Until recently, results of *in vitro* stimulation experiments have generally been unsatisfactory. B cells often become unstable to fusion after *in vitro* stimulation, and successful hybrids are usually IgM secretors (O5). However, Takahashi *et al.* recently demonstrated successful production of IgG class monoclonal antibodies by optimizing antigen concentration in the presence of mitogen- and thymocyte-conditioned media and lengthening the culture time before fusion to 8 days (T2). Further enhancement of the IgG response is obtained by combining purified lymphoblast cells, primed *in vivo,* with *in vitro* immunization techniques (E8). Muramyl dipeptide has been effectively employed in culture medium to enhance fusion frequency (B11). Thus, considerable progress has been made recently in developing *in vitro* B cell stimulation methodologies and in controlling the class of monoclonal antibody produced *in vitro* by stimulated B cells.

A number of laboratories are working on improving hybridoma screening methods. Care must be taken to minimize overgrowth by competitive nonsecreting clones (G9), and thus culture density is a critical parameter. Hadas and Theilen have shown that hybridoma seeding density is extremely important for improving the yield of ovalbumin-specific hybridoma clones (H1). They observed an inhibitory effect on hybridoma production due to nonproducing cells or residual cell fragments on ELISA screening protocols. Maximum yield of antibody-secreting clones was obtained when one cell was present per well. New screening methods are also being developed to select high-affinity monoclonal antibodies by directed fusion methods. Reason *et al.* have demonstrated a method for generating high-affinity antibodies by complexing antibody-producing B cell with antigen-coupled avidin *in vitro,* followed by fusion with biotinylated myeloma cells using polyethylene glycol (R2). The biotin- and avidin-directed fusion technique has been used successfully for several different antigens and provides an excellent selection method for high-affinity ($K_d > 10^{-8}$ mol/liter) antibodies. It is also important to note that the recent application of computer-aided techniques for monoclonal generation is decreasing the time necessary to identify desired antibodies by facilitating screening protocols. Data base software for screening against several antigens using multiple assay formats is becoming more common, particularly when complex screening procedures are required to select antibodies with complex physical or biological properties (H5).

2.2. Quadroma Generation

When hybridoma cells that recognize two different antigens are fused, hybrid hybridoma, or quadroma, progeny are produced that secrete hybrid antibodies originating from both parental lines (as outlined in Fig. 3). Hybrid hybridomas codominantly express H and L chains from both parental hybridoma lines. If the H and L chain polypeptides from both parental

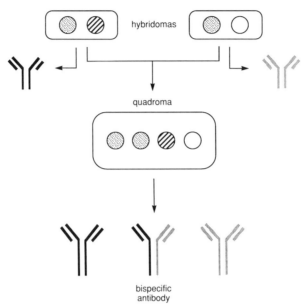

FIG. 3. Scheme for generation of hybrid hybridomas or quadromas. Hybridomas, each producing antibodies that recognize two different antigens, are fused. Hybrid hybridomas are generated by a combination of heavy and light chains from both parental lines. Although 10 separate antibody species are theoretically possible, quadroma cells usually express only three major antibody species. Two types are homologous to the original parental cells and bispecific antibodies that carry one binding site for each of the two different antigens recognized by the parental lines.

lines combine randomly after translation, ten different IgG molecules are theoretically possible. In practice, however H–L chain combination in hybrid hybridomas seems to be restricted, so that after translation the H–L chains associate preferentially and form half-molecules. Thus, three kinds of IgG antibodies are codominantly expressed—two species with identical phenotype to each respective parental line, and a hybrid molecule which is bispecific and composed of half-molecules from each parental line. Each IgG type is expressed in hybrid hybridoma cells in roughly equal proportions (S25). Separation can be accomplished by successive affinity separations or by conventional chromatographic methods.

Generation of hybrid hybridoma cells can be quite labor intensive since additional selection procedures have to be employed to separate the parental hybridoma cells from progeny hybrid hybridoma lines. For example, Staerz has generated a hybrid hybridoma to focus T cell effector activity to Thy-1.1 antigen expressing tumor cells (S19). A T cell-specific hybridoma sensitive to HAT media was first isolated and made ouabain resistant. Selection of hybrid hybridoma T cell–Thy-1.1 cells was accomplished in HAT–ouabain selection. Another approach involves selection of hybrid hybridomas by treating both hybridoma cell lines with irreversible inhibitors immediately prior to fusion. Suresh *et al.* developed antiperoxidase–antisubstance P hybrid hybridomas by treating one parental cell with a protein inhibitor (emetine) and the other with an inhibitor of RNA synthesis (actinomycin D) (S25). The resultant hybrid hybridomas were isolated in an actinomycin- and ouabain-containing selection media. Direct selection of hybrid hybridomas may also be accomplished by labeling one parental hybridoma cell with fluorescein isothiocyanate and the other with rhodamine isothiocyanate, followed by fusion and isolation of heterofluorescent hybrid cells using a fluorescence-activated cell sorter (K2).

Although heterohybrid methods of bispecific antibody production are more common, bispecific antibodies can also be generated chemically by mild reduction to monovalent IgG half-molecules followed by reoxidation. When monoclonal IgG half-molecules produced by two different hybridoma cells are mixed and reoxidized, bivalent IgG molecules reform randomly, resulting in bispecific antibodies. Chemical reduction often results in loss of binding activity and denaturation, which can be a limitation (K2). It is important to note, however, that new chemical methods for producing bispecific antibodies have been very successful (P4). Arsenite can be used as a dithiol-complexing agent to prevent intramolecular disulfide formation and 5,5-dithiobis(2-nitrobenzoic acid) can be used to direct the assembly of IgG half-molecules with good retention of binding activity. Thus, controlled chemical procedures could become the method of choice for bispecific antibody generation, and complex hybridization techniques may not be necessary.

2.3. Human Monoclonal Antibody Production and Difficulties

Developing procedures to immortalize human B cells producing antibodies represents an important and expanding research area. For those interested in more detail on the subject of human monoclonal antibody production with an emphasis on the difficulties involved, James and Bell have recently published a thorough review (J1). A recent book edited by Engleman *et al.* is also recommended reading (E4). The book contains a compendium of articles by key researchers in this area and a useful appendix of techniques for those interested in specific methodologies.

The methods used to immortalize human B cells are similar to those used for mouse B cell immortalization. Human myeloma cell lines, Epstein–Barr virus (EBV)-transformed human lymphoblastoid cells, or mouse myeloma cells (making heterohybridomas) are often employed as fusion partners. Immortalization by direct transformation using lymphotrophic viruses such as EBV is also common. EBV transformation has proved one of the most effective immortalization techniques to date (C1). Steinitz *et al.* were the first to immortalize human B cells and point out the importance of human monoclonal antibodies for medical and research applications (S22). They immortalized donor lymphocytes with a natural antibody titer to a synthetic hapten molecule (4-hydroxy-3,5-dinitrophenacetic acid) by direct EBV transformation.

However, despite a number of successes, immortalization of human B cells is an evolving technology fraught with difficulties. A major difficulty is that immune B cells are not readily available in humans. Peripheral blood lymphocytes are generally employed with *in vitro* immunization or sensitization procedures to generate adequate numbers of immunocompetent B cells. Procedures for *in vitro* stimulation and immunization of human B cells are still poorly defined. Many parameters are under study including (1) the duration of use and concentration of mitogen and antigen used for stimulation; (2) the addition of immunoregulatory molecules; (3) control of the differentiation state of immune B cells; (4) removal of supressor cells or cytotoxic T cells which inhibit human B cell survival; and (5) improvement of cultivation procedures after stimulation.

Human B cell immortalization protocols are also far from perfected. Most procedures are 100 to 1000 times less efficient than those used for murine hybridoma generation. Human myeloma cell lines such as the robust lines employed for mouse B cell immortalization are not yet available. Human myeloma lines generally grow poorly in culture or in selective media and are often killed by standard PEG fusion techniques. Electrofusion procedures under development may facilitate human B cell immortalization by enhancing fusion frequencies (E6). In addition to the problem of low fusion frequencies, immortalized B cells often lose the ability to produce antibody. This instability has been attributed to a variety of factors, including lack of certain growth and differentiation factors (W4), absence of important receptor signals (M6), structural or regulatory defects as a result of gene translocation or loss of heavy or light chain genes (T4, S8), and deficiencies in antibody secretory pathways (G1, K11).

Immortalized human B cells in culture are difficult to clone and select due to their inherent instability and to the low levels of antibody secreted. Antibody-producing cells also tend to grow much more slowly than murine hybridomas and exhibit doubling times of 40–60 h in cell culture. Production and scale-up of human monoclonal antibodies can be difficult. Concentrations of human monoclonal antibodies achieved in cell culture or nude mouse ascites are typically 10 to 100 times less than for murine hybridomas employing comparable cultivation methods. Understanding the mechanisms of this instability and generating stable human antibody-producing cells are research areas of key importance to the future of this technology.

2.4. Hybridoma Cultivation

A variety of different techniques are available for the cultivation of human and murine antibody-producing cells. For example, research quantities of murine or human monoclonal antibodies are often produced in the laboratory by simple expansion of stationary culture, roller bottles, or small spinner cultures (G2). *In vivo* growth in the peritoneal cavity of mice is also a reliable means of producing research quantities of mouse monoclonal antibodies. Most often within several weeks, 3–20 mg/ml of monoclonal antibody in 10 ml of peritoneal ascites fluid can be obtained from a pristane-primed mouse (E6). Production of human monoclonal antibodies is more difficult. Human–mouse heterohybrid lines appear to grow better in mouse ascites than do cell lines generated by transfection or other immortalization methods (J1), and nude mice are required for production of human monoclonal antibodies in ascites (B10). Usually, even after X-irradiation to destroy natural killer (NK) and macrophage cytotoxic activity, only 50% of human B cell immortalized lines produce *in vivo* (J1). Thus, most human monoclonal antibodies are produced by cultivation and expansion of immortalized lines in cell culture.

Although gram quantities of monoclonal antibodies can be conveniently and inexpensively produced in mouse ascites fluid for research purposes, commercial production of human or mouse monoclonal antibodies *in vivo* has several potential disadvantages:

1. Thousands of mice are required to generate kilogram quantities of antibody.
2. Caution must be taken to prevent rodent viral or bacterial infections; tissue-specific xoonotic viruses can be passed from rodents to man (C3), and outbreaks of viral disease have been reported even in barrier-maintained mouse colonies (B7).
3. It is difficult and expensive to maintain sterile mouse colonies.
4. Synchronizing ascites production from thousands of animals is labor intensive and expensive.
5. Replacement of a contaminated mouse colony is considerably more difficult than sterilizing several large fermenters.
6. Monoclonal antibodies from ascites fluid can be contaminated with up to 20% endogenous murine antibodies.

In vitro methods of hybridoma cultivation are relatively simple to scale-up and large quantities of antibody can be produced for commercial applications. Experience has shown that mammalian cell products, including antibodies, can be produced and purified safely and efficiently using simple batch fermentation techniques (F3). If oxygen tension and pH are carefully controlled, mouse hybridoma cells grow to a level of $1-4 \times 10^6$ cell/ml. Production rates of up to 40 µg per 10^6 cells per day are not unusual. Once media nutrients are depleted and waste products accumulate, cell viability drops precipitously. Perfusion techniques can provide further nutrients and remove some waste products, leading to higher cell densities and improved antibody quality. Oxygen tension and pH are the two parameters most carefully controlled during *in vitro* production. Hybridoma cells may also be immobilized by encapsulation or entrapment in inert supports to increase cell density and antibody production. Immobilization is accomplished with soft gels such as agarose or calcium alginate (N5), polyethyleneimine, or poly(lysine) (L11), and entrapment densities of 10^8 cells per ml have been achieved (L14). The encapsulation process must be carefully controlled. If the size of encapsulated microspheres is greater than 80–100 µm in diameter, cells in the center can become starved for oxygen (F6). Microsphere technology is still under development and producing large numbers of microspheres with a predefined size for mass culture can be an expensive proposition. Hybridoma cells may also be grown in hollow fiber (H9, K7, L10) or ceramic cartridge reactors (P12) in a media perfusion mode. The fibers, typical ultrafiltration cartridges engineered for cell growth, contain capillary channels which entrap cells. The capillaries have pores large enough to allow exchange of low-molecular-weight media constituents, but small enough to exclude antibody passage. Capillary-entrapped cells can grow to very high densities and be sustained for several months by media perfusion. Similar observations have been made with ceramic cartridges. With the ceramic device developed by Corning Glass Corporation, hybridoma cells adhere to or are entrapped by a ceramic support channeled with large pores in direct contact with perfusion media. Scale-up of the length of a hollow fiber or ceramic device to increase antibody production is limited by inlet and outlet gradients of pH, oxygen, nutrients, and waste products. Thus, new reactor methodologies may be needed before this technology is commonly employed by industry.

2.4.1. *Serum-Free Media*

A number of laboratories are working to develop low-protein or protein-free defined media for hybridoma cultivation (C6, C7, J2). Where large amounts of monoclonal antibody are required for clinical or therapeutic purposes, inclusion of serum for production has several disadvantages. The growth-promoting components in serum are not well defined and reproducible lots are often not available in the quantities needed for large-scale production endeavors. Serum may also be contaminated by adventitious viruses such as bovine diarrhea virus, or may contain factors which inhibit antibody secretion (G7). Use of serum can also complicate a purification process by adding contaminants that are difficult to remove (such as bovine IgG) and unwanted proteases that compromise product quality. Thus, defined media formulations for hybridoma growth represent an important alternative for commercial production.

Serum-free media consists of a basal salt solution, such as a mixture of Ham's F-12 and Iscove's modified Dulbecco's medium, containing carbohydrates, amino acids, vitamins, and minerals along with various growth supplements (C6). Common protein additives that may be employed are the iron carrier protein transferrin, the hormone insulin, and bovine serum albumin (C4). Others additives that have been investigated include hormones such as 3,3,5-triiodo-L-thyronine, ethanolamine, β-estradiol, hydrocortisone, and α-thioglycerol, along with certain trace elements such as selenium (M10). In the future it may be possible to support hybridoma growth in chemically defined media without any added proteins. Cole *et al.* recently described a human hybridoma cell line that grows in a culture medium completely free of added proteins and supplemented only with ethanolamine (C7).

Several companies offer commercial serum-free media formulations that are also pyrogen free for monoclonal antibody production. Although a number of excellent serum-free media formulations are on the market, a generic serum-free formulation is not yet available. Hybridoma lines originating in serum-containing media must be adapted for serum-free growth. Unfortunately, adaptation is relatively time consuming. A better alternative is to prepare a myeloma cell bank preadapted to serum-free conditions and then to generate, clone, and select hybridoma cells in serum-free media, eliminating the requirement for further adaptation. Also, commercially available serum-free media formulations are currently more expensive than serum-supplemented media due to the inclusion of relatively rare growth supplements (M10), such as relaxin or prolactin. This cost can generally be justified if high-purity monoclonal antibody preparations are required.

3. Monoclonal Antibody Purification

Many monoclonal antibody applications require purified protein, and reliable purification techniques are needed for both research and clinical purposes. Monoclonal antibodies employed for *in vivo* imaging, passive immunization, and other therapeutic purposes have to be carefully prepared utilizing safe and reproducible recovery procedures. Patients must be protected from the danger of adventitious viral or microbial pathogens, and from the potential untoward effects of immunization with irrelevant proteins. Fortunately, excellent antibody purification and characterization techniques are available today to provide such protection. Government regulatory and safety considerations for therapeutic monoclonal antibody products will be discussed in this section, along with several purification methodologies.

3.1. COMMON PURIFICATION TECHNIQUES

A number of techniques have been established for the purification of murine or human monoclonal antibodies from cell harvest or ascites fluids, as listed in Table 1. Human and mouse monoclonal antibodies can be purified using basically equivalent techniques. For research applications,

TABLE 1
COMMON MONOCLONAL ANTIBODY PURIFICATION METHODS

Category	Method	Reference
Precipitation	Ammonium or sodium sulfate	H4
	Polyethylene glycol	C16
	Caprylic acid	M4
	Caprylic acid + ammonium sulfate	R4
	Rivanol	F9
Ion-exchange chromatography		
Anion	DEAE	F1, G4, G9, J3
	QAE	B9
Cation	Sulfonic acid	C2
Size-exclusion chromatography	Sepharose	F1
	Superose	P7
	Sephacryl	P7
	TSK-4000	P6
Other chromatography methods	DEAE Affi-gel blue	B13
	Hydroxylapatite	J7, S20
	Baker ABx resin	N2
Affinity methods	Protein A resins	E9, G8, L12, P8
	Protein G resins	A4, B9
	Immobilized anti-antibodies	B4, P10
	Immobilized antigen	W3

simple enrichment by precipitating agents is often sufficient. Commonly described precipitation steps include salting out with sodium or ammonium sulfate (H4) and precipitation with Rivanol (2-ethoxy-6,9-diaminoaridine lactate) (F9), polyethylene glycol (C16), and caprylic acid (M4). Precipitation methods are often employed when large numbers of monoclonal antibody samples need to be characterized and moderate purity is sufficient. Reik *et al.* have recently described a two-step precipitation method for the purification of monoclonal antibodies from mouse ascites fluid (R4). In the first step, caprylic acid is used to precipitate albumin and other non-IgG proteins followed by ammonium sulfate precipitation. Relatively high-purity monoclonal antibodies result. It is important to note that some monoclonal antibodies can be denatured by salt treatment (B14). Thus, salt precipitation should be judiciously practiced.

When the application requires greater purity, chromatographic separations utilizing cation or anion exchangers (B15, C2, F1, G4, G9, J3), hydroxylapatite (S20), size exclusion (F1), Affi-gel blue (B13), or affinity supports are commonly employed (Table 1). With anion-exchange methods, antibody and acidic contaminants are bound between pH 7.0–9.0. Monoclonal antibody culture fluid is often concentrated and adjusted to low ionic strength before chromatography. Purification from more tightly bound acidic contaminants is accomplished by salt or pH gradient elution. Cation-exchange resins (e.g., sulfonic acid-substituted resins) may also be used to purify monoclonal antibodies (C2). An advantage of utilizing cation-exchange resins is that the monoclonal antibody can be concentrated and purified directly from hybridoma culture fluids after pH adjustment (4.0–6.0). Acidic protein contaminants (including albumin) flow unbound through the column. Monoclonal antibodies are then eluted by salt or pH gradient elution. In this way, purification and concentration can be performed as a single operation. Albumin and transferrin commonly contaminate antibodies purified by ion-exchange methods, and can be removed by size-exclusion chromatography (J3, P7). Size-exclusion chromatography is also often utilized also for the purification of IgM class monoclonal antibodies (J3, P7), because of the unusually high molecular weight of IgM (>900,000 Da).

A number of high-performance and fast-protein liquid chromatography (HPLC- and FPLC-configured) purification protocols have been proposed for monoclonal antibody purification (Table 1). Several mechanically stable bonded phases have had demonstrated utility, including ion exchange (B15, G4), hydroxylapatite (J7, S20), size exclusion (P6, S2), affinity methods (B4, P10), and mixed ion-exchange bonded-phase ABx proprietary supports (N2). Currently, most monoclonal antibody purification work performed by HPLC is at an analytical scale. High-performance techniques can be employed to purify and characterize small quantities of monoclonals within several minutes. There is considerable interest in configuring high-performance techniques for the production of larger amounts of antibody for clinical or therapeutic applications. With appropriately sized HPLC columns, pumps, and automated operation, kilogram quantities of antibody could be rapidly purified.

3.1.1. Affinity Purification

Affinity purification (Table 1) of monoclonal antibodies has several advantages relative to other purification techniques. Affinity chromatography steps are easier to implement, automate, and scale-up. Large quantities of antibody can be purified in a single step employing simple column loading, washing, and elution procedures. High-purity (>98%) monoclonal antibody preparations can be reproducibly obtained from cell culture, ascites fluid, or other complex feedstocks. Affinity purification procedures are relatively generic. For example, immobilized protein A can be employed to purify hundreds of different human or mouse monoclonal antibodies with very few procedural changes. Ion exchange, on the other hand, would require optimizing conditions each time a monoclonal antibody is purified.

Protein A is a *Staphyloccocus aureus*-derived membrane protein that selectively binds a site between the C2 and C3 constant region H chain domains of antibody molecules from many species. If a particular monoclonal antibody is of a class and species bound by protein A, protein A chromatography is an excellent purification method (E9, G8, P8). Immobilized protein A resins are commercially available (Pharmacia, BioRad, Fermentech, Repligen, and others) and have a high capacity for many human and mouse monoclonal antibodies. Monoclonal antibodies are usually desorbed under relatively gentle conditions (pH 3.0–6.0), depending on the species and subclass purified. The mouse IgG_1 subclass, and IgA, IgM, and IgE do not bind well to protein A (G8). Mouse IgG_1 represents over 80% of the antibodies in mouse serum, and is the most common mouse monoclonal antibody subclass. The low affinity of protein A for mouse IgG_1 has limited its application for many mouse monoclonal antibodies. With recent advances in protein A chromatography, this limitation has been virtually eliminated (P8). Chromatography buffers with high pH and ionic strength (1.5 mol/liter glycine, 3.0 mol/liter NaCl, pH 8.9) dramatically increase the capacity of protein A resins for the IgG_1 subclass. It is important to note that protein G, another membrane-derived bacterial protein that binds selectively to antibody Fc regions, is currently being investigated and may provide an alternative approach. Protein G exhibits a broader specificity for antibody classes and subclasses for immunoglobulins from many species. This fact is being exploited for the purification of less common antibody classes (A4, B9) and eliminates the need for a high salt high pH buffer.

Protein A-based purification techniques may be superior to a number of other chromatography techniques employed for the manufacture of monoclonal antibodies. As an affinity technique, loading and washing steps can be performed at relatively high ionic strength, so DNA contamination is minimized, and no salt gradient is required. Protein A resins are also stable to cycles of reuse without significant losses in binding capacity or selectivity. It has been reported that protein A–Sepharose can be reused over 200 times without noticeable loss in binding capacity (L12). Thus, the initial high cost of a protein A resin ($5000 to 10,000 per liter) is relatively easy to justify in terms of antibody production cost after repeated cycles of use. The protein A molecule is also stable to exposure to high temperatures, low pH, and denaturing agents (e.g., guanidine–HCl) (L12), which makes protein A resins easily amenable to sterilization prior to or during manufacturing.

It is important to keep in mind that monoclonal antibodies purified with protein A or G may be contaminated with trace levels of bovine antibodies from cell culture harvest fluids or with nonspecific endogenous antibodies present in ascites fluids. Antimonoclonal antibody approaches can minimize contamination by heterologous antibodies, since the binding specificity is controlled by the specificity of antimonoclonal antibodies employed for purification (P10). For example, Bazin *et al.* described a procedure to purify ascites fluid-derived LOU rat monoclonal antibodies against a dinitrophenyl (DNP) hapten employing immobilized κ chain, allotype-specific, monoclonal antibodies (B4). The monoclonal antibody was selected for desorption under mild elution conditions to preserve anti-DNP binding activity. DNP-specific IgG_1, IgG_{2a}, and IgG_{2c} antibodies were purified from either serum or ascites utilizing the same allotype-specific antibody and employing virtually the same immunoaffinity protocol. Other anti-antibody approaches are also commonly employed. For example, subclass-specific antibodies can be prepared and immobilized so that contamination from heterologous antibody subclasses is minimized. Chromatography employing immobilized antiidiotypic antibodies carrying an "internal image" of the antigen might offer an even higher degree of specificity, based on the fact that such antibodies would be binding-site directed (J5). Antimonoclonal antibody affinity chromatography can eliminate contamination from endogenous ascites antibodies or bovine antibodies present in cell culture media. Unfortunately, antimonoclonal antibody reagents, particularly antiidiotypic antibodies, are relatively difficult and time consuming to prepare, isolate, and immobilize. Despite these difficulties, anti-antibody methods are often employed for research purposes.

Antigen-specific affinity chromatography is potentially the best way to purify monoclonal antibodies. Because binding occurs based on antigen specificity, heterologous antibody contaminants from the same or different species would not be expected. However, antigen-specific techniques have several common limitations. For rare or novel proteins there is usually not enough protein available to prepare a suitable antigen-specific matrix. Further, many proteins lose the ability to bind antibody upon immobilization, or are unstable even to relatively mild elution conditions. Some antigens bind antibody with such high affinity that dissociation can only be accomplished with denaturing solvents such as 3.5 mol/liter thiocyanate, 8 mol/liter urea, 6 mol/liter guanidine HCl, or 1 mol/liter propionic acid, which often leads to loss of antibody-binding potency. For these reasons antigen-based affinity methods, although powerful, are often difficult to implement (W3).

3.2. Monoclonal Antibodies for Medical Applications

In the United States, therapeutically administered monoclonal antibody products and *in vitro* diagnostic products used to test blood are regulated by the Federal Drug Administration (FDA), by either the Center for Drugs and Biologics or the Office of Biologics Research and Review. Other *in vitro* diagnostic products are regulated by the FDA Center for Devices and Radiologic Health under the Medical Device Amendment of the Food, Drug, and Cosmetics Act. The mission of these agencies is to protect public health by establishing requirements for licensing medical products. For therapeutic monoclonal antibody products, licensing proceeds in stages. First, permission is requested to test unlicensed biological products in a small number of human volunteers in clinical trials by filing of a Notice of Claimed Investigational Exemption for a New Drug (IND). Permission is granted based on a review of data detailing the potential benefit of a product for treating disease and relevant toxicology and scientific data that establish the relative risk in therapy. Obtaining regulatory approval requires clinical trials and is expensive and time consuming. Most monoclonal antibody biologicals administered in clinical trials to date have been of mouse origin. Only two therapeutic monoclonal antibody products have been licensed. Both products are monoclonal antibodies of mouse origin (see Section 4.3). The development of a new drug generally proceeds in stages, culminating after a period of 6 to 12 yrs with a license to manufacture a product for sale. At each stage the product's safety and efficacy is defined more stringently and larger scale clinical trials are performed.

3.2.1. Regulatory and Manufacturing Considerations

Monoclonal antibodies are complex molecules and it can be difficult to establish reproducibly efficacious preparations without rigorous standardization of manufacturing procedures. Schauf recently summarized basic government procedures and guidelines for licensure of therapeutic monoclonal antibody products and reviewed regulatory and safety considerations as well (S4). Documents which point out major concerns with respect to production, purity, and safety of monoclonal products have been published by the FDA and are available for review (F4, F7). Also, the Code of Federal Regulations (21 CFR Parts 200–299 and 600–680) deals with government guidelines for the production of biologicals and are broadly relevant.

Monoclonal antibody products cannot be administered safely, or prepared reproducibly without the development of standard operating procedures for production, purification, and characterization. Typically a manufacturer's working cell bank is prepared containing cells with a well-documented history, free of viruses, and of a defined passage number. Tests are routinely performed on antibody lots and at various stages of the production process to demonstrate freedom from viruses. Consistent molecular characteristics must also be established. Qualities like antibody stability, specificity, purity, sterility, and molecular integrity are usually carefully assessed. SDS–polyacrylamide gel electrophoresis, light scattering, protein concentration, isoelectric-focusing gels, and HPLC analysis may be employed to evaluate physical parameters likely to alter the therapeutic value of the administered monoclonal antibodies. Common issues of concern include cross-reactivity with irrelevant human tissues, antibody aggregation, denaturation, fragmentation, and charge heterogeneity—factors which might alter the molecular binding potency or toxicity after administration. Other issues of concern include removal of residual components such as protein A, bovine IgG, pristane, and penicillin. Monoclonal antibody potency is carefully monitored. It may be sufficient to measure potency by performing a simple binding assay. However, if animal models exist, an *in vivo* estimate of antibody potency is preferred. Final product vials must also be tested for sterility, general safety, and pyrogenicity. The product must be consistent with respect to molecular and pharmacological properties, most often based on a comparison to an acceptable reference standard.

Monoclonal antibodies employed for *in vitro* diagnostic applications or for the purpose of purifying other pharmaceutical products by immunoaffinity chromatography require similar, although less stringent, standard procedures for production and purification. In the case of *in vitro* diagnostic assays, shelf life and reproducibility are key issues. If monoclonal antibodies are used for immunoaffinity purification, the implications of employing monoclonal antibodies in the purification process are carefully considered. Procedures must be in place to guarantee that the affinity-purified product is made reproducibly and that product safety is not compromised by contamination with mouse viruses, mouse DNA, monoclonal antibody, or other possible adventitious contaminants.

3.2.2. *Removal of Viral Contaminants*

Viral contamination is a major concern for therapeutically administered antibody products. The perceived risk of pathogenic viral contamination is greatest for human monoclonal antibody products. Proposed safety, efficacy, and reproducibility issues for monoclonal antibodies are similar to those for human immune serum-derived heterologous antibody products. Detailed studies have to be performed on hybridoma cell banks and purified monoclonal antibody to establish freedom from virus contamination. As reviewed by Carthew (C3), many rodent viruses are known to replicate in human and monkey cells in culture (Table 2), and certain viruses, such as Hantaan (D4), lymphocytic choriomeningitis (B8), and rat rotavirus-like agent (E2), have been shown to transmit disease from animal to man in infected rodent colonies. The perceived risk is even greater for human monoclonal antibody products because of the potential propagation of human viruses. Transmission of blood-borne viral diseases such as human parvovirus infection (T1, W2) are well known among human serum-derived products. Clearly, EBV-immortalized human B cell products must be carefully checked for active virus particles (C14). The possible contamination of monoclonal-producing human B cell lines by virus is an important product safety issue.

The safety risk with respect to viral contaminants must be established before monoclonal antibodies are approved for therapeutic use. Fortunately, a variety of excellent tests are available for the detection of endogenous mouse viruses, including the frankly pathogenic viruses. If murine cells are employed, detailed testing of murine cells for viruses, including polyoma viruses, reoviruses, lymphocytic choriomeningitis, and murine leukemia viruses, is essential. Recommended tests include mouse or rat antibody production (MAP or RAP) tests to identify murine virus contamination in manufacturers' banks or purified product. Fermenter derived monoclonal antibody feedstocks, serum-free or serum-containing media supplements, or ascites fluids must all be considered as a possible

TABLE 2
MOUSE ANTIBODY PRODUCTION (MAP) TEST: POTENTIAL PATHOGENICITY OF MOUSE VIRUSES[a]

MAP test viruses	Replication in human or monkey cell cultures[b]	Isolated from primate tissues or the cause of disease in humans (H) or monkeys (M)[b]
Hepatitis virus	No (P3)	—
Adenovirus	Yes (S11)	—
Pneumonia virus	Yes (B6)	—
Ectromelia	Yes (P12)	—
Polyoma	Yes (D9)	—
Sendai	Yes (N7)	Yes, M, disease (J4)
Minute virus	No (C15)	
Reovirus-3	Yes (B5)	Yes, H, tissues (R9)
Pneumonitis (K virus)	No (R10)	
Lymphocytic choriomeningitis		Yes, H, disease (B8)
Encephalomyelitis	Yes (H6)	Yes, H, disease (D7)

[a]Based on replication in human or monkey cells or on rodent viruses known to cause primate disease. Table and examples adapted from a review by Carthew (C3), with permission.
[b]References are given in parentheses.

source of viral contamination. Examples of viruses detected in a MAP test are summarized in Table 2. A number of companies provide a battery of sensitive tests to establish the safety of therapeutically administered antibody products with respect to endogenous murine viruses or other pathogens.

There is no guarantee that a particular assay will be sensitive enough to detect certain viruses. Furthermore, some pathogenic viruses are undoubtedly yet to be discovered. Thus, viral assays are helpful but do not guarantee safety with respect to endogenous viruses. Since freedom from viral contamination can not be assured by viral assays alone, the process employed for antibody purification may be designed to provide a further margin of safety. Steps are included to inactivate viruses by physical or chemical treatments. Most rodent viruses are inactivated by a 1- to 2-h heat treatment at 50-60°C (A7, C3). However, certain rodent viruses such as encephalomyocarditis virus and Kilham rat virus are known to be resistant to heat inactivation and survive even after a 2 hour treatment at 80°C (A7). Chemical treatments such as pH extremes, detergents, urea, or chaotropic agents may also be evaluated (A8). Unfortunately, it can be difficult to find a chemical or physical treatment that inactivates a specific virus without compromising antibody integrity.

It is also important to note that chromatography steps employed in a purification process are likely to be quite protective. Affinity chromatographic procedures (protein A chromatography, for example) have been shown to provide over log six clearance with respect to type C retroviruses (L8) and are also effective for the clearance of DNA and RNA contaminants. Retroviruses are of particular concern with respect to murine hybridoma cells, since Bartal *et al.* have reported that hybridoma cells carry C-oncorna and A-oncorna retroviruses (B1). Similar observations have been made with mouse plasmacytoma cells (L7). To establish that a particular immunoglobulin preparation will be free of contamination, it may be necessary to demonstrate the removal of viruses in "spiking" experiments. In such experiments, known viruses are added at each stage of the purification process and clearance is then assessed by appropriate viral assays. Spiking experiments can be used to validate physical or chemical inactivation steps or the clearance of viruses at different stages of a purification process. Spiking experiments are routinely employed in manufacture of biological products (C14, F5, L8) to provide further evidence for safety.

3.2.3. *Heterogeneity*

Each monoclonal antibody has a characteristic fingerprint of isoelectric bands (referred to as a spectrotype) that characterizes the molecule with respect to charge heterogeneity. The spectrotype for several different IgG_1 mouse monoclonal antibodies specific for recombinant tissue plasminogen activator are shown in Fig. 4 (lanes A–H, left panel). Each of the monoclonal antibodies shown focuses as 3–10 closely spaced isoelectric bands. Monclonal antibody heterogeneity often results from posttranslational modifications (mechanisms are described below). The large pI shifts shown for the different tissue plasminogen activator (tPA)-specific monoclonal antibodies are likely due to carbohydrate alterations or variable region differences, since each of the monoclonal antibodies is from the same subclass.

The spectrotype of a particular monoclonal antibody may also change depending on cultivation methods. For example, the purified monoclonal antibody (5B6) preparations shown in Fig. 4 (lanes I and J) exhibit different spectrotypes when produced from ascites or serum-free media; however, both (5B6) preparations are indistinguishable based on SDS–polyacrylamide gel analysis (lanes A and B, right panel). Heterogeneity with respect to molecular weight has been observed recently by Manil *et al.*, who have characterized monoclonal antibody molecules by two-dimensional gel electrophoresis and observed light chain molecular-weight differences (M1). Changes in antibody specificity may also occur as a result of changing cultivation methods. For example, Underwood and Bean isolated mouse

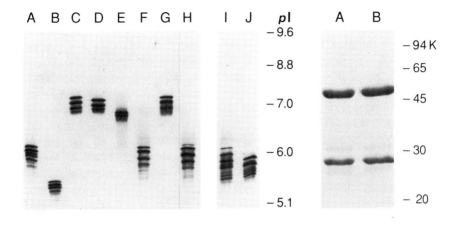

FIG. 4. Isoelectric-focusing gels illustrating heterogeneity of purified mouse IgG$_1$ monoclonal antibodies recognizing recombinant tissue plasminogen activator (left panel, lanes A–H). Each monoclonal antibody molecule has a particular spectrotype or fingerprint of isoelectric bands due to posttranslational modifications. Major pI shifts shown among the IgG$_1$ monoclonal antibodies result from carbohydrate or V region differences. The isoelectric-focusing (IEF) spectrotype may also vary with production methods. Different IEF spectrotypes are obtained from hybridoma cell line (5B6) cultivated in serum-free cell culture (lane I) or as ascites fluid (lane J). The right panel shows that the ascites (lane A) and serum-free cell culture-produced (5B6) antibodies (lane B) were not distinguishable by SDS–polyacrylamide gel electrophoretic analysis.

hybridoma cells for 26 different influenza subtypes (U1) and found that passage in ascites alters specificity with respect to several of the subtypes. They also found changes in specificity dependent on storage methods. Standardized methods of monoclonal antibody production, purification, formulation, and storage are therefore important, particularly for development of a manufacturing process.

The reasons for the significant heterogeneity observed for monoclonal antibodies in response to variations in growth conditions are not clear. Possible explanations are (1) posttranslational modification, including changes in carbohydrate composition or proteolysis (B13, P5); (2) selection of an unexpressed genotype from a mixed population not sufficiently cloned (C8); and (3) genome alterations involving point, frameshift, deletion and somatic mutation mechanisms (A9, C9), as well as class (K4) or subclass (B3) switching. It is important to mention that heterogeneity may also result through postcultivation mechanisms, such as deamination during purification and storage or carbamylation during IEF analysis using urea (P5). During scale-up of hybridoma cultivation, the potential for genetic change is high. Immortalized antibody-producing cell lines are predisposed to genetic changes. Isotype switching (among IgG_1, IgG_{2a}, IgG_{2b}, and IgG_3 molecules) is relatively frequent in hybridoma lines and occurs at a rate of approximately 10^{-5} per generation. And although the mechanism of somatic hypermutation is still not understood, codon changes in the hypervariable regions of V region genes also accumulate at a fairly high rate (approximately $0.3-1.0 \times 10^{-4}$ point mutations per codon per generation) (W1). This is at least 10,000 times more frequent than the mutation rate estimated for the mouse genome (W1).

Heterogeneity with respect to carbohydrates is another parameter which should be considered for the development of effective monoclonal antibody products. Inefficient glycosylation or the variant structure of attached oligosaccharides may be therapeutically important. Nose and Wigzell studied the effect of carbohydrate alterations on mouse IgG_{2a} molecules (N8) and showed that removal of asparagine-linked carbohydrate inhibits antibody clearance from circulation, antigen-dependent cellular cytotoxicity, Fc receptor binding on macrophages, and complement activation. Human IgG molecules contain at least 16 different complex-type asparagine-linked carbohydrate chains (H3). Since human monoclonal antibodies have species-specific carbohydrates, the structure or distribution on monoclonal antibody molecules is likely to change with different methods of B cell immortalization. The mouse component of heterohybrid cells, for example, is likely to express a unique set of glycosyltransferases (M9) that would alter the carbohydrate structure of heterohybrid-produced human antibodies. It remains to be determined, however, whether such carbohydrate differences occur and whether these differences will alter the therapeutic value of administered monoclonal antibody products.

Aggregation can also compromise the safety of antibody preparations administered therapeutically (O1). Human-derived serum antibodies have been used therapeutically for many years, and the safety of intravenous heterologous immunoglobulin preparations has been thoroughly studied. The degree of aggregation of immunoglobulin preparations administered therapeutically has been correlated closely with the onset of uncomfortable and sometimes dangerous side effects, including chills, nausea, blushing, abdominal cramps, wheezing, and local pain upon infusion (I1). Considerable evidence suggests that these side effects are predominantly caused by complement-mediated inflammatory responses (R7) through activation of the properdin pathway. The properdin pathway is activated by antibody aggregates and is Fc dependent. Inflammatory side effects can be minimized or eliminated by appropriate procedures, including (1) removal of antibody Fc regions by proteolytic digestion (K8); (2) chemical inactivation of Fc binding sites by reduction followed by alkylation (S6); and (3) appropriate attention to formulation buffers. Glycine buffers containing 5 to 10% maltose in glycine buffer, for example, can substantially reduce aggregation and side effects after intravenous injection (O2). Aggregation of human or mouse monoclonal antibody preparations prior to administration is an important concern and should be carefully considered when developing purification and formulation protocols.

4. Monoclonal Antibody Technology

Monoclonal antibodies are beginning to change the way diseases are diagnosed, monitored, and treated. It has been estimated that the market for monoclonal products was $155 million in 1985, mostly due to the development of new diagnostic assay products. Assay kits are being configured in new and imaginative ways, and many former polyclonal antibody-based assays are being replaced with superior monoclonal antibody-based products. The market for monoclonal antibody products is expected to exceed $4.0 billion in the 1990s. Over half of the projected increase relates to the development of *in vivo* therapeutic and imaging products (D8). A number of therapeutic and *in vivo* diagnostic assay products are under development. These include products for the treatment of cancer, imaging of tumors and blood clots, passive immunization against viral and bacterial pathogens, protection against toxins and poisons, and treatment of allergic disorders and organ transplant rejection. Although great progress is being made, commercialization of administered monoclonal antibody products is slow, due in part to clear technological and scientific limitations and also to the great cost and time required to license safe and effective therapeutic products.

The power of monoclonal antibodies to purify, quantitate, characterize, and identify is contributing to the development of new recombinant protein products for biotechnology, and is also stimulating the advance of biological research. The use of monoclonal antibodies has increased scientific knowledge of the immune system, hormone receptors, histocompatibility antigens, T cell differentiation factors, virus pathogenicity, tumor markers, enzymes, and a number of basic biological processes. A number of recent reviews have emphasized the important contribution of monoclonal antibody reagents in veterinary and human medicine and in biological research (E7, N3, V6). Examples of a number of applications are summarized in Table 3. This section discusses several key diagnostic and therapeutic uses and illustrates some limits of the current technology. Several new ways to utilize and generate monoclonal antibodies are also described, with a look toward future developments.

TABLE 3
CURRENT AND PROSPECTIVE MONOCLONAL ANTIBODY PRODUCTS/APPLICATIONS

Product/application	Example(s)
Diagnostics	
In vitro	See Table 4
In vivo	Imaging glioma (P9), bronchial (S15), and other tumors; locating blood clots, infarction, abscesses, parasites, etc.
Therapy	
Cancer treatment	Lymphatic cancer (M5, M7), melanoma (H11), gastrointestinal (S9)—direct tumor killing or targeting of antitumor drugs or cytotoxic T cells against tumor antigens
Passive immunization	For viral, bacterial, and parasitic diseases
Immunization with idiotypic antibodies	Vaccines (D10), or against tumor antigens as a strategy in cancer therapy (K10)
Drug, poison, and toxin removal	Mouse IgG product administered as antidote for digitoxin and digoxin poisoning; for endotoxin-induced septic shock (B10, L3); snake antivenom (L3)
Inhibition of graft rejection	Mouse IgG Mab to CD-3 cluster determinant on immune T cells—marketed product to prevent renal graft rejection (T3)
Control of allergic reactions	Inhibition of IgE degranulation of mast cells; prevention of autoimmune diseases (L2)
Fertility modification	Selection of X or Y spermatocytes; hormone regulation to improve fertility or for birth control (L2)
Pharmaceutical purification	Manufacture of immune interferon (S18), hepatitis B (M3), interleukin-2 (K6)

4.1 *In Vitro* DIAGNOSTIC ASSAYS

Many new assays configured using monoclonal antibodies are being developed for research and medical purposes and applied for the detection of a broad range of biomolecules. Common assay configurations include (1) assays where antibodies are attached to particles which agglutinate in the presence of antigen; (2) direct attachment of antibody (covalently tagged with fluorescent dye, radioactively labeled, or enzyme conjugated) to antigen where antigen–antibody reaction is directly visualized in tissue smears, on cells, or on proteins employing Western blot methodologies (T5); (3) competition assays using tagged antigen; and (4) ELISA (B18) methodologies; for example, indirect ELISA methods where bound monoclonal antibody is immobilized to capture soluble antigen, then bound antigen is detected using a second nonhomologous antibody that is enzyme conjugated for signal development.

When a group of animals is immunized, typically only a few will produce the quality of antisera desired for assay purposes. When the immune system encounters antigen, a heterologous antibody response develops and a variety of antigenic sites are recognized. Antibodies of differing isotypes, affinities, and specificities are produced. Antibodies to trace contaminants in a protein preparation are also common. Immune sera therefore represent a complex mixture of antibodies that can lead to unwanted cross-reactivity, making reproducible polyclonal assay reagents quite difficult to generate. Monoclonal antibodies circumvent many of these problems. Several monoclonal antibody-based diagnostic assays have been commercialized or are in advanced stages of development (Table 4). These include tests (1) to predict pregnancy or ovulation and protect the newborn from anti-

TABLE 4
MONOCLONAL ANTIBODIES FOR in Vitro DIAGNOSTIC APPLICATIONS[a]

Product category	Company
Fertility	
Ovulation kit	Monoclonal Antibodies, Mountain View, CA
Pregnancy kit	Ortho Pharmaceutical Corporation, Raritan, NJ
Anti-Rh factor kit	Immunotech SA, Marseilles, France
Sexually transmitted diseases	
Herpes virus kit	Genetic Systems, Seattle, WA
Chlamydia kit	Genetic Systems, Seattle, WA
Blood screening	
Hepatitis non-A, non-B kit	Centocor, Philadelphia, PA
AIDS virus kit	Centocor, Philadelphia, PA
Cancer detection and monitoring	
Gastrointestinal cancer kit	Abbott Laboratories, North Chicago, IL
Prostate cancer kit	Hybritech, San Diego, CA
Leukemia phenotyping kit	Becton Dickinson, Mountain View, CA
Pancreatic carcinoma kit	Biogenex Laboratories, Dublin, CA
Drug and hormone monitoring	
Digoxin Mab	Chemicon International Inc., Segundo, CA
Parathyroid hormone Mab	Meloy Life Sciences, Springfield, VA
Theophylline Mab	Ventrex Laboratories, Portland, ME
Tissue or cell typing	
HLA antigen kit	Biogenex, Dublin, CA
OKT4/OKT8 ratio kit	Becton Dickinson, Mountain View, CA

[a] Adapted with permission from an article originally published by Tami *et al.* (T3).

Rh hemolytic disease; (2) to diagnose sexually transmitted diseases, including *Chlamydia* bacteria and herpes virus infections; (3) to protect the blood supply from AIDS and hepatitis virus contamination; and (4) for early detection and monitoring of cancer, utilizing monoclonal antibody-specific tumor antigens such as carcinoembryonic antigen or prostatic acid phosphatase; and (5) to detect or monitor for drugs or other small molecules. Monoclonal antibody-configured assays are typically more sensitive and rapid, and often give a very low incidence of false positives compared to when heterologous antibodies are employed. In addition, an unlimited and reproducible supply of assay reagents is virtually guaranteed. It has been pointed out that home pregnancy tests marketed by Warner–Lambert Company and Ortho Pharmaceutical Corporation take only 30 min to develop and are 99% accurate (T3). Pregnancy assessment is based on measurement of human chorionic gonadotropin (hCG) levels in urine. Previous pregnancy tests based on polyclonal reagents were slow in comparison, and cross-reactivity led to a higher incidence of false positives. Centocor's blood-screening assay for hepatitis is two to four times more sensitive than previous assays utilizing polyclonal sera. Centocor is also marketing a monoclonal kit that has been described as 98.5% accurate for the early detection of stomach, pancreas, liver, and colorectal cancers using a test specific for gastrointestinal tumor antigens (V1). Abbott Laboratories has developed a diagnostic kit utilizing a monoclonal antibody to carcinoembryonic antigen (CEA). High CEA levels are common in patients with lung, ovary, pancreas, and gastrointestinal tract tumors, and the assay is employed to monitor regression and prognosis of the disease (C5). Other effective monoclonal antibody tests include assays for histocompatability antigens for organ transplantation and assays to monitor T cell phenotype. OKT4 and OKT8 T cell monoclonal antibody reagents have been very important for the diagnosis of AIDS (D6). New assays are also used to assess clearance of drugs or poisons. For example, tests monitoring digoxin, phenobarbital, and theophylline based on monoclonal antibodies are available (R5).

It should be mentioned that monoclonal antibodies can be too specific. Since monoclonal antibodies bind only to one epitope, they can be less reliable than heterologous antibodies for the quantitation of complex antigens subject to genetic drift. For example, hog cholera virus and bovine viral diarrhea strains have been missed in assays employing monoclonal reagents generated to viral subtypes (V4). Such issues are also of particular concern for current diagnostic assays for the AIDS virus. Since multiple epitopes are recognized, polyclonal antibody-based assays may sometimes be superior for detection of rapidly changing pathogens or subtypes, although monoclonal antibody cocktails may also be employed to account for genetic drift.

4.2 Immunoaffinity Purification

Monoclonal antibody affinity techniques are being broadly applied for research purposes and in pharmaceutical development, and are part of the arsenal of purification techniques employed for high-value recombinant proteins in biotechnology. In the late 1970s, polyclonal rabbit antibodies began to be used for affinity purification of proteins at a micropreparative or an analytical scale. The availability of monoclonal antibody reagents has expanded the use of immunoaffinity techniques in manufacturing programs and research labs. Many natural proteins and enzymes have been isolated employing monoclonal immunoaffinity methods, including *Bordetella pertussis* proteins for vaccine development (L4), polio and other vaccines in manufacturing operations (V3), plasminogen activators for characterizing the enzymes present in human urine (K1), oxidases for studies of monoamine oxidase isoenzymes in the liver (D3), phosphodiesterases for characterization of the enzyme from bovine heart (H2), and insulin-like growth factor I from human serum for studies of its role as a growth factor (L5).

The use of immunoaffinity steps for the purification of high-value recombinant proteins by bioindustry is also becoming more common. Immunoaffinity methods have several advantages over classical protein purification approaches (such as protein precipitation or ion exchange, and size-exclusion chromatography). Immunoaffinity methods are simple to implement, rapid, and easy to automate (A6). Proteins can be isolated from complex mixtures in a single-process operation with high yield and purity. Conventional approaches require multiple purification operations to obtain comparable purity, and this can often result in poor product yield. The manufacturing process for *Escherichia coli*-derived recombinant immune interferon, for example, employs an immunoaffinity operation that increases purity over 800-fold in a single step with greater than 90% recovery of activity (S18). Other examples of successfully purified recombinant proteins include hepatitis B surface antigen from yeast (M3), bacteria-derived transforming growth factor α (W5), interleukin-2 (K6), and bovine somatotropin (K13).

It is important to point out that monoclonal antibodies used for preparation of an immunoaffinity column for a pharmaceutical process must be carefully selected and that the cost of producing an immunoabsorbent can be rather high. In addition to regulatory issues (discussed in Section

3.2.1), several factors are commonly considered. First, an appropriate monoclonal antibody must be identified. Considerable effort may be required to generate and select a suitable antibody-producing cell line. The monoclonal antibody must be stable to process solvents, proteases in product feed streams, and to the chemical coupling agents used for immobilization. The monoclonal antibody must also release product under relatively mild desorption conditions that maintain product potency and avoid denaturation. Second, the purification operation must be configured to circumvent the tendency of coupled monoclonal antibody to degrade or denature in process solvents. Finally, the binding capacity of the column and the column reuse capability must make sense in terms of overall process economics.

4.3. Mouse Monoclonal Antibody Therapeutic Products

Several therapeutic roles have been proposed for monoclonal antibody products (Table 3). These include passive administration for viral and bacterial diseases, cancer therapy, treatment of common allergies, autoimmune diseases, control of graft transplant rejection, cardiac and tumor imaging, receptor modulation, and human antiidotypic vaccines. Clinical trials are underway to evaluate products in many of these areas (D8, L2, T3). To date, two mouse-derived monoclonal antibody products have been licensed by the FDA for therapy. One, from Wellcome Biotechnology, Ltd. (Kent, United Kingdom), is employed for the clearance of the heart drugs digoxin and digitoxin. The antibody is administered in many hospital emergency rooms as an antidote for lethal effects of digoxin and digitoxin poisoning. Small children are the most common victims. The other product is from Ortho Pharmaceutical Corporation (Raritan, New Jersey) and is an antibody directed against human cell surface antigen CD-3 (a lymphocyte cell differentiation antigen), which is administered for treatment of human renal allograft rejection. The antibody therapy serves as an adjunct to immunosuppressive therapy and is effective in reversing acute renal graft rejections where steroids or therapeutic human polyclonal antibody preparations have failed.

4.3.1. *Cancer Therapy*

One of the most important potential applications of monoclonal antibodies is in cancer therapy. Over half of the $4 billion projected market for monoclonal antibody products in the 1990s is based on the assumption that monoclonal antibodies will become key products for cancer treatment and imaging. Unfortunately, except for a few remarkable cases, patients treated with mouse monoclonal antibodies for T cell lymphomas (M7), B cell leukemia (M5), gastrointestinal cancers (S9), and melanoma (H11) show no significant decrease in tumor burden. As has been reviewed by Smith and Teng (S17), the general lack of success has been attributed to several factors:

1. The tumor burden may be too high, or in the case of solid tumors, antibody penetration may not be possible.
2. Genetic drift may lead to changes in tumor antigens.
3. Antigen shedding may block available antibody sites, shielding the tumor.
4. Antibodies employed for therapy may not be specific enough, have high enough affinity, or be of the correct subtype.
5. The injected antibody may be cleared too fast or the dosage may not be sufficient.
6. The patient may develop a human antimouse antibody or suffer from some fundamental immunodeficiency.

Results of cancer therapeutic trials employing mouse monoclonal antibodies raise serious questions about their therapeutic value. Innovations are likely to be necessary. New approaches may involve treatments with antiidiotypic antibodies that would contain internal images of a specific tumor antigen. Antiidiotypic antibodies appear to have led to partial or even complete regression in patients treated for leukemias (K10). New approaches involving the targeting of toxins, drugs, or radionuclides to tumor cells employing monoclonal antibodies and utilizing bispecific antibodies to capture circulating chemotherapeutic drugs are also under investigation (see Section 4.5.2) and may be more successful therapeutic agents than mouse monoclonal antibodies administered directly.

4.4. HUMAN MONOCLONAL ANTIBODY PRODUCTS

Human serum-derived antibody has proved useful for the treatment of viral and bacterial infections in unimmunized, congenitally immunodeficient individuals, and also to treat normal patients with antibody levels temporarily depleted by systemic infection. Conditions treated include Rh (D) hemolytic disease of the newborn, viral diseases such as rubella, rabies, hepatitis A or B (also non-A, non-B), and varicella-zoster virus (C17). Other applications include treatment of gram-negative sepsis, use as a snake antivenom, reversal of drug overdose, and fertility control (L2, L3). Many serum-derived antibody products are likely to be replaced by superior human monoclonal antibody products as human monoclonal antibody technology develops. As described in Section 2.3, human B cell immortalization is a difficult and evolving technology. Activated human

B cells against specified antigens are difficult to obtain, immortalization techniques are still poorly defined, and immortalized human B cells often produce very low levels of antibody and are frequently unstable. Despite these limitations, human monoclonal antibodies have been generated to a large variety of antigens of interest for therapeutic and research purposes. These include human monoclonal antibodies to viral, bacterial, and parasite antigens, cell-specific or histocompatibilty antigens, tumor antigens, DNA, sperm, and cytoskeletal protein (Table 5). No human monoclonal antibody products have yet been approved for sale. A human IgG_1 monoclonal antibody from Celltech (United Kingdom) is being tested for treatment of anti-Rh disease to prevent hemolytic episodes in the newborn (L2). Another is being tested for imaging of glioma (P9) and bronchial (S15) cancers.

The interest in human monoclonal antibody development is expanding because of the importance of human monoclonal antibodies as therapeutic products and as research tools. Human monoclonal antibodies are perceived to have a number of advantages. One reason for this is that the human B cells are likely to respond to human antigens differently than mouse lymphocytes. This is because the human B cell repertoire is quite different. Mouse monoclonal antibodies, for example, seem to favor common HLA framework epitopes and have limited usefulness for tissue typing (self, nonself discrimination). Human monoclonal antibodies are expected to better discriminate among HLA histocompatibility antigens (E7, H12, P2). The therapeutic value of human monoclonal antibodies is also likely to be different than for mouse monoclonal antibodies. Human species-specific carbohydrate differences have been reported which effect cellular cytotoxicity, complement activation, and phagocytosis (L2, N6). Changes in any effector properties would be expected to limit the value of mouse-derived therapeutic products. A more serious problem is the immunogenicity of mouse monoclonal antibodies in humans. There is evidence that the effectiveness of mouse monoclonal antibodies administered for cancer therapy may be limited by immunogenicity. An anti-antibody response has been observed in patients treated for melanoma (H11), gastrointestinal cancer (S9), leukemia, and lymphoma (L9). Most of the antiglobulin response appears to be directed against H chain C2 and C3 constant regions (S7). Some of this antigenicity may be avoided by the development of chimeric human–mouse monoclonal antibodies (see Section 4.5.3) or by the utilization of human monoclonal antibodies for cancer treatment. In the case of human monoclonal antibodies, an idiotypic or allotypic immune response against administered antibody is expected, but should be well tolerated based on experience gained with human serum antibodies in therapeutic situations. In fact, human idotypic monoclonal antibodies carrying an "internal image" for viral, bacterial, or parasitic antigens are likely to be an important source of future vaccines (D10).

TABLE 5
HUMAN MONOCLONAL ANTIBODIES[a]

Example	Reference
Bacterial	
Tetanus toxoid	G5
Haemophilus influenzae (type B)	G6
Pseudomonas aeruginosa	S3
Chlamydia	R8
Viral	
Hepatitis B	B17
Rabies, rubella, hepatitis A	V2
Herpes VZV	S24
Parasite	
Plasmodium falciparum	S5
Cell-specific or histocompatibility antigens	
Neurons	S16
Islet cells	E3
Erythrocytes	S13
Rhesus D	C14
HLA	H12
Blood group A	R1
Tumor antigens	
Lung carincoma antigen	H8
Gastric and colorectal antigens	H7
Glioma	P9
Bronchial	S15
Melanoma	I2
Lymphoid cancers	A1
Other	
DNA (ss and ds)	W4
The hapten 4-hydroxy-3,5-dinitro-phenacetic acid	S22
Sperm	K14
Cytoskeletal protein	C12

[a] Adapted with permission from a review article by James and Bell (J1).

Although there is no clear consensus, it has recently been pointed out (J1) that generation of heterohybrids using mouse myeloma cells for immortalization may be the method of choice for human monoclonal antibody production. Mouse myeloma fusion has been employed for a range of antigens, including lung carcinoma antigen (H8), tetanus toxin (G4), *Haemophilus influenzae* B (G6), and *Pseudomonas aeruginosa* (S3). In general, the mouse myeloma cells have been reported to be easier to clone and fuse and the parental heterohybrid cells secrete higher levels of antibody and are more stable (up to 30 months) than when other methods of immortalization are employed. Recently, Sugano *et al.* generated "stable" human antibody-producing heterohybrids for herpes VZ virus by fusion with mouse myeloma cells (S24). Donor lymphocytes were taken from thrombocytopenic patients requiring splenectomy and were stimulated *in vitro* utilizing viral antigens and pokeweed mitogen. The heterohybrids produced IgG_1-neutralizing antibodies for herpes VZ virus that cross-react with herpes simplex virus strains and are 1000 times more potent than normal human serum. The fusion frequency was higher than one hybridoma per 5×10^4 lymphocytes. The antibody-producing cells have been followed for 9 months and have shown no loss in antibody-producing capability. Also, the heterohybrid cells secreted antibody at levels comparable to mouse hybridoma cells (30–40 µg per 10^6 cells per day). It is therefore likely that many of the technical limitations facing human monoclonal antibody production will be overcome in the near future, utilizing mouse fusion partners.

4.5. Developing Areas of Monoclonal Antibody Technology

Several interesting and new areas of monoclonal antibody technology are considered briefly here with respect to their practical and theoretical implications (Table 6). First, monoclonal antibody methodologies are being combined with advances in engineering, leading to the development of electronically based assay devices termed immunosensors. Second, techniques to routinely produce monoclonal antibodies with dual specificities or bispecific antibodies are being perfected. Third, recombinant techniques are being perfected to generate man-made antibodies. The implementation of recombinant DNA techniques to engineer antibody molecules is likely to lead to an entirely new generation of antibody reagents. Finally, although the combining sites of antibodies are normally not catalytic, research is underway to understand and selectively modify the combining regions of antibody molecules so that enzymatically active sites are created. Such monoclonal antibodies are referred to as catalytic antibodies or "abzymes."

4.5.1. *Immunosensors*

Immunosensors are simple, probelike devices engineered and electronically configured to detect the concentration of biomolecules. It is hoped that immunosensors can be designed that will read out antigen concentration within a few minutes or seconds. Applications proposed for immunosensors include precise quantitation of biochemicals for medical and

TABLE 6
DEVELOPING AREAS OF MONOCLONAL ANTIBODY TECHNOLOGY

Immunosensors
  Piezoelectric immunosensors: detect and quantitate human IgG and influenza type A virus (R6)
  Potentiometric immunosensors: detect and quantitate hCG (Y1) and *Candida albicans* (K3)
  Indirect immunosensors: detect and quantitate albumin, hCG, tumor antigen, thyroxin, and hepatitis B surface antigen (A2)

Bispecific antibodies
  Diagnostic assays: increase sensitivity in immunohistochemical studies (S25); simplify ELISA (S25) and immunoblot assays (L6)
  Targeting techniques: target cytotoxic T cells to Thy-1.1-expressing tumor cells, parasites, viruses, or other pathogens (L1, S19); target vindesine to CEA-expressing tumor cells (C10)

Recombinant antibodies
  Humanizing monoclonal antibodies: mouse variable domains combined with human constant domains (J6, O3)
  Altering specificity or affinity: altering antibody binding site specificity and affinity by controlled mutagenesis (S10)

Catalytic antibodies
  Generate enzymatically active antibodies: use of transition-state analog immunogens to generate monoclonal antibodies that catalyze ester (T6) and carbonate (P11) hydrolysis

veterinary immunodiagnostic purposes, food contamination monitoring, diagnosis of human and animal disease, industrial effluent monitoring, and fermentation and purification process control and monitoring (N6). Consequently, there is a growing interest in the development of efficient immunosensors.

Immunosensors consist of antibodies coupled to a stable support and a transducing element that senses antigen–antibody interaction and sends a proportional electrical signal to a measuring device. There are two classes of immunosensors—one detects antigen directly, the other indirectly. Examples of direct immunosensors under study are: (1) electromechanical or piezoelectric-based immunosensors, where changes in electric charge are encountered in response to mechanical stress as antibody interacts with antigen at the surface of a sensor; (2) potentiometric systems, where the effect of antigen–antibody charge is measured directly by potentiometric devices; and (3) field-effect transistors (FET), where changes in electrical field strength are measured. Piezoelectric immunosensor devices for human IgG and influenza type A virus have been constructed that recognize nanogram quantities of antigen within seconds (K3, R6). Potentiometric devices have been used successfully to detect hCG in the urine of pregnant women (Y1) and for measuring *Candida albicans* in blood specimens (K3).

Indirect immunosensors differ in that signal transduction is achieved indirectly. Typically, indirect immunosensors measure competitive reactions between antigen and antibody. For example, competition may be measured at the surface of a Clarke oxygen electrode by employing catalase-labeled antigen (A3). Competition by unlabeled antigen lowers oxygen production. Optical fiber sensors have also been employed with indirect immunosensors utilizing fluorescently tagged antigen to produce a signal (T7). Indirect immunosensors for human serum albumin, human immunoglobulin, human chorionic gonadotropin, tumor antigen α-fetoprotein, thyroxin, and hepatitis B surface antigen have been described (A2).

Immunosensor technology currently has several limitations. Relatively expensive devices are needed to transduce and record antibody–antigen signals. Another consideration is sensitivity. The sensitivity for antigen detection is often less than with comparable ELISA methods. Finally, the lifetime of probes is often short and subject to limited reuses due to antibody stability and probe sterility concerns. As has been pointed out (N6), many of these limitations are the subject of active investigation, and considerable progress is being made toward commercialization.

4.5.2. Bispecific Antibodies

Bispecific antibodies have dual binding specificity (see Section 2.2), which provides new opportunities for application in therapy and diagnosis. For example, with one binding site directed to a signal enzyme such as horseradish peroxidase and the other toward antigen, ELISA (K2) and immunoblot-type assays (L6) can be performed in a single step. Suresh *et al.* studied the advantages of bispecific antibodies as assay reagents and noted an increase in sensitivity and decrease in signal-to-noise ratio in immunohistochemical studies of spinal cord slices visualized by microscopy, and in competitive immunoassays involving tissue-specific antigens (S25). Bispecific antibodies are also being evaluated for therapeutic uses in targeting cytostatic drugs or cytotoxic T cells to tumor cells or pathogens. For example, vinca alkaloid drugs have been targeted to human colorectal cancers using bispecific mouse monoclonal antibodies to vindesine and carcinoembryonic antigen (C10). Cytotoxic T cells with allotypic determinants recognizing one antibody site have been efficiently targeted to Thy-1.1-expressing tumor cells. Several stable bispecific antibodies have also been employed to selectively target cytotoxic T cells to ovary carcinoma cells, *Toxoplasma gondii,* and HLA antigens (L1,

S19). Therefore, T cell targeting by bispecific antibodies might be used therapeutically to alter histocompatibility, destroy parasites, and kill tumor cells.

4.5.3. Recombinant Monoclonal Antibodies

The assembly of man-made monoclonal antibodies is now possible using recombinant DNA techniques. For a review of this subject, a discussion by Oi and Morrison is relevant (O3). Immunoglobulin genes coding for mouse variable region domains have been combined with human constant region domains and expressed in several different host cells, including myelomas (J6). Site-directed mutagenesis has also been used to alter specificity and idiotype of mouse monoclonal antibodies (S10).

Although the technology is still evolving, it may soon be possible to devise antibody vectors and develop a generic host for high-level production of recombinant monoclonal antibodies. Recombinant DNA methods under development have the potential of circumventing many of the drawbacks of conventional methods for monoclonal antibody production. Antibody diversity might be generated *in vitro* without the need for *in vivo* immunization. Bispecific antibodies and enzyme-fused antibodies may also be prepared employing recombinant procedures. Difficulties in the production of human antibodies might be eliminated by constructing "humanized" antibodies grown in well-defined host cells adapted to a serum-free medium. Mouse variable region genes against human antigens could be expressed with different functionalities depending on the human constant domains employed. Such chimeric mouse–human monoclonal antibodies may be rendered nonimmunogenic by the substitution of human constant region genes. It is clear recombinant techniques will be extremely important in the development of "humanized" and multifunctional antibodies for a variety of clinical and diagnostic applications.

4.5.4. Catalytic Antibodies

Recent work suggests that appropriately selected antibody molecules can function as enzymes. Catalytic antibodies are currently produced by selecting antibodies against transition-state analogs. Antibody-catalyzed hydrolysis has been demonstrated for specific esters (T6) and carbonates (P11) with kinetic rate accelerations of approximately 1000- to 17,000-fold over the uncatalyzed reaction rate. In addition to increasing the understanding of catalytic processes, catalytic antibodies may have important applications in science and medicine. A major technological goal is to produce catalytic antibodies that can function as selective proteases, which modify proteins in the same way as DNA is modified by restriction enzymes (M2). Many believe catalytic antibodies may also have pharmaceutical potential (B2). Binding and enzymatic functions might be incorporated into catalytic antibodies to search and destroy viruses and other pathogens, remove atherosclerotic plaque from blood vessel walls, or for many other medical purposes. No one has demonstrated the transfer of such complex catalytic functions as plaque or cell lysis to antibody molecules. It may soon be possible to fuse enzyme active sites to antibody molecules using genetic engineering techniques. This would make the preparation of such complex catalytic antibodies more feasible.

References

A1. Abrams, P., Knost, J., Clarke, G., Wilburn, S., Oldham, R., and Foon, K., Detection of the optimal human cell lines for development of human hybridomas. *J. Immunol.* **131**, 1201–1204 (1983).

A2. Aizawa, M., Molecular recognition and chemical amplification of biosensors. *In* Proceedings of the International Meeting on Chemical Sensors. *Anal. Chem., Symp. Ser.* **17**, 683–692 (1983).

A3. Aizawa, M., Morioka, A., Matsuoka, H., Suzuki, S., Nagamura, Y., and Shinohara, R., An enzyme immunosensor for IgG. *J. Solid-Phase Biochem.* **1**, 319–328 (1976).

A4. Akerstrom, B., and Bjork, L., A physiological study of protein G, a molecule with unique immunoglobulin binding properties. *J. Biol. Chem.* **261**, 10240–10247 (1986).

A5. Amit, A., Mariuzza, R., Phillips, S., and Poljak, R., Three-dimensional structure of an antigen-antibody complex at 6 Å resolution. *Nature (London)* **313**, 156–158 (1985).

A6. Anderson, N., Willis, D., Holladay, D., Caton, J., Holleman, J., Eveleigh, J., Attrill, J., Ball, F., Frances, L., and Anderson, N., Analytical techiques for cell fractions: Automatic system for cyclic chromatography. *Anal. Biochem.* **66**, 159–174 (1975).

A7. Andrews, C., and Pereira, H., "Viruses of Vertebrates," 2nd ed. Bailliere, Tindall & Cassell, London, 1967.

A8. Andrews, C., Harris, R., Coleman, P., and Morahan, P., Stability of minute virus of mice to chemical and physical agents. *Appl. Microbiol.* **28**, 351–354 (1974).

A9. Argon, A., Burrone, O., and Milstein, C., Molecular characterization of a nonsecreting myeloma mutant. *Eur. J. Immunol.* **13**, 301–305 (1983).

B1. Bartal, A., Feit, C., Erlandson, R., and Hirshaut, Y., The presence of viral particles in hybridoma clones secreting monoclonal antibodies. *N. Engl. J. Med.* **306**, 1423 (1982).

B2. Baum, R., Catalytic antibodies open up new strategy for protein engineering. *Chem. Eng. News* **65**(14), 30–33 (1987).

B3. Baumhaeckel, H., Liesegang, B., Radbruch, B., Rajewsky, K., and Sablitzky, F., Switch from NIP specific IgG_3 to IgG_1 in the mouse hybridoma cell line S24/63/63. *J. Immunol.* **128**, 1217–1220 (1982).

B4. Bazin, H., Cormont, F., and Clercq, L., Rat monoclonal antibodies. II. A rapid and efficient method of purification from ascitic fluid or serum. *J. Immunol. Methods* **71**, 9–16 (1984).

B5. Bell, T., and Ross, M., Persistently latent infection of human embryonic cells with reovirus type 3. *Nature (London)* **212**, 412–414 (1966).

B6. Berthiaume, L., Joncas, J., and Pavilanis, V., Comparative structure, morphogenesis and biological characteristics of the respiratory syncytial (RS) virus and the pneumonia virus of mice (PVM). *Arch. Virol.* **45**, 39–51 (1974).

B7. Bhatt, P., and Jonas, A., An epizootic of Sendai infection with mortality in a barrier-maintained mouse colony. *Am. J. Epidemiol.* **100**, 222–229 (1974).
B8. Biggar, R., Schmidt, T., and Woodall, J., Lymphocytic choriomeningitis in laboratory personnel exposed to hamsters inadvertently infected with LCM virus. *J. Am. Vet. Med. Assoc.* **171**, 829–832 (1977).
B9. Bjork, L., and Kronvall, G., Purification and some novel properties of streptococcal protein G, a novel IgG-binding reagent. *J. Immunol.* **133**, 969–974 (1984).
B10. Bogard, W., Hornberger, E., and Kung, P., Production and characterization of human monoclonal antibodies against Gram negative bacteria. *In* "Human Hybridomas and Monoclonal Antibodies" (E. Engleman, S. Foung, J. Larrick, and A. Raubitschek, eds.), pp. 95–112. Plenum, New York, 1985.
B11. Boss, B., An improved *in vitro* immunization procedure for the production of monoclonal antibodies against neural and other antigens. *Brain Res.* **291** 193–196 (1984).
B12. Brennand, D., Danson, M., and Hough, D., A comparison of ELISA screening methods for the production of monoclonal antibodies against soluble protein antigens. *J. Immunol. Methods* **93**, 9–14 (1986).
B13. Bruck, C., Portetelle, D., Glineur, C., and Bollen, A., One-step purification of mouse monoclonal antibodies from ascitic fluid by DEAE Affi-gel blue chromatography. *J. Immunol. Methods* **53**, 313–319 (1982).
B14. Burchiel, S., Purification and analysis of monoclonal antibodies by high-performance liquid chromatography. *In* "Methods in Enzymology" (J. Langone and H. Van Vunakis, eds.), Vol. 121, pp. 596–615. Academic Press, Orlando, Florida, 1986.
B15. Burchiel, S., Billman, J., and Alber, T., Rapid and efficient purification of mouse monoclonal antibodies from ascites fluid using high performance liquid chromatography. *J. Immunol. Methods* **69**, 33–42 (1984).
B16. Burnet, F., A modification of Jerne's theory of antibody production using the concept of clonal selection. *Aust. J. Sci.* **20**, 67–69 (1957).
B17. Burnett, K., Leung, J., and Martinis, J., Human monoclonal antibodies to defined antigens: Toward clinical applications. *In* "Human Hybridomas and Monoclonal Antibodies" (E. Engleman, S. Foung, J. Larrick, and A. Raubitschek, eds.), pp. 113–133. Plenum, New York, 1985.
B18. Butler, J., The amplified ELISA: Principles of an application for the comparative quantitation of class and subclass antibodies and the distribution of antibodies and antigen biochemical separates. *In* "Methods in Enzymology" (J. Langone and H. Van Vunakis, eds.), Vol. 73, pp. 482–523. Academic Press, New York, 1981.
C1. Campling, B., Cole, C., Atlaw, T., Roder, J., and Kozbor, D., The EBV hybridoma technique and its applications. *In* "Human Hybridomas: Diagnostic and Therapeutic Applications" (J. Strelkauskas, ed.), pp. 3–22. Dekker, New York and Basel, 1987.
C2. Carlsson, M., Herdin, M., Inganas, B., Harfast, B., and Blomberg, F., Purification of *in vitro* produced mouse monoclonal antibodies. A two step procedure utilizing cation exchange chromatography and gel filtration. *J. Immunol. Methods* **79**, 89–98 (1985).
C3. Carthew, P., Is rodent virus contamination of monoclonal antibody preparations for use in human therapy a hazard? *J. Gen. Virol.* **67**, 963–974 (1986).
C4. Chang, T., Steplawski, Z., and Koprowski, H., Production of monoclonal antibodies in serum free medium. *J. Immunol. Methods* **39**, 369–375 (1980).
C5. Chisholm, R., On the trail of the magic bullet: Monoclonal antibodies promise perfectly targeted chemicals. *High Technol.* **3**(1), 57–63 (1983).
C6. Cleveland, W., Wood, I., and Erlanger, B., Routine large-scale production of monoclonal antibodies in a protein free culture medium. *J. Immunol. Methods* **56**, 221–234 (1983).

C7. Cole, S., Vreeken, S., Mirski, S., and Campling, B., Growth of human × human hybridomas in protein-free medium supplemented with ethanolamine. *J. Immunol. Methods* **97**, 29–35 (1987).

C8. Coller, H., and Coller, B., Statistical analysis of repetitive subcloning by the limiting dilution technique with a view toward ensuring hybridoma monoclonality. *Hybridoma* **2**, 91–96 (1983).

C9. Cook, W., Rudikoff, S., Giusti, A., and Scharff, M., Somatic mutation in cultured mouse myeloma cells affects antigen binding. *Proc. Natl. Acad. Sci. U.S.A.* **79**, 1240–1244 (1982).

C10. Corvalan, J., Smith, W., Gore, V., and Brandon, D., Specific *in vitro* and *in vivo* drug localization to tumor cells using a hybrid–hybrid monoclonal antibody recognizing both carcinoembryonic antigen and vinca alkaloids. *Cancer Immunol. Immunother.* **24**(2), 133–137 (1987).

C11. Cory, S., Tyler, B., and Adams, J., Sets of immunoglobulin Vκ genes homologous to ten cloned Vκ sequences: Implications for the number of germline Vκ genes. *J. Mol. Appl. Genet.* **1**. 103–116 (1981).

C12. Cote, R., Morrissey, D., Houghton, A., Beattie, E., Oettgen, H., and Old, L., Generation of human monoclonal antibodies reactive with cellular antigens. *Proc. Natl. Acad. Sci. U.S.A.* **80**, 2026–2030 (1983).

C13. Cotton, R., and Milstein, C., Fusion of two immunoglobulin producing myeloma cells. *Nature (London)* **244**, 42–43 (1973).

C14. Crawford, D., Barlow, M., Mulholland, N., McDougall, D., Zanders, E., Tippett, R., and Huehns, E., The production and characterization of a human monoclonal antibody to rhesus D antigen. *Proc. Br. Blood Transfus. Soc.* **1**, 113–121 (1984).

C15. Crawford, L., A minute virus of mice. *Virology* **29**, 605–612 (1966).

C16. Creighton, W., Lambert, P., and Miescher, P., Detection of antibodies and soluble antigen-antibody complexes by precipitation with polyethylene glycol. *J. Immunol.* **3**, 1219–1227 (1973).

C17. Cunningham-Rundles, C., Appropriate uses of human immunoglobulin in clinical practice: memorandum from an IUIS/WHO meeting. *Bull. W.H.O.* **60**(1) 43–47 (1982).

D1. Dangl, J., Parks, D., Oi, V., and Herzenberg, L., Rapid isolation of cloned isotype switch variants using fluorescence activated cell sorting. *Cytometry* **2**(6), 395–401 (1982).

D2. De Blas, A., Ratnaparkhi, L., and Mosimann, J., Estimation of the number of monoclonal hybridomas in a cell fusion experiment. Effect of post-fusion cell dilution on hybridoma survival. *J. Immunol. Methods* **45**, 109–115 (1981).

D3. Denney, R., Fritz, R., Patel, N., and Abell, C., Human liver MAO-A and MAO-B separated by immunoaffinity chromatography with MAO-B-specific antibody. *Science* **215**, 1400–1403 (1982).

D4. Desmyter, J., Johnson, K., Deckers, C., Leduc, J., Brasseur, F., and Van Ypersele De Strihou, C., Laboratory rat associated outbreak of haemorrhagic fever with renal syndrome due to Hantaan-like virus in Belgium. *Lancet* **2**, 1445–1448 (1983).

D5. D'Eustachio, P., Bothwell, A., Takaro, T., Baltimore, D., and Ruddle, F., Chromosomal location of structural genes encoding murine immunoglobulin lambda light chains. *J. Exp. Med.* **153**, 793–800 (1981).

D6. DeVita, V., Hellman, S., and Rosenberg, S., eds., "AIDS: Etiology, Diagnosis, Treatment, and Prevention." Lippincott, Philadelphia, Pennsylvania, 1985.

D7. Dick, G., Best, A., Haddow, A., and Smithburn, K., Mengo encephalomyelitis, a hitherto unknown virus affecting man. *Lancet* **2**, 286–289 (1948).

D8. Dixon, B., Words of caution as monoclonals mature. *Bio/Technology* **4**, 604 (1986).

D9. Dmochowski, L., Viruses and tumours. *Bacteriol. Rev.* **23**, 18–40, (1959).

D10. Dreesman, G., and Kennedy, R., Anti-idiotype antibodies: Implications of internal image-based vaccines for infectious diseases. *J. Infect. Dis.* **151**, 761–765 (1985).
E1. Early, P., and Hood, L., Allelic exclusion and nonproductive immunoglobulin gene rearrangements. *Cell (Cambridge, Mass.)* **24**, 1–3 (1981).
E2. Eiden, J., Vonderfecht, S., and Yolken, R., Evidence that a novel rotavirus-like agent of rats can cause gastroenteritis in man. *Lancet* **2**, 8–10 (1985).
E3. Eisenbarth, G., Linenbach, A., Jackson, R., Scearce, R., and Croce, C., Human hybridomas secreting anti-islet autoantibodies. *Nature (London)* **300**, 264–267 (1982).
E4. Engleman, E., Foung, S., Larrick, J., and Raubitschek, A., eds., "Human Hybridomas and Monoclonal Antibodies." Plenum, New York, 1985.
E5. Engvall, E., Enzyme immunoassay ELISA and EMIT. *In* "Methods in Enzymology" H. Van Vunakis and J. Langone, eds.), Vol. 70, pp. 419–439. Academic Press, New York, 1980.
E6. Epstein, A., and Epstein, M., The hybridoma technology. I. Production of monoclonal antibodies. *Adv. Biotechnol. Processes* **6**, 179–218 (1986).
E7. Epstein, N., and Epstein, M., The hybridoma technology. II. Applications of hybrid cell products: Monoclonal antibodies and lymphokines. *Adv. Biotechnol. Processes* **6**, 219–251 (1986).
E8. Erkman, L., Soldati, G., James, R., and Kato, A., Partial purification of lymphoblasts after *in vitro* immunization increases the yield in Ig-producing hybridomas. *J. Immunol. Methods* **98**, 43–52 (1987).
E9. Ey, P., Prowse, S., and Jenkin, C., Isolation of pure IgG_1, IgG_{2a}, and IgG_{2b} Igs from mouse serum using protein A-sepharose. *Immunocytochemistry* **15**, 429–436 (1978).
F1. Fahey, J., and Terry, E., Ion exchange chromatography and gel filtration. *In* "Handbook of Experimental Immunology: Immunochemistry" (D. Weir, ed.), Vol. 1, pp. 8.1–8.16. Blackwell, Oxford, 1978.
F2. Fazekas de St. Groth, S., and Scheidegger, D., Production of monoclonal antibodies: Strategy and tactics. *J. Immunol. Methods* **35**(1), 1–21 (1980).
F3. Feder, J., and Tolbert, W., The large-scale cultivation of mammalian cells. *Sci. Am.* **248**(1), 36–43 (1983).
F4. Federal Register, Docket No. 83N-0070, Licensing of a biological monoclonal antibody product prepared by hybridoma technology. *Fed. Regist.* **48**; 50795250796 (1983).
F5. Finter, N., and Fantes, K., The purity and safety of interferons prepared for clinical use: The case for lymphoblastoid interferon. *In* "Interferon 2" (I. Gresser, ed.), pp. 65–80. Academic Press, London, 1981.
F6. Fleischaker, R., Applying large-scale culture techniques to the *in vitro* production of monoclonal antibodies. *Pharm. Manuf.* **3**(1), 17–21 (1986).
F7. Food and Drug Administration, "Points to Consider in the Manufacture and Testing of Monoclonal Antibody Products for Human Use." Office of Biologics Research and Review: Center for Drugs and Biologics, FDA, Washington, D.C., 1987.
F8. Fox, P., Berenstein, E., and Siraganian, R., Enhancing the frequency of antigen-specific hybridomas. *Eur. J. Immunol.* **11**, 431–434 (1981).
F9. Frantisek, F., Purification of IgG monoclonal antibodies from ascitic fluid based on rivanol precipitation. *In* "Methods in Enzymology" (J. Langone and H. Van Vunakis, eds.), Vol. 121, pp. 631–638. Academic Press, New York, 1986.
G1. Gaffar, S. A., Surh, C. D., and Glassy, M. C., Variations in secretion of monoclonal antibodies by human–human hybridomas. *Hybridoma* **5**, 93–105 (1986).
G2. Galfré, G., and Milstein C., Preparation of McABs: Strategies and procedures. *In* "Methods in Enzymology" (J. Langone and H. Van Vunakis, eds.), Vol. 73, pp. 3–46. Academic Press, New York, 1981.

G3. Gally, J., and Edelman, A., The genetic control of immunoglobulin synthesis. *Annu. Rev. Genet.* **6,** 1–46 (1972).
G4. Gemski, M., Doctor, B., Gentry, M., Pluskal, M., and Strickler, M., Single step purification of monoclonal antibody from murine ascites and tissue culture fluids by anion exchange high performance liquid chromatography. *Bio Techniques* **3**(5), 378–384 (1985).
G5. Gigliotti, F., and Insel, R., Protective human hybridoma antibody to tetanus toxin. *J. Clin. Invest.* **70,** 1306–1309 (1982).
G6. Gigliotti, F., Smith, L., and Insel, R., Reproducible production of human monoclonal antibodies by fusion of peripheral blood lymphocytes with a mouse myeloma cell line. *J. Infect. Dis.* **149,** 43–47 (1984).
G7. Glassey, M., Gaffar, S., Peters, R., and Royston, I., Human monoclonal antibodies to human cancer cells. *In* "Monoclonal Antibodies and Cancer Therapy" (R. Reisfeld and S. Sell, eds.), Vol. 27, pp. 97–110. Alan R. Liss, New York, 1985.
G8. Goding, J., Use of staphylococcal protein A as an immunological reagent *J. Immunol. Methods* **20,** 241–253 (1978).
G9. Goding, J., Antibody production by hybridomas (review article). *J. Immunol. Methods* **39**(4), 285–308 (1980).
G10. Goding, J., "Monoclonal Antibodies: Principles and Practice," 2nd ed. Academic Press, London, 1986.
G11. GraveKamp, C., Bol, S., Hagemeijer, A., and Bolhuis, R., Production of human T-cell hybridomas by electrofusion. *In* "Human Hybridomas and Monoclonal Antibodies" (E. Engleman, S. Foung, J. Larrick and A. Raubitschek, eds.), pp. 323–339. Plenum, New York, 1985.
H1. Hadas, E., and Theilen, G., Production of monoclonal antibodies: The effect of hybridoma concentration on the yield of antibody-producing clones. *J. Immunol. Methods* **96,** 3–6 (1987).
H2. Hansen, R., and Beavo, J., Purification of two calcium/calmodulin-dependent forms of cyclic nucleotide phosphodiesterase by using conformation-specific monoclonal antibody chromatography. *Proc. Natl. Acad. Sci. U.S.A.* **79,** 2788–2792 (1982).
H3. Harada, H., Masugu, K., Tokumoto, Y., Seiko Y., Koyama, F., Kochibe, N., and Korata, A., Systematic fractionation of oligosaccharides of human immunoglobulin G by serial affinity chromatography on immobilized lectin columns. *Anal. Biochem.* **164,** 374–381 (1987).
H4. Heide, K., and Schwick, H., Salt fractionation of Igs. *In* "Handbook of Experimental Immunology: Immunochemistry" (D. Weir, ed.), Vol. 1, pp. 7.1–7.11. Blackwell, Oxford, 1978.
H5. Heinsohn, R., and Poschmann, A., Computer-aided hybridoma screening. Production of monoclonal antibodies against membrane fractions from mammary tumour tissue. *J. Immunol. Methods* **96,** 179–184 (1987).
H6. Helwig, F., and Schmidt, E., A filter-passing agent producing interstitial myocarditis in anthropoid apes and small animals. *Science* **102,** 31–33 (1945).
H7. Hirohashi, S., Shimosato, Y., and Ino, Y., The *in vitro* production of tumour-related human monoclonal antibody and its immunohistochemical screening with autologous tissue. *Gann* **73,** 345–347 (1982).
H8. Hirohashi, S., Shimosato, Y., and Ino, Y., Antibodies from EB-virus-transformed lymphocytes of lymph nodes adjoining lung cancer. *Br. J. Cancer* **46,** 802–805 (1985).
H9. Hopkinson, J., Hollow fiber cell culture systems for economical cell-product manufacturing. *Bio/Technology* **3,** 225–230 (1985).

H10. Horibata, K., and Harris, A., Mouse myelomas and lymphomas in culture. *Exp. Cell Res.* **60**, 621–677 (1970).
H11. Houghton A., Mintzer, D., Cordon-Cardo, C., Weltz, S., Fliegel, B., Vadhan, S., Corswell, E., Melamed, M., Oettgen, H., and Old, L., Mouse monoclonal IgG_3 antibody detecting GD3 ganglioside: A phase I trial in patients with malignant melanoma. *Proc. Natl. Acad. Sci. U.S.A.* **82**, 1242-1246 (1985).
H12. Hulette, C., Effros, R., Dillard, L., and Walford, R., Production of human monoclonal antibody to HLA by human-human hybridoma technology. *Am. J. Pathol.* **121**, 10–14 (1985).
11. Imbach, P., Barandun S., Hirt A., and Wagner, H., Intravenous immunoglobulin for idiopathic thrombocytopenic purpura (ITP) in childhood. *In* "Immunochemotherapy: A Guide to Immunoglobulin Prophylaxis and Therapy" (U. Nydegger, ed.), pp. 291–297. Academic Press, London, 1981.
12. Irie, R., Sze, L., and Saxton, R., Human antibody to OFA-1, a tumour antigen produced *in vitro* by Epstein-Barr virus-transformed human B-lymphoid cell lines. *Proc. Natl. Acad. Sci. U.S.A.* **79** 5666–5670 (1982).
J1. James, K., and Bell, T., Human monoclonal antibody production: Current status and future prospects (review article). *J. Immunol. Methods* **100**, 5–40 (1987).
J2. Jayme, D., and Blackman, E., Culture media for propagation of mammalian cells, viruses, and other biologicals. *Adv. Biotechnol. Processes* **5**, 1–30 (1985).
J3. Jehanli, A., and Hough, D., A rapid procedure for the isolation of human IgM myeloma proteins. *J. Immunol. Methods* **44**, 199–204 (1981).
J4. Jensen, K., Minuse, E., and Ackermann, W., Serologic evidence of American experience with pneumonitis virus (type Sendai). *J. Immunol.* **75**, 71–77 (1955).
J5. Jerne, N., Roland, J., and Cazenave, P., Recurrent idiotopes and internal images. *EMBO J.* **1**, 243–247 (1982).
J6. Jones, P., Dear, P., Foote, J., Neuberger, M., and Winter, G., Replacing the complementarity determining regions in a human antibody with those from a mouse. *Nature (London)* **321**, 522–525 (1986).
J7. Juarez-Salinas, H., Engelhorn, S., Bigbee, W., Lowry, M., and Stanker, L., Ultrapurification of monoclonal antibodies by high-performance hydroxylapatite (HPHT) chromatography. *Bio Techniques* **2**(8), 164–169 (1984).
K1. Kaltoft, K., Nielsen, L., Zerthen, J., and Dan, K., Monoclonal antibody specificity inhibits a human M_r 52,000 plasminogen-activating enzyme. *Proc. Natl. Acad. Sci. U.S.A.* **79** 3720–3723 (1982).
K2. Karawajew, L., Micheel, O., Bersing, O., and Gaestel, M., Bispecific antibody-producing hybrid hybridomas selected by fluorescence activated cell sorter. *J. Immunol. Methods* **96**, 265–270 (1987).
K3. Karube, I., and Suzuki, M., Novel immunosensors. *Biosensors* **2**, 343–362 (1986).
K4. Kataoka, T., Kawakam T., Takahashi N., and Honjo T., Rearrangement of immunoglobulin gamma-1-chain gene and mechanism for heavy-chain class switch. *Proc. Natl. Acad. Sci. U.S.A.* **77**, 919–923 (1980).
K5. Kearney, J., Radbruch, A., Liesegang, B., and Rajewsky, K., A new mouse myeloma cell line that has lost immunoglobulin expression but permits construction of antibody-secreting hybrid cell lines. *J. Immunol.* **123**, 1548–1550 (1979).
K6. Kiso, Y., Okamoto, K., Makiyama, M., Shimokura, M., Hirai, Y., Kawai, K., and Kikuishi, H., Preparation of a monoclonal antibody to a synthetic peptide related to human interleukin-2 (IL-2, T-cell growth factor) and affinity-purification of IL-2 produced by a recombinant gene. *Pept. Chem.* **22**, 103–108 (1985).

K7. Knazek, R., Kohler, P., and Dedrick, R., Cell culture on artificial capillaries applied to tissue growth *in vitro. Science* **178**, 65–67 (1972).

K8. Koblet, H., Barandun, S., and Diggelmann, H., Turnover of standard gammaglobulin, pH-4-gammaglobulin, and pepsin disaggregated gammaglobulin and clinical implications. *Vox Sang.* **13**, 93–102 (1967).

K9. Kohler, G., and Milstein, C., Continuous cultures of fused cells secreting antibody of predefined specificity. *Nature* **256**, 495–497 (1975).

K10. Koprowski, H., Herlyn, D., Lubeck, M., DeFreitas, E., and Sears, H., Human anti-idiotype antibodies in cancer patients: Is the modulation of the immune response beneficial for the patient? *Proc. Natl. Acad. Sci. U.S.A.* **81**, 216–219 (1984).

K11. Kozbor, D., and Croce C., Fusion partners for production of human monoclonal antibodies. *In* "Human Hybridomas and Monoclonal Antibodies"(E. Engleman, S. Foung, J. Larrick, and A. Raubitschek, eds.), pp. 21–36. Plenum, New York, 1985.

K12. Kozbor, D., and Roder, J., Requirements for the establishment of high titered human monoclonal antibodies against tetanus toxoid using the EBV techniques. *J. Immunol.* **127**(4), 1275–1280 (1981).

K13. Krivi, G., and Rowold, E., Monoclonal antibodies to bovine somatotropin: Immunoadsorbent reagents for mammalian somatotropins. *Hybridoma* **3**(2), 151–162 (1984).

K14. Kyurkchiev, S., Shegata, M., Koyama, K., and Isojima, S., A human-mouse hybridoma producing monoclonal antibody against human sperm coating antigen. *Immunology* **57**, 489–492 (1986).

L1. Lanzavecchia, A., Hybrid antibodies as a tool to direct T lymphocytes against target cells. *Meet. Abstr. Biotech, Int. Symp. Monoclonals DNA Probes Diagn. Prev. Med., 1986* p. 53 (1986).

L2. Larrick, J., and Bourla, J., Prospects for the therapeutic use of human monoclonal antibodies. *J. Biol. Response Modif.* **5**, 379–393 (1986).

L3. Larrick, J., and Buck, D., Practical aspects of human monoclonal antibody production. *Bio Techniques* **1**, 6–14 (1984).

L4. Larsen, F., Selmer, J., and Hertz, J., Purification of *Bordetella pertussis* antigens using monoclonal antibodies. *Acta Path. Microbiol. Immunol. Scand., Sect. C* **92C**, 271–277 (1984).

L5. Laubli, U., Baier, W., Binz, H., Celio, M., and Humbel, R., Monoclonal antibodies directed to human insulin-like growth factor. I. Use for radioimmunoassay and immunopurification of IGF. *FEBS Lett.* **149**(1), 109–112 (1982).

L6. Leong, M., Milstein, C., and Pannell, R., Luminescent detection method for immunodot, western and southern blots. *J. Histochem. Cytochem.* **34**(12), 1645–1650 (1986).

L7. Levy, J., Mouse plasmacytoma cells produce infectious type C viruses. *Lancet* **2**, 522 (1983).

L8. Levy, J., Kawahata, T., and Spitler, L., Purification of monoclonal antibodies from mouse ascites eliminates contaminating infectious mouse type C viruses and nucleic acids. *Clin. Exp. Immunol.* **56**, 114–120 (1984).

L9. Levy, R., and Miller, R., Tumor therapy with monoclonal antibodies. *Fed. Proc., Fed. Am. Soc. Exp. Biol.* **42**(9), 2650–2656 (1983)

L10. Lewis, C., Tolbert, W., and Feder, J., Large scale perfusion culture system used for production of monoclonal antibodies: Abstracts from the Third Annual Congress for Hybridoma Research. *Hybridoma* **3**(1), 74 (1984).

L11. Lim, F., and Sun, A., Microencapsulated islets as bioartificial endocrine pancreas. *Science* **210**, 908–910 (1980).

L12. Lindmark, R., Thoren-Toolling, K., and Sjöquist, J., Binding of immunoglobulins to protein A and immunoglobulin levels in mammalian sera (review article). *J. Immunol. Methods* **62**, 1–13 (1983).

L13. Littlefield, J., Selection of hybrids from matings of fibroblasts *in vitro* and their presumed recombinants. *Science* **145**, 709–710 (1964).

L14. Littlefield, S., Gilligan, K., and Jarvis, A., Growth and monoclonal antibody production from rat × mouse hybridomas: A comparison of microcapsule cultures with conventional suspension culture: Abstract from the Third Annual Congress for Hybridoma Research, February 22, 1984. *Hybridoma* **3**(1), 75 (1984).

L15. Lovgren K., Lindmark, J., Pipkorn, R., and Morein, B., Antigenic presentation of small molecules and peptides conjugated to a preformed iscom as carrier. *J. Immunol. Methods* **98**, 137–143 (1987).

M1. Manil, L., Motte, P., Pernas, P., Troalen, F., Bohuon, C., and Bellet, D., Evaluation of protocols for purification of mouse monoclonal antibodies. *J. Immunol. Methods* **90**, 25–37 (1986).

M2. Marx, J., Making antibodies work like enzymes. *Science* **234**, 1497–1498 (1986).

M3. McAleer, W., Buynak, E., Maigetter, R., Wampler, D., Miller, W., and Hilleman, M., Human hepatitis B vaccine from recombinant yeast. *Nature (London)* **307**, 178–180, 1984.

M4. McKinney, M., and Parkinson, A., A simple, non-chromatographic procedure to purify immunoglobulins from serum and ascites fluid. *J. Immunol. Methods* **96**, 271–278 (1987).

M5. Meeker, Y., Lowder, J., Maloney, D., Miller, R., Thielemans, R., and Levy, R., A clinical trial of anti-idiotype therapy for B-cell malignancy. *Blood* **65**, 1349–1363 (1985).

M6. Melamed, M., Gordon, J., Ley, S., Edgar, D., and Hughes-Jones, N., Senescence of a human lymphoblastoid clone producing anti-Rhesus (D). *Eur. J. Immunol.* **15**, 742–746 (1985).

M7. Miller, R., Oseroff, A., Stratte, P., and Levy, R., Monoclonal antibody therapeutic trials in seven patients with T-cell lymphoma. *Blood* **62**, 988–995 (1983).

M8. Milstein, C., Monoclonal antibodies. *Sci. Am.* **243**, 66–74 (1980).

M9. Mizuochi, M., Taniguchi, T., Shimizu, A., and Kobata, A., Structural and numerical variations of the carbohydrate moiety of immunoglobulin G. *J. Immunol.* **129**, 2016–2020 (1982).

M10. Murakami, H., Masui, H., Sato, G., Sueoka, N., Chow, T., and Kano-Sueoka, T., Growth of hybridoma cells in serum free medium: Ethanolamine is an essential component. *Proc. Natl. Acad. Sci. U.S.A.* **79**, 1158–1162 (1982).

N1. Najbauer, J., Tigyi, G., and Nemeth, P., Antigen-specific cell adherence assay: A new method for separation of antigen-specific hybridoma cells. *Hybridoma* **5**(4), 361–370 (1986).

N2. Nau, D., A unique chromatographic matrix for rapid antibody purification. *Bio/Chromatography* **1**(2), 82–94 (1986).

N3. Nielsen, K., Henning M., and Duncan, J., Monoclonal antibodies in veterinary medicine. *Biotechnol. Genet. Eng. Rev.* **4**, 311–353 (1986).

N4. Nilsson, B., Svalander, P., and Larsson, A., Immunization of mice and rabbits by intrasplenic deposition of nanogram quantities of protein attached to Sepharose beads or nitrocellulose paper strips. *J. Immunol. Methods* **99**, 67–75 (1987).

N5. Nilsson, K., Scheirer, W., Merten, O., Ostberg, L., Liehl, E., and Mosbach, K., Entrapment of animal cells for production of monoclonal antibodies and other biochemicals. *Nature (London)* **304**, 629–630 (1983).

N6. North, J., Immunosensors: Antibody-based biosensors. *Trends Biotechnol.* **3**(7), 180–185 (1985).
N7. Northrop, R., and Walker, D., A stable human cell line as a host system for Sendai virus. *Proc. Soc. Exp. Biol. Med.* **118**, 698–703 (1965).
N8. Nose, M., and Wigzell, H., Biological significance of carbohydrate chains on monoclonal antibodies. *Proc. Natl. Acad. Sci. U.S.A.* **80**, 32–36 (1983).
O1. Ochs, H., Alving, B., and Finlayson, J., eds., "Immunoglobulins: Characteristics and Uses of Intravenous Preparations," DHEW Publ. No (FDA) 80–9005, pp. 9–13. U.S. Govt. Printing Office, Washington, D.C., 1980.
O2. Ochs, H., Buckley, R., Pirofsky, B., Fischer, S., Rousell, R., Anderson, C., and Wedgwood, R., Safety and patient acceptability of intravenous immune globulin in 10% maltose. *Lancet* **2**, 1158–1159 (1980).
O3. Oi, V., and Morrison, S., Chimeric antibodies. *Bio Techniques* **4**(3), 214–221 (1986).
O4. Oi, V., Vuong, T., Hardy, P., Reidler, J., Dangl, J., Herzenberg, L., and Stryer, L., Correlation between segmental flexibility and effector function of antibodies. *Nature (London)* **307**, 136–140 (1984).
O5. Ossendorp, F., De Boer, M., Al, B., Hilgers, J., Bruning, P., and Tager, J., Production of murine monoclonal antibodies against human thyroglobulin using *in vitro* immunization procedure in serum-free medium. *J. Immunol. Methods* **91**, 257–264 (1986).
P1. Padlan, E., Davies D., Pecht, I., Givol, D., and Wright, C., Model-building studies of antigen-binding sites: Hapten-binding site of MOPC-315. *Cold Spring Harbor Symp. Quant. Biol.* **41**, 627–637 (1976).
P2. Parham, P., Androlewicz, M., Brodsky, F., Holmes, N., and Ways, J., Monoclonal antibodies: Purification, fragmentation and application to structural and functional studies of class I MHC antigens (review article). *J. Immunol. Methods* **53**(2), 133–173 (1982).
P3. Parker, J., Tennant, R., Ward, T., and Rowe, W., Virus studies with germfree mice: I. Preparation of serological diagnostic reagents and survey of germfree and monocontaminated mice for indigenous murine viruses. *J. Natl. Cancer Inst. (U.S.)* **34**, 371–380 (1965).
P4. Paulus, H., Preparation and biomedical application of bispecific antibodies. *Behring Inst. Mitt.* **78**, 118–132 (1985).
P5. Pearson, T., and Anderson, L., Use of high resolution two-dimensional gel electrophoresis for analysis of monoclonal antibodies and their specific antigens. *In* "Methods in Enzymology" (J. Langone, and H. Van Vunakis, eds.), Vol. 92 Part E, pp. 196–220. Academic Press, New York, 1983.
P6. Perry, G., Jackson, J., McDonald, T., Crouse, D., and Sharp, J., Purification of monoclonal antibodies using high performance liquid chromatography (HPLC). *Prep. Biochem.* **14**(5), 431–437 (1985).
P7. Pharmacia Laboratory Separation Division, Monoclonal antibody purification, *Sep. News* **13**(4) (1986).
P8. Pharmacia Laboratory Separation Division, Specific monoclonal antibody purification techniques: The use of Protein-A Sepharose to purify murine IgG monoclonal antibodies. *Sep. News* **13**(5) (1986).
P9. Phillips, J., Sikora, M., and Watson, J., Localization of glioma by human monoclonal antibody. *Lancet* **2**, 1214–1215 (1982).
P10. Phillips, T., More, N., Queen, T., Holohan, N., Kremerr, N., and Thompson, A., High-performance affinity chromatography: A rapid technique for the isolation and quantitation of IgG from cerebral spinal fluid. *J. Chromatogr.* **317**, 173–179 (1984).

P11. Pollack, S., Jacobs, J., and Schultz, P., Selective chemical catalysis by an antibody. *Science* **2**, 1570–1573 (1986).
P12. Porterfield, J., and Allison, A., Studies with pox viruses by an improved plaque technique. *Virology* **10**, 233–244 (1960).
P13. Putnam, J., Wyatt, D., Pugh, G., Noll, L., and Lydersen, B., Maximizing antibody production using the Opticell culture system. *Hybridoma* **4**(1), 63 (1985).
R1. Raubitschek, A., Senyk, G., Larrick, J., Lizak, G., and Foung, S., Human monovalent antibodies against group A red blood cells. *Vox Sang.* **48**, 305–308 (1985).
R2. Reason, D., Carminati, J., Kimura, J., and Henry, C., Directed fusion in hybridoma production. *J. Immunol. Methods* **99**, 253–257 (1987).
R3. Rees, A., and de la Paz, P., Investigating antibody specificity using computer graphics and protein engineering. *Trends Biochem. Sci.* **11**(3), 144–148 (1986).
R4. Reik, L., Maines, S., Ryan, D., Levin, W., Bandiera, S., and Thomas, P., A simple, non-chromatographic purification procedure for monoclonal antibodies: Isolation of monoclonal antibodies against cytochrome P450 isozymes. *J. Immunol. Methods* **100**, 123–130 (1987).
R5. Reimer, C., Immunofluorometric techniques. *In* "Monoclonal Antibodies in Clinical Diagnostic Medicine" (D. Gordon, ed.), pp. 5–18. Igaku-Shoin, Tokyo, 1985.
R6. Roederer, J., and Bastiaans, G., Microgravimetric immunoassay with piezoelectric crystals. *Anal. Chem.* **55**, 2333–2336 (1983).
R7. Romer, J., Spath, P., Skvaril, F., and Nydegger, U., Characterization of various immunoglobulin preparations for intravenous application. II. Complement activation and binding to staphylococcus protein A. *Vox Sang.* **42**, 74–80 (1982).
R8. Rosen, A., Persson, K., and Klein, G., Human monoclonal antibodies to a genus-specific chlamydial antigen produced by EBV transformed B cells. *J. Immunol.* **130**, 2899–2907 (1983).
R9. Rosen, L., Hovis, J., Mastrota, F., Bell, J., and Huebner, R., Observations on a newly recognized virus (Abney) of the reovirus family. *Am. J. Hyg.* **71**, 258–265 (1960).
R10. Rowe, W., and Capps, W., A new mouse virus causing necrosis of the thymus in newborn mice. *J. Exp. Med.* **113**, 831–844 (1961).
S1. Sakano, H., Kurosawa, Y., Weigert, M., and Tonegawa, S., Identification of nucleotide sequence of a diversity DNA segment (D) of immunoglobulin heavy chain genes. *Nature (London)* **290**, 562–565 (1981).
S2. Sampson, I., Hodgen, A., and Arthur, I., The separation of immunoglobulin M from human serum by fast protein liquid chromatography. *J. Immunol. Methods* **69**, 9–15 (1984).
S3. Sawada, S., Kawamura, T., Masuho, Y., and Tomibe, K., Characterization of a human monoclonal antibody to lipopolysaccharides of *Pseudomonas aeruginosa* serotype 5: A possible candidate as an immunotherapeutic agent for infection with *P. aeruginosa*. *J. Infect. Dis.* **152**, 965–970 (1985).
S4. Schauf, V., Perspectives on regulatory considerations for human monoclonal antibodies for use in man. *In* "Human Hybridomas: Diagnostic and Therapeutic Applications" (A. Strelkauskas, ed.), pp. 269–277. Dekker, New York and Basel, 1987.
S5. Schmidt, U., Brown, J., Whittle, H., and Lin, P., The human-human hybridomas secreting monoclonal antibodies to the M.W. 195,000 *Plasmodium falciparum* blood stage antigen. *J. Exp. Med.* **163**, 179–188 (1986).
S6. Schroeder, D., Tankersley, D., and Lundblad, J., A new preparation of modified immune serum globulin (human) suitable for intravenous administration. I. Standardization of the reduction and alkylation reaction. *Vox Sang,* **40**, 373–382 (1981).

S7. Schroff, R., Foon, K., Beatty, S., Oldham, R., and Morgan, A., Human anti-murine immunoglobulin responses in patients receiving monoclonal antibody therapy. *J. Cancer Res.* **45**, 879–885 (1985).
S8. Schwaber, J., Posner, M., Schlossman, S., and Lazarus, H., Human-human hybrids secreting pneumococcal antibodies. *Hum. Immunol.* **9**, 137–143 (1984).
S9. Sears, H., Herlyn, D., Steplewski, Z., and Koprowski, H., Effects of monoclonal antibody immunotherapy on patients with gastrointestinal adenocarcinoma. *J. Biol. Response Modif.* **3**, 138–150 (1984).
S10. Sharon, J., Gefter, M., Manser, T., and Ptashne, M., Site directed mutagenesis of an invariant amino acid residue at the variable diversity segments junction of an antibody. *Proc. Natl. Acad. Sci. U.S.A.* **83**, 2628–2631 (1986).
S11. Sharon, N., and Pollard, M., Propagation of mouse adenovirus on cell lines of human origin. *Nature (London)* **202**, 1139–1140 (1964).
S12. Shelman, M., Wilde, C., and Kohler, G., A better cell line for making hybridomas secreting specific antibodies. *Nature (London)* **276**, 269–270 (1978).
S13. Shoenfeld, Y., Hsu-Lin, S., Gabriels, J., Silberstein, L., Furie, B. C., Stoellar, B., and Schwartz, R., Production of autoantibodies by human-human hybridomas. *J. Clin Invest.* **70**, 205–208 (1982).
S14. Sigal, N., and Klinman, N., The B-cell clonotype repertoire. *Adv. Immunol.* **26**, 255–328 (1978).
S15. Sikora, K., Nethersell, A., Ritson, A., and Smedley, H., Tumour localization by human monoclonal antibodies. *Med. Oncol. Tumor Pharmacother.* **2**(2), 77–86 (1985).
S16. Simpson, J., Boyd, J., Yates, C., Cristie, J., Fink, G., James, K., and Gordon, A., Autoantibodies to Alzheimer and normal brain structures from virus transformed lymphocytes. *J. Neuroimmunol.* **13**, 1–8 (1986).
S17. Smith, L., and Teng, N., Applications of human monoclonal antibodies in oncology. *In* "Human Hybridomas: Diagnostic and Therapeutic Applications" (A., Strelkauskas, ed.), pp. 121–158. Dekker, New York and Basel, 1987.
S18. Staehelin, T., Hobbs, D., Hsiangfu, K., Lai, C., and Pestka, S., Purification and characterization of recombinant human leukocyte interferon (IFLrA) with monoclonal antibodies. *J. Biol. Chem.* **256**(9), 9750–9754 (1981).
S19. Staerz, U., and Bevan, M., Hybrid hybridoma producing a bispecific monoclonal antibody that can focus effector T cell activity. *Proc. Natl. Acad. Sci. U.S.A.* **83**, 1453–1457 (1986).
S20. Stanker, L., Vanderlaan, M., and Juarez-Salinas, H., One-step purification of mouse monoclonal antibodies from ascites fluid by hydroxylapatite chromatography. *J. Immunol. Methods* **76**, 157–169 (1985).
S21. Steele, E., and Pollard, J., Hypothesis: Somatic hypermutation by gene conversion via the error prone DNA-RNA-DNA information loop. *Mol. Immunol.* **24**(6), 667–673 (1987).
S22. Steinitz, M., Kelin, G., Koskimies, S., and Mäkelä, O., EB virus-induced B lymphocyte cell lines producing specific antibody. *Nature (London)* **269**, 420–422 (1977).
S23. Strosberg, A., Collins, J., Black, P., Malamud, D., Wilbert, S., Bloch, K., and Haber, E., Transformation by simian virus 40 of spleen cells from hyperimmune rabbit: Demonstration of production of specific antibody to the immunizing antigen. *Proc. Natl. Acad. Sci U.S.A.* **71**, 263–264 (1974).
S24. Sugano, T., Matsumoto, Y., Miyamoto, C., and Masuho, Y., Hybridomas producing human monoclonal antibodies against varicella-zoster virus. *Eur. J. Immunol.* **17**, 359–364 (1978).

S25. Suresh, M., Cuello, A., and Milstein, C., Advantages of bispecific hybridomas in one-step immunocytochemistry and immunoassays. *Proc. Natl. Acad. Sci. U.S.A.* **83,** 7989–7993 (1986).
T1. Tabor, E., "Infectious Complications of Blood Transfusions." Academic Press, New York, 1982.
T2. Takahashi, M., Fuller, S., and Hurrell, J., Production of IgG-producing hybridomas by *in vitro* stimulation of murine spleen cells. *J. Immunol. Methods* **96,** 247–253 (1987).
T3. Tami, J., Parr, M., Brown, S., and Thompson, J., Monoclonal antibody technology. *Am. J. Hosp. Pharm.* **43,** 2816–2825 (1986).
T4. Teng, N., Reyes, B., Bieber, G., Fry, M., Lam, K., and Herbert, J., Strategies for stable human monoclonal antibody production: Construction of heteromyelomas *in vitro* sensitization and molecular cloning of human immunoglobulin genes. *In* "Human Hybridomas and Monoclonal Antibodies" (E. Engleman, S. Foung, J. Larrick, and A. Raubitschek, eds.), pp. 71–91. Plenum, New York, 1985.
T5. Towbin, H., Staehelin, T., and Gordon, J., Electrophoretic transfer of proteins from polyacrylamide gels to nitrocellulose sheets: Procedure and some applications. *Proc. Natl. Acad. Sci. U.S.A.* **76,** 4350–4354 (1979).
T6. Tramontano, A., Janda, K., and Lerner, R., Catalytic antibodies. *Science* **234,** 1566–1569 (1986).
T7. Tromberg, B., Sepaniak, M. J., Vo-Dinh, T., and Griffin, G., Fiber-optic chemical sensors for competitive binding fluoroimmunoassay. *Anal. Chem.* **59**(8), 1226–1230 (1987).
U1. Underwood, P., and Bean, P., The influence of methods of production, purification and storage of monoclonal antibodies upon their observed specificities. *J. Immunol. Methods* **80,** 187–197 (1985).
V1. Van Brunt, J., Centocor: Cashing in on serendipity. *Bio Technology* **3,** 126–128 (1985).
V2. Van Meel, F., Steenbakkers, P., and Oomen, J., Human and chimpanzee monoclonal antibodies. *J. Immunol. Methods* **80,** 267–276 (1985).
V3. Van Wezel, A., and Van der Marel, P., The application of immunoadsorption on immobilized antibodies for large scale concentration and purification of vaccines. *In* "Affinity Chromatography and Related Techniques" (T. Gribnau, J. Visser, and R. Nivard, eds.), pp. 283–292. Elsevier, Amsterdam, 1982.
V4. Van Zaane, D., Use of monoclonal antibodies in virus diagnosis. *In* "Recent Advances in Virus Diagnosis" (M. McNulty and J. McFerran, eds.), pp. 145–156. Martinus Nijhoff Publishers, Boston, Massachusetts, 1984.
V5. Vienken, J., and Zimmermann, U., Electric field induced fusion: Electrohydraulic procedure for production of heterokaryon cells in high yield. *FEBS Lett.* **137**(1), 11–13 (1982).
V6. Vora, S., Monoclonal antibodies in enzyme research: Present and potential applications. *Anal. Biochem.* **144,** 307–318 (1985).
W1. Wabl, M., Burrows, P., von Gabain, A., and Steinberg, C., Hypermutation at the immunoglobulin heavy chain locus in a pre-B-cell line. *Proc. Natl. Acad. Sci. U.S.A.* **82,** 479–482 (1985).
W2. Whitley, R., Parvovirus infection (editorial). *N. Engl. J. Med.* **313,** 111–112 (1985).
W3. Wilchek, M., Miron, T., and Kohn, J., Affinity chromatography. *In* "Methods in Enzymology" (W. Jakoby, ed.), Vol. 104, Part C, pp. 3–55. Academic Press, New York, 1984.
W4. Winger, L., Winger, C., Shastry, R., Russel, A., and Longenecker, M., Efficient generation *in vitro* from human peripheral blood cells of monoclonal EB-virus trans-

formants producing specific antibody to a variety of antigens without prior deliberate immunization. *Proc. Natl. Acad. Sci. U.S.A.* **80,** 4484–4488 (1983).

W5. Winkler, M., Bringman, T., and Marks, B., The purification of fully active recombinant transforming growth factor alpha produced in *Escherichia coli. J. Biol. Chem.* **261,** 13838–13843 (1986).

Y1. Yananitim, B., Nagasawa, Y., Shuto, S., Tsubomura, H., Sawai, M., and Okumura, H., Antigen-antibody reaction investigated with use of a chemically modified electrode. *Clin. Chem. (Winston-Salem, N.C.)* **26,** 1569–1572 (1980).

MONOCLONAL ANTIBODIES: CLINICAL APPLICATIONS

Rudolph Reckel*

Immunology Development,
Immunomedics, Inc.,
Newark, New Jersey

1. Introduction... 355
   1.1. Monoclonal versus Polyclonal Antibodies........................... 356
   1.2. General Considerations 356
2. Neoplastic Disease... 357
   2.1. Diagnostic Applications 358
   2.2. Therapeutic Applications.. 371
3. Hormones and Proteins 376
   3.1. Hormones.. 376
   3.2. Enzyme Research 379
4. Cell Surface Antigens 381
   4.1. Leukocyte Antigens 382
   4.2. Histocompatibility Antigens 384
   4.3. Erythrocyte Antigens 386
5. Miscellaneous Clinical Applications..................................... 387
   5.1. Infectious Diseases 387
   5.2. Clinical Rheumatology 389
6. Conclusions... 391
   References... 391

1. Introduction

Since the hallmark paper of Kohler and Milstein (K16, K17), increasing numbers of applications for monoclonal antibodies (Mabs) have been reported (e.g., M17, M28, S7, S12, T16, Y2, Y3). The use of Mabs has applications in many areas of biological research, clinical medicine, and new drug delivery systems. Slowly, as the data warrant, Mabs are beginning to replace conventional polyclonal antibodies (Pabs), and commercial companies are introducing them in a variety of immunoassay kits. In this review, an attempt will be made to summarize those applications which

*Deceased.

best serve the medical community. Particular emphasis has been placed on the uses of Mabs in neoplastic disease, where many *in vitro* and *in vivo* diagnostic and therapeutic applications have been developed and refined for practical use. Additional areas of current Mab research have been selected for discussion, because it is felt that they have advanced beyond fundamental studies and have progressed into the working laboratory. The actual techniques for the preparation of Mabs and their purification are covered elsewhere (see Vetterlein, this volume).

1.1. Monoclonal versus Polyclonal Antibodies

When an animal is injected with an immunogen, the animal responds by producing an enormous diversity of antibody structures directed against different antigens and different determinants; different antibody structures may be directed against the same determinant. Once these antibodies are produced, they are released into the circulation and it is virtually impossible to separate all the individual components present in serum. Each antibody, however, is made by individual cells, and the immortalization of specific antibody-producing cells by somatic cell fusion, followed by cloning of the appropriate hybrid derivative, allows the perpetual production of each of the antibodies in separate culture vessels. These cells can usually be either injected into animals to develop myeloma-type tumors, forming ascites (which are rich in Mabs), or grown in an assortment of sophisticated fermentors designed to produce gram or kilogram amounts of Mab. Monoclonal antibodies only require an initial immunization with an antigen of reasonable purity, since one can select the antibody of desired specificity from the many cross-reactive types made. The only concerns initially are that the selected antigen is imunogenic in the host animal and that an adequate screening test is available to the desired antibody once it is produced (B4, D22, F5, L3, R9, S41). Homogeneity, specificity, and availability make Mabs attractive for future expansion in immunoassays, immunoaffinity chromatography, and areas still to be discovered.

1.2. General Considerations

A physical listing of the myriad of immunoassays described in the literature would be impractical in a review of this nature, thus the reader is referred to reviews which summarize these (e.g., S14). Each of the ensuing sections will summarize those Mabs which have made useful contributions to the field of immunodiagnostics under discussion. In Section 2 monoclonal antibodies will be discussed as both diagnostic and therapeutic tools in neoplastic disease. The traditional immunoassays are replacing Pabs with Mabs where justifiable [radioimmunoassay (I2); enzyme immunoassay (S39, U2); immunofluorescence (D7)]; and others (D8, R9, S28). Affinity chromatography (C17) is an exciting area for Mab application (K4, Q1, S13). The immunometric assay (W15) has made enormous inroads using Mabs. Assays such as the forward and reverse two-step methods, and the simultaneous method (M22), take advantage of the newly created Mabs which offer reproducible reactants to multiple epitopes on a single molecule making two-site assays a reality. However, investigators have discovered that advantages offered by Mabs are at times accompanied by disappointment and challenge.

Most Pabs, particularly those which are IgG, are quite stable as liquid reagents. Purified Mabs with excellent binding properties, however, are sometimes unstable in solution at physiological pH (6–8) or in certain media (e.g., low ionic strength) due to characteristics of the Mabs structure (poor solubility, intermolecular hydrophobic interactions, etc.). Also, some Mabs become useless for applications where Mab modifications (i.e., radioiodination, fragmentation, and enzyme conjugation) result in inactivation of the antigen-binding portions of the antibody molecule due to reactive sites being located in the variable parts of the F(ab) region.

Cross-reactivity of new Mabs to epitopes in unrelated tissue has been observed. This can be a serious diagnostic problem for the pathologist. Cross-reactions such as that between mammalian neurones and *Typanosoma cruzi* parasites as reported by Wood *et al.* (W13) have been reported. Inaccurate diagnosis can result if a Mab does not recognize an important isoenzyme because a functionally active genetic variant lacks the epitope. Consequently, care must be always taken when dealing with these new diagnostic tools.

Some concern has been expressed over the possibility that the large-scale production of Mabs in rodents will result in material contaminated by rodent viruses. This potential problem has been reviewed by Carthew (C6), who discusses the hazard posed, particularly in human therapy, and the safeguards that can be taken.

2. Neoplastic Disease

In 1984, approximately 6 million new cases and 4.5 million deaths were attributed to cancer world wide (W16). In the United States, the four most common cancers are carcinoma of the lung (126,000 deaths, 144,000 new cases), colon/rectum (60,000 deaths, 138,000 new cases), breast (38,000 deaths, 119,000 new cases), and prostate (26,000 deaths, 86,000 new cases). In industrial countries, lung, breast, and colorectal cancers are most common, while in developing countries cervical, oral, and liver cancers predominate.

Immunological reactions to malignant disease have been under study since the turn of the century. In the 1940s, animal models were developed (F8, G40) which showed that malignant cells could invoke an immune response. Antisera to animal xenografts to other tumor types followed (D6, P8). The advent of monoclonal antibodies (Mabs) has allowed us to make many new discoveries since the problems asociated with polyclonal nonspecificity were overcome. In the following sections we will review many clinical applications in which Mabs have been used in both diagnostic and therapeutic areas of human malignant disease.

2.1. Diagnostic Applications

The differential diagnosis of cancer is difficult because cancer exists as many diseases (over 200 types) and the symptoms are not always specific. Once located, the tumor must still be classified as either benign or malignant. The first diagnostic priority ideally is to screen the entire symptomatic population or high-risk groups for early disease. The next diagnostic criterion is the accurate diagnosis of symptomatic patients. Next, tests to monitor disease progress enable evaluation for purposes of determining therapy or for detecting recurrences. Finally, indicators of disease prognosis are necessary to plan treatment programs. *In vitro* testing by standard histochemical stains (histopathology of biopsy specimens) confirms most cancers today. Tumor marker immunoassays and occult blood tests comprise most of the remaining *in vitro* monitoring and screening procedures. Noninvasive *in vivo* diagnosis is possible by traditional palpation (e.g., breast cancer), visualization by endoscopy (e.g., colorectal cancer), and imaging with ultrasound [computerized axial tomography (CAT) scans] and nuclear magnetic resonance (NMR). However, Mabs have introduced newer possibilities with parenterally introduced labeled antibodies to tumors to locate and treat cancers *in vivo*.

One of the major problems in treating cancer patients is the inadequate assessment of disease. Only a few tumors secrete circulating markers which accurately reflect the presence of active tumor and tumor load. In germ cell tumors, α-fetoprotein (AFP) and human chorionic gonadotropin (hCG) have been shown to have clinical value (A4). Many solid tumors, which comprise up to 90% of all cancers, do not have convenient markers. The advent of Mabs has greatly improved the quality of existing tests of known markers [i.e., carcinogenic embryonic antigen CEA)]. Novel new markers may be assayed using Mabs, i.e., assays for oncogene products, and these may become the next index of tumor burden (H3).

2.1.1. Human Tumor Marker Assays

Considerable debate has existed whether specific tumor markers exist. Early animal experiments with virus-induced or chemically induced tumors demonstrated the presence of transplantation antigens when these tumor cells were put into syngeneic hosts (K11, O4). Antibodies to various carcinomas could also be found circulating in patient sera (E3, H1, S19).

The advent of Mabs made it possible to develop antibodies to tumors of many different histological types. Several of these Mabs to tumor markers (antigens) have been shown to have great utility as diagnostic indicators and are discussed below.

2.1.1.1. α-*Fetoprotein.* AFP is a glycoprotein and one of most extensively studied tumor markers. Initially found in serum of mice with chemically induced hepatomas (A3, K10, O3), it is an example of an oncofetal protein and is synthesized by liver cell during fetal development (A2). Immunofluorescent studies showed excellent correlations between circulating serum levels and ontogenesis (A2, B25, S20, S21, S35). After reaching a maximum level during the first trimester, serum levels range from 1 to 4 ng/ml after the first year of life (R23, R24). It was found to be useful for diagnosing or monitoring conditions other than cancer. Elevated AFP levels in the amniotic fluid or maternal serum indicated the presence of fetal neural tube defects (e.g., spina bifida). Partial hepatectomy or chemical (carbon tetrachloride) poisoning causes liver regeneration resulting in AFP synthesis and elevated serum levels that last a week before normalizing (T2). Neoplasms reported to contain AFP include hepatoma, germinal teratocarcinoma, embryonal cell carcinoma, mixed seminoma, ovarian endodermal sinus tumor, serous cystadenocarcinoma, gastric carcinoma, colonic carcinoma, and pancreatic carcinoma (G26, K7, M13, O2, S5). Monitoring of serum AFP levels in primary hepatocellular tumors (elevated in 70–80% of patients) and yolk sac-derived tumors of the ovary and testes (elevated in 55–80% of males) has been demonstrated (A1, A6, N6, R25, W1). Moderate and transient elevated AFP serum levels also occur in approximately 10% of patients with acute and chronic hepatitis, cirrhosis, and liver metastases (B22, D13, R17). Two AFP Mabs were described by Uotila *et al.* (U1, U2), who used them in a two-site serum enzyme-linked immunosorbent assay (ELISA). Bellet *et al.* (B10) have developed a sensitive Mab radioimmunoassay which has high sensitivity and specificity for primary hepatomas.

Commercial kits are marketed for the dual purpose of fetal neural tube defect screening and cancer management, including polyclonal two-site ELISA assays (e.g., Abbott Laboratories, Behring Diagnostics), monoclonal immunoradiometric assays (IRMAs) (e.g., Hybritech, Inc.), and a monoclonal IRMA and enzyme-enhanced luminescent immunoassay (ELIA) (Amersham, Inc.).

2.1.1.2. *Carcinoembryonic Antigen.* CEA is a ubiquitous glycoprotein antigen associated primarily with epithelial tumors in man and was probably the first antigen recognized as a possible tumor-specific marker. It is a complex antigen composed of several epitopes and is generally regarded as an antigen of the gastrointestinal system. It is present in the serum of normal individuals at a concentration of about 2.5 ng/ml and is found elevated in smokers and in about 20% of persons with ulcerative colitis, Crohn's disease, pancreatitis, and liver or lung infections. When levels exceed 20 ng/ml, advanced malignancy is suspected; levels 100 times normal have been seen in some cancer patients. Antigens from other organ systems, however, have been isolated which are cross-reactive with CEA isolated from the gastrointestinal region. These include the nonspecific cross-reacting antigen (NCA) and a meconium antigen (MA), found in colonic tumors, normal colonic tissue, spleen, lung, and plasma (B48, K12, K14). Cultured human cell lines producing CEA have been reported from colon (E1, G18), liver (R17), lung (E5, N4), prostate (W11), pancreas, thyroid, cervix, breast, testes (N4), and melanomas (N4, R17). The molecular weight of CEA isolated from colonic tumors is about 200,000 daltons while CEA from other tumors may have molecular weights as high as 370, 000 Da (H14, P8). These molecular-weight discrepancies appear to be related to variations in the carbohydrate region of the glycoprotein, with carbohydrate:protein ratios varying from 1:1 to 1:5 among tumor types (B3, R22). Histological studies revealed CEA to be on the plasma membrane (G13) and electron microscopy further showed it in the glycocalyx as an associated antigen and not an integral part of the cellular membrane (G15). CEA has been shown to be present in gastric, colonic, pancreatic, biliary, bronchogenic, mammary, uterine, ovarian, prostatic, and urinary bladder carcinoma (G19). There have been reports of a great many monoclonal antibodies to various epitopes on CEA (A4, B23, B45, B47, G13, G14, I1, K13, K33, M14, N5, S5). In colonic cancer, elevated CEA levels relate to Dukes' (D25) classification as follows: stage A, 20–40%; stage B, 50–70%; stage C, 70–80%; and metastases, 80–90% (M14). When colonic metastases of the liver are present, CEA levels often rise rapidly and are valuable in differentiating these from other liver diseases (S5, S27). The use of monoclonal antibodies in immunoassays has helped discriminate between small differences in the CEA molecule (R22). A solid-phase enzyme immunoassay was shown to have good correlation with conventional antisera (B39), but cross-reactivity with normal sera was observed. Primus *et al.* (P14) described four Mabs (NP-1, NP-2, NP-3, and NP-4) that recognized at least four different determinants of CEA.

Three of these (NP-2, NP-3, and NP-4) did not cross-react with NCA. NP-1 cross-reacted with NCA and a meconium antigen. NP-2 and NP-3 recognize epitopes that are shared with CEA and MA but not NCA. The NP-4 Mab recognizes an epitope only expressed on CEA. Another enzyme immunoassay described by Hedin et al. (H8) uses two noncrossreactive Mabs, and when compared to conventional Pabs shows greater specificity for carcinomas. Predictive values of 60–90% (pancreatic cancer), 60–65% (lung cancer), and 50–65% (breast cancer) have been reported. Pinto (P7) assayed CEA levels in 14 pericardial effusions of suspected malignant pericarditis patients using a commercial Mab ELISA with a sensitivity of 0.5 ng/ml (serum) and 1.0 ng/ml (pleural fluid).

When compared to a cytological examination, the method was 100% specific and sensitive (100% specific, 90% sensitive by cytological exam). Many commercial kits have been marketed as cancer immunoassays. These include Mab-based, two-site ELISA tests with a sensitivity of 0.3–50 ng/ml (e.g., Roche, Abbott, Hybritech), Mab IRMAs with sensitivities between 0.5 and 50 ng/ml (e.g., Immunomedics, Hybritech, Amersham), and polyclonal RIAs and ELISAs with sensitivities between 2.5 and 250 ng/ml (e.g., Roche, Pharmacia, Diagnostic Products, Boehringer Mannheim, Behring).

2.1.1.3. *Pancreatic Oncofetal Antigen.* This oncodevelopmental marker was first described by Banwo et al. in 1974 (B5). It was demonstrated to be immunologically distinct from carcinoembryonic antigen, α-fetoprotein, ferritin, and all acute-phase reactants. It had a molecular weight between 800,000 and 900,000 and has been shown to be present in pancreatic, biliary tract, gastric, colonic, breast, and bronchogenic carcinomas (G5–G7, H20). The highest serum pancreatic oncofetal antigen (POA) levels are found in carcinomas of the pancreas, with cross-reactivity observed in biliary tract and bronchogenic carcinomas. Poorly differentiated large tumors have lower serum levels than smaller well-differentiated tumors and this may suggest the POA serum concentration is not a good immunodiagnostic test (H16).

2.1.1.4. *Prostatic Acid Phosphatase and Prostate-Specific Antigen.* Prostatic acid phosphatase (PAP) was first described in the 1970s and at first was thought to be a specific marker for adenocarcinoma of the prostate (C11, G44). Fewer than 50% of malignancies confined to the prostate are detectable but almost 95% of metastasized prostatic carcinomas are detectable (B38, F10, M6, Q2). PAP (isoenzyme II) is a glycoprotein of approximately 100,000 Da and has a unique organ-specific epitope located in the protein region of the molecule. The antigens originate in the ductal epithelium of the prostate, and while they are shed in greater quantities in cancer than in benign prostatic hypertrophy, the presence of PAP in benign disease and its occasional absence in cancer patients limits its usefulness as a cancer diagnostic. It is still useful for monitoring defined cases. Monoclonal antibodies have been made making specific immunoassays possible (C10, G20). Pab- and Mab-based commercial kits have been marketed for this purpose. These include two-site Mab and Pab ELISA kits (e.g., Hybritech, Abbott, Behring, Cetus), IRMAs (e.g., Hybritech), and RIAs (e.g., Diagnostic Products, Amersham).

Prostate-specific antigen (PSA) has been shown more recently to have promise as a marker for monitoring the progress and therapy of adenocarcinoma of the prostate (W5). A two-site ELISA (Hybritech) is being marketed currently, but "for research only" kits are available (Cetus and Diagnostic Products).

2.1.1.5. *Melanoma Antigen p97 and Mab 9.2.27.* Melanomas have been extensively studied by immunological means. Melanoma antigens have been detected in the cell cytoplasm (B43, M35, W12) as well as on the cell membrane (G33, L13, M21, R16). Antibodies from patients with malignant melanoma and hyperimmune sera from immunized animals were the reagents used initially to develop assays (G31, M21, V3). Conventional immunofluorescent and complement fixation tests using Pabs have been used to detect serum antibodies to malignant melanoma (G41, G42, L13, M38, R16). Recently, a melanoma-specific antigen (p97) and a Mab (9.2.27) to a melonoma antigen have been described (B34–B37, M31, W12, W14).

The p97 antigen is a 97,000-Da iron-binding glycoprotein related to transferrin and forms an integral part of the cell membrane (B35, B37). This antigen, which was initially thought to be tumor specific, is expressed in low amounts in normal tissues and at high levels in embryonic colon and intestine (B36). Most melanomas have from 50,000 to 500,000 p97 molecules per cell as compared to 10,000 p97 molecules per cell in normal tissue (B36). Mabs to p97 have been used as histochemical stains to aid in diagnosing primary tumors and to aid in the selection of Mabs useful in binding tumors, and can be used for tumor localization for therapy. The Mab 9.2.27 is reactive with the glycoprotein melanoma-associated chondroitin sulfate proteoglycan complex (MPG) (B41, B42, K33). The levels of serum antibodies diminish as the tumor progresses (C1, L15, M34) and immune complexes between soluble circulating antigen and antibody may form (J10, L14). Mabs to melanoma antigen thus far have been used primarily as histochemical reagents and in therapy, and serological assays are still under development. This Mab and related antigens have been reviewed by Reisfeld (R8).

2.1.1.6. *Human Chorionic Gonadotropin.* Human chorionic gonadotropin was first isolated from the human placenta and described by Hirose

(H15). Analysis of the hCG structure shows that it contains two peptide chains, the α and β subunits, which are held together by disulfide bridges. The α subunit is common to the other human glycoprotein hormones, follicle-stimulating hormone (FSH), luteinizing hormone (LH), and thyroid-stimulating hormone (TSH). The β subunit is specific for hCG and contains unique epitopes. Following the observations by Aschheim and Zondek that the urine of pregnant women contained a substance that could stimulate ovaries of test mammals (A10), gonadotropins were found in the urines of patients with tropoblastic neoplasms (B33, H9, Z2). Human chorionic gonadotropin has been also demonstrated in nontrophoblastic neoplasms (A5, M7, S23), spermatozoa (A9), and in the following human malignancies: germinal cell tumors, trophoblastic disease, and ovarian, pancreatic, biliary tract, gastric, hepatocellular, pulmonary, and breast carcinomas (B30, B31, D19, G1, G27, L2, S3, S18, S34, S40, T9). In 1972, Vaitukaitis (V1) described a specific radioimmunoassay which used highly specific Pabs capable of measuring hCG in the presence of cross-reacting human luteinizing hormone. In 1973, the linear amino acid sequences of the β subunits of hCG were reported (B11). The need to differentiate normal gestation and gestational trophoblastic disease (GTD) led to the development of sensitive immunoassays which had been made β-subunit specific and which could accurately measure elevated hCG levels in the first trimester of pregnancy and monitor the return of these values to normal at the termination of the pregnancy (B29). If the hCG did not decline to normal levels it indicated GTD (B28, D12). Mabs have been developed to the β-hCG subunit which are not cross-reactive with subunits. The role of hCG in monitoring choriocarcinoma (T17) and testicular cancer (J7) has been studied extensively. Two groups (B24, T3) have described Mabs to β-hCG with low cross-reactivity to luteinizing hormone (principal cross-reacting gonadotropin) and show that their assays correlate well with conventional assays.

These are many quantitative and qualitative tests for β-hCG. The qualitative tests are primarily for pregnancy testing while the quantitative assays are used for both pregnancy testing and cancer monitoring. The commercial kits include two-site Mab ELISAs and IRMAs with sensitivities of 1.5–1.7 mIU/ml (e.g., Hybritech, Roche), Mab ELISAs and RIAs with sensitivities of 0.6 mIU/ml (Amersham), and a Pab EIA with a sensitivity of 1 mIU/ml (Abbott).

2.1.1.7. *Ca 19-9*. Koprowski *et al.* (K21), using colorectal cancer cells as the immunogen, succeeded in isolating a hybridoma antibody which bound specifically with colorectal cancer cells. This Mab also reacted with sera from patients with colorectal cancer and not with normal sera (K20). It was subsequently shown that the antigen involved is a monosialoganglioside (M5). Elevated serum levels were shown in patients with colorectal cancer (64%), gastric cancer (72%), and pancreatic cancer (92%), and it may be specific for gastrointestional cancer (H11, K24). The Mab to Ca 19-9 has also been shown to cross-react with human Lewis b blood antigen (K19). It is currently used primarily as a marker for pancreatic cancer even though the prognosis for this disease is poor. Commercial kits are available using this Mab.

2.1.1.8. *Ca 125.* Following immunization with a serous ovarian cell line, Bast and co-workers (B7) were able to isolate a hybridoma with a Mab reacting to an ovarian cancer antigen (Ca 125). The antibody reacted against 12 of 20 ovarian cancer biopsies. Subsequently these workers developed an RIA for Ca 125 (K15), which showed normal serum values of 11.2 ± 5.4 U/ml. A cutoff value of 35 U/ml was selected and 82% of ovarian cancer patients showed values >35 U/ml, with 6% of normals also elevated. At a cutoff of 65 U/ml, this changed to 73% elevated for ovarian cancer, with 2% elevated in normal subjects. While elevated values of Ca 125 are potentially diagnostic for ovarian cancer, cross-reactivity to the endometrial, cervical, and large intestinal carcinomas has been shown. A good correlation has been shown with rates of tumor recurrence and with responses to chemotherapy (C2). A Mab RIA for this marker has recently been marketed (Centocor) for monitoring patients with ovarian cancer.

2.1.1.9. *Placental Alkaline Phosphatase.* Placental alkaline phosphatase (PLAP) has been associated with carcinoma of the testes, ovaries, lung, stomach, and pancreas (S33, S40), but assays using polyclonal antisera could not differentiate the different isoenzymes of alkaline phosphatase, making these assays of little use. The recent development of Mabs to PLAP showed specificity and usefulness in monitoring tumor progression (E8). Studies comparing PLAP with AFP and hCG are being considered for the monitoring of testicular cancer.

2.1.2. *Tumor Localization*

A great deal of research has gone into the development of Mabs directed to tumor antigens; when these Mabs are radiolabeled and injected into patients with suspected cancers, they can be used for the purpose of imaging the malignancies. Imaging with radioactive antibodies or radioimmunodetection (RAID) has been under development for at least 35 yr. Several reviews have been written on this subject (D11, G16, G17, K5, K15, R1, R8). Recent progress has resulted from several advances, namely, identification of suitable tumor-associated antibodies, suitable human xenograft models (G10), knowledge that circulating tumor antigens do not block tumor localization, and development of computer-assisted and biological methods for reducing nontarget background radioactivity. The advent of new Mabs to tumor antigens came with great expectancies, but it was soon realized that some of the initial asumptions concerning Mabs were incorrect. The great specificity to antigen over that offered by polyclonal antibodies has at times turned out to be a disappointment, since only single epitopes are reacted with by Mabs while several epitopes are reacted with with Pabs. Thus for tumor visualization, Mabs offer in some instances less or sometimes inadequate imaging due to fewer reactive sites on the cell surface. Additional factors which would enhance tumor uptake of antibody (e.g., antibody binding constant) need to be considered. Heterogeneity became an issue when the specificities of many Mabs for tumor epitopes were shown to cross-react with normal tissues. At times this was not a major concern, e.g., when the relative level of antibody reactivity with tumor was many times more reactive than with normal cells. Such events have led to the speculation that cells in transition from normal to malignant states may pass through a series of program changes (L7). Other forms of heterogeneity have arisen with respect to antigenic properties, drug response, receptors, etc. (K6, W7). A Mab to one specific antigen may fail to react with histologically identical tumors in another individual, or even with metastases of the tumor from which the Mab was derived. Such loss of antigen expression has invoked other approaches (i.e., Mab/Mab or Mab/Pab mixtures). Additionally, an antigen can be expressed on the cell surfaces of a number of tissues and tumors (F4). Several Mabs to adenocarcinoma of the colon have been developed (S31). Some of these have shown binding to melanoma, breast carcinoma, and blood group substance B. Other Mabs to ovarian carcinoma and melanoma have been demonstrated (B24). Many other such cross-reactivities have been noted (C18, M10, T4).

The role of Mabs in diagnostic imaging, through much research, has generated several opinions (B2, C12, E7, E8, F2, G17, H7, H18, L10, M2, M3, M36, M37, P5, P6, S26, S32). Improved tumor-to-blood and tumor-to-normal tissue ratios have been documented in studies using Mab rather than Pab (M37). Mach *et al.* (M1, M3) showed that if the affinity constants were the same, Pab and Mab uptake on the same tumor antigen were equivalent. In similar studies, Primus *et al.* (P11, P13) evaluated the tumor concentration of Mabs against CEA with Pabs made in the goat. Three monoclonal anti-CEAs with reactivity to different epitopes on the CEA molecule (60–80% immunoreactivity following radiolabeling), and the goat polyclonal anti-CEA (70% immunoreactive), when injected into animal tumor models, all gave quite similar tumor-to-tissue ratios. This demonstrates that polyclonal and monoclonal antibodies against the same soluble antigen will localize equivalently.

The imaging of tumors with radiolabeled antibodies requires computer processing. When major sites of the blood pool are involved, conventional nuclear medicine collimators are not very applicable to the use of I-labeled antibodies for cancer imaging. Potent, high-affinity antibodies against tumor-associated antigens cannot overcome the relatively low concentration of labeled antibody on the tumor compared to that present in the background tissues. Highly vascular areas (i.e., heart, liver) interfere with suitable images. The inadequate concentration of radiolabeled antibody in tumor with respect to the background tissue, resulting in poor imaging, has been well described (D11, G21, G25, P4).

Ballou, Goldenberg, and others (B2, G23, G24) have reported several techniques which are used to enhance tumor imaging. The techniques for improving target-to-nontarget ratios which have been successfully demonstrated include the delaying of imaging to reduce background, the administration of radiolabeled compounds such as ^{99m}technetium (^{99m}Tc)-labeled albumin followed by a normalization and subtraction method, the administration of a second antibody directed to the radiolabeled primary tumor antibody, and the use of radiolabeled immunoglobulin fragments [i.e., F(ab')$_2$, F(ab')], which are cleared from the vascular system more rapidly.

The simple delay in imaging to allow reduction of background is clinically unwieldy, since delays can be as long as 7 days with variable results. Also, results are required more rapidly than this.

Forty-eight hours after the intravenous injection of radiolabeled antibody, average target-to-nontarget ratios of 2.3:1 can be shown. This, however, is not usually acceptable for routine planar (two-dimensional) imaging. The use of a radiolabeled tracer to mimic the nonspecific background distribution, followed by a normalization and computer-assisted subtraction, is attractive. This technique of subtraction (selective removal of nontarget radioactivity) with a second radiolabeled tracer as introduced by Kaplan et al. (K3) has been extended to remove nontarget radioactivity obscuring the lesion (D2, D3). This procedure has been used by Goldenberg in over 800 patients receiving ^{131}I-labeled anticancer antibodies. The basic requirements for a second radiotracer in the subtraction procedure are that the principal energy of the γ-ray emission from the second agent must be lower than that of the radionuclide label on the tumor, and that the two energies are far enough apart to be resolved in the dual-channel analyzers without overlap in the windows in which each nuclide is being collected (i.e., ^{99m}Tc-labeled albumin and ^{99m}Tc-labeled pertechnetate mixtures for imitating intravascular and extravascular components). It is also possible to use an organ-specific tracer (i.e., ^{99m}Tc–sulfur colloid) for liver.

With this method, simultaneous images for ^{131}I and ^{99m}Tc are stored in the computer, and after selection of an area of interest, the ^{99m}Tc background is subtracted from the ^{131}I activity. The result is an image with increased contrast and an increase in target:nontarget ratio of twofold to threefold. This technique is explained in detail elsewhere (D14).

The use of a second antibody directed against the primary radiolabeled anticancer antibody is still in its experimental stage. Sharkey *et al.* (S17) showed fourfold reduction of primary antibody radioactivity within 2 h after second-antibody administration. This technique has also been reported by Bradwell *et al.*, (B27) and Goodwin *et al.* (G30).

The use of antibodies which have been enzymatically digested with pepsin or papain to remove the Fc regions of these immunoglobulins is another technique which has been demonstrated. These fragments, either the F(ab)$_2$, or Fab fragments, can be produced and radioiodinated without significant loss of immunoreactivity. These radiolabeled antibodies are cleared from the circulation more rapidly than the intact IgG antibodies, since the elimination of the Fc portion of IgG antibodies reduces the nonspecific binding to many existing tissue Fc receptors and results in lowered background radiation. Buchegger *et al.* (B40) found fragments superior to intact IgG Mab in detecting cancer grafted into mice. In patients receiving fragments, 61% showed positive scans as opposed to 51% positive scans for those receiving intact IgG (M4). Additionally, the fragments, without the Fc piece attached, are much less immunogenic to the host. Fragments may also be hybridized, yielding binary molecules with different target specificities, which may increase tumor saturation.

Berche *et al.* (B14), using tomographic scintigraphy with improved image specificity and planar imaging with background subtraction detected lesions in the 18-cm^3 (3.5 cm) size range. Recently, Goldenberg's group (G22) examined 13 patients with a history of primary liver carcinoma. They were given either ^{131}I-labeled Pab or Mab against α-fetoprotein and were scanned with a γ camera. ^{99m}Tc imaging agents were used for tumor image enhancement by computer-assisted subtraction. A sensitivity of 91% (primary site), 50% (lungs), 33% (chest area), or 75% (abdomen, pelvis) was achieved with specificity of >94%. Figure 1 shows typical results of such a subtraction obtained by the Goldenberg group.

Recent progress with Mabs in RAID have introduced new dimensions. Recently, liposomes with second antibody have been used by Begent *et al.* (B9) as another technique to reduce background radioactivity. Radioiodinated primary goat anti-CEA was administered followed by a second antibody (horse anti-goat) entrapped in liposomes (B9). Accelerated clearance was demonstrated in four of five patients and improved γ camera

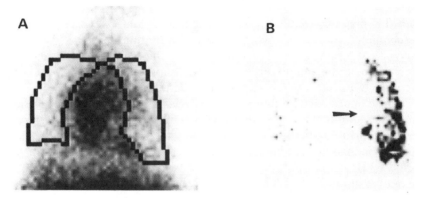

FIG. 1. Anterior chest view of a patient with colonic carcinoma metastatic to the left lung. (A) ^{131}I murine monoclonal anti-CEA IgG scan at 48 h prior to dual-isotope subtraction; (B) after subtraction of blood pool background radioactivity. Diffuse radioactivity throughout left lung.

imaging was shown in three of five patients. One drawback to this approach is that the liver sequesters liposomes and raises the radioactive levels in that organ, making imaging in this area less sensitive.

There has also been an increased use of other isotopes. Otsuka and Welch have recently reviewed the various radionuclides available (07) for labeling monoclonal antibodies. These include ^{131}I, ^{123}I, ^{77}Br (radiohalogens), and ^{111}In and ^{99m}Tc (radioactive metals) attached to antibody with bifunctional chelates [i.e., derivatives of ethylenediaminetetraacetic acid (EDTA) or derivatives of diethylenetriaminepentaacetic acid (DTPA)]. ^{131}I has been used in most of the immunolocalization studies published. This isotope's energy level is too high for conventional γ cameras and there is some risk to the patient. Both ^{111}In and ^{123}I have been investigated primarily because of lower energy and shorter half-life. The choices of which label to use depend on whether a suitable means is available to attach the isotope to the antibody (or antibody fragment) without destroying immunoreactivity, whether the isotope has a suitable decay rate during the clinical study, or whether the uptake of radiolabeled metabolites will interfere with imaging. This also assumes that the radiolabel remains attached to the antibody after administration and is not exchanged with other proteins or tissues (one major concern with chelated metals). Animal models have been described to evaluate labeled antibodies (B17, J2, K6, S9) as well as other nonanimal models (K18, 08).

2.1.3. *Histological Tumor Cell Typing*

Diagnostic immunohistology has become one of the well established areas in which Mabs have been used with great success. This has been important in the pathology laboratory where homogeneous reagents with precise specificity are invaluable. Two widely used techniques, the peroxidase–antiperoxidase method and the avidin–biotin–phosphatase complex method, have seen many Mab applications and this area of study has been recently reviewed by Falini and Taylor (F1). Since many tumor cells can be cultured readily, they can be used as immunogens for the generation of new Mabs. Some of these will react with the immunizing cell line but not with other tumors of the same organ, limiting their applicability. Characterization of these antibodies often reveals cross-reactivity with normal tissues. The discovery of "identical determinants" can be of great interest to the research biologist, since it can reveal previously unknown areas of homology between diverse macromolecules, whereas to the pathologist it can result in misleading diagnostic interpretations. Many sensitive tumor markers have been described, including those to tumors of the skin, breast, brain, lung, stomach, pancreas, colon/rectum, ovary, testes, bladder, kidney, and prostate and to lymphoma, melanoma, osteosarcoma (A5). In several malignant diseases, they have had particular importance (e.g., malignant lymphoma, melanoma, breast cancer, gastrointestinal cancer).

2.1.3.1. *Malignant Lymphoma.* The different lymphoid cells that exist and their different biological functions (i.e., helper or suppressor cells) can now be differentiated by certain cell surface markers using panels of select monoclonal antibodies. Malignancies of the lymphoid cells lead to their transformation and rapid multiplication. However, using monoclonal antibodies, these cells can still be typed as to their lymphoid cell origins and a proper assessment of the malignancy can be made. Neoplastic B cells in follicular lymphomas can be recognized by unique immunoglobulin light chains identifiable by a Mab (W6). Similarly, cutaneous T cell lymphomas can be differentiated (K32). The correct selection of monoclonal antibodies can be more accurate than conventional techniques (J4). There are instances when differentiation on morphological grounds is not possible while examination with monoclonal antibodies of cell surface markers and epithelial surface antigens is diagnostic. Gatter *et al.* (G4) examined 23 carcinomas and 22 non-Hodgkin's lymphomas and showed all the carcinomas to be positive for at least one epithelial antigen (four were positive for leukocyte antigens), and all the lymphomas to be positive for two leukocyte antigens (none positive for epithelial antigens). Four of five tumors which were morphologically equivical could be correctly identified with Mabs. Such differentiation could be invaluable in determining prognosis, chemotherapy, etc.

2.1.3.2. *Melanoma.* Melanoma cell lines have been used to develop Mabs (C4, K22, N6). Koprowski identified three clones specific for melanoma, and several others partially reactive or cross-reactive with normal and colonic cancer cells (K22). A common antigen in melanoma and the neuroectoderm has been revealed and further studies may uncover more information useful in the study of the behavior of this cell line.

2.1.3.3. *Breast Cancer.* Several studies have shown the utility of monoclonal antibodies in the examination of breast tumors. Mabs to human milk-fat globules have been produced which revealed different staining patterns in normal tissue and papillomas (T4). Other Mabs to epithelial surface antigens may indicate premalignant tissue (H9, M8). The development of Mabs to breast estrogen receptors (G36) brings the possibility of better methods for identifying these sites.

2.1.3.4. *Gastrointestinal Cancer.* Conventional histology in gastrointestinal tract tumors is still diagnostic, and Dukes' histopathological classification is a reliable guide for prognosis (D25). The value of antigenic characterization, however, is still unclear. Burtin and Gold (B46) have established the relationship between gastrointestinal cancer and CEA. It has been shown that CEA is a family of substances (R2) and many monoclonal antibodies have been developed to sort out the heterogeneity of CEA. Primus *et al.* describe a Mab which reacts with 70% of primary and secondary colorectal tumors examined (P11). Koprowski *et al.* (H12, K21) developed Mabs to 17-1A and Ca 19-9 (non-CEA antigens). One Mab, 17-1A, reacts with membrane antigens which are not shed, while Ca 19-9 reacts with shed membrane antigen. Ca 19-9 recognizes a gastrointestinal-associated carcinoma antigen (GICA), which is present on 100% of tissue sections of gastric carcinoma and 82% of pancreatic cancers.

2.1.3.5. *Intermediate Filament Proteins.* All mammalian cells contain in their cytoskeleton a number of filaments 7 to 11 nm in size; these are known as intermediate filaments (IFs). These are significant because different cell types possess immunologically distinct types of filaments which have been in turn identified with various tumors.

The five IFs and their normal tissue and tumor associations are as follows: keratin (epithelia, carcinomas, mesotheliomas), desmin (muscle, leiomyosarcomas), vimentin (sarcomas, lymphomas, melanomas), neurofilaments (neurons, neuroblastomas, oat cell carcinomas), and glial fibrillary acidic protein (gliomas). Panels of Pabs and/or Mabs can be used by the pathologist in the differential diagnosis of tumors. Additionally, they can be used to determine the origin of poorly differentiated carcinomas whose primary site is unknown. These highly specific Mabs may one day play a major role in the classification of neoplasms. An excellent review of Mabs that have been developed and their uses has been detailed by Osborn *et al.* (O6).

2.2. THERAPEUTIC APPLICATIONS

The therapeutic concept of antibodies as weapons against neoplastic disease (magic bullets) was introduced many years ago by Ehrlich (E2). This was followed by the concept of immunologic surveillance (B44). There is evidence that the normal host immunologic apparatus, under some circumstances, is capable of mounting a defense against cancer, e.g., the presence of antitumor antibodies and immune complexes in patient serum, infiltration of tumors by mononuclear cells, tumor development in immunodeficiency, and immunosuppression.

In the following sections we review some approaches that have been used to treat malignancies immunologically, principally with Mabs that have been tagged with radiolabels or toxins.

2.2.1. Radiolabeled Antibodies

There have scattered reports of successful outcomes with patients treated therapeutically with anticancer antibodies. This has been accomplished with both Pabs and more recently with Mabs. Basically, radioimmunotherapy (RAIT) involves the delivery and subsequent deposition of therapeutic doses of radiolabeled antibody on cancer cells. Antibody selection and delivery, tolerable doses of radioactivity (or other toxins) as well as the density of tumor antigen, noncancer cells in the host with similar targets, and accessibility of the antigen sites to circulating antibody are factors that must be carefully considered before therapy is attempted. Some reports estimate antigen density on a cell to range from hundreds to 1 million per cell (D14, J1, T12). With a labeling ratio of one atom of radioactive iodine per IgG molecule, 270–27,000 mCi of ^{131}I constitutes the maximum quantity of ^{131}I delivered to 1 g of tissue during its physical decay (8-day half-life). Thus, 32,400 to 3,240,00 rad can theoretically be delivered to a tumor (D14). This assumes the presence of 10^9 cell/g or cm^3 of homogeneous cancer are present and that 10^{13}–10^{15} antibodies can be accumulated in 1 g of cancer tissue. In actuality, few antigen sites are probably available, distribution of administered antibody to other tissues will occur, neutralization of antibody by circulating antibody is possible, or host reacting with the foreign protein delivered will reduce the theoretically delivered radiolabeled dose considerably. Despite this, favorable reports with ^{131}I have been reported.

Antiferritin and anti-CEA labeled with ^{131}I have been administered by the Order group (E9). Treated patients were restricted to those with primary hepatic cancers in a protocol including chemotherapy. A recent review of this work shows remissions in 11 of 28 patients treated with antiferritin (hepatomas not secreting AFP) and in 4 to 6 patients with biliary cancer treated with anti-CEA (O5). Larson and co-workers (L6) have treated melanoma patients with ^{131}I-labeled Mab, or Mab F(ab) fragments to p97 (B35, C3), and high-molecular weight antigen (HMWA) (H10), showing transient, partial tumor regression in two of three patients. A phase I trial of radioimmunotherapy with ^{131}I-labeled anti-CEA and AFP was recently reported by Goldenberg *et al*, with similar encouraging data (D9). Still other approaches are being explored. Human monoclonals have been generated by infection of anibody-secreting B cells with Epstein–Barr virus (S29, T14, Z3). More recently, new mouse/human heteromyeloma fusion partners (K32, T6) and lymphoblastoid fusion partners (G36, K26) have been developed, as well as some of these in combination with EBV B cell transformations (K27, L5). These developments have recently been reviewed (E9, L4, S37). Recombinant DNA technology and oncogenes have been used in which cloned mouse heavy and light chains have human constant region genes substituted (M33). Ultimately, with this approach, many useful mouse monoclonals could be switched to human immunoglobulins.

2.2.2. Toxic Antibodies

The idea of using toxins on Mabs instead of radioisotopes has been under consideration for some time (G8, G39, R2). Naturally occurring toxins consist generally of an A chain, which is toxic, and a B chain, which binds to the cell, and the chains are connected by means of disulfide bridges. Examples of these are ricin, abrin, modeccine, and diphtheria toxin.

The normal mechanism of action once bound to the cell involves internalization followed by blockade of protein synthesis or 60 S ribosomes. This toxicity is nonspecific since B chains have an affinity for galactose residues. Mabs can be used to deliver the A chains to the target cell, or intact toxins can be used which have the galactose binding of the B chains blocked. Both of these methods have been applied to ricin. Conjugates have been made to the antibody to common acute lymphocytic leukemia antigen (CALLA) (R3), anticarcinoembryonic antigen (G11, L9), anti-T101 (C7), and anti-p97 (J5). The studies *in vitro* have shown conjugates to A chain alone less cytotoxic than whole ricin conjugates. These whole conjugates, using lactose to block the B chain, have been studied *in vitro* by Youle and Neville (Y4) using anti-Thy-1.2. Trials using animal tumor models have been conducted with ricin A-chain conjugates to test for possible toxicity of the compounds. Whole ricin conjugate studies in mice suggest a dose–response correlation of toxicity (F7) and other studies with various other animal models have given somewhat mixed data (J6, K29, T11) and some evidence that dissociation of the ricin A chain from conjugate *in vivo* (F7) occurs, thus posing problems similar to those previously cited for Mab chelate complexes with isotopes. Cross-reactivity has also been shown with anti-CALLA and renal tubular epithelial cells (T10). Other areas of antibody–toxin application has been in autologous bone marrow transplantation. The technique of ablative systemic therapy followed by autologous bone marrow transplantation (Z1) could be useful in tumors such as lymphomas, leukemias, and small cell carcinoma of the lung, where use of available systemic antitumor agents is limited by suppression of bone marrow precursors. Patient bone marrow could be harvested after systemic therapy to reduce tumor load, then could be treated *in vitro* with a Mab–toxin conjugate to eliminate malignant cells and returned to the patient following lethal doses of systemic radiation and chemotherapy to restore normal marrow function. This approach has been established *in vitro* (B8, D14, J5, R15) with favorable results reported in a T cell lymphoma by Kaiser *et al.* (K1), and in acute leukemia by Ritz *et al.* (R13). The use of *in vitro* Mab treatment in autologous transplant programs is appealing, but obstacles remain, including high costs of treatment, support necessary for large numbers of patients in periods of marrow aplasia, and tolerence of older patients to aggresive therapy regimens. For solid tumors without bone metastases, ablative systemic therapy plus autologous marrow transplantation is encouraging (D4, G12). What remains for carcinoma of lung, colon, breast, and prostrate are more advances in systemic treatment regimens.

There has been considerable work ongoing using conjugates of antibody and chemotherapeutic agents (G8, G39). Successful conjugates have been made with chlorambucil (D15, G9), adriamycin (H21), daunomycin (L12), methotrexate (G8), and vindecine (R18). Animal studies have shown that antibody–drug conjugates are more effective than antibody alone or drug–nonspecific immunoglobulin alone (G8, T1). This was accomplished with Pabs. The availability of Mabs will allow an increase in tumor cytotoxicity without a significant increase in reactivity to normal cells. Initial reports in animals with adriamycin conjugates are encouraging (P12) and large-scale trials in humans are anticipated.

2.2.3. *Passively Administered Antibodies*

A variety of animal models have been used to demonstrate the antitumor potential of unlabeled Mabs directed at tumor-associated antigens. Using AKR/J mice, Bernstein *et al.* (B18, B19) studied spontaneous leukemia and soft tissue leukemia/lymphoma models. An anti-Thy-1.1 Mab (IgG$_{2a}$)

prolonged survival compared to anti-Thy-1.1 (IgM) and IgG_3 controls. This was accomplished even though the Mab reacted with normal thymocytes and T cells.

While the tumors themselves were not eradicated by this treatment, cure was possible by successful surgery 10 days later followed by another course of anti-Thy-1.1 plus rabbit complement. A complement-mediated cytotoxicity (CMC) was demonstrated with all antibodies used, but only IgG_{2a} and IgG_3 showed antibody-dependent cell-mediated cytotoxicity (ADCC). IgG_{2a} was shown to be superior to IgG_3. Serotherapy against T cell antigens Thy-1.2 and Lyt-2.2 was performed using ASL.1 leukemia injected into A/Thy-1.1 cogenic hosts and ERLD and EL-4 tumors in C57BL/6/Lyt-2.1 congenic mice as models. In the ASL.1 model, an IgG_3 Mab prevented tumor development when given in less than 24 h but an IgM Mab actually prolonged survival. In the ERLD and EL-4 tumor models, an IgG_{2a} Mab prolonged survival with ERLD, but decreased survival in EL-4 (K23). The possibility of an Ig subclass being a factor and the possibility of tumor enhancement have been suggested (K2).

Many other animal models have been studied (F9, H13, S8, S36, Y5), and these studies suggest that Mab inhibition of tumor cells occurs *in vivo* without involving complement activation, that the Mab Ig isotype may be important, and that the antitumor effect on established tumors is minimal, all suggesting that larger Mab doses or use under minimum tumor burden are required and that with widely disseminated tumor, Mab therapy is not effective. The ease of obtaining peripheral blood, bone marrow, and lymph nodes has led to the quick development of Mabs to hematopoietic cells (K31, L11, R13, R19). Therefore, most of the clinical applications have been in hematopoietic malignancies. A number of these published studies have been summarized by Dillman (D15). For the most part, these experiences in humans have brought focus on several key areas to be addressed in future efforts.

2.2.3.1. *Toxicity.* Reactions observed have included fever, chills, bronchospasm, and anaphylactoid reactions. Dillman *et al.* (D18) reported several patients with significant toxicity when treating chronic lymphocytic leukemia (CLL) patients with Mabs. Ritz *et al.* (R14) have not reported such reactions. No reports of definite serum sickness have been reported, however.

2.2.3.2. *Target Cell Binding.* It has been possible to demonstrate binding of administered Mabs to peripheral cells by isolating them and demonstrating them with techniques such as immunofluorescence. However, demonstration of administered Mabs to solid tumor has been poorly documented. Some reports have shown evidence of Mab binding to skin of patients with chronic T cell leukemia (CTCL) and melanoma.

2.2.3.3. *In Vivo Specificity.* Many reports show that Mab cross-reactivity occurs *in vivo* with other cell types not demonstratable *in vitro*. Infusions of antimelanoma and antihuman T cell Mabs did not reduce platelets or granulocytes, but significantly reduced T lymphocytes with T101 but not with anti-p240 antimelanoma Mab (D17). The J-5 Mab cross-reacts with renal tubular epithelium *in vitro* (M27), but Ritz *et al.* (R14) saw no nephrotoxicity in acute lymphocytic leukemia (ALL) patients after treatment with anti-CALLA (J-5). It is important to establish antigen specificity prior to administration to humans. Some of the unexplained cross-reactions may be due to reactivity of intact antibody with Fc receptors *in vivo*.

2.2.3.4. *Free Antigen.* With some tumors, considerable tumor antigen is shed into the circulation. The potential problems of overcoming the effects of large levels of circulating antigen will have to be dealt with on an individual basis, since not all Mabs react the same in this regard. Antigens detected by anti-Leu-1 (T101 and J-5) and anti-P-240 (p97) do not appear to shed significantly. Nadler *et al.* (N2), using Mab AB89, had to use large doses of antibody to overcome circulating antigen. Possibly antigen is released by cells after treatment (M27).

2.2.3.5. *Endogenous Antimouse Antibody.* There has been long-standing feeling that administration of murine antibodies would result in the host making neutralizing doses of human antimouse antibody (HAMA). Long-term treatment of patients has confirmed that this occurs, but the extent that it is a problem is still being assessed. Miller and Levy (M26) found that a patient with cutaneous T cell lymphoma (CTCL), previously successfully treated, suddenly showed a progression of tumor. This could have been the result of Mab neutralization by HAMA. Several additional studies have acknowledged HAMA and its frequent occurrence (D16, P4, P12, S11, T6, T8). Primus *et al.* (P12) described a Mab ELISA for CEA in serum using murine anti-CEA 31C5A4 as the solid-phase capture antibody and peroxidase-conjugated anti-CEA as the probe, resulting in higher levels due to HAMA reacting with the murine capture antibody (See Fig. 2). An ELISA assay for HAMA for resolving these HAMA-elevated CEA assays is described.

2.2.3.6. *Clinical Efficacy.* The real goal of any anticancer therapy program should be a clinical benefit to the patient. We should be encouraged by some reports showing dramatic responses to Mab serotherapy (M26). Many of these successes are only moderately sustained, however. The overall outcome of passive antibody administration shows that the use of

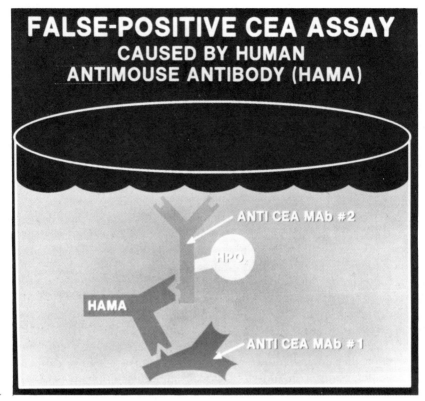

FIG. 2. Illustration of a false-positive CEA assay caused by human antimouse antibody (HAMA).

Mab conjugates offers more promise for the goals in mind, namely, sustained benefits to the patient.

3. Hormones and Proteins

Significant advances have been made in the areas of hormone and protein research because of the unique specificities offered by Mabs. In the following section we review progress in the areas of steroid hormone receptor, protein hormone, and enzyme research.

3.1. HORMONES

The recent advances in the development of Mabs has made it possible to study hormones and their receptors in those areas where considerable cross-reactivity was experienced when using polyclonal antibodies and when analytical studies between complex structures was involved.

3.1.1. Steroid Hormone Receptors

The study of receptors for estrogen and progesterone are important particularly in the area of human breast cancers. The studies began in the early 1960s when little was known of their chemical structure, their biosynthesis, or exactly where they were localized on a cellular level. Initially, polyclonal antibodies were prepared to estrogen receptors (estrophilin) by Greene et al. in 1977 (G38); others subsequently prepared antibodies to progesterone receptors and glucocorticoid receptors (G32, L17). Antibody impurity in Pabs prevented the direct detection of receptors either by immunocytochemical or immunoblot methods. Contaminating components upon immunization of the appropriate animals often made antibody responses which exceeded the antibody levels generated to the receptors themselves. The first Mabs were to the receptors to estrogen (G36), progesterone (L19), and glucocorticoid (O1, W10). These Mabs were generally prepared using steroid antigens purified by methods using the affinity of the desired activated receptor for nuclei or DNA (G34, L19, O1, W10).

The breast cancer MCF-7 cell and calf uterine estrogen receptors were purified by affinity chromatography (G35, M25). Following immunization of rats and mice, hybrids were selected by an ELISA screening method and then were subjected to rigorous examination to prove that the Mabs were specific for steroid hormone receptors. Two types of artifacts, namely, those due to chemical impurities and those due to the stickiness of steroid hormone receptors to proteins, had to be dealt with. Reactions with only weak interactions were usually nonspecific and therefore density gradient sedimentation profiles were used to verify receptor–IgG formation and high ionic strength was used to abolish weak interactions (L17).

The estrogen receptor Mabs were used for immunochemical studies on the human estrogen receptors. The receptor was subjected to proteolytic cleavage (G37) with mercuripapain, chymotrypsin, and trypsin, and it was shown that most of the Mabs recognized a region of the receptor close to the estradiol-binding site. However, one Mab recognized the DNA-binding site and another a site unrelated to estradiol or DNA. The antigenic determinants that were conserved in different species were those to the steroid- or DNA-binding sites. A two-site ELISA sandwich assay was developed for the estrogen receptor based on the work of Greene and Jensen (G36). One Mab to receptor is bound to beads (solid phase) and this binds receptor if present. A second enzyme-linked Mab (probe) is added and quantifies estrogen receptor when present.

Immunocytological studies with Mab to human estrogen receptors by King and Greene (K8) localized estrogen receptors in human breast cancer and rabbit uterus. The receptor was found to be intranuclear. The receptor is extractable from nuclei and is probably loosely attached. The same Mabs were used by Press and co-workers (P9, P10) to localize estrogen receptor in human endometrium and human breast tumors (K9, M25). This observation is important since estrogen receptor determinations are used to guide the treatment and to determine the prognosis of breast cancers. A benefit of this technique is that immunocytochemical determinations require only a small tumor fragment.

Glucocorticoid-specific Mabs have been used in immunoblotting techniques to study rat liver glucocorticoid receptor (G2, O1, W10). The same band was detected and estimated as having a molecular weight of 90,000 (W10), 94,000 (O1), and 95,000 (G2). The glucocorticiod receptor has been immunopurified by Gametchu and co-workers by elution with 3–5 mol/ liter NaSCN, pH 5 (E4, G3, H4).

The progesterone receptor of chick oviduct consists of two steroid-binding subunits (B21). Subunit A (79,000 Da) binds DNA, and subunit B (110,000 Da) binds chromatin. This model has been extended to the human breast cancer progesterone receptor (L8). The Mabs to progesterone receptor on immunoblotting showed both the 110,000- and 79,000-Da bands. The Mabs also recognized not only the subunits, but also the entire assembled steroid receptor.

Immunopurification was achieved by Logeat *et al.* (L18). An IgG$_{2a}$ Mab was cross-linked to protein A–Sepharose through the Fc portion. Following prefiltration of the cytosol through protein A–Sepharose to remove extraneous protein, the receptor was separated on a Mab Mi60-10 column and eluted in stable form at pH 10.5–11.

The production of Mabs to partially purified steroid receptors has been important in the study of these regulatory proteins. Immunoaffinity chromatography with Mabs ultimately allows complete separation of these molecules in high yield.

3.1.2. *Protein and Glycoprotein Hormones*

A considerable effort has gone into the development of Mabs to polypeptide and glycoprotein hormones. The polypeptide hormones include parathyroid, adenocorticotropic, and growth hormones; the glycoprotein hormones include thyrotropin, follicle-stimulating hormone, luteinizing hormone, and chorionic gonadotropin. The development of Mabs to polypeptide hormones has helped enormously in the standardization of these hormones. The lack of specificity in assays for glycoprotein hormones comes about since they each have a common subunit, resulting in considerable cross-reactivity. In the following section we trace the development of Mabs that have greatly resolved many of these issues.

3.1.2.1. *Parathyroid Hormone.* Mabs have been developed to the biologically active amino-terminal sequence of parathyroid hormone (PTH) 1-34 (N7). The PTH 1-34 Mab appears to be quite specific to the 1-34 region and shows poor reactivity with a radioiodinated PTH 1-84 preparation.

3.1.2.2. *Growth Hormone.* A Mab highly specific for human growth hormone (hGH) has been reported by Ivanyi (I2). This Mab was one of four generated and showed no cross-reactivity to human placental lactogen (hPL). Additional Mabs were generated (S38) for the development of radioimmunoassays. One Mab had an affinity constant of 4.4×10^{10} liters/mol with only 3% cross-reactivity with hPL; an assay for hGH in serum could thus be developed.

3.1.2.3. *Thyrotropin.* A Mab to the β subunit of thyrotropin (thyroid-stimulating hormone, TSH) was described by Ridgeway *et al.* (R11). It had a 1% cross-reactivity with hFSH and hLH, but was useful as an immobilized antibody in affinity chromatography and in immunofluorescent identification of TSH-containing cells.

3.1.2.4. *Chorionic Gonadotropin.* It is necessary to obtain a high degree of specificity and sensitivity with human chorionic gonadotropin assays in order to differentiate low serum hCG levels and serum hLH and hFSH in patients with early intra- or extrauterine pregnancies (B12). One Mab developed against hCG is to the unique 32-amino acid carboxy-terminus of the β subunit of hCG. It does not cross-react with the β subunits of TSH, FH, and LH, and has greater precipitating capacity than do polyclonal antibodies (W4).

3.2. ENZYME RESEARCH

In this section we focus on some of the advances made in basic and clinical research as a result of Mabs directed to enzymes and isozymes.

3.2.1. *Human Phosphofructokinase*

It has been established that human phosphofructokinase (PFK) is under multigenic control (V5), and this was accomplished using Mabs. It was shown that the three structural loci encode the muscle (M), liver (L), and platelet (P) subunits, which are expressed in various tissues. Biochemical studies revealed that the M_4, L_4, and P_4 homotetramers could be separated readily by ion-exchange chromatography, while the random polymers, which are easily formed, could be only poorly separated. To study the various isozymes, Mabs had to be generated because the cross-reactivity between M, L, and P subunits made this approach unwieldy. Mabs to M and L were readily made. While the Mabs did not cross-react with the human M and L subunits, a vertebrate panel of PFKs showed that the Mabs to L subunit cross-reacted. This suggests that duplication and subsequent divergence of ancestral PFK had occurred (V8). The biochemical and genetic data gained using the Mabs permitted the placement of PFK subunits M and L on chromosomes 1 and 21, respectively (V10). Inherited deficiency of PFK M produces glycogenesis type VII, resulting in exertional myopathy with hemolysis. The use of the Mabs enabled verification of this isoenzyme deficiency in several patients (V9). The weak immunogenicity of PFK P did not make it possible to generate a Mab, which is one of the weaknesses in this technique, as pointed out in an earlier section. However, using the Pab generated in earlier work, and with the aid of the Mabs to the M and L subunits, the Pab could be rendered functionally monospecific (V7). This led to the assignment of PFK P to chromosome 10.

Many Mabs have been made to enzymes or isoenzymes from many sources. The majority (60%) are to mammalian enzymes. The Commission on Biochemical Nomenclature of the International Union of Biochemistry defines isoenzymes operationally. When an enzyme preparation can be resolved into several molecular species, these forms are designated isoenzymes (regardless of whether the molecular model of multiplicity is known). Also, once the molecular basis is known, those which are encoded in different genes are designated isoenzymes while the rest can be designated multienzyme systems (M9). This classification allows eight classes of isoenzymes as summarized by Vora *et al.* (M9, V6). Each of these are discussed in brief and Mab applications are listed when clinically relevant.

1. Class I. Polymeric and monomeric isoenzymes, e.g., phosphofructokinase, prostatic acid phosphatase, and alkaline phosphatase; many Mabs are made to PAP and placental alkaline phosphatase. PAP is secreted into the plasma and is a marker for prostatic adenocarcinoma. It can also be elevated in other diseases (Y1). These Mabs are available commercially (D5, L7, L16). In immunocytochemical studies, pooled Mabs appear more effective than single Mabs (N1, N3). The antibodies detected three nonoverlapping determinants on the PAP molecule clustering on a single proteolytic fragment of the enzyme. Three alkaline phosphatases, placental, intestinal, and liver–bone–kidney types are considered encoded by separate genes (A8). Placental alkaline phosphatase is secreted into plasma from the eleventh week of gestation. The use of polyclonal sera did not eliminate the cross-reactivity with the six common phenotypes. Mabs were developed to placental alkaline phosphatase which distinguished between it and intestinal isoenzymes (A8, M16, M23, S36, W9), between those that cross-reacted with all six phenotypes (M16) or detected them specifically (M24), and between those that cross-reacted with it and intestinal isoen-

zymes (W10). Immunoradiometric assays and antigen EIA were two types of assays set up with Mabs to further study these isoenzymes (J8, S22).

2. Class II. Soluble and particulate enzymes encoded in separate genes, e.g., tyrosine hydroxylase.

3. Class III. Isoenzymes encoded by allelic genes, e.g., glucose 6-phosphate dehydrogenase. Mabs to normal human B-type isoenzyme and rat liver enzyme have been used to define the mechanism of inherited enzyme deficiency (D1).

4. Class IV. Isoenzymes resulting from a series of polymers of a single subunit, e.g., acetylcholinesterase.

5. Class V. Isoenzymes produced by partial proteolysis of the original polypeptide, e.g., plasminogen-activating enzyme.

6. Class VI. Posttranslational modification of the original polypeptide, e.g., pyruvate kinase.

7. Class VII. Conformational isoenzymes produced by the presence of an additional substance, e.g., Ca^{2+} and calmodulin-dependent cyclic nucleotide phosphodiesterase.

8. Class VIII. Multimeric enzymes sharing peptides (structural isoenzymes), e.g., RNA and DNA polymerases.

The major applications of Mabs in basic research will be immunoaffinity purification. Purified enzymes in turn will open up areas of investigations (immunochemical, biochemical, and genetics) now made possible by the availability of homogeneous enzyme preparations. In somatic cell genetics and molecular genetics of enzymes, Mabs allow recognition of enzymes which ultimately can lead to gene localization. In clinical medicine, many Mabs can be used as tumor markers, e.g., PLAP, PAP, and terminal deoxynucleotidetransferase (TdT).

Other potential markers can perhaps be established, e.g., urokinase, xanthine oxidase, or γ-cystathionase. Some of these may even be used to treat malignancies. γ-Cystathionase antibodies could deplete L-cysteine as a treatment in leukemia. Mabs appropriately tagged to xanthine oxidase may be useful in selective chemotherapy in breast cancer. Animal models for the study of human disease can utilize Mabs, as is the case for myasthenia gravis, where depletion of acetylcholine brings on experimental myasthenia gravis in animals (R10).

4. Cell Surface Antigens

The development of Mabs has had major impact on the study of the cell surface antigens on both the white and the red blood cells. In the following sections we attempt to review how Mabs have contributed to our current understandings of these cell types and what directions they are taking us. Again, these fields cannot be reviewed in depth, and when definitive reviews are available they will be cited.

4.1. Leukocyte Antigens

The recent advances in Mabs to leukocyte antigens and the advances in flow cytometry instrumentation have gone hand in hand in the identification of leukocyte cell types and subtypes. In-depth descriptions of flow cytometric instrumentation and their many applications in the quantitation of heterogeneous leukocyte populations have been written by Melamed *et al.* and others (B15, M18, S16). The list of Mabs developed for study of leukocyte antigens is large and bibliographies including more than 1000 studies have been compiled by the major producers of these antibodies (Ortho Diagnostic Systems, Inc., Raritan, New Jersey; Becton Dickinson, Mountain View, California) (see K30, R6, R7). Additionally, studies on Mabs to human T cell antibodies can be found for *in vitro* cellular interactions (W9), phenotyping lymphomas and leukemias (H19), and lymphocyte subpopulations in disease (S25). This section confines discussion to how these recent developments have influenced our understanding of human T cell differentiation and the development of its functional program.

4.1.1. *Human Cell Surface Antigens*

Classically, human T lymphocytes can be identified by their binding characteristics to sheep erythrocytes to form E-rosettes. Variations in raw materials and technique have posed many problems, many of which have been solved by Mabs. The First International Workshop on Leukocyte Differentiation Antigens met in 1982; the workshop participants compared the data on 160 Mabs submitted by an international group and were able to reduce the many reactions to 18 patterns of anti-T cell reactivity (B16). Seven major T cell clusters were resolved from this (four total T, one helper–inducer T, one suppressor–cytotoxic T, and one corticothymic T). Commercial sources of Mabs have been made available, chief among them the following Mabs:

1. Anti-Leu-5,* OKT11, T11 (Total T: 40–50K antigen).
2. Anti-Leu-1, OKT1, T101 (Total T: 67K antigen).
3. Anti-Leu-4, OKT3 (Total T: 19–30K antigen).
4. Anti-Leu-3, OKT4, T4 (Helper T: 55–65K antigen).
5. Anti-Leu-2, OKT8, T8 (Suppressor T: 32–43K antigen).

*Anti-Leu (Becton Dickinson), OKT (Ortho Diagnostic Systems), T4–T8 (Coulter Immunology), T101 (Hybritech).

The Mabs that identify the various T cell subpopulations allow for certain distinctions to be made. OKT1-positive lymphocytes define T cells, since the Mab is only to T cells and not to normal B cells. However, this Mab is reactive with B cell chronic lymphocytic leukemia cells (W9). Both T helper and T suppressor cells as defined by the above Mabs are heterogeneous and thus a T helper cell for antibody synthesis comprises only a fraction of the total OKT4-positive cells (C15).

Less is known about the Mabs to granulocytes and B cell subsets, and they will not be discussed here. However, the best marker for quantitating B cells may still be surface immunoglobulin.

4.1.2. Cell Surface Analysis: Clinical Uses

The following discussions include clinical uses to which Mabs to T cell subgroups have been applied with particular success.

4.1.2.1. *Lymphoid Malignancies.* The identification of immunologic phenotype aids in the classification of patients with lymphoid malignancies. Surface markers define both biological and prognostic characteristics of the leukemias (W3). In acute lymphocytic leukemia (ALL), knowing the T, B, and non-T phenotypes has been prognostic (T15). Further subclassifications using Ia or CALLA may be of benefit. In a study of lymphoma patients, Sallan *et al.* (S2) showed 12% of T cells positive, with 88% positive for Ia. The Ia-positive patients were 84% positive for common acute lymphocytic leukemia antigen. The Ia-positive, CALLA-negative patients had a poorer prognosis than did patients with non-T cell lymphomas. The prognostic implications of the various Mab typings are still under development.

4.1.2.2. *Acquired Immunodeficiency Syndrome.* The epidemic nature of acquired immunodeficiency syndrome (AIDS) prompted an examination of many of the immunological functions in the patients with this disease (e.g., antibodies to HIV, Ig levels, immune complexes, cell-mediated immunity). The most common immunological abnormality was quickly shown to be lymphocytopenia with disproportionate decrease in the T helper cells. Further studies have revealed that this retrovirus selectively infected this T cell population. A T helper/T suppressor cell ratio (OKT4$^+$/OKT8$^+$), therefore, can readily demonstrate onset of the infection, which correlates well with detection of antibody to HIV in early disease (U3).

4.1.2.3. *Renal Transplantation.* One of the most exciting applications of Mabs has been in the area of renal transplantation. Following transplantation, renal failure episodes are associated with inverted ratios of T helper cells to T suppressor cells (e.g., OKT4$^+$/OKT8$^+$). Immunosuppressive therapy is employed as treatment (C14, V2). Using abnormal T cell ratios to predict rejection crises and intervene with immunosuppressive therapy has been disappointing (E6). The Mabs to total T cells (e.g., OKT3) were used to treat renal rejection. Despite some problems with antigenic modulation and production of HAMA, the results using OKT3 (Ortho Pharmaceuticals) have been exciting and this Mab has been commercialized for this purpose (B49, C8, C9, C14).

4.1.2.4. *Systemic Lupus Erythematosus.* This autoimmune disease is characterized be the production of multiple autoantibodies and other disorders. Alterations in the immunoregulatory cells of systemic lupus erythematosus (SLE) patients has been proposed. Initial studies with Mabs found decreased numbers of T suppressor cells (M32). Bakke *et al.* (B1) found a decrease in T helper cells in a study of 28 patients. Smolen *et al.* (S25) found increases in both subsets in 32 patients. These data suggest that SLE may be a collection of similar syndromes, but the hope is that these approaches may prove of great help in providing prognostic information.

4.1.2.5. *Multiple Sclerosis.* Immunoregulatory disturbances have been observed during the clinical changes found in multiple sclerosis. Consistantly, a decrease in T suppressor cells has been noted in patients with active or chronic progressive disease (W8). Normal subjects showed a 3% elevated T helper/T suppressor ratio. Patients with inactive disease had an 8% elevated ratio, patients with active disease had a 68% elevated ratio, and 41% of patients with chronic active disease were abnormal. The mechanisms behind this phenomenon is unknown, and while T subset analysis is helpful in categorizing disease activity at this time, it cannot be considered currently as a diagnostic test.

4.2. Histocompatibility Antigens

There are three distinct classes of human leukocyte antigens (HLAs).

1. *Class 1 Antigens.* These antigens are detected on most nucleated leukocytes as a 43,000-Da glycopeptide associated with β_2-microglobulin. There are three allogeneic determinants on the 43,000-Da piece designated A, B, and C. By the use of conventional HLA serology, approximately 20 different HLA-A, 30 different HLA-B, and 8 different HLA-C determinants have been recognized.

2. *Class 2 Antigens.* These antigens consist of two chains (α and β) and are 34,000 and 28,000 Da, respectively. The β chain carries the polymorphic determinants termed HLA-DR.

3. *Class 3 Antigens.* These antigens are the C2, C4, and Bf components of the complement system and are located on the same chromosome.

The main source of reagents for HLA tissue typing was traditionally antisera obtained from multiparous women. Many limitations were encountered due to the character of the raw materials (i.e, low antibody titer, ability to fix complement, poor specificity, and low supply). Despite this, much of the HLA system was unraveled and the complexity of the system was worked out in several successive histocompatability workshops (T7). To minimize these problems, Mabs were made in many laboratories. The immunogen used was generally whole cells or purified proteins from Epstein–Barr virus-transformed B lymphoblastoid cell lines. Peripheral blood lymphocytes and leukemic cell lines were also used. The resultant Mabs to Class 1 antigens had finer specificity than that revealed by human polyclonal sera. Most of the Mabs generated are mouse or rat/mouse hybrids (M15). Brodsky *et al.* have shown that most Mabs produced do not recognize polymorphic determinants of the molecule (B32) and instead recognize only antigens identical in every individual of the species (i.e., the HLA products from chromosome 6 and the products of chromosome 15, such as β_2-microglobulin (B32, T13). There have been other Mabs made to monomorphic determinants on Class 2 antigens; these have been used to define the products of three loci associated with HLA-DR. Mabs to three β and three α chains have been generated (D10). In 1982 a registry of 121 Mabs to HLA antigens was set up (C13). There were 54 to the Class 1 antigens, 12 of these monomorphic, 15 with broad polymorphic specificity, and 27 with narrow polymorphic specificity. The specificities were classified further into seven groupings with increasing broadness of reactivity:

1. Splits of a classic serotype, e.g., A2, A28.
2. Perfect fits of known antigens, e.g., B7.
3. Classic cross-reactions, e.g., B7/B40.
4. Nonclassic cross-reactions.
5. Interlocus A, B, and C polymorphic specificities, e.g., A2/B17.
6. α and β locus specificity.
7. Broad polymorphic specificity.

Mabs to Class 2 antigens were also summaried and 53 were identified. None of these had narrow specificity to HLA-DR antigens, but 18 had polymorphic specificity to DC, MT, MB, and SB gene products, 5 had broad polymorphic specificity, and 30 had monomorphic specificity.

The appearance of Mabs has led to a review of the techniques for tissue typing. Many new Mab techniques involve binding assays rather than the traditional (National Institutes of Health) tissue typing method based on complement-mediated cytotoxicity. Krakauer *et al.* (K28) summarized the activities of a workshop on methodologies and this study recommended the enhanced cytotoxicity method, which requires the appropriate polyclonal antimouse IgG as being most appropriate for routine screening because of its simplicity and universality and similarity to routine tissue typing.

4.3. Erythrocyte Antigens

The traditional production of reagents for the typing of human erythrocytes for ABO antigens uses pools of high-titered human sera obtained by plasmaphoresis from human donors who have high titers of polyclonal antibody of the desired specificity. These individuals are frequently immunized with a blood group substance of the appropriate specificity to keep the antibody concentrations elevated. Enormous volumes of sera were needed to make anti-A, anti-B, and associated subtyping reagents of sufficient potency to properly type erythrocytes. The government specifications required high-potency (high-titer) reagents that were usually manufactured with minimum dilution. It is therefore not surprising that Mabs were considered as ideal sources of these antisera. In the following sections we describe some of the advances in this field to accomplish this goal.

4.3.1. ABO Typing Sera

The majority of the first Mabs were directed to high-frequency antigens, which resulted in their agglutinating virtually all erythrocytes tested (A7, P1). These had little commercial value but proved invaluable in the reference laboratory, e.g., assignment of Wr^a antigen to the major erythrocyte sialoglycoprotein (R12). One of the first Mabs to the polymorphic red cell antigen from blood group A was reported by Voak et al. (V4). This was the result of an immunization of mice with human colon carcinoma cells. Subsequently, many examples of potent Mabs to A and B erythrocytes have been demonstrated using intact erythrocytes, secretor substances, and synthetic oligosaccharides as immunogens (e.g., L20, S1). Additionally, Mabs have been generated to ABO groups subgroups, i.e., A_1, A_2B, AB, and H (D20, M19, M29). There have also been many Mabs generated to several polymorphic groups outside the ABO group, i.e., M, N, Le^a, and Le^b (e.g., B20, F6, M20, Y6). Examples of EBV-transformed human lymphoblastoid cell lines secreting anti-D (Rh_o) have been reported (B26, K25) and these lines have been cloned for the production of anti-D (C16, D24).

In addition to Mabs to erythrocyte antigens, there have been Mabs made to immunoglobulins and complement components, which are of interest to immunohematologists. The antihuman globulin (AHG) reagent usually consists of a blend of anti-IgG and anti-C3 polyclonal antibodies. AHG reagents using Mabs to these proteins have been evaluated (B6, D23). Many problems have been encountered in producing an adequate substitute to the polyclonal AHG reagents due chiefly to lack of a universal anti-IgG Mab. However, there has been considerable success with a compromise, which uses the established anti-IgG polyclonal antibodies into which is blended murine anti-C3d Mab (H17). Much work needs to be done if Mabs are to substitute completely for all of the human polyclonal antibodies that exist. One of the chief hurdles thus far has been the limited range of Mabs to polymorphic erythrocyte determinants; most Mabs are directed to carbohydrate-associated erythrocyte antigens. It will remain for us to see if these challanges will be met.

5. Miscellaneous Clinical Applications

The following sections selectively review some of the major advances made in several areas of clinical immunology where Mabs have been shown to have tremendous utility.

5.1. INFECTIOUS DISEASES

The diagnosis of infectious diseases was a fertile area for Mabs, because there were numerous reagents of conventional design which needed the immaculate specificity of Mabs to be improved. Major inroads have been made in areas such as the development of cloned and synthetic vaccines. The final assessment of the large array of diagnostic reagents, developed in the initial burst of activity following the discovery of hybridoma technology, is still under evaluation with respect to long-term impact. The literature abounds with Mab applications to the diagnosis of viruses, bacteria, and parasites. An excellent summary of this literature has been prepared by McDade (M11).

Infectious diseases are traditionally diagnosed in one of four ways, namely, microscopic examination of specimen or exudate with visual identification of bacterium, virus, or parasite; culture methods with selective media; immunolgical identification of specific antigens associated with pathogens; and/or measurement of circulating antibodies to infectious agents. This area of medical science is so complex that a single methodology could not prove ideal for all situations. Usually laboratories set up a combination of two or more of these techniques to handle the large variety of assays they are confronted with. Limitations of various techniques, i.e., inability of microscopes to differentiate viruses and difficult and time-consuming culture methods, have led to the need to supplement immunological methods, and Mabs have allowed that to happen. Mab techniques now abound for viruses, bacteria, and parasites.

One principal area of interest has been in the field of sexually transmitted diseases. The impact on the detection of three pathogens, *Neisseria gonorrhoeae, Chlamydia trachomatis,* and herpes simplex virus (HSV), type 2, of which more than 10 million cases are reported annually, will be reviewed below.

5.1.1. *Neisseria gonorrhoeae*

More than one million cases annually are reported to the Centers for Disease Control in Atlanta, Georgia (M30), representing a major public health problem. The development of specific Mabs that do not cross-react with the other *Neisseria* species has been difficult. Mabs were prepared using a principal outer-membrane protein PrI (34,000–37,000 Da) as the immunogen (T13), but these only reacted with a subset of the *N. gonorrhoeae* reference strains. Several Mabs to PrI were screened as discrete pools. Two subsets (PtIA and PrIB) were identified when 16 Mabs were screened (S4) as pools. Three Mabs (4-G5, 2-H1, and 3-C8) were eventually pooled and together detected 99.6% of 719 *N. gonorrhoeae* isolates and showed no cross-reactivity with other species of *Neisseria*. This pool could be developed satisfactorily into a radioimmunoassay.

5.1.2. *Chlamydia trachomatis*

The infectious form of *Chlamydia* is an extracellular element (300 nm diameter) that is transmissible from one cell to another (S6, S24). Diagnosis of chlamydia is usually done in culture with mammalian cells, taking approximately 6 days. Mabs were developed to the major membrane protein of *C. trachomatis*. Comparing fluorescein-conjugaated Mab staining of cells with traditional iodine staining for inclusion bodies at 18, 24, 48, and 72 h, it was clearly shown that Mab staining started in 18 h while the iodine method began in 48 h. Mab staining generally revealed from 8 to 11 times more inclusion bodies than iodine, and four times more with Giemsa staining of infected cells (S30). A 30-min direct test where urethral or cervival smears are examined by Mab immunofluorescence was successfully developed.

5.1.3. *Herpes Simplex Virus*

The initial differentiation of HSV, types 1 and 2, led to the general concept that HSV-1 infections were confined to the upper body while HSV-2 infections were located principally in the genital areas. However, it has been shown that HSV-1 infections can occur in the genital areas, making the differential diagnosis of these two types of clinical importance. Mabs to the five major glycoproteins of the virus (gC, gB, gA, gE, and gD) have been studied (G29). Four Mabs were developed which distinguished HSV-1 from HSV-2. Mab 3-G11 reacts with an 80,000-Da glycoprotein complex (HSV-1), Mab 6-A6 reacts with a 140,000-Da protein (HSV-2), Mab 6-E12 reacts with a 55,000-Da protein (HSV-2), and Mab 60H11 reacts with a 38,000-Da protein (HSV-2). Testing of 122 isolates showed 100% concordance with the restriction endonuclease analysis of viral DNA. Polyclonal typing sera detected only 66% of the 122 isolates (P3). Staining could also be directly obtained using scraping from lesions of infected patients with either HSV-1 or HSV-2. Similar findings have been shown by Pereira *et al.* (P2).

Mabs have also been useful in the direct detection of a variety of pathogens. The differentiation of viral hepatitis A from B is accomplished by using sensitive assays for the detection of the hepatitis B surface antigen (HB_sAg). A sandwich assay using IgM capture Mab on beads and IgM Mab as the probe was described by Wands (W2) and was found superior to tests using IgG Mabs. The sensitivity of this assay is about 100 pg/ml. Shafritz *et al.* showed that HB_sAg in circulating immune complexes could be detected by this assay (S15). Mab to hepatitis B core antigen has also been successfully developed and demonstrated (T5).

Mabs have been used to serve as unchanging reagents of limitless supply in certain applications. The diagnosis of streptococcal group B infection in the newborn has become increasingly important as bacteremia and meningitis due to this pathogen has increased. Mabs to this pathogen with good specificity and sensitivity have been developed (R21).

5.2. CLINICAL RHEUMATOLOGY

Several rheumatic diseases have an immunological or autoimmune basis. Mabs have helped unravel some of the mysteries surrounding these issues. In the following sections we review some of the applications made in this area.

5.2.1. *Lymphoid Cell Phenotypes*

Many relationships between lymphoid cells and autoimmune disease have been uncovered. The relationship of Epstein–Barr virus to rheumatoid arthritis (RA) has been actively pursued. Lymphoid cells from RA patients do not limit an autologous B cell infection with EBV *in vitro* as do normal lymphocytes (H5). Mabs have been responsible for localizing this defect to a T cell subset, T4, which is functionally defective in the production of interferon (H6). Systemic lupus erythematosus is considered a prime example of an autoimmune disease. SLE is a disease of B cell hyperreactivity and T suppressor cell dysfunction (B13, F3). These relationships to disease state or etiology are unclear, and it is not, as is the case with a heterogeneous disease such as SLE, time to commit these phenotyping techniques to the clinical management of patients.

5.2.2. Idiotypes and Autoantibodies

One of the most important functions of our immune system is to discriminate self from nonself, and to extend this to the recognition and elimination of foreign material introduced and recognized as nonself. Autoimmune clones exist and under the proper conditions can be induced in normal individuals (G43).

The major control of cellular or humoral response rests with T cells (inducer–helper), which recognize antigen in concert with the major histocompatibility complex (MHC) (B20, F9, G28, R5). The T cells induce other B cells or cytotoxic T cells to perform their antigen-specific immunologic functions. Antiidiotype regulation (J9) involves generation of an antibody to an antibody already formed (anti-antibody); this regulates production of the original antibody by interaction with idiotype receptors of B or T cells (J3). Mab technology has made it possible to examine the fine specificity and idiotypes of autoantibodies. Mabs with antinuclear and rheumatoid factor (anti-IgG) activity have been studied. These Mabs appear polyspecific with examples showing cross-reactivity between antibodies such as anti-IgG and anti-histone, or anti-DNA and anti-cardiolipin (R20, S6, S10). Some of these cross-reactions are no doubt due to similar epitopes, but in others they are too diverse to be accounted for in this way. Mabs have led to the understanding of multiple reactivity, where this phenomenon is common. This is helpful in the study and understanding of autoantibodies (H6, L1).

Mabs have had an impact on the tests performed in the clinical laboratory. HLA typing has had major applications in diseases such as RA and SLE (F3) and Mabs have been introduced for this function. In immune complex testing, which is used extensively in rheumatic diseases, two ELISA tests have been commercialized which use unique monoclonal anti-C1q and anti-C3d antibodies to capture complement-containing complement immune complexes in serum containing endogenous C1 and C3. A monoclonal anti IgG:Fc–HRP conjugate reactive with all human IgG subgroups is used as the probe (R4). Similar applications to rheumatoid factor testing using Mabs to IgG have been described (C5, M12).

6. Conclusions

Monoclonal antibody technology, in the course of 10 yr, has impacted on virtually every aspect of the biological world. In the research lab, it is now possible to engineer protocols for isolation and discovery never before possible. Immunodiagnostic assays and methods abound, and our experiences with Mab shortfalls (such as unwanted and unexpected cross-reactivities) have left us smarter, particularly in the way we are addressing the utilization of these new diagnostic tools. In our attack on cancer, we have been able to sort out an ocean full of Mab candidates and utilize increasingly, and with great benefit, those remaining for rapid serodiagnosis, radioimmunodetection, and radioimmunotherapy. The increasing number of clinical trials underway are witness to this. We are aware that transitions must be made based on our acquired knowledge. In radioimmunotherapy these changes have been addressed by Halpern and Dillman (H2). They envision a metamorphosis from intact radioiodinated murine antibody to less immunogenic Mab fragments with more efficient chelated metals attached, and ultimately predominant utilization of human Mabs. Of course there will be great strides in the areas of new drug delivery systems, new toxins for the Mabs, and inroads in genetically engineered products. All these exciting prospects will require many years of dedicated research and development. However, is this not the type of challenge that we enjoy?

REFERENCES

A1. Abelev, G. I., Alpha-fetoprotein as a marker of embryo-specific differentiations in normal and tumor tissues. *Transplant. Rev.* **20,** 3–37 (1974).

A2. Abelev, G. I., Alpha-fetoprotein as a model for studying reexpression of embryonic antigens in neoplasia. *In* "Immunodiagnosis of Cancer" (R. B. Herberman and K. R. McIntire eds), pp. 76–77. Dekker, New York, 1979.

A3. Abelev, G. I., Perova, S. D., and Khramkova, Z., Production of embryonal serum alpha globulin by the transplantable mouse hepatomas. *Transplantation* **1,** 174–180 (1963).

A4. Accolla, R. S., Carrel, S., and Mach, J., Monoclonal antibodies specific for carcinoembryonic antigen and produced by two hybrid cell lines. *Proc. Natl. Acad. Sci. U.S.A.* **77,** 563–566 (1980).

A5. Acevedo, H. F., Campbell-Acevedo, E. A., and Pardo, M., Immunohistochemical localization of chorionic gonadotropin-like antigen in animal malignant cells. *In* "Tumor Imaging: The Radioimmunochemical Detection of Cancer" (S. Burchill and B. A. Rhodes, eds.), pp. 73–88. Masson, New York, 1981.

A6. Albert, E., Hershberg, R., and Schur, P. H., Alpha-fetoprotein in human hepatoma: Improved detection in serum, and quantitative studies using a new sensitive technique. *Gastroenterology* **61,** 137–143 (1971).

A7. Anstee, D. J., and Edwards, P. A. W., Monoclonal antibodies to human erythrocytes. *Eur. J. Immunol.* **12**, 228–232 (1982).
A8. Arklie, J., Trowsdale, J., and Bodmer, W. F., A monoclonal antibody to intestinal alkaline phosphatase made against D98/AH-2 (HeLa) cells. *Tissue Antigens* **17**, 303 (1981).
A9. Asch, R. H., Fernandez, E. O., and Siler-Khodr, T. M., Presence of hCG-like substance in human sperm. *Am. J. Obstet. Gynecol.* **135**, 1041–1047 (1979).
A10. Aschheim, S., and Zondek, B., Hypophysenvorderlappenhormon und ovarialhormon in harn von schwangeren. *Klin. Wochenschr.* **6**, 1322 (1927).
B1. Bakke, A. C., Kirkland, P. L., Kitridon, R. C., Quismorio, F. P., Jr., Rea, T., Ehresmann, G. M., and Horwitz, D. A., T lymphocyte subsets in systemic lupus erythematosus. Correlations with corticosteroid therapy and disease activity. *Arthritis Rheum.* **26**, 745–750 (1983).
B2. Ballou, B., Levine, G., and Hakala, T. R., Tumor localization detected with radioactive labeled monoclonal antibody and external scintigraphy. *Science* **206**, 844–847 (1979).
B3. Banjo, C., Shuster, J., and Gold, P., Intermolecular heterogeneity of the carcinoembryonic antigen. *Cancer Res.* **34**, 2114–2121 (1973).
B4. Bankert, R. B., DesSoye, D., and Powers, L., Screening and replicating of anti-hapten hybridomas with a transfer template hemolytic spot assay. *J. Immunol. Methods* **35**, 23–32(1981).
B5. Banwo, O., Versey, J., and Hobbs, J. R., New oncofetal antigen for human pancreas. *Lancet* **1**, 643–645 (1974).
B6. Barker, J. M., Serologic comparison of a monoclonal anti-C3d with commercial antiglobulin reagents. *Transfusion (Philadelphia)* **22**, 507–510 (1982).
B7. Bast, R. C., Feeney, M., Lazarus, H., Nadler, L. M., Colvin, R. B., and Knapp, R. C., Reactivity of a monoclonal antibody with human ovarian carcinoma. *J. Clin. Invest.* **68**, 1331–1337 (1981).
B8. Bast, R. C., Ritz, J., Lipton, J. M., Feeney, M., Sallan, S. E., Nathan, D. G., and Schlossman, S. F., Elimination of leukemic cells from human bone marrow using monclonal antibody and complement. *Cancer Res.* **43**, 1389–1394 (1986).
B9. Begent, R. H. J., Keep, P. A., Green, A. J., Searle, F., Bagshawe, J. D., Tewkes, R. F., Jones, B. F., Barratt, G. M., and Ryman, B. E., Liposomally entrapped second antibody improves tumor imaging with radiolabeled (first) antitumor antibody. *Lancet* **2**, 739–742 (1982).
B10. Bellet, D. H., Wands, J. F., Isselbacher, K. J., and Bohnon, C., Serum alpha-fetoprotein levels in human disease: Perspective from a highly specific monoclonal radioimmunoassay. *Proc. Natl. Acad. Sci. U.S.A.* **81**, 3969–3873 (1984).
B11. Bellisario, R., Carlson. R. B., and Bahl, O. P., Human chorionic gonadotropin, Linear amino acid sequence of the subunits. *J. Biol. Chem.* **248**, 6796–6809 (1973).
B12. Bellve, A. R., and Moss, S. B., Monoclonal antibodies as probes of reproductive mechanisms. *Biol. Reprod.* **28**, 1–26 (1983).
B13. Benaceraf, B., Role of MHC gene products in immune regulation *Science* **212**, 1229–1238 (1981).
B14. Berche, C., Mach, J.-P., Lumbroso, J.-D., Langlais, C., Aubry, F., Buchegger, F., Carel, S., Rougier, P., Parmenter, C., and Tubiani, M., Tomoscintigraphy for detecting gastrointestinal and medullary thyroid cancers: First clinical results using radiolabeled monoclonal antibodies against carcinoembryonic antigen. *Br. Med. J.* **285**, 1447–1451 (1982).
B15. Bernard, A., Boumsell, L., and Dausett, eds., J., "Leucocyte Typing. Human Leu-

cocyte Differentiation Antigens Detected by Monoclonal Antibodies: Specification, classification, and Nomenclature." Springer-Verlag, New York, 1984.

B16. Bernard, A., Boumsell, L., and Hill, C., Joint report of the first international workshop on human leukocyte differentiation antigens by the investigators of the participating laboratories. In "Leukocyte Typing. Human Leukocyte Differentiation Antigens Detected by Monoclonal Antibodies" (A. Bernard, L. Boumsell, and J. Dausset. eds.), p. 9. Springer-Verlag, New York, 1984.

B17. Bernhard, M. I., Hwang, K. M., Foon, K. A., Keenan, A., M., Kessler, R. M., Frincke, J. M., Tallum, D. J., Hanna, M. G., Jr., Peters, L., and Oldham, R. K., Localization of Indium-111-and I-125-labeled monoclonal antibody in guinea pigs bearing line 10 hepatocarcinoma tumors. Cancer Res. 43, 4429 (1983).

B18. Bernstein, I. D., and Nowinski, R. C., Monoclonal antibody treatment of transplanted and spontaneous murine leukemia. In "Hybridomas in Cancer Diagnosis and Treatment" (M. S. Mitchell and H. F. Oetigen, eds.), p. 97. Raven Press, New York, 1982.

B19. Bernstein, I. D., Tam, M. R., and Nowinski, R. C., Mouse leukemia therapy with monoclonal antibodies against a thymus differentiation antigen. Science 207, 68 (1980).

B20. Bigbee, W. L., Vandervlaan, M., Fong, S. S. N., and Jensen, R. H., Monoclonal antibodies specific for the M- and N-forms of human glycophorin A. Mol. Immunol. 20, 1353–1362 (1983).

B21. Birnbaumer, M., Schrader, W. T., and O'Malley, B. W., Assessment of structural similarities in chick oviduct progesterone receptor subunits by partial proteolysis of photo-affinity-labeled proteins. J. Biol. Chem. 258, 7331–7337 (1983).

B22. Bloomer, J. R., Waldmann, T. A., and McIntire, K. R., Serum alpha-fetoprotein in non-neoplastic hepatic disorders. JAMA, J. Am. Med. Assoc. 233, 38–41 (1975).

B23. Booth, S. N., King, J, and Leonard, J., Carcinoembryonic antigens in the management of colorectal carcinoma. Br. Med. J. 4, 183–187 (1974).

B24. Bosch, A. M. G., Stevens, W., and Schumn, A., Characteristics of monoclonal antibodies against human chorionic gonadotropin. Protides Biol. Fluids 29, 837 (1981).

B25. Bosteris, B., Localisation tissuiares comparatives de l'alpha-fetoprotéine et de l'albumine par immunofluoresence au cours de l'ontogénèse et pendant la période post natale chez le rat. Tesis, Marseille, 1976.

B26. Boylston, A. W., Gardiner, B., Anderson, R. L., and Hughes-Jones, N. C., Production of human IgM anti-D in tissue culture by EB-virus transformed lymphocytes. Scand. J. Immunol. 12, 355–358 (1980).

B27. Bradwell, A. R., Vaughn, A., and Fairweather, D. S., Improved radioimmunodetection of tumors using second antibody. Lancet 1, 247 (1983).

B28. Braunstein, G. D., Use of human chorionic gonadotropin as a tumor marker in cancer. In "Immunodiagnosis of Cancer" (R. B. Herbermann and K. R. McIntire, eds.), pp. 383–409. Dekker, New York, 1979.

B29. Braunstein, G. D., Rasor, J., and Adler, D., Serum human chorionic gonadotropin levels thoughout normal pregnancy, Am. J. Obstet. Gynecol. 126, 678–681 (1976).

B30. Braunstein, G. D., Vaitukaites, J. L., and Carbone, P. P., Ectopic production of human chorionic gonadotropin by neoplasms. Ann. Intern. Med. 78, 39–45 (1973).

B31. Broder, L. E., Weintraube, B. D., and Rosen, S. W., Placental proteins and their subunits as tumor markers in prostatic carcinoma. Cancer (Philadelphia) 40, 211–216 (1977).

B32. Brodsky, F. M., Parham, P., Barnstable, C. J., Crumpton, M. J., and Bodmer, W. F., Monoclonal antibodies for analysis of the HLA system. Immunol. Rev. 47, 3 (1979).

B33. Brouha, L., and Hinglais, H., Le diagnostic de la grossesse par la réaction de Brouha-Hinglais-Simmonet. Gynecol. Obstet. 24, 42–56 (1931).

B34. Brown, J. P., Wright, P. W., and Hart, C. E., Protein antigen of normal and malignant human cells identified by immunoprecipitation with monoclonal antibodies. *J. Biol. Chem.* **255**, 4980–4983 (1980).
B35. Brown, J. P., Nishiyama, K., Hellström, I., and Hellström, K. E., Structural characteristics of human melanoma-associated antigen p97 using monoclonal antibodies. *J. Immunol.* **127**, 539–546 (1981).
B36. Brown, J. P., Woodbury, R. G., Hart, I., Hellström, I., and Hellström, K. E., Quantitative analysis of melanoma-associated antigen p97 in normal and neoplastic tissues. *Proc. Natl. Acad. Sci. U.S.A.* **78**, 539–543 (1981).
B37. Brown, J. P., Hewick, R. M., Hellström, I., Hellström, K. E., Doolittle, R. F., and Dreyer, W. J., Human melanoma antigen p97 is structurally and functionally related to transferrin. *Nature (London)* **296**, 171–173 (1982).
B38. Bruce, A. W., Mahan, D. E., and Bellville, W. D., The role of radioimmunoassay for prostatic acid phosphatase in prostatic carcinoma, *Urol. Clin. North Am.* **7**, 645–652 (1981).
B39. Buchegger, F., Phan, M., and Rivier, D., Monoclonal antibodies against carcinoembryonic antigen (CEA) used in a solid phase enzyme immunoassay: First clinical results. *J. Immunol. Methods* **49**, 129 (1982).
B40. Buchegger, F., Haskell, C. M., Schreyer, C. M., Scazziga, B. A., Raudin, S., Carel, S., and Mach, J. P., Radiolabeled fragments of monoclonal antibodies against carcinoembryonic antigen for localization of colon carcinoma grafted into nude mice. *J. Exp. Med.* **15**, 413–427 (1983).
B41. Bumol, T. F., and Reisfeld, R. A., Unique glycoprotein proteoglycan complex defined by monoclonal antibody on human melanoma cells. *Proc. Natl. Acad. Sci. U.S.A.* **79**, 1245–1249 (1982).
B42. Bumol, T. F., Walker, L. E., and Reisfeld, R. A., Biosynthetic studies on proteoglycans in human melanoma cells with monoclonal antibody to a core glycoprotein of chondroitin sulfate proteoglycans. *J. Biol. Chem.* **259**, 12733–12741 (1984).
B43. Burgoin, J., Jr., and Bourgoin, A., Cytoplasmic antigens in human malignant melanoma cells. *In* "Pigment Cell Conference" (S. Basel, ed.), pp. 366–371. Karger, Basel Switzerland, 1973.
B44. Burnet, F. M., The concept of immunologic surveillance. *Prog. Exp. Tumor Res.* **13**, 1 (1970).
B45. Burtin, P., and Escribano, M. J., The carcinoembryonic antigen and its cross-reacting antigens. *In* "Oncodevelopmental Markers" (W. H. Fishman, ed.), pp. 316–317. Academic Press, New York, 1983.
B46. Burtin, P., and Gold, P., Carcinoembryonic antigen. *Scand. J. Immunol.* **8**, Suppl., 27 (1978).
B47. Burtin, P., Calmettes, C., and Fondaneche, M. C., CEA and non-specific cross-reacting antigens (NCA) in medullary carcinomas of the thyroid *Int. J. Cancer* **23**, 741–745 (1979).
B48. Burtin, P., Chavenel, G., and von Kleist, S., Demonstration in normal human plasma of an antigen that crossreacts with the carcinoembryonic antigen of digestive tract tumors. *J. Natl. Cancer Inst. (U.S.)* **49**, 1727–1729 (1972).
B49. Burton, R. C., Cosimi, A. B., Colvin, R. B., Rubin, R. H., Delmonico, F. L., Goldstein, G., and Russell, P. S., Monoclonal antibodies to T cell subsets: Use for immunological monitoring and immuno suppression in renal transplantation. *J. Clin. Immunol.* **2**, 1425–1478 (1982).
C1. Canevari, S., Fossati, G., and Porta, G., Humoral cytotoxicity in melanoma patients and its correlation with the extent and course of the disease. *Int. J. Cancer* **16**, 722–729 (1975).

C2. Canney, C. A., Moore, M., James, R. A., and Wilkinson, P. M., Initial results with ovarian cancer antigen Ca 125. *Br. J. Cancer* **50**, 261 (1984).
C3. Carrasquillo, J. A., Krohn, K. A., and Beaumier, P., Diagnosis of and therapy for solid tumors with radiolabeled antibodies and immune fragments. *Cancer Treat. Rep.* **68**, 317–328 (1984).
C4. Carrel, S., Schreyer, M., and Schmidt-Kessen, A., Use of monoclonal antibodies for the histological localization of melanoma-associated antigen on fresh tumor material. *Protides Biol. Fluids* **31**, 821 (1983).
C5. Carson, D. A., Lawrence, S., and Catalano, M. A., Radioimmunoassay of IgG and IgM rheumatoid factors reacting with human IgG. *J. Immunol.* **119**, 295–300 (1977).
C6. Carthew, P., Is rodent virus comtamination of monoclonal antibody preparations for use in human therapy a hazard? *J. Gen. Virol.* **67**, 963–974 (1986).
C7. Cassalas, P., Blythman, H. E., Gros, O., Richer, G., and Jansen, F. K., *Protides Biol. Fluids*, p. 359 (1983).
C8. Chatenoud, L., Chkoff, N., Kreis, H., and Bach, J. F., Interest and limitations of the use of monoclonal anti-T cell antibodies for the follow-up of renal transplant patients. *Transplantation* **36**, 45–50 (1983).
C9. Chatenoud, L., Baudrihaye, M. F., Kreis, H., Goldstein, G., Schindler, J., and Bach, J. F., Human *in vivo* antigenic modulation induced by the anti-T cell OKT3 monoclonal antibody. *Eur. J. Immunol.* **12**, 979–983 (1982).
C10. Choe, B.-K., Lillehoy, H. S., Dong, M. K., Gleason, S., Barron, N., and Rose, N. R., Characterization of antigenic sites of human prostate acid phosphatase. *Ann N.Y. Acad. Sci.* **390**, 16–20 (1982).
C11. Chu, T. M., Wang, M. C., and Lee, C. L., Phosphate acid phosphatase in human prostate cancer. *In* "Biochemical Markers for Cancer" (T. M. Chu, ed.). Dekker, New York, 1982.
C12. Colcher, D., Esteban, J. M., Carrasquillo, J. A., Sugarbaker, P., Reynolds, J. C., Bryant, G., Larson, S. M., and Schlom, J., Quantitative analyses of selective radiolabeled monoclonal antibody localization in metastatic lesions of colorectal patients. *Cancer Res.* **47**, 1185–1189 (1987).
C13. Colombani, J., Dausset, J., Lepage, V., Degos, L., Kalil, J., and Fellous, M., HLA monoclonal antibody registry: A proposal. *Tissue Antigens* **20**, 161–171 (1982).
C14. Colvin, R. B., Cosimi, A. B., and Burtin, R. C., Circulating T cell subsets in 72 human renal allograft recipients: The OKT4+/OKT8+ cell ratio correlates with reversibility of graft injury and glomerulopathy. *Transplant. Proc.* **15**, 1166–1169 (1983).
C15. Corte, G., Mingari, C., Moretta, A., Damiani, G., Moretta, L., and Bargellisi, A., Human T-cell sub-populations defined by a monoclonal antibody. I. A small subset is responsible for proliferation to allogenic cells or to soluble antigens and for helper activity for B-cell differentiation. *J. Immunol.* **128**, 16–19 (1982).
C16. Crawford, D. H., Barlow, M. J., Harrison, J. F., Winger, L., and Huehns, E. R., Production of human monoclonal antibody to Rhesus D antigen. *Lancet* **1**, 386–388 (1983).
C17. Cuatrecasas, P., and Anfinsen, C. B., Affinity chromatography. *Annu. Rev. Biochem.* **40**, 259–278 (1971).
C18. Cuttita, F., Rosen, S., Gazdar, A. F., and Minna, J. D., Monoclonal antibodies that demonstrate specificity for several types of lung cancer. *Proc. Natl. Acad. Sci. U.S.A.* **78**, 4591–4595 (1981).
D1. Damiani, G., Frascio, M., Benatti, U., Zocchi, E., Bargellisi, A., and deFlora, A., Monoclonal antibodies to human erythrocyte glucose-6-phosphate dehydrogenase. *FEBS Lett.* **119**, 169–173 (1980).

D2. Damron, J. R., Beihn, R. M., and DeLand, F. H., Detection of upper abdominal abscesses by radionuclide imaging. *Radiology (Easton, Pa.)* **120,** 131–134 (1976).
D3. Damron, J. R., Beihn, R. M., and Selby, J. B., Gallium technetium substraction scanning for the localization of subphrenic abscesses. *Radiology (Easton, Pa.)* **113,** 117–122 (1974).
D4. Davies, D. A., and O'Neill, G. J., In vivo and in vitro effects of tumor specific antibodies with chlorambucil. *Br. J. Cancer* **28,** 286 (1973).
D5. Davies, S. N., and Gochman, N., Evaluation of a monoclonal-antibody based immunoradiometric assay for prostatic acid phosphatase. *Am J. Clin. Pathol.* **79,** 114–119 (1983).
D6. Day, E. D., Pleninsek, K., Korngold, L., and Pressman, D., Tumor localizing antibodies purified from antisera against Murphy rat lymphosarcoma. *J. Natl. Cancer Inst. (U.S.)* **17,** 517 (1956).
D7. deCock, W., deCree, J., and Verhaegen, H., Enumeration of human peripheral T lymphocytes with E-rosettes and OKT3.PAN monoclonal antibody; a close correlation. *Thymus* **2,** 133–137 (1980).
D8. deCock, W., deCree, J., and Verhaegen, H., An enzyme immunoassay for the enumeration of peripheral human T-lymphocytes with OKT#.PAN monoclonal antibody. *J. Immunol. Methods* **43,** 131–134 (1981).
D9. DeJager, R. L., Primus, F. J., Ford, E. H. *et al.,* Phase I clinical trial of radiotherapy with I-131 labeled anti-CEA or anti-AFP in solid tumors producing CEA or AFP. *Proc. Am. Soc. Clin. Oncol.* **4,** 224, c-871 abstr. (1985).
D10. deKretser, T. A., Crumpton, M. J., Bodmer, J. G., and Bodmer, W. F., Two dimensional gel analysis of the polypeptides precipitated by a polymorphic HLA-DR1, w6 monoclonal antibody; evidence for a third locus. *Eur. J. Immunol.* **12,** 600–606 (1982).
D11. DeLand, F. H., and Goldenberg, D. M., Diagnosis and treatment of neoplasms with radio-nuclide labeled antibodies. *Semin. Nucl. Med.* **15,** 2–11 (1985).
D12. Delfs, E., Quantitative chorionic gonadotropin. Prognostic value in hydatidiform mole and chorionepithelioma. *Obstet. Gynecol.* **9,** 1–24, (1957).
D13. Delmont, J., Kermarec, J., and Lafon, J., Radioimmunoassay of alpha-fetoprotein. II. Studies in patients suffering from hepatic diseases. *Digestion* **10,** 29–39 (1974).
D14. DeNardo, G. L., Raventos, A., Hines, H. H., Scheibe, P. O., Macey, D. J., Hays, M. T., and DeNardo, S. J., Requirements for a treatment planning system for radioimmunotherapy. *Int. J. Radiat. Oncol., Biol. Phys.* **11,** 335–348 (1985).
D15. Dillman, R. O., Monoclonal antibodies in the treatment of cancer. *CRC Crit. Rev. Oncol. Hematol.* **1,** 357–383 (1985).
D16. Dillman, R. O., Beauregard, J. C., Halpern, S. E., and Clutter, M., Toxicities and side effects associated with intravenous infusions of murine monoclonal antibodies. *J. Biol. Response Modif.* **5,** 73–84 (1986).
D17. Dillman, R. O., Beauregard, J. C., Shawler, D. L., and Royston, I., Results of early trials using murine monoclonal antibodies as anti-cancer therapy. *Protides Biol. Fluids,* p. 353 (1983).
D18. Dillman, R. O., Shawler, D. L., Collins, H., A., Wormsley, S. B., and Royston, I., Murine monoclonal antibody therapy in two patients with chronic lymphocytic leukemia. *Blood* **59,** 1036 (1982).
D19. Dodson, M. G., Klegerman, M. E., and Menon, M., Establishment and characterization of a squamous cell carcinoma of the vulva in tissue culture and immunological evaluation of the host. *Am J. Obstet. Gynecol.* **131,** 606–619 (1978).
D20. Doinel, C., Edelman, L., Rouger, P., LeBlanc, J., Reviron, J., Bach, J. F., and Salmon,

C., A murine monoclonal antibody against blood group H type-1 and type-2 structures. *Immunology* **50**, 215–221 (1983).
D21. Douer, D., Champlin, R. E., Ho, W. E., Sarna, G. P., Grase, P. R., and Gale, R. P., High-dose combines modality therapy and autologous bone-marrow transplantation in resistant cancer. *Am. J. Cancer* **71**, 973 (1981).
D22. Douilllard, J. Y., Hoffman, T., and Herberman, R. B., Enzyme-linked immunosorbant assay for screening monoclonal antibody production. Use of intact cells as antigen. *J. Immunol. Methods* **39**, 309–316 (1980).
D23. Downie, D. M., Voak, D., and Jarvis, J., The use of monoclonal antibodies to human IgG in blood transfusion serology. *Biotest Bull.* **1**, 348–352 (1983).
D24. Doyle, A., Jones, T. J., Bidwell, L. W., and Bradley, B. A., *In vitro* development of human monoclonal antibody-secreting plasmacytomas. *Hum. Immunol.* **13**, 199–209 (1985).
D25. Dukes, C. E., The classification of cancer of the rectum. *J. Pathol. Bacteriol.* **35**, 322 (1935).
E1. Egan, M. I., and Todd, C. W., Carcinoembryonic antigen: Synthesis by a continuous line of adenocarcinoma cells, *J. Natl. Cancer Inst. (U.S.)* **49**, 887–889 (1972).
E2. Ehrlich, P., Uber den jetzigen stand der karzinimforschung. *In* "The Collected Works of Paul Ehrlich," Vol. 2. Pergamon, Oxford, 1957.
E3. Eilber, F. R., and Morton, D. L., Sarcoma-specific antigens. Detection by complement fixation with serum from sarcoma patients. *J. Natl. Cancer Inst. (U.S.)* **44**, 651–656 (1970).
E4. Eisen, L. P., Reichman, M. E., Gametchu, B., Harrison, R. W., and Eisen, H. J., Immunocytochemical purification of affinity-labeled glucocorticoid receptor. *Proc. Int. Congr. Endocrinol. 7th, 1984* p. 374 (1984).
E5. Ellion, M. L., Lamb, D., and Rwett, J., Quantitative aspects of carcinoembryonic antigen output by human lung carcinoma cell line. *J. Natl. Cancer Inst. (U.S.)* **59**, 309–312 (1977).
E6. Ellis, T. M., Lee, H. M., and Mohanakumar, T, Alterations in human regulatory T lymphocyte subpopulations after renal allografting. *J. Immunol.* **127**, 2199–2203 (1981).
E7. Epenetos, A. A., Nimmon, C. C., and Arklie, J., Detection of human cancer in an animal model using radiolabeled tumor-associated monoclonal antibodies. *Br. J. Cancer* **36**, 1–8 (1982).
E8. Epenetos, A. A., Travers, P., Gatter, K. C., Oliver, R. D., Mason, D. Y., and Bodmer, W. F., An immunohistological study of testicular germ cell tumors using two different monoclonal antibodies against placental alkaline phosphatase. *Br. J. Cancer* **49**, 11 (1984).
E9. Ettinger, E. S., Order, S. E., Wharam, M. D., Parker, M. K., Klein, J. L., and Leichner, P. K., Phase 1–11 study of isotope immunoglobulin therapy for primary liver cancer. *Cancer Treat. Rep.* **66**, 289–297 (1982).
F1. Falini, B., and Taylor, C. R., New developments in immunoperoxidase techniques and their applications. *Arch. Pathol. Lab. Med.* **107**, 105–117 (1983).
F2. Farrands, P. A., Pimm, M. V., and Embleton, M. J., Radioimmunodetection of human colorectal cancers by an anti-tumor monoclonal antibody. *Lancet* **2**, 397–400 (1982).
F3. Fauci, A. S., Lane, H. C., and Volkam, D. J., Activation and regulation of human immune responses: Implications in normal and disease states. *Ann. Intern. Med.* **99**, 61–75 (1983).
F4. Fink, L. M., and Clarke, S. M., Monoclonal antibodies as diagnostic reagents for the identification and characterization of human tumor diagnosis. *Hum. Pathol.* **13**, 121–133 (1982).

F5. Fisher, A. G., and Brown, G., A rapid method for determining whether monoclonal antibodies react with the same or different antigens on the cell surface. *J. Immunol. Methods* **39**, 377–385, (1980).
F6. Fletcher, A., and Harbour, C., An interesting monoclonal anti-N produced following immunization with human group O, NN erythrocytes. *J. Immunogenet.* **11**, 121–126 (1984).
F7. Fodstad, O., and Phil, A., Doxorubicin and ricin, a strongly synergistic combination in mouse leukemia. *Cancer Treat. Rep.* **64**, 1375 (1980).
F8. Foley, E. J., Antigenic properties of methylcholanthrene induced tumors in mice of the strain of origin. *Cancer Res.* **13**, 835 (1953).
F9. Foon, K. A., Bernhard, M. I., and Oldham, R. K., Monoclonal antibody therapy: Assessment by animal tumor models. *J. Biol. Response Modif.* **1**, 277 (1982).
F10. Foti, A. G., Cooper, J. F., Herschman, H., *et al.*, Detection of prostatic cancer by solid phase radioimmunoassay of serum prostatic acid phosphatase. *N. Engl. J. Med.* **297**, 1357–1361 (1977).
G1. Gailani, S., Chu, T. M., and Nussbaum, A., Human chorionic gonadotropin (hCG) and CEA in nontropoblastic neoplasms. Assessment of abnormalities of hCG and CEA in broncogenic and digestive neoplasms. *Cancer (Philadelphia)* **38**, 1684–1686 (1974).
G2. Gametchu, B., and Harrison, R. W., Characterization of a monoclonal antibody to the rat liver glucocorticoid receptor. *Endocrinology (Baltimore)* **114**, 274–279 (1984).
G3. Gametchu, B., and Harrison, R. W., Use of an immunoaffinity column prepared from a monoclonal antibody in the purification of the glucocortcoid receptor. *In* "Proceedings of the 75th Annual Meeting of the American Society of Biological Chemists and the 68th Annual Meeting of American Association of Immunologists, St. Louis," Vol 43. F.A.S.E.B., Bethesda, Maryland, 1984.
G4. Gatter, K. C., Abdulatiz, Z., Beverly, P., Corvalan, J. R., Ford, C., Lane, E. B., Mota, M., Nash, J. R., Pulford, K., Stein, H., Taylor-Papadimitriou, J., Woodhouse, C., and Mason, D. Y., Use of monoclonal antibodies for the histopathologic diagnosis of human malignancy. *J. Clin. Pathol.* **35**, 1253–1267 (1982).
G5. Gelder, F. B., Mackie, C. R., and Cooper, M. J., Identification, isolation and clinical application of a pancreatic oncofetal antigen. *Cancer Res.* **38**, 312–324 (1978).
G6. Gelder, F. B., Moosa, A. R., and Hunter R. L., Identification and partial characterization of a pancreatic oncofetal antigen (POA). *Prev. Detect. Cancer [Proc. Int. Symp.], 3rd, 1976* pp. 587–595 (1977).
G7. Gelder, F. B., Reese, C., and Moosa, A. R.. Pancreatic oncofetal antigen. *In* "Immunodiagnosis of Cancer" (R. B. Herberman and K. R. McIntire, eds.), Part 1, pp. 357–368. Dekker, New York, 1979.
G8. Ghose, T., and Blair, A. H., Antibody-linked cytotoxic agents in the treatment of cancer: Current status and future prospects. *JNCI, J. Natl. Cancer Inst.* **61**, 657 (1978).
G9. Ghose, T., Gucul, A., and Tai, J., Suppression of an AKR lymphoma by antibody and chlorambucil. *J. Natl. Cancer Inst. (U.S.)* **55**, 1353 (1975).
G10. Ghose, T., Ferrone, S., Imai, K., Norvell, S. T., Luner, S. J., Martin, R. H., and Blair, A. Imaging of human melanoma xenografts in nude mice with a radiolabeled monoclonal antibody. *JNCI, J. Natl. Cancer Inst.* **69**, 823–826 (1982).
G11. Gilliland, D. G., Stepelewski, Z., Collier, R. H., Mitchell, K. F., Chang, T. H., and Koprowski, H., Antibody-directed cytotoxic agents: Use of monoclonal antibody to direct the action of toxin A chains to colorectal carcinoma cells. *Proc. Natl. Acad. Sci. U.S.A.* **77**, 4539 (1980).

G12. Glode, L. M., Robinson, W. A., Hartmann, D. W., Klein, J. J., Thomas, M. R., and Moorton, N., Autologous bone marrow transplantation in the therapy of small cell carcinoma of the lung. *Cancer Res.* **42,** 4270 (1982).
G13. Gold, P., Carcinoma of the colon and antecedent epithelium. *In* "Carcinoma of the Colon" (W. J. Burdette, ed.), pp. 131–142. Thomas, Springfield, Illinois, 1970.
G14. Gold, P., and Freedman, S. O., Demonstration of tumor-specific antigens in human colonic carcinoma by immunological tolerance and absorption techniques. *J. Exp. Med.* **121,** 439–462 (1962).
G15. Gold, P., Krupey, J., and Ansari, H., Position of the carcinoembryonic antigen of the human digestive system in ultrastructure of human tumor cell surface. *J. Natl. Cancer Inst. (U.S.)* **45,** 219–225 (1975).
G16. Goldenberg, D. M., and DeLand, F. H., History and status of tumor imaging with radiolabeled antibodies. *J. Biol. Response Modif.* **1,** 121–136 (1983).
G17. Goldenberg, D. M., Goldenberg, H., and Primus, F. J., Cancer diagnosis and therapy with radiolabeled antibodies. *In* "Immunoconjugates. Antibody Conjugates in Radioimaging and Therapy" (C. W. Vogel, ed.), Chapter 13, pp. 259–280. Oxford Univ. Press, New York, 1987.
G18. Goldenberg, D. M., Pavia, R. A., and Hansen, J. H., Synthesis of carcinoembryonic antigen *in vitro. Nature (London), New Biol.* **239,** 189–190 (1972).
G19. Goldenberg, D. M., Primus, F. J., and DeLand, F. H., Tumor detection and localization with purified antibodies to carcinoembryonic antigen. *In* "Immunodiagnosis of Cancer" (R. B. Herberman and K. R. McIntire, eds.), Part 1, pp. 265–304. Dekker, New York, 1979.
G20. Goldenberg, D. M., Sharkey, R. M., and Primus, F. J., Carcinoembryonic antigen in histopathology: Immunoperoxidase staining of conventional tissue sections. *J. Natl. Cancer Inst. (U.S.)* **57,** 11–22 (1976).
G21. Goldenberg, D. M., Preston, D. F., Primus, F. J., and Hansen, H. J., Photoscan localization of GW-39 tumors in hamsters by radiolabeled, heterospecific antibody to carcinoembryonic antigen. *Cancer Res.* **34,** 1–9 (1974).
G22. Goldenberg, D. M., Goldenberg, H., Higginbotham-Ford, E., Schochat, D., and Ruoslahti, E., Imaging of primary and metastatic liver cancer with I-131 monoclonal and polyclonal antibodies against alpha-fetoprotein. *J. Clin. Oncol.* (in press).
G23. Goldenberg, D. M., Kim, E. E., Bennett, S. J., Nelson, M. O., and DeLand, F. H., Carcinoembryonic antigen radioimmunodetection in the evaluation of colorectal cancer and the detection of occult neoplasms. *Gastroenterology* **84,** 524–532 (1983).
G24. Goldenberg, D. M., Kim, E. E., DeLand, F. H., Bennett, S. J., and Primus, F. J., Radioimmunodetection of cancer with radioactive antibodies to carcinoembryonic antigen. *Cancer Res.* **40,** 2984–2992 (1980).
G25. Goldenberg, D. M., DeLand, F. H., Kim, E. E., Bennett, S. J., Primus, F. J., and Rayburn, P., Use of radiolabeled antibodies to carcinoembryonic antigen for the detection and localization of diverse cancers by external photoscanning. *N. Engl. J. Med.* **298,** 1384–1388 (1978).
G26. Goldenberg, D. M., Kim, E. E., DeLand, F. H., Spremulli, E., Nelson, J. R., Jr., Gockerman, J. P., Primus, F. J., Corgan, R. I., and Alpert, P., Clinical studies on the radioimmunodetection of tumors containing alpha-fetoprotein, *Cancer Res.* **45,** 2500–2505 (1980).
G27. Goldstein, D. P., Kosasa, T. S., and Sharin, A. T., The clinical application of a specific radioimmunoassay for human chorionic gonadotropin in trophoblastic and nontrophoblastic tumors. *Surg., Gynecol. Obstet.* **138,** 747–751 (1974).

G28. Goldstein, G., Lifter, J., and Mittler, R., Immunoregulatory changes in human disease detected by monoclonal antibodies to T lymphocytes. In "Monoclonal Antibodies" (A. J. McMichael and J. W. Fabre, eds.), pp. 39–70. Academic Press, London, 1982.

G29. Goldstein, L. C., Corey, L., McDougal, J. K., Tolentino, E., and Nowinski, R. C., Monoclonal antibodies to herpes simplex viruses: Use in antigenic typing and rapid diagnosis. *J. Infect. Dis.* **147**, 829–837 (1983).

G30. Goodwin, D. A., Meares, C., and Diamanti, C., Use of specific antibody for rapid clearance of circulating blood background from radiolabeled tumor imaging proteins. *Eur. J. Nucl. Med.* **9**, 209–215 (1984).

G31. Goodwin, D. P., Hornung, M. O., and Leong, S. P. L., Immune responses induced by human malignant melanoma in the rabbit. *Surgery (St. Louis)* **72**, 737, 743 (1972).

G32. Govindan, M. V., and Sekeres, C. E., Purification of two dexamethasone-binding proteins from rat liver cytosol. *Eur. J. Biochem.* **89**, 95–104 (1978).

G33. Gray, B., Mehigan, J. F., and Morton, D. L., Demonstration of antibodies in melanoma patients cytotoxic to human melanoma cells. *Proc. Am. Assoc. Cancer Res.* **123**, 79 (1971).

G34. Greene, G. L., Fitch, F. W., and Jensen, E. V., Monoclonal antibodies to estrophilin: Probes for the study of estrogen receptors. *Proc. Natl. Acad. Sci. U.S.A.* **77**, 157–161 (1980).

G35. Greene, G. L., King, W. J., and Press, M. F., Monoclonal antibodies as probes of estrogen receptor. *Proc. Int. Congr. Endocrinol., 7th, 1984* p. 117. (1984).

G36. Greene, G. L., Nolan, C., Engler, J. P., and Jensen, E. V., Monoclonal antibodies to human estrogen receptor. *Proc. Natl. Acad. Sci. U.S.A.* **77**, 5115–5119 (1980).

G37. Greene, G. L., Sobel, N. B., King, W. J., and Jensen, E. V., Immunochemical studies of estrogen receptors. *J. Steroid Biochem.* **20**, 51–56 (1984).

G38. Greene, G. L., Closs, L. E., Fleming, H., DeSombre, E. R., and Jensen, E. V., Antibodies to estrogen receptor: Immunochemical similarity between estrophilin from various mammalian species. *Proc. Natl. Acad. Sci. U.S.A.* **74**, 3681–3685 (1977).

G39. Gregoriadis, G., Targeting of drugs, implications in medicine. *Lancet* **2**, 241 (1981).

G40. Gross, L., Intradermal immunization of C3H mice against a sarcoma that originated in an animal of the same line. *Cancer Res.* **3**, 326 (1943).

G41. Gupta, R. K., and Moreton, D. L., Suggestive evidence for *in vivo* binding of specific antitumor antibodies of human melanomas. *Cancer Res.* **35**, 58–62 (1975).

G42. Gupta, R. K., Irie, R. F., and Morton, D. L., Antigens on human tumor cells assayed by complement fixation with allogenic sera. *Cancer Res.* **38**, 2573–2580 (1978).

G43. Gurlbert, B., Dighicro, G., and Avrameas, S., Naturally occurring antibodies against nine common antigens in normal humans. I. Detection, isolation and characterization. *J. Immunol.* **128**, 2779–2787 (1982).

G44. Gutman, E. B., Sproul, E. B., and Gutman, A. B., Significance of increased phosphatase activity of bone at site of osteoclastic metastases secondary to carcinoma of prostate gland. *Am. J. Cancer* **28**, 485–495 (1936).

H1. Hakala, T. R., Castro, A. E., and Elliot, A. Y., Humoral cytotoxicity in human renal cell carcinoma. *Invest. Urol.* **11**, 405–410 (1974).

H2. Halpern, S. E., and Dillman, R. O., Problems with radioimmunodetection and possibilities for future solutions. *J. Biol. Response Modif.* **6**, 235–262 (1987).

H3. Hamlyn, P. H., and Sikora, K., Oncogenes. *Lancet* **2**, 326–331 (1983).

H4. Harrison, R. W., and Gametchu, B., Mouse glucocorticoid receptor immunoaffinity purification using a monoclonal antibody. *Proc. Int. Congr. Endocrinol., 7th, 1984* p. 512 (1984).

H5. Hasler, F., Bluestein, H. G., and Zvaifler, N. J., Analysis of the defects responsible for rheumatoid arthritis lymphocytes. I. Diminished gamma interferon production in response to autologous stimulation. *J. Exp. Med.* **157,** 173–188 (1983).
H6. Haspel, M. V., Onodera, T., Prabhakar, B. S. *et al.,* Multiple organ-reactive monoclonal autoantibodies. *Nature (London)* **304,** 73–76 (1983).
H7. Hedin, A., Wahren, B., and Hammarström, S., Tumor localization of CEA-containing human tumors in nude mice by means of monoclonal anti-CEA antibodies. *Int. J. Cancer* **30,** 547–552 (1982).
H8. Hedin, A., Carlsson, L., Berglund, A., and Hammarström, S., A monoclonal antibody enzyme immunoassay for serum carcinoembryonic antigen with increased specificity for carcinomas. *Proc. Natl. Acad. Sci. U.S.A.* **80,** 3470 (1983).
H9. Heidrich, L., Feis, E., and Mathias, E., Testiculares chorinoephitheliom mit gynakomastic und mit einigen schwangerschaftserscheinungen gleichzeitig ein beitrag zur pathologie der hormonal gewachse. *Beitr. Z. Klin. Chir.* **150,** 349–384 (1930).
H10. Hellström, I., Garrigues, J., Cabasco, L., Mosely, G. H., Brown, J. P., and Hellström, K. E. Studies of a high-molecular weight human melanoma-associated antigen. *J. Immunol.* **130,** 1467–1472 (1983).
H11. Herlyn, M., Blaszczyk, M., and Koprowski, H., Immunodiagnosis of human solid tumors. *Contrib. Oncol.* **19,** 160 (1984).
H12. Herlyn, M. Steplewski, Z., Herlyn, D., and Koprowski, H., Colorectal carcinoma specific antigen: Detection by means of monoclonal antibodies. *Proc. Natl. Acad. Sci. U.S.A.* **76,** 1438 (1979).
H13. Herlyn, D. M., Steplewski, Z., Herlyn, M. F., and Koprowski, H., Inhibition of growth of carcinoma in nude mice by monoclonal antibody. *Cancer Res.* **40,** 717–721 (1980).
H14. Herzog, B., Hendrick, J. C., and Franchimont, P., Heterogeneity of carcinoembryonic antigen (CEA) in human serum. *Eur. J. Cancer* **12,** 657–658 (1976).
H15. Hirose, T., Exogenous stimulation of corpus luteum formation in the rabbit; influence of extracts of human placenta, decidua, fetus, hydatid mole, and bovine corpus luteum on the rabbit gonad. *J. Jpn. Gynecol. Soc.* **16,** 1065–1060 (1920).
H16. Hobbs, J. R., Knapp, M. L., and Branfoot, A. C., Pancreatic oncofetal antigen (POA): Its freqency and localization in humans. *Oncodev. Biol. Med.* **1,** 37–48 (1980).
H17. Holt, P. D. J., Donaldson, C., Judson, P. A., Johnson, P., Parsons, S. F., and Anstee, D. J., NBTS/BRIC 8. A monoclonal anti-C3d antibody. *Transfusion (Philadelphia)* (1985).
H18. Houston, H. H., Nowinski, R. C., and Bernstein, I. D., Specific *in vivo* localization of monoclonal antibodies directed against the Thy 1.1 antigen. *J. Immunol.* **125,** 837–843 (1980).
H19. Hunnighake, C. W., and Crystal, R. G., Pulmonary sarcoidosis: A disorder mediated by excess helper T lymphocytes. *N. Engl. J. Med.* **305,** 429–434 (1981).
H20. Hunter, R., Gelder, F. B., and Moosa, A. R., Pancreatic oncofetal antigen (POA). In "Compendium of Assays for Immunodiagnosis of Human Cancer" (R. B. Herberman, ed.), pp. 247–249. Elsevier/North-Holland, New York, Amsterdam, 1979.
H21. Hurwitz, E., Levy, R., Maron, R., Wilchek, M., Arnon, R., and Sela, M., The covalent binding of daunomycin and adriamycin to antibodies. *Cancer Res.* **35,** 1175 (1975).
I1. Ishikawa, N., and Hamada, S., Association of medullary carcinoma of the thyroid with carcinoembryonic antigen. *Br. J. Cancer* **34,** 111–115 (1976).
I2. Ivanyi, J., Competition and affinity assay of monoclonal antibodies against human growth hormone. *Protides Biol. Fluids* pp. 471–478 (1980).
I3. Ivanyi, J., Paratope specificity of monoclonal antibodies to human growth hormone.

In "Monoclonal Antibodies and T-Cell Hybridomas: Perspectives and Technical Advances" (G. J. Hammerling, U. Hammerling, and J. F. Kearney, eds.), pp. 349–356. Elsevier, Amsterdam, 1981.

J1. Jackson, D. C., Howlett, G. J., and Nestorowicz, A., The equilibrium constant for the interaction between a monoclonal Fab fragment and an influenza virus neuraminidase. *J. Immunol.* **130,** 1313–1316 (1983).

J2. Jakowatz, J. G., Beatty, B. G., Vlahos, W. G., Porudominsky, D., Philben, V. J., Williams, L. E., Paxton, R. J., Shively, J. E., and Beatty, J. D., High-specific-activity In-111-labeled anticarcinoembryonic antigen monoclonal antibody: Biodistribution and imaging in nude mice bearing human colon cancer xenografts. *Cancer Res.* **45,** 5700 (1985).

J3. Secarz, E. E., Janeway, C., and Wigzell, H., eds. "Immunoglobulin Idiotypes." Academic Press, New York, 1981.

J4. Janossy, G., Thomas, J. A., Pizzolo, G., Granger, S. M., McLaughlin, J., Habeshaw, J. A., Stanfeld, A. G., and Sloane, J., Immunohistological diagnosis of lymphoproliferative diseases by selected combinations of antisera and monoclonal antibodies. *Br. J. Cancer* **42,** 224 (1980).

J5. Jansen, F. K., Blythman, H. E., Carriere, D., Cassalas, P., Gros, O., Gros, P., Laurent, J. C., Paolucci, F., Pau, B., Poncelet, P., Richer, C., Vidal, H., and Voicin, G. A., Immunotoxins: Hybrid molecules combining high specificity and potent cytotoxicity. *Immunol. Rev.* **62,** 71 (1981).

J6. Jansen, F. K., Blythman, H., Carriere, D., Casallas, P., Hellström, I., Hellström, K. E., Gros, O., Gros, P., Laurent, J. C., Poncelet, P., Richer, P., Royston, I., Vidal, H., and Voison, G. A., Immunotoxins showing high specific cytotoxicity in four different tumor model systems. *Proc. Int. Cancer Conr., 13th, 1982* p. 585 (1983).

J7. Javadpour, N., The role of biologic tumor markers in testicular cancer. *Cancer (Philadelphia)* **45,** 1755 (1980).

J8. Jemmerson, R., and Fishman, W. H., *Anal. Biochem.* **124,** 286 (1982).

J9. Jerne, N. K., Towards a network theory of the immune system. *Ann. Immunol. (Paris)* **125,** 373–389 (1974).

J10. Jerry, L. M., Rowden, G., and Cano, P. O., Immune complexes in human melanoma: A consequence of deranged immune regulation. *Scand. J. Immunol.* **5,** 845–859 (1976).

K1. Kaiser, H., Levy, R., Broval, C., Civan, C. I., Fuller, D. J., Hsu, H. H., Leventhat, B. G., Miller, R. A., Milvanan, E. S., Santos, G. W., and Waram, M. D., Autologous bone marrow transplantation in T-cell malignancies: A case report involving *in vitro* treatment of marrow with a pan-T-cell monoclonal antibody. *J. Biol. Response Modif.* **1,** 233 (1982).

K2. Kaliss, N., Immunologic enhancement of tumor homografts in mice: A review. *Cancer Res.* **18,** 922 (1958).

K3. Kaplan, E., Beth-Porath, M., and Fink, S., Elimination of liver interference from the selenomethionine pancreas scan. *J. Nucl. Med.* **7,** 807–816 (1966).

K4. Katzman, J. A., Nesheim, M. E., Hibbard, L. S., and Mann, K., Isolation of functional human coagulation Factor V by using hybridoma antibody. *Proc. Natl. Acad. Sci. U.S.A.* **78,** 162–166 (1981).

K5. Keenan, A. M., Harbert, J. C., and Larson, S. M., Monoclonal antibodies in nuclear medicine. *J. Nucl. Med.* **26,** 531–537 (1985).

K6. Khaw, B. A., Fallon, J. T., Srauss, H. W., and Haber, E., Myocardial infarct imaging of antibodies to canine cardiac myosin with Indium-111-diethylenediaminepentaacetic acid. *Science* **209,** 295 (1980).

K7. Kim, E. E., DeLand, F. H., and Nelson, M. O., Radioimmunodetection of cancer with radiolabeled antibodies to alpha-fetoprotein. *Cancer Res.* **40**, 3008–3012 (1980).
K8. King, W. J., and Greene, G. L., Monoclonal antibodies localize oestrogen receptor in the nuclei of target cells. *Nature (London)* **307**, 745–749 (1984).
K9. King, W. J., DeSombre, E. R., Jensen, E. V., and Greene, G. L., Comparison of immunocytochemical and steroid-binding assays for estrogen receptor in human breast tumors. *Cancer Res.* **45**, 293–304 (1985).
K10. Klein, G., Tumor-specific transplantation antigens, G. H. A. Clowes Memorial Lecture. *Cancer Res.* **28**, 625–635 (1968).
K11. Klein, G., Klein, E., and Haughton, G., Variation of antigenic characteristics between different mouse lymphomas induced by the Moloney virus. *J. Natl. Cancer Inst. (U.S.)* **36**, 607–621 (1966).
K12. Kleist, S. von, and Burtin, P., Identification of a normal antigen from human colonic tumors. *Cancer Res.* **29**, 1961–1964 (1969).
K13. Kleist, S. von, and Burtin, P., Antigens cross reacting with CEA. *In* "Immunodiagnosis of Cancer" (R. B. Herberman and K. R. McIntire, eds.), pp. 322–342. Dekker, New York, 1979.
K14. Kleist, S. von, Chavanel, G., and Burtin, P., Identification of a normal antigen that cross-reacts with the carcinoembryonic antigen. *Proc. Natl. Acad. Sci. U.S.A.* **69**, 2492–2494 (1972).
K15. Klug, T. L., Bast, R. C., Jr., Niloff, J. M., Knapp, R. C., and Zurawski, Z. R., Jr., Monoclonal antibody radioimmunometric assay for an antigenic determinant (CA 125) associated with human epithelial ovarian carcinomas. *Cancer Res.* **44**, 1048–1053 (1984).
K16. Kohler, G., and Milstein, C., Continuous cultures of fused cells secreting antibody of predefined specificity. *Nature (London)* **256**, 495–497 (1975).
K17. Kohler, G., Howe, C. S., and Milstein, C., Fusion between immunoglobulin secreting and nonsecreting myeloma cell lines. *Eur. J. Immunol.* **6**, 292–295 (1976).
K18. Koizumi, M., Endo, K., Sakahara, H., Nakashima, T., Kunumatsu, M., Ohta, H., Konishi, J., and Torizuka, K., A model system for the evaluation of radioimmunoimaging of tumors. *J. Nucl. Med.* **26**, P26 (1985).
K19. Koprowski, H., Blaszczyk, M., and Steplewski, Z., Lewis blood-type may affect the incidence of gastrointestinal cancer. *Lancet* **1**, 1332 (1982).
K20. Koprowski, H., Herlyn, D., and Steplewski, Z., Specific antigen in serum of patients with colon carcinoma. *Science* **212**, 53 (1981).
K21. Koprowski, H., Steplewski, Z., and Mitchell, K., Colorectal carcinoma antigens detected by hybridoma antibodies. *Somatic Cell Genet.* **5**, 957 (1979).
K22. Koprowski, H., Steplewski, Z., Herlyn, D., and Herlyn, M., Study of antibodies against human melanoma produced by somatic cell hybrids. *Proc. Natl. Acad. Sci. U.S.A.* **75**, 3405–3409 (1978).
K23. Koprowski, H., Steplewski, Z., Herlyn, D., and Herlyn, M., Inhibition of growth of colorectal carcinoma in nude mice by monoclonal antibody. *Cancer Res.* **40**, 717 (1980).
K24. Koprowski, H., Herlyn, D., Herlyn, M., Steplewski, Z., and Sears, H. F., Specific antigen in the serum of patients with colon carcinoma. *Science* **212**, (1981).
K25. Koskimies, S., Human lymphoblastoid cell line producing specific antibody against Rh-antigen D. *Scand. J. Immunol.* **11**, 73–77 (1980).
K26. Kozbor, D., and Croce, C., Fusion partners for production of human monoclonal antibodies. *In* "Human Hybridoma Antibodies and Monoclonal Antibodies" (E. Engleman, S. Foung, J. Larrick, and A. Raubitschek, eds.), pp. 21–36. Plenum, New York, 1986.

K27. Kozbor, D., Lagarde, A., and Roder, J., Human hybridomas constructed with antigen-specific EBV transformed lines. *Proc. Natl. Acad. Sci. U.S.A.* **79,** 6651–6655 (1982).

K28. Krakauer, H., Hartzman, R. J., and Johnson, A. H., Monoclonal antibodies specific for human polymorphic cell surface antigens. 1. Evaluation of methodologies. Report on a workshop. *Hum. Immunol.* **4,** 167–181 (1982).

K29. Krolick, K. A., Uhr, J. W., Slavin, S., and Vitetta, E. S., In vivo therapy of a B cell tumor (BCL_1) using antibody-ricin A chain immunotoxins. *J. Exp. Med.* **155,** 1797 (1982).

K30. Kung, P. C., Goldstein, G., and Reinherz, E. L., Monoclonal antibodies defining distinctive human T-cell surface antigens. *Science* **206,** 347–349 (1979).

K31. Kung, P. C., Goldstein, G., Reinhertz, E. L., and Schlossman, S. F., Monoclonal antibodies defining distinctive human T cell surface antigens. *Science* **206,** 347 (1979).

K32. Kung, P. C., Burger, C. L., Goldstein, G., LoGerfo, P., and Edelson, R. L., Cutaneous T cell lymphoma: Characterization by monoclonal antibodies. *Blood* **57,** 261 (1981).

K33. Kupchik, H. Z., Dissection and characterization of human tumor-associated macromolecules using monoclonal antibodies. *In* "Monoclonal Antibodies and Cancer" (G. L. Wright, Jr., ed.), p. 299. Dekker, New York, 1984.

L1. Lane, D., and Koprowski, H., Molecular recognition and future of monoclonal antibodies. *Nature (London)* **296,** 200–202 (1982).

L2. Lange, P. H., McIntire, K. R., and Waldman, T. A., Serum alpha-fetoprotein and human chorionic gonadotropin in the diagnosis and management of nonseminomatous germ-cell testicular cancer. *N. Engl. J. Med.* **295,** 1237–1240 (1976).

L3. Lansdorp, P. M., Astaldi, G. C. B., Oosterhof, F., Janssen, M. D. C., and Zeijlemaker, W. P., Immunoperoxidase procedure to detect monoclonal antibodies against cell surface antigens. Quantitation of binding and staining of individual cells. *J. Immunol. Methods* **39,** 393–405 (1980).

L4. Larrick, J. W., and Bourla, J. M., Prospects for the therapeutic use of human monoclonal antidodies. *J. Biol. Response Modif.* **5,** 379–393 (1986).

L5. Larrick, J. W., Hart, S., Lippman, D., Glembourtt, M., Hsu, Y.-P., and Moss, R., Generation and characterization of human monoclonal anti-*Pseudomonas aeruginosa* antibodies. *In* "Human Hybridomas: Diagnostic and Therapeutic Applications" (A. Strelkelkaus, ed.). Dekker, New York, 1986.

L6. Larson, S. M., Carrasquillo, J. A., Krohn, K. A., McGuffin, R. W., and Williams, D. L., Diagnostic imaging of malignant melanoma with radiolabeled antitumor antibodies. *JAMA, J. Am. Med. Assoc.* **249,** 811 (1983).

L7. Lee, C. L., Li, C. Y., Jou, Y. H., Murphy, G. P., and Chu, T. M., Immunochemical characterization of prostatic acid phosphatase with monoclonal antibodies. *Ann. N.Y. Acad. Sci.* **390,** 52–61 (1982).

L8. Lessey, B. A., Alexander, P. S., and Horwitz, K. B., The subunit structure of human breast cancer progesterone receptors. Characterization by chromatography and photoaffinity labeling. *Endocrinology (Baltimore)* **112,** 1267–1274 (1983).

L9. Levin, L., Griffin, T. W., Haynes, L. R., and Sedor, C. J., Selective cytotoxicity for a colorectal carcinoma cell line by a monoclonal anti-carcinoembryonic antigen antibody coupled to the A chain of ricin. *J. Biol. Response Modif.* **1,** 149 (1982).

L10. Levine, G., Ballou, B., Reiland, J. *et al.,* Loacalization of I–131 labeled tumor-specific monoclonal antibody in the tumor-bearing BALB/c mouse. *J. Nucl. Med.* **21,** 570–573 (1980).

L11. Levy, R., Dilley, J., Fox, R. I., and Warnke, R., A human thymus-leukemia antigen defined by hybridoma monoclonal antibodies. *Proc. Natl. Acad. Sci. U.S.A.* **76,** 6552 (1979).

L12. Levy, R., Hurwitz, E., Maron, R., Arnon, R., and Sela, M., The specific cytotoxic effects of daunomycin conjugated to antitumor antibodies. *Cancer Res.* **35**, 1182 (1975).
L13. Lewis, M. G., and Phillips, T. M., The specificity of surface membrane immunofluorescence of human malignant melanoma. *Int. J. Cancer* **10**, 105–111 (1971).
L14. Lewis, M. G., Hartmann, D., and Jerry, L. M., Antibodies and anti-antibodies in human malignancy: An expression of deranged immune regulation. *Ann. N.Y. Acad. Sci.* **276**, 316–327 (1976).
L15. Lewis, M. G., Ikonopisov, R. L., and Nairn, R. C., Tumor specific antibodies in malignant melanoma and their relationship to the extent of the disease. *Br. Med.* **3**, 547–552 (1969).
L16. Lillehoj, H. S., Choe, B. K., and Rose, N. R., Monoclonal anti-human prostatic acid phosphatase antibodies. *Mol. Immunol.* **19**, 1199 (1983).
L17. Logeat, F., Vu Hai, M. T., and Milgrom, E., Antibody to rabbit progesterone receptor: Cross reaction with human receptor. *Proc. Natl. Acad. Sci. U.S.A.* **78**, 1426–1430 (1981).
L18. Logeat, F., Pamphile, R., Loosfeld, H., Jolivet, A., Fournier, A., and Milgrom, E., One step immunoaffinity purification of active progesterone receptor. Further evidence in favor of the existence of a single steroid binding subunit. *Biochemistry* **24**, 1029–1035 (1985).
L19. Logeat, F., Vu Hai, M. T., Fournier, A., Legrain, P., Buttin, G., and Milgrom, E., Monoclonal antibody to rabbit progesterone receptor: Cross reaction with other mammalian progesterone receptors. *Proc. Natl. Acad. Sci. U.S.A.* **80**, 6456–6459 (1983).
L20. Lowe, A. D., Lennox, E. S., and Voak, D., A new monoclonal anti-A. Culture supernatants with the performance of hyperimmune human reagents. *Vox Sang.* **46**, 185–194 (1984).
M1. Mach, J.-P., Carrel, S., Forni, M., Ritchard, J., Donath, A., and Alberto, P., Tumor localization of radiolabeled antibodies against carcinoembryonic antigen in patients with carcinoma: A critical evaluation. *N. Engl. J. Med.* **330**, 5–10 (1980).
M2. Mach, J.-P., Buchegger, F., Girardet, C., Forni, M., Ritsheard, J., Accollo, R. S, and Carrel, S., Use and limitations of monoclonal anti CEA antibodies in immunoassays and in tumour localization by immunoscintigraphy: A discussion. *In* "Markers for Diagnosis and Monitoring of Human Cancer" (M. I. Colnaghi, G. L. Buraggi, and M. Ghione, eds.), Vol. 2, p. 189. Academic Press, New York, 1982.
M3. Mach, J.-P, Buchegger, F., Forni, M., Ritshcard, J., Berche, C., Lumbruso, J. D., Schreyer, M., Girardet, C., Accola, R. S., and Carrel, S., Use of radiolabeled monoclonal anti-CEA antibodies for the detection of human carcinomas by external scanning and tomoscintigraphy. *Immunol. Today* **2**, 239–249 (1981).
M4. Mach, J.-P., Chatal, J. F., Lumbroso, J. D., Buchegger, F., Forni, M., Ritshcard, J., Berche, C., Doullard, J. Y., Carel, S., and Herlyn, D., Tumor localization in patients by radiolabeled monoclonal antibodies to colon carcinoma. *Cancer Res.* **43**, 5593 (1983).
M5. Magnani, J. L., Brockhaus, M., Smith, D. F., Ginsburg, V., Blaszczyk, M., Mitchell, K. F., Steplewsli, Z., and Koprowski, H., A monosialoganglioside is a monoclonal antibody defined antigen of colon carcinoma. *Science* **212**, 53 (1981).
M6. Mahan, D. E., and Doctor, B. P., A radioimmunoassay for human prostatic acid phosphatase in prostatic disease. *Clin. Biochem. (Ottawa)* **12**, 10–17 (1979).
M7. Malkin, A., Keller, J. A., and Raviczky, M. I., Immunohistochemical detection of ectopic hormones in experimental rat tumors. *In* "Carcino-Embryonic Proteins" (F.-G. Lehman, ed.), Vol. 2, pp. 751–758. Elsevier/North-Holland, New York, 1979.
M8. Mansel, R. E., Edwards, D. P., and Sewell, B., Monoclonal antibody immunohistochemistry identifies "high risk" antigen in benign breast biopsies. *Br. J. Surg.*

M9. Markert, C. L., Current topics of biological medical research. *In* "Isoenzymes" (M. C. Rattazzi, J. G. Scandalios, and G. S. Whitt, eds.), p. 1. Alan R. Liss, New York, 1977.

M10. Mazauric, T., Mitchell, K. F., Letchworth, G. J., III, Koprowski, H., and Steplewski, Z., Monoclonal antibody-defined human lung cell surface protein antigens. *Cancer Res.* **42**, 150–154 (1982).

M11. McDade, J. E., Diagnostic applications of monoclonal antibodies: Infectious disease diagnosis. *In* "Monoclonal Antibodies in Clinical Diagnostic Medicine" (D. S. Gordon, ed.), pp. 138–139. Igaku-Shoin, New York, 1985.

M12. McDougal, J. S., Kennedy, M. S., Hubbard, M., McDuffie, F. C., and Moore, D. D., Immunoassay for IgG rheumatoid factor using a murine monoclonal anti-Fd antibody. *J. Lab. Clin. Med.* **106**, 80–87 (1986).

M13. McIntire, K. R., Waldman, T. A., and Moertel, C. G., Serum alpha-fetoprotein in patients with neoplasms of the gastrointestinal tract. *Cancer Res.* **35**, 991–996 (1975).

M14. McKay, I. R., Carcinoembryonic antigen: Use in screening. *In* "Immunodiagnosis of Cancer" (R. B. Herberman and K. R. McIntire, eds.), pp. 255–264. Dekker, New York, 1979.

M15. McKearn, T. J., Smilek, D. E., and Fitch, F. W., Rat–mouse hybridomas and their application to studies of the major histocompatability complex. *In* "Monoclonal Antibodies, Hybridomas: A New Dimension in Biological Analyses" (R. H. Kennet, T. J. McKearn, and K. B. Bechtel, eds.). Plenum, New York, 1980.

M16. McLaughlin, P. J., Cheng, M. H., Slade, M. B., and Johnson, P. M., Expression on cultured human tumour cells of placental trophoblast membrane antigen and placental alkaline phosphatase defined by monoclonal antibodies. *Int. J. Cancer* **30**, 21–26 (1982).

M17. McMichael, A. J., and Fabre, J. W., eds., "Monoclonal Antibodies in Clinical Medicine." Academic Press, New York, 1982.

M18. Melamed, M., Mullaney, P., and Mendelson, M., Flow cytomerization of cutaneous T-cell lymphoma. Use of monoclonal antibodies to compare with other malignant T-cells. *N. Engl. J. Med.* **304**, 1319–1323 (1981).

M19. Messeter, L., Brodin, T., Chester, M. A., Low, B., and Lundblad, A., Mouse monoclonal antibodies with anti-A, anti-B and anti-A,B specificities; some superior to human polyclonal ABO reagents. *Vox Sang.* **46**, 185–194 (1984).

M20. Messeter, L., Brodin, T., Chester, M. A., Karlsson, K. A., Zopf, D., and Lundblad, A., Immunochemical characterization of a monoclonal anti-Le^b blood grouping reagent. *Vox Sang.* **46**, 66–74 (1984).

M21. Metzgar, R. S., Bergoc, P. M., and Moreno, M. A., Melanoma specific antibodies produced in monkeys by immunization with human melanoma cell lines. *J. Natl. Cancer Inst. (U.S.)* **50**, 1065–1068 (1973).

M22. Miles, L. E. M., *in* "Handbook of Radioimmunoassay" (G. Abraham, ed.), pp. 131–154. Dekker, New York, 1977.

M23. Millan, J. L., Beckman, G., Jeppsson, A., and Stigbrand T., Genetic variations of placental acid phosphatase as detected by a monoclonal antibody. *Hum. Genet.* **60**, 145–149 (1982).

M24. Millan, J. L., Stigbrand, T., Rouslahti, E., and Fishman, W. H., Characterization and use of an allotype-specific antibody to placental alkaline phosphatase in the study of cancer related phosphatase polymorphism. *Cancer Res.* **42**, 2444–2449 (1982).

M25. Miller, L. S., and Konrath, J. G., An immunocytochemical assay for estrogen receptors in breast tumors using monoclonal antibodies. *J. Steroid Biochem.* **20**, 1633 (1984).

M26. Miller, R. A., and Levy, R., Response of cutaneous T-cell lymphoma therapy with hybridoma monoclonal antibody. *Lancet* **2**, 226 (1981).

M27. Miller, R. A., Maloney, D. G., McKillop, J., and Levy, R., In vivo effects of hybridoma monoclonal antibody in a patient with T-cell leukemia. *Blood* **58**, 78 (1981).
M28. Milstein, C., Monoclonal antibodies from hybrid myelomas. *Proc. R. Soc. London, Ser. B.* **211**, 393–412 (1981).
M29. Moore, S., Chirnside, A., Micklem, L. R., McClelland, D. B., and James, K., A mouse monoclonal antibody with anti-A, (B) specificity which agglutinates A_x cells. *Vox Sang.* **47**, 427–434 (1984).
M30. Morbidity and Mortality Weekly Report, Progress toward achieving the national 1990 objectives for sexually transmitted diseases. *Morbid. Mortal. Wkly. Rep.* **36**, 173–176 (1987).
M31. Morgan, A. C., Jr., Galloway, D. R., and Reisfeld, R. A., Production and characterization of monoclonal antibody to a melanoma specific glycoprotein. *Hybridoma* **1**, 27–36 (1981).
M32. Morimoto, C., Reinherz, E. L., Schlossman, S. F., Schur, P. H., Mills, J. A., and Steinberg, A. D., Alterations in immunoregulatory T cell subsets in active systemic lupus erythematosus. *J. Clin. Invest.* **66**, 1171–1174 (1980).
M33. Morrison, S., Transfectomas provide novel chimaeric antibodies. *Science* **229**, 1202–1207 (1985).
M34. Morton, D. L., Immunological studies with human neoplasms. *RES, J. Reticuloendothel. Soc.* **10**, 137–160 (1971).
M35. Morton, D. L., Malmgren, D. A., and Holmes, E. C., Demonstration of antibodies against human malignant melanoma by immunofluorescence. *Surgery (St. Louis)* **64**, 2333–2340 (1968).
M36. Moshakis, V., McIlhinney, R. A. J., Raghaven, D., and Neville, A. M., Localization of human tumor xenografts after i.v. administration of radiolabeled monoclonal antibodies. *Br. J. Cancer* **44**, 91–99 (1981).
M37. Moshakis, V., McIlhinney, R. A. J., Raghaven, D., and Neville, A. M., Monoclonal antibodies to detect human tumours: An experimental approach. *J. Clin. Pathol.* **34**, 314–319 (1981).
M38. Muna, N. M., Marcus, S., and Smart, C., Detection by immunofluorescence of antibodies specific for human malignant melanoma cells. *Cancer (Philadelphia)* **23**, 88–93 (1969).
N1. Nadji, M., and Morales, A. R., Immunochemistry of prostatic acid phosphatase. *Ann. N.Y. Acad. Sci.* **390**, 133 (1982).
N2. Nadler, L. M., Stashenko, P., Hardy, R., Kaplan, W. D., Dutton, L. N., Kufe, D. W., Antman, K. H., and Schlossman, S. F., Serotherapy of a patient with a monoclonal antibody directed against a human lymphoma-associated antigen. *Cancer Res.* **40**, 3147 (1980).
N3. Naritoku, W. Y., and Taylor, C. Y., A comparative study of the use of monoclonal antibodies using three different monoclonal and polyclonal antibodies against human prostatic acid phosphatase. *J. Histochem. Cytochem.* **30**, 253 (1982).
N4. Neuvald, P. D., Anderson, C., Salivar, W. O., Andenderfer, P. H., Dermody, W. C., Weintraub, B, D., Rosen, S. W., Nelson-Rees, W. A., and Rudden, R. W., Expression of oncodevelopmental gene products by human tumor cells in culture. *JNCI, J. Natl. Cancer Inst.* **64**, 447–459 (1980).
N5. Newell, K. D., Goldenberg, D. M., and Primus, F. J., Identification and differentiation of carcinoembryonic antigen determinants by monoclonal antibodies. *Fed. Proc., Fed. Am. Soc. Exp. Biol.* **41**, 411 (abstr.) (1982).
N6 Norgaard-Pederson, B., Albrechtsen, R., and Teilum, G., Serum alpha-fetoprotein as a marker for endodermal sinus tumor (yolk sac tumor) or a vitelline component of 'teratocarcinoma.' *Acta Pathol. Microbiol. Scand., Sect. A* **83A**, 573–589 (1975).

N7. Nussbaum, S. R., Rosenblatt, M., and Mudgett-Hunter, M., Monoclonal antibodies directed against the biologically active of parathyroid hormone. In "Monoclonal Antibodies in Endocrine Research" (R. E. Fellows and E. S. Eisenbarth, eds.), pp. 181–192. Raven Press, New York, 1981.
O1. Okret, S., Wikström, A. C., Wrange, O., Andersson, B., and Gustafsson, J. A., Monoclonal antibodies against the rat liver glucocorticoid receptor. *Proc. Natl. Acad. Sci. U.S.A.* **81,** 1609–1613 (1984).
O2. Okuda, K., Kubo, Y., and Obata, H., Serum alpha-fetoprotein in the early stages of heptocellular carcinoma and its relationship to gross anatomical types. *Ann. N.Y. Acad. Sci.* **259,** 248–252 (1975).
O3. Old, L. J., Immunology: The search for specificity, G. H. A. Clowes Memorial Lecture. *Cancer Res.* **41,** 361–375 (1981).
O4. Old, L. J., Boyse, E. A., and Clark, D. A. Antigenic properties of chemically induced tumors. *Ann. N.Y. Acad. Sci.* **101,** 80–106 (1962).
O5. Order, S. E., Radioimmunoglobulin therapy of cancer. *Compr. Ther.* **10,** 9–18 (1984).
O6. Osborn, M., Altmannsberger, M., Debus, E., and Weber, K., Differentiation of the major human tumor groups using conventional and monoclonal antibodies specific for intermediate filament proteins. *Ann. N.Y. Acad. Sci.* **455,** 649–668 (1985).
O7. Otsuka, F. L., and Welch, M. J., Methods to label monoclonal antibodies for use in tumor imaging. *Nucl. Med. Biol.* **14,** 243–249 (1987).
O8. Otsuka, F. L., Welch, M. J., McElvany, K. D., Nicolotti, R. A., and Fleischman, J. B., Development of a model system to evaluate methods for radiolabeling monoclonal antibodies. *J. Nucl. Med.* **25,** 1343 (1984).
P1. Parsons, S. F., Monoclonal antibodies to human erythrocyte antigens. *Biotest Bull.* **1,** 315–317 (1983).
P2. Pereira, L., Dondero, D. V., and Gallo, D., Serological analysis of herpes simplex types 1 and 2 with monoclonal antibodies. *Infect. Immunol.* **35,** 363–367 (1982).
P3. Peterson, E., Schmidt, G. W., Goldstein, L. C., Nowinski, C., and Corey, L., Typing of clinical herpes simplex virus isolates with mouse monoclonal antibodies to herpes simplex virus types 1 and 2: Comparison with type-specific rabbit anti-sera and restriction endonuclease analysis of viral DNA. *J. Clin. Microbiol.* **17,** 92–96 (1983).
P4. Pimm, M. V., Perkins, A. C., Armitage, N. C., and Baldwin, R. W., The characteristics of blood-borne radiolabels and the effect of anti-mouse IgG antibodies on localization of radiolabeled monoclonal antibody in cancer patients. *J. Nucl. Med.* **26,** 1011–1023 (1985).
P5. Pimm, M. V., Jones, J. A., Price, M. R., Middle, J. G., Embleton, M. J., and Baldwin, R. W., Tumor localization of monoclonal antibody against a rat mammary carcinoma and suppression of tumor with adriamycin-antibody conjugates. *Cancer Immunol. Immunother.* **12,** 125–134 (1982).
P6. Pimm, M. V., Embleton, M. J., Perkins, A. C., Price, M. R., Robins, R. A., Robinson, G. R., and Baldwin, R. W., In vitro localization of anti-oestrogenic sarcoma 79IT monoclonal antibody in oestrogenic sarcoma xenographs. *Int. J. Cancer* **30,** 75–85 (1982).
P7. Pinto, M. M., Carcinoembryonic antigen in pericardial effusion. *Lab Med.* **18,** 671–672 (1987).
P8. Pletch, Q., and Goldenberg, D. M., Molecular size of carcinoembryonic antigen in plasma of patients with malignant disease. *J. Natl. Cancer Inst. (U.S.)* **53,** 1201–1204 (1974).
P9. Press, M. F., and Green, G. L., Methods in laboratory investigation. An immunocytochemical method for demonstrating estrogen receptor in human uterus using monclonal antibodies to human estrophilin. *Lab. Invest.* **50,** 480–486 (1984).

P10. Press, M. F., Nousek-Goebl, N., King, W. J., Herbst, A. L., and Greene, G. L., Immunohistochemical assessment of estrogen receptor distribution in the human endometrium throughout the menstrual cycle. *Lab. Invest.* **51**, 495–503 (1984).
P11. Primus, F. J., Kuhns, W. J., and Goldenberg, D. M., Immunological heterogeneity of carcinoembryonic antigen: Immunohistochemical detection of carcinoembryonic antigen determinants in colonic tumors with monoclonal antibodies. *Cancer Res.* **43**, 693 (1983).
P12. Primus, F. J., Kelley, E. A., Hansen, H. J., and Goldenberg, D. M., Sandwich immunoassay for carcinoembryonic antigen in patients receiving murine monoclonal antibodies for diagnosis and therapy. *Clin. Chem. (Winston-Salem, N.C.)* (in press).
P13. Primus, F. J., MacDonald, R., Goldenberg, D. M., and Hansen, H. J., Localization of GW-39 human tumors in hamsters by affinity-purified antibody to carcinoembryonic antigen. *Cancer Res.* **37**, 1544–1547 (1977).
P14. Primus, F. J., Newell, K. D., Blue, A., and Goldenberg, D. M., Immunological heterogeneity of carcinoembryonic antigen: Antigenic determinants on CEA distinguished by monoclonal antibodies. *Cancer Res.* **43**, 686 (1983).
Q1. Quaranta, V., Walker, L. E., Pelligrino, M. A., and Ferrone, S., Purification of immunologically functional subsets of human Ia-like antigens on a monoclonal antibody (Q5/13) immunoadsorbant. *J. Immunol.* **125**, 1421–1425 (1980).
Q2. Quinones, G. R., Rohner, T. J., Drago, J. R., and Demers, L. M., Will prostatic acid phosphatase determination by radioimmunoassay increase the diagnosis of early prostatic carcinoma? *J. Urol.* **125**, 361–364 (1981).
R1. Rankin, E. M., and McVie, J. G., Radioimmunodetection of cancer: Problems and potential. *Br. Med. J.* **287**, 1402–1404 (1983).
R2. Raso, V., Antibody mediated delivery of toxic molecules to antigen bearing target cells. *Immunol. Rev.* **63**, 93 (1982).
R3. Raso, V., Ritz, J., Bassala, M., and Schlossman, S. F., Monoclonal antibody-ricin A chain conjugate selectively cytotoxic for cells bearing the common acute lymphoblastic leukemia antigen. *Cancer Res.* **42**, 457 (1982).
R4. Reckel, R. P., Harris, J., Botsko, E., Wellerson, R., and Varga, S., The detection of circulating immune complexes containing C1Q and IgG using a new rapid ELISA test system. *Diagn. Immunol.* **2**, 228–237 (1984).
R5. Reinherz, E. L., and Schlossman, S. F., The characterization and function of human immunoregulatory T lymphocyte subsets. *Immunol. Today* **1**, 67–75 (1981).
R6. Reinherz, E. L., Kung, P. C., and Goldstein, G., Separation of functional subsets of human T cells by a monoclonal antibody. *Proc. Natl. Acad. Sci. U.S.A.* **76**, 4061–4065 (1979).
R7. Reinherz, E. L., Kung, P., C., and Goldstein, G., A monoclonal antibody with selective reactivity with functionally mature human thymocytes and all peripheral human T cells. *J. Immunol.* **123**, 1312–1317 (1979).
R8. Reisfeld, R. A., Immunochemical characterization of human tumor antigens. *Semin. Oncol.* **13**, 153–164 (1986).
R9. Richardi, P., Amoroso, A., Crepaldi, T., and Curtoni, E. S., Cytolytic activity of monoclonal antibodies strongly depends on rabbit complement used. *Tissue Antigens* **17**, 368–371, (1981).
R10. Richman, D. P., Gomez, C. M., Berman, P. W., Burres, S. A., Fitch, F. W., and Wainer. *Nature (London)* **286** (1980).
R11. Ridgeway, E. C., Ardisson, L. J., Meskell, M. J., and Midgett-Hunter, M., Monoclonal antibody to human thyrotropin. *J. Clin. Endocrinol. Metab.* **55**, 44–48 (1982).
R12. Ridgewell, K., M. J. A., and Anstee, D. J., The Wr^b antigen, a receptor for *Plasmodium*

falciparum malaria, is located on a helical region of the major sialoglycoprotein of human red blood cells. *Biochem. J.* **209**, 615–619 (1983).

R13. Ritz, J., Pesando, J., Notis-McConarty, J., Lazarus, H., and Schlossman, S. F., A monoclonal antibody to human acute lymphoblastic leukemia antigen. *Nature (London)* **283**, 583 (1980).

R14. Ritz, J., Pesando, J. M., Sallan, S. E., Clavell, L. A., Notis-McConarty, J., Rosenthal, P., and Schlossman, S. F., Serotherapy of acute lymphoblastic leukemia with monoclonal antibody. *Blood* **58**, 141 (1981).

R15. Ritz, J., Bast, R. C., Clavell, L. A., Hercent, T., Sallan, S. E., Lipton, J. M., Feeney, M., Nathan, D. G., and Schlossman, S. F., Autologous bone-marrow transplantation in CALLA-positive acute lymphoblastic leukemia after *in vitro* treatment with J5 monoclonal antibody and complement. *Lancet* **2**, 60 (1982).

R16. Romsdahl, M., and Cox, I. S., Human malignant antibodies demonstrated by immunofluorescence. *Arch. Surg. (Chicago)* **100**, 491–497 (1970).

R17. Rosen, S. W., Weintraub, B. D., and Aaronson, S. A., Nonrandom ectopic protein production by malignant cells: Direct evidence *in vitro*. *J. Biol. Chem.* **254**, 9409–9415 (1979).

R18. Rowland, G. F., Simmonds, R. G., Corvalan, R. F., Marsden, C. H., Johnson, R. R., Woodhouse, C. S., Ford, C. H. G., and Neuman, C. C., The potential use of monoclonal antibodies in drug targeting. *Protides Biol. Fluids* p. 921 (1982).

R19. Royston, I., Majda, J. A., Baird, S., Meserve, B. L., and Griffiths, J. C., Human T-cell antigens defined by monoclonal antibodies: The 65,000-dalton antigen of T-cells (T65) is also found on chronic lymphocytic leukemia cells bearing surface immunoglobulin. *J. Immunol.* **125**, 725 (1980).

R20. Rubin, R. L., Balderas, R. S., and Tan, E. M., Multiple autoantigen binding capabilities of mouse monoclonal antibodies selected for rheumatoid factor activity. *J. Exp. Med.* **159**, 1429–1440 (1984).

R21. Ruch, F. E., Jr., and Smith, L., Monoclonal antibody to streptococcal group B carbohydrate: Applications in latex agglutination and immunoprecipitation assays. *J. Clin. Microbiol.* **16**, 145–162 (1982).

R22. Rule, A. H., and Golesky-Reilly, C., Carcinoembryonic antigen (CEA): Separation of CEA reacting molecules from tumor, fetal gut, meconium and normal colon. *Immunol. Commun.* **2**, 213–226 (1973).

R23. Ruoslahti, E., and Seppela, M., Studies of carcinofetal proteins. III. Development of a radioimmunoassay for alpha-fetoprotein in serum of healthy adults. *Int. J. Cancer* **8**, 374–383 (1971).

R24. Ruoslahti, E., and Seppala, M., Alphafetoprotein in cancer and fetal development. *Adv. Cancer Res.* **29**, 275 (1979).

R25. Ruoslahti, E., Salaspuro, M., and Pihko, H., Serum alpha fetoprotein: Diagnostic significance in liver disease. *Br. Med. J.* **2**, 527–529 (1974).

S1. Sacks, S., and Lennox, E. S., Monoclonal anti-B as a new blood typing reagent. *Vox Sang.* **40**, 99–104. (1981).

S2. Sallan, S. E., Ritz, J., Pesando, J., Gelber, R., O'Brien, C., Hitchcock, S., Coral, F., and Schlossman, S. F., Cell surface antigens: Prognostic implications in childhood acute lymphocytic leukemia. *Blood* **55**, 395–402 (1980).

S3. Samaan, N. A., Smith, J. P., and Rutledge, F. N., The significance of measurement of placental lactogen, human chorionic gonadotropin, and carcinoembryonic antigen in patients with ovarian carcinoma. *Am. J. Obstet. Gynecol.* **126**, 186–189 (1976).

S4. Sandstrom, E., and Danielsson, D., Serology of *Neiserria gonorrheae:* Classification by co-agglutination. *Acta Pathol. Microbiol. Scand., Sect. B* **88 B**, 27–38 (1981).

S5. Scardino, P. T., Cox, H. D., and Waldman, T. A., The value of serum tumor markers in the staging and prognosis of germ cell tumors of the testes. *J. Urol.* **118**, 994–999 (1979).
S6. Schachter, J., and Caldwell, D. H., Chlamydiae. *Annu. Rev. Microbiol.* **34**, 285–309 (1980).
S7. Scharff, M. D., Roberto, S., and Thammana, P., Monoclonal antibodies. *J. Infect. Dis.* **143**, 364–351 (1981).
S8. Scheinberg, D. A., and Strand, M., Leukemic cell targeting and therapy by a monoclonal antibody in a mouse model system. *Cancer Res.* **42**, 44 (1982).
S9. Scheinberg, D. A., Strand, M., and Gansow, O. A., Tumor imaging with radioactive metal chelates conjugated to monoclonal antibodies *Science* **215**, 1511–1513 (1982).
S10. Schoenfeld, Y., Rauch, J., and Massicote, H., Polyspecificity of monoclonal lupus autoantibodies produced by human–human hybridomas. *N. Engl. J. Med.* **308**, 414–420 (1983).
S11. Schroff, R. W., Foon, K. A., and Beatty, S. M., Human anti-murine immunoglobulin responses in patients receiving monoclonal antibody therapy. *Cancer Res.* **45**, 879–885 (1985).
S12. Secher, D. S., Monoclonal antibodies by cell fusion. *Immunol. Today* July, pp. 22–26 (1980).
S13. Secher, D. S., and Burke, D. C., A monoclonal antibody for large scale purification of human leukocyte interferon. *Nature (London)* **285**, 446–450 (1980).
S14. Sevier, E. D., David, G. S., Martinis, J., Desmond, W. J., Batholemew, R. M., and Wang, R., Monoclonal antibodies in clinical immunology. *Clin. Chem. (Winston-Salem, N.C.)* **27**, 1797–1807 (1981).
S15. Shafritz, D. A., Lieberman, H. M., and Isselbacher, K. J., Monoclonal immunoassays for hepatitis B surface antigen. Demonstration of hepatitis B virus DNA or related sequences in serum and viral epitopes in immune complexes. *Proc. Natl. Acad. Sci. U.S.A.* **79**, 5675–5679 (1982).
S16. Shapiro, H. M., Multistation multiparameter flow cytometry: A critical review and rationale. *Cytometry* **4**, 11–19 (1983).
S17. Sharkey, R. M., Primus, F. J., and Goldenberg, D. M., Second antibody clearance of radiolabeled antibody in cancer radioimmunodetection. *Proc. Natl. Acad. Sci. U.S.A.* **81**, 2843–2846 (1984).
S18. Sheth, N. A., Saruiya, J. N., and Ranadive, K. J., Ectopic production of human chorionic gonadotropin by human breast tumors. *Br. J. Cancer* **30**, 566–570 (1974).
S19. Shiku, H., Takahashi, T., and Oettgen, H. F., Cell surface antigens of human malignant melanoma. II. Serological typing with immune adherence assays and definition of two new surface antigens. *J. Exp. Med.* **144**, 873–881 (1976).
S20. Shipova, L. Y., Goussev, A. L., Alpha-fetoprotein in the liver of embryonic and newborn rats. *Ontogenez* **7**, 392–395 (1976).
S21. Shipova, L. Y., Goussev, A. I., and Englehardt, N. V., Immunohistochemical study of alpha-fetoprotein and serum albumin in the early postnatal period in mice. *Ontogenez* **5**, 53–60 (1974).
S22. Slaughter, C. A., Coseo, M. C., Cancro, M. P., and Harris, H., *Proc. Natl. Acad. Sci. U.S.A.* **78**, 1124 (1981).
S23. Slifkin, M., Acevedo, H. F., and Pardo, M., Human chorionic gonadotropin in cancer cells. II. Ultrastructural localization. *In* "Prevention and Detection of Cancer" (H. E. Nieburgs, ed.), Part 2, Vol. I, pp. 965–979. Dekker, New York, 1978.
S24. Smith, T. F., Brown, S. D., and Weed, L. A., Diagnosis of *Chlamydia trachomatis* infections by cell culture and serology. *Lab. Med.* **13**, 92–100 (1982).

S25. Smolen, J. S., Chused, J. M., Leiserson, W. M., Reeves, J. P., Alling, D., and Steinberg, D. A., Heteregeneity of immunoregulatory T cell subsets in systemic lupus erythematosus. Correlations with clinical features. *Am. J. Med.* **72**, 783–790 (1982).
S26. Solter, D., Ballou, B., and Reilan, J., Radioimmunodetection of tumors using monoclonal antibodies. *Prog. Cancer Res. Ther.* **21**, 241–244 (1982).
S27. Steele, L., Cooper, E. H., and MacKay, A. M., Combination of carcinoembryonic antigen and gamma glutamyl transpeptidase in the study of the evaluation of colorectal cancer. *Br. J. Cancer* **30**, 319–324 (1974).
S28. Steensgaard, J., Jacobsen, C., Lowe, J., Hardie, D., Ling, N. R., and Jeffris, R., The development of difference turbidimetric analysis for monoclonal antibody to human IgG. *Mol. Immunol.* **17**, 1315–1318 (1980).
S29. Steinitz, M., Klewin, G., Koskimes, S., and Mäkelä, O., EB virus-induced B lymphocyte cell lines producing specific antibody. *Nature (London)* **269**, 420–422 (1977).
S30. Stephens, R. S., Kuo, C. C., and Tam, M. R., Sensitivity of immunofluorescence with monoclonal antibodies for detection of *Chlamydia trachomatis* inclusions in cell culture. *J. Clin. Microbiol.* **16**, 4–7 (1982).
S31. Steplewski, Z., and Koprowski, H., Monoclonal antibody development in the study of colorectal carcinoma associated antigens. *Methods Cancer Res.* **20**, 286–316 (1982).
S32. Stern, P. H., Hagan, P., and Halpern, S., The effect of radiolabel on the kinetics of monoclonal anti-CEA in a nude mouse–human colon tumor model. *Prog. Cancer Res. Ther.* **21**, 245–253 (1982).
S33. Stolbach, L. L., Krant, M. J., and Fisherman, W. J., Ecotopic production of alkaline phosphate isoenzyme in patients with cancer. *N. Engl. J. Med.* **281**, 757 (1969).
S34. Stone, M., Bagshawe, K. D., and Kardena, Beta human chorionic gonadotropin and carcinoembryonic antigen in the management of ovarian carcinoma. *Br. J. Obstet. Gynaecol.* **84** (1977).
S35. Stratel, P., Dolejalova, V., and Feit, J., Localization of alphafetoprotein in liver tissue of rat during postnatal development. Comparison of the immunofluoresence and autoradiographic methods. *Neoplasma* **23**, 1–10 (1976).
S36. Stratte, P. T., Miller, R. A., Amyx, H. L., Asher, D. M., and Levy, R., *In vivo* effects of murine monoclonal anti-T cell antibodies in subhuman primates. *J. Biol. Response Modif.* **1**, 137 (1982).
S37. Strelkelkaus, A., ed., "Human Hybridomas: Diagnostic and Therapeutic Applications." Dekker, New York, 1986.
S38. Stuart, M. C., Walichnowski, C. M., Underwood, P. A., Hussain, S., Harmon, D. F., Rathjen, D. A., and von Sturmer, S. R., The production of high affinity monoclonal antibodies to human growth hormone. *J. Immunol. Methods* **61**, 33–42 (1983).
S39. Suddith, R. L., Townsend, C. M., and Thompson, J. C., Detection of monoclonal antibodies against synthetic human gastrin and pentagastrin by an indirect enzyme-linked immunosorbent assay (ELISA). *Surg. Forum* **31**, 185–186 (1980).
S40. Sufrin, G., Mirand, E. A., and Moore, R. H., Hormones in renal cancer. *J. Urol.* **117**, 433–438 (1977).
S41. Suter, L., Bruggern, J., and Sorg, C., Use of an enzyme-linked immunosorbant assay (ELISA) for screening of hybridoma antibodies against cell surface antigens. *J. Immunol. Methods* **39**, 407–411 (1980).
T1. Tai, J., Blair, A., and Ghose, T., Tumor inhibition by chorambucil linked to antitumor globulin. *Eur. J. Cancer* **15**, 1357 (1979).
T2. Taketa, K., Watanabe, A., and Kosaka, K., Different mechanisms of increased alphafetoprotein production in rats following CCl_4 intoxication and partial hepatectomy. *Ann. N.Y. Acad. Sci.* **259**, 80–84 (1975).

T3. Tanner, P., Stenman, U. H., Seppala, M., and Schroder, J., Sensitive and specific RIA for HCG using monoclonal antibodies. *Protides Biol. Fluids* **29**, 843 (1981).
T4. Taylor-Papadimitriou, J., Peterson, J. A., Arklie, J., Burchell, J., Ceriani, R. L., and Bodmer, W. F., Monoclonal antibodies to epithelium-specific of the human milkfat globule membrane: Production and reactions of cells in culture. *Int. J. Cancer* **28**, 17–21 (1981).
T5. Tedder, R. S., Guarascio, P., Yao, J. L. *et al.*, Production of monoclonal antibodies to hepatitis surface and core antigens, and the use in the detection of viral antigens in liver biopsies. *J. Hyg.* **90**, 135–142 (1983).
T6. Teng, N., Lam, K., Calvo Riera, F., and Kaplan, H., Construction and testing of mouse–human heteromyelomas for human monoclonal antibody production. *Proc. Natl. Acad. Sci. U.S.A.* **80**, 7308–7312 (1983).
T7. Terasaki, P., *in* "Histocompatibility Testing 1980." UCLA Tissue Typing Laboratory Publ., University of California, Los Angeles, 1980.
T8. Thompsin, R. H., Jackson, A. P., and Langlois, N., Circulating antibodies to mouse monoclonal immunoglobulins in normal subjects. Incidents, species specificity, and effect on the two site assay for creatinine kinase-MB isoenzyme. *Clin. Chem. (Winston-Salem, N.C.)* **32**, 476–481 (1986).
T9. Tormey, D. C., Waalkes, T. P., and Ahmann, D., Biological markers in breast carcinoma. I. Incidence of abnormalities of CEA, HCG, 3 polyamines and 3 nucleotides. *Cancer (Philadelphia)* **35**, 1095–1100 (1975).
T10. Trowbridge, I. S., Cancer monoclonals. *Nature (London)* **294**, 204 (1981).
T11. Trowbridge, I. S., and Domingo, D. L., Anti-transferrin receptor monoclonal antibody and toxin-antibody conjugates affect growth of human tumour cells. *Nature (London)* **294**, 171 (1981).
T12. Trucco, M., and dePetris, S., Determination of equilibrium binding parameters of monoclonal antibodies specific for cell surface antigens. *In* "Immunological Methods" (I. Lefkovitz and R. Pernis, eds.), Vol. 2, pp. 1–26. Academic Press, New York, 1981.
T13. Trucco, M. M., Garotta, G., Stocker, J. W., and Ceppillini, R., Murine monoclonal antibodies against HLA structures. *Immunol. Rev.* **47**, 219 (1979).
T14. Tsuchiya, S., Yokoyama, S., Yoshie, O., and Ono, Y., Production of diptheria antitoxin antibody in Epstein–Barr virus induced lymphoblastoid cell lines. *J. Immunol.* **124**, 1970–1976 (1980).
T15. Tsukimoto, I., Wong, K. Y., and Lampkin, B. C., Surface markers and prognostic factors in acute lymphocytic leukemia. *N. Engl. J. Med.* **294**, 245–248 (1976).
T16. Turkin, D., and LaPointe, J., Hybridomas and monoclonal antibodies. *Ligand Q.* **3**, 31–33 (1980).
T17. Tyrey, L., Human chorionic gonadotropin: Structural, biologic and immunologic aspects. *Semin. Oncol.* **9**, 163 (1983).
U1. Uotila, M., Engvall, E., and Ruoslahti, E., Monoclonal antibodies to human alphafetoprotein. *Mol. Immunol.* **17**, 741–748 (1980).
U2. Uotila, M., Ruoslahti, E., and Engvall, E., Two-site sandwich ELISA with monoclonal antibodies to human AFP. *J. Immunol. Methods* **42**, 11–15 (1981).
U3. U.S. Centers for Disease Control, Update: Acquired immunodeficiency syndrome (AIDS)—United States. *Morbid, Mortal. Wkly. Rep.* **32**, (1984).
V1. Vaitukaitis, J. L., Braunstein, G. D., and Ross, G. T., A radioimmunoassay which specifically measures human chorionic gonadotropin in the presence of human luteinizing hormone. *Am. J. Obstet. Gynecol.* **113**, 751–758 (1972).
V2. Van Es, A., Tanke, H. J., Baldwin, W. M., Oljans, P. J., Ploem, J. S., and Vanes, L. A., Ratios of T-lymphocyte subpopulations predict survival of cadaveric renal al-

lografts in adult patients with low dose corticosteroid therapy. *Clin. Exp. Immunol.* **52,** 13–20 (1983).
V3. Visa, D., and Phillips, J., Identification of an antigen associated with malignant melanoma. *Int. J. Cancer* **16,** 312–317 (1975).
V4. Voak, D., Sacks, S., Alderson, T., Takei, F., Lennox, E., Jarvis, J., Milstein, C., and Darnborough, J., Monoclonal anti-A from a hybrid myeloma: Evaluation as a blood grouping reagent. *Vox Sang.* **39,** 134–140 (1980).
V5. Vora, S., Isozymes of human phosphofructokinase in blood cells and cultured cell lines: Molecular and genetic evidence for a trigenic system. *Blood* **57** (1981).
V6. Vora, S., Monoclonal antibodies in enzyme research: Present and potential applications. *Anal. Biochem.* **144,** 307–318 (1985).
V7. Vora, S., Miranda, A., Hernandez, E., and Franke, U., Regional assignment of the human gene for platelet type phosphofructokinase (PFKP) to chromosome 10p: Novel use of a polyspecific rodent antisera to localize human enzymic genes. *Hum. Genet.* **63,** 3774 (1983).
V8. Vora, S., Wims, L. A., Durham, S., and Morrison, S. L., Production and characterization of monoclonal antibodies to the subunits of human phosphofructokinase: New tools for the immunochemical and genetic analyses of isozymes. *Blood* **58,** 823–829 (1981).
V9. Vora, S., Corash, L., Engel, W. K., Durham, S., Seaman, C., and Piomelli, S., The molecular mechanism of the inherited phosphofructokinase deficiency associated with hemolysis and myopathy. *Blood* **55,** 629 (1980).
V10. Vora, S., Davidson, M., Seaman, C., Miranda, A. F., Noble, N. A., Tanaka, K. R., and DiMauro, S., Heterogeneity of the molecular lesions in inherited phosphofructokinase deficiency. *J. Clin. Invest.* **72,** 1995–2006 (1983).
W1. Waldmann, T. A., and McIntire, K. R., The use of a radioimmunoassay for alpha-fetoprotein in the diagnosis of malignancy. *Cancer (Philadelphia)* **34,** 1510–1515 (1974).
W2. Wands, J. R., Carlson, R. I., Schoemaker, H., *et al.*, Immunodiagnosis of hepatitis B with high affinity IgM monoclonal antibodies. *Proc. Natl. Acad. Sci. U.S.A.* **78,** 1214–1218 (1981).
W3. Wang, C. Y., Good, R. A., and Ammirati, P., Identification of p69, 71 complex expressed on human T-cells sharing determinants with B-type chronic lymphatic leukemia cells. *J. Exp. Med.* **151,** 1539–1544 (1980).
W4. Wang, L., Rahamin, N., and Harpon, N., Monoclonal antibodies against immunodeterminants associated with the alpha and beta subunits of human chorionic gonadotropin. *Hybridoma* **1,** 293–302 (1982).
W5. Wang, M. C., Valenzuela, L., and Murphy, G. P., Purification of a human prostate specific antigen. *Invest. Urol.* **17,** 159–163 (1979).
W6. Warnke, R., and Levy, R., Immunopathology of follicular lymphoma: A model of B lymphocyte homing. *N. Engl. J. Med.* **298,** 481 (1978).
W7. Warren, L., Buck, C. A., and Tuszynski, G. P., Glycopeptide changes and malignant transformation. A possible role for carbohydrate in malignant behavior. *Biochim. Biophys. Acta* **516,** 97–127 (1978).
W8. Weiner, H. L., Hafler, D. A., Fallis, R. J., Johnson, D., Ault, K. A., and Hauser, S. L., Altered blood T-cell subsets in patients with multiple sclerosis. *J. Neuroimmunol.* **6,** 75–84 (1984).
W9. Welte, K., Platzer, E., Wange, C. Y., Rinnoosy, Kan E. A., Moore, M. A., and Mertelsmann, R., OKT8 antibody inhibits OKT3-induced IL-2 production and proliferation in OKT8+ cells. *J. Immunol.* **131,** 2356–2361 (1983).

W10. Westphal, H. M., Molderhauer, G., and Beato, M., Monoclonal antibody to the rat liver glucocorticoid receptor. *EMBO J.* **1**, 1467–1471 (1982).
W11. Williams, R. D., Bronson, D. L., and Ellio, A. Y., Production of carcinoembryonic antigen in human prostate epithelial cells *in vitro. J. Natl. Cancer Inst. (U.S.)* **58**, 1115–1116 (1977).
W12. Wood, G. W., and Barth, R., Immunofluorescent studies of the serologic reactivity with malignant melanoma against tumor-associated cytoplasm antigens. *J. Natl. Cancer Inst. (U.S.)* **53**, 309–316 (1974).
W13. Wood, J. N., Hudson, J., and Jessel, T. M., A monoclonal antibody defining antigenic determinants on subpopulations of mammalian neurones and *Trypanosoma cruzi* parasites. *Nature (London)* **296**, 34–38 (1982).
W14. Woodbury, R. G., Brown, J. P., Meh, M.-Y., Hellström, I., and Hellström, K. E., Identification of a cell surface protein, p97, in human melanomas and certain other neoplasms. *Proc. Natl. Acad. Sci. U.S.A.* **77**, 2183–2187 (1980).
W15. Woodhead, J. S., Addison, G. M., and Hales, C. N., The immunoradiometric assay and related techniques. *Br. Med. Bull.* **30**, 44–49 (1974).
W16. World Health Organization, *WHO Bull.* **62** (1984).
Y1. Yam, L. T., Li, C. Y., and Lam, K. D., *in* "Male Accessory Sex Glands" (E. Springhile and E. S. E. Hatez, eds.), p. 183. Elsevier/North-Holland, Amsterdam, 1980.
Y2. Yelton, D, E., and Scharff, M. D., Monoclonal antibodies. *Am. Sci.* **4**, 510–516 (1980).
Y3. Yelton, D. E., Roberts, S. B., and Scharff, M. D., Hybridomas and monoclonal antibodies. *Lab. Manage.* January, pp. 19–24 (1981).
Y4. Youle, R. J., and Neville, D. M., Anti-thy 1.2 monoclonal antibody linked to ricin is a potent cell-type specific toxin. *Proc. Natl. Acad. Sci. U.S.A.* **77**, 5483 (1980).
Y5. Young, W. W., and Hakomori, S., Therapy of mouse lymphoma with monoclonal antibodies to glycolipid: Selection of low antigenic variance *in vivo. Science* **211**, 487 (1981).
Y6. Young, W. W., Johnson, H. S., Tamura, Y., Karlsson, K. A., Larson, G., Parker, J. M., Khare, D. P., Spohr, U., Baker, D. A., Hindsgaul, O., and Lemieux, R. U., Characterization of monoclonal antibodies specific for the Lewis A human blood group determinant. *J. Biol Chem.* **258**, 4890–4894 (1983).
Z1. Zoler, M. L., New marrow transplant technique aids some cancers. *JAMA, J. Am. Med. Assoc.* **248**, 2213 (1982).
Z2. Zondek, B., Hypophysenvorderlappen und schwangershaft. *Endokrinologie* **5**, 425–434 (1929).
Z3. Zurawski, V., Black, P., and Haber, E., Continuously proliferating human cell lines synthesizing antibody of predetermined specificity. *In* "Monoclonal Antibodies" (R. Kennett, T. McKearn, and K. Bechtol, eds.), pp. 19–33. Plenum, New York, 1980.

INDEX

A

Abrin, in cancer therapy, 372
Acetaminophen, 203
Acetate, role in acid–base balance, 240
Acetazolamide, 214, 236, 243, 245
Acetylcholinesterase, 381
Acetylsalicylic acid, 10
Acid–base monitoring, in intensive care patients, 227–249
   computer programs, 249–261
   metabolic acidosis, 235–238
   metabolic alkalosis, 240–243
   mixed disorders, 245–247
   reference intervals, 227
   respiratory acidosis, 239–240
   respiratory alkalosis, 243–244
   role of ions, 232–235
   therapeutic calculations, 247–249
Acid–base status disorders, 228–249
Acidemia, 229–231, 237, 239, 240, 249
Acidosis, 19
   metabolic, 209, 228–239, 245–249
   and potassium balance, 215
   respiratory, 228–231, 239–240, 245–248
Acid phosphatase, serum, 19
Acquired immunodeficiency syndrome
   assay, using monoclonal antibodies, 331, 332
   cell surface analysis, 383
   effect on neopterin levels, 107–109
Acute catabolism, as cause of hyperosmolality, 204
Addison's disease, hypoosmolality in, 206
Adenovirus, 325
β-Adrenergic blockers, 246
Adriamycin, in cancer therapy, 373
Affinity purification, of monoclonal antibodies, 320–322
Age effects
   in neopterin values, 97, 98
   reference values, 22, 25
AIDS, see Acquired Immunodeficiency Syndrome
Alanine aminotransferase, serum, 20, 21, 40
Albumin
   and acid–base balance, 234, 235
   immunosensors, 340
   serum, 19, 40
Alcohols, see specific compounds
Aldosterone, 210, 221, 224, 259
Alkalemia, 215, 229, 234, 239, 240, 244, 245, 247, 249
Alkaline phosphatase
   as human tumor marker, 364
   monoclonal antibodies, 380
   serum, 19, 21, 39
Alkalosis
   metabolic, 228–235, 239–243, 246–249
   respiratory, 228–232, 234, 239, 244–246
Allograft rejection
   neopterin levels in, 100–104
   treatment, with monoclonal antibodies, 334
Altitude, effect on hemoglobin, 21
Amanita phaloides, and kidney failure, 217
Amiloride, 243
Amino acid monooxygenases, 85
Aminoglycosides, and kidney failure, 217, 220
2-Amino-4-oxo-3,4-dihydropteridine, see Pterin
Ammonium chloride, 243
Ammonium intoxication, 241
S-Amylase, 19
Anderson–Darling test, 51, 54, 57
Aneurysma, vascular, electrolyte balance in, 221
Anion gap, 234–236, 241, 245
Anthranilic acid 3-monooxygenase, 94
Antibiotics, and kidney function, 217, 218, 222
Antibodies
   blood, 39

to hepatitis B virus, 150–151, 165–167, 172
monoclonal, see Monoclonal antibodies
Anti-diuretic hormone, 203, 204, 210, 213, 221, 235
Antiferritin, in cancer therapy, 371–372
Antigens
    cell-surface, 381–387
    to hepatitis B virus, 150–151, 165–168, 172, 178
    as human tumor markers, 360–364
    leukocyte, 382–384
    in tumor cell typing, 368–370
    in tumor localization, 364–368
α_1-Antitrypsin, serum, 39
Apis, 82
Arginine hydrochloride, 243
Arthrosis, 4
Aseptic meningoencephalitis, neopterin levels in, 117–118
Aspartate aminotransferase, serum, 20, 21, 40
Assays
    for hepatitis B virus enzymes, 174–177
    for human tumor markers, 359–364
    monoclonal antibodies in, 330–333
Autoimmune disease
    monoclonal antibodies and, 384
    neopterin levels in, 112–116
Autoradiogram, of DNA probes, 157
Avidin, 158
Avidin–biotin–phosphatase complex, 369

B

Bacterial infection, neopterin levels in, 110–112
Bacteriemia, neopterin production in, 111
Barbiturates, 203, 213, 239
Bartter's syndrome, 243
Base deficiency, 247
Base excess, 227, 228, 230, 247–248
Basophils, serum, 40
Bicarbonate
    reference intervals, 227
    role in acid–base balance, 227–230, 233–236, 238–247
Biliary carcinoma, markers for, 360, 361, 363
Bilirubin, 19, 21
    as marker for hepatitis B virus, 150
Biological variation, factors in, 18, 25–26

Biopterin, 87, 88, 95
    in AIDS patients, 107
    catabolism, 91
    measurement, 95
Biotin, as label for DNA probes, 158
Bispecific antibodies, 340–341
Bladder tumor, effect on neopterin levels, 120–121
Blood donors, screening, 106–107
Bone marrow
    hepatitis B virus DNA in, 173
    transplant
        monoclonal antibodies in, 373
        neopterin levels in, 104
Breast cancer
    cell typing, 370
    incidence rate, 357
    markers for, 360, 361, 363
    neopterin levels in, 122
    steroid hormone receptors and, 377–378
    therapy, 373
Bronchial asthma, neopterin levels in, 111
Bronchogenic carcinoma, markers for, 360, 361
Buffer bases, 232–233, 245, 246
Burns, and electrolyte balance, 204, 210, 217, 220

C

Ca 19-9 antigen, as human tumor marker, 363–364
Ca 125 antigen, as human tumor marker, 364
Caffeine, pharmacological effects, 21, 25, 26
Calcium
    role in acid–base balance, 234, 237, 249
    serum, 19, 40
    urinary, 116
Cancer, see also specific cancers
    effect on pteridines, 82–83, 86–87, 92–93, 118–123
    monoclonal antibodies and, 331, 332, 357–376
        in diagnosis, 358–370
        therapy, 371–376
        in tumor cell typing, 368–370
        in tumor localization, 364–368
        tumor marker assays, 359–364
Candida albicans, measurement, using monoclonal antibodies, 340

Carboanhydrase inhibitors, 246
Carbonate, and acid–base balance, 240
Carbon dioxide, role in acid–base balance, 228–230, 238–240, 244–248
Carbon tetrachloride, poisoning, 359
Carcinoembryonic antigen, as human tumor marker, 360–361
Cardiac arrhythmia, 245
Cardiotonics, and electrolyte balance, 213, 218
Cardiovascular disease, 3
S-Carotene, 19
Catecholamines, 21, 237
Celiac disease, neopterin levels in, 116–117
Cell surface antigens, 381–387
Cerebrospinal injury, osmolarity disorders in, 204
Ceruloplasmin, serum, 21
Cervical cancer, 358
  neopterin levels in, 120
Chemotactics, 87
Chicken pox, effect on neopterin levels, 104–106
Chlamydia, monoclonal antibodies, 331, 332, 337, 388
Chlorambucil, in cancer therapy, 373
Chloride
  and acid–base balance, 232–236, 239, 241–243, 246
  serum, 19, 40
Chlorohydrocarbons, and kidney failure, 217
Chlorpropamide, 203
Cholesterol, serum, 19, 21, 22, 40
Choriocarcinoma, markers for, 363
Chorionic gonadotropin, 379
  as human tumor marker, 362–363
  immunosensors, 340
Chronobiological rhythms, and reference values, 20–21, 25, 60, 61
Circadian variations, in neopterin levels, 98–99
Cirrhosis, liver, 145, 165, 182, 214, 359
Citrate, role in acid–base balance, 270
Clones, as DNA probes, 155–156
Cobalamin, 19
Coefficient-based tests of Gaussian distribution, 51–53
Colonic carcinoma
  markers for, 359–361, 370
  neopterin levels in, 122
  therapy, 373

Colorectal cancer
  incidence rate, 357
  markers for, 363–364
Coma, hyperosmolality and, 204
Computers
  for ion and acid–base monitoring, 249–261
    approach, 254–261
    data inputs, 254–255
    development of, 250–254
    program output, 256–257
    sodium and potassium doses, 259–261
  in organization and management, 269–301
    access, 274–275
    access, prevention, 276–277
    applications, 270–271
    audit trails, 275
    branching, 273–274
    check digits, 277
    computer architecture, 272
    cost accounting, 299–300
    data entry methods, 283
    data presentation and retrieval, 289–292
    electronic mail, 278
    emergency printing, 291
    graphic output, layouts, 291–292
    help functions, 274
    human resource management, 296–297
    inquiries, 280
    instrument interfaces, 286, 288
    label generation, 280–281
    laboratory quality control, 285–286
    labor productivity, 298–299
    labor reduction, 275
    long-term data storage, 292
    management information systems, 297
    management quality control, 294–296
    management tools, 293–294
    menus, 273
    patient charting, 289–290
    patient registration, 278–279
    patient security, 275–276
    report generation, 284–285
    screen formats, 273
    service level analysis, 300
    specimen accessioning, 280
    specimen collection, 281–282
    test orders, 279–280
    tumor registry, 296
    utilization reporting, 297–298
    workload allocation, 282–283
    workload reports, 286–288

Confidence intervals, in reference limits, 41–43, 46, 57–58
Contraceptives, oral, 21, 25
Corticoids, urinary, 21
Cortisol, serum, 21, 25, 40, 60, 61
Cortisone, serum, 60
Cramér–von Mises test, 51, 54
C-Reactive protein, 113, 114
Creatine kinase, serum, 20, 40
Creatinine
  renal clearance rates, 218–220, 223–227, 259
  serum, 19, 21, 25, 40, 96–100, 103, 218, 223
Crithidia fasciculata, 85, 95
Crohn's disease, 360
  neopterin levels in, 114–115
Cyclic nucleotide phosphodiesterase, 381
Cyclosporine A, 102–103
γ-Cystathionase, in leukemia therapy, 381
Cytomegalovirus, 104–106, 177

D

Data management, 66–67
  computer-assisted, *see* Computers, in organization and management
Data rounding, correction for, 53–54
Daunomycin, in cancer therapy, 373
Deamination, 87
Dehydration
  and metabolic alkalosis, 241
  with natremia, 209–213
Delta virus, 147
Demeclocydine, 221
Deoxyribonucleic acid, *see* Hepatitis B virus, deoxyribonucleic acid
Desmin, 370
Diabetes insipidus, and electrolyte balance, 203, 204, 210, 221, 227
Diabetes mellitus, 11
  neopterin levels in, 115–116
Diagnosis, 5–6
  monoclonal antibodies in, *see* Monoclonal antibodies, clinical applications
Dictyostelium discoideum, 87
Digitalis, 240
Digitoxin, monoclonal antibody treatment, 334
Digoxin, monoclonal antibody treatment, 334
Dihydrobiopterin, 88

Dihydrobiopterin synthetase, 86, 100
7,8-Dihydro-6-hydroxylumazine, 83
7,8-Dihydroneopterin, 88–91
Dihydroneopterin triphosphate, 87, 90, 91, 94, 100
Dihydropteridine reductase, 86
7,8-Dihydropterin, 89, 92, 95
7,8-Dihydroxanthopterin, 89, 92
2,4-Dioxo-1,2,3,4-tetrahydropteridine, *see* Lumazine
Diphtheria toxin, in cancer therapy, 372
Disease, as concept, 17
Disseminated autonomy, neopterin levels in, 115
Distributions
  Gaussian, *see* Gaussian distributions
  normal, *see* Normal values
  reference, *see* Reference values, distributions
Diuresis
  computer estimates of, 255–258
  role in electrolyte balance, 217, 219–222, 225, 227
  volume, in intensive care patients, 223, 259
Diuretics, and electrolyte balance, 209, 210, 212–215, 220, 221, 243, 245, 246
DNA, *see* Deoxyribonucleic acid
DNA polymerase, from hepatitis B virus, *see* Hepatitis B virus, DNA polymerase
Dopamine, 85, 203, 218
Dot hydridization, of hepatitis B virus DNA, 158–159
Dyspnea, 237

E

Ectromelia virus, 325
Edema, and electrolyte balance, 205, 213, 214, 237, 239, 243, 244, 246
Ehrlich ascites tumor, 83
Electroencephalogram, in hyper- and hypoosmolality, 205
Electrolyte monitoring, in intensive care patients, 201–267
  computer programs, 249–261
  osmolality, 202–206
  potassium, 215–217
  renal function and, 217–227
  sodium, 206–215
Embryonal cell carcinoma, 359
Encephalitis, electrolyte balance in, 221, 244

Encephalomyelitis, 325
Encephalomyocarditis virus, 325
Encephalopathy, metabolic, 204–205
Enzymes, research, monoclonal antibodies in, 379–381
Eosinophils, blood, 40
Epidemiology, of hepatitis B virus, 144–145
Epstein–Barr virus
  inhibition by phosphonoformate, 177
  neopterin levels in, 105
  in rheumatoid arthritis, 389
Erythrocytes
  antigens, 386–387
  monoclonal antibodies, 337
  sedimentation rate, 114, 120
Escherichia coli, 87, 93, 155
Estradiol, plasma, 22
Estrogen receptors, 377–378
Estrogens, 21
Ethacrynic acid, 218
Ethanol
  and acid–base balance, 236, 246
  hyperosmolality and, 204, 206
  pharmacological effects, 21, 25, 26
Ethylene glycol, 217, 236
Exercise, effects on reference values, 25

F

Fasting, effect on specimen collection, 18–19
α-Fetoprotein, as tumor marker, assay for, 359–360, 367
Folic acid, 84, 86, 87
  biosynthesis, 90–91
  metabolic products, 91
Formamide, in deoxyribonucleic acid hybridization, 152
Fractiles, in confidence intervals, 43, 57–58, 63–64, 67
Furosemide, 214, 218, 220

G

β-Galactosidase, 158
Gastric carcinoma, markers for, 359–361, 363, 364, 370
Gastrointestinal cancer
  cell typing, 370
  neopterin levels in, 122

Gaussian distribution, 3, 39, 43–58, 63
  tests of, 48–54
Genitourinary tract tumors, neopterin levels in, 120–121
Germinal cell tumors, markers for, 359, 363
α-Gliadin, intolerance to, 116
Gliomas, 370
Glomerular filtration rate, 217–219, 222–225, 236, 242
Glucocorticoid receptors, 377–378
Glucocorticoids, 241
Glucose
  blood, 20
  and electrolyte balance, 215, 224, 243
  role in natremia, 207, 208
Glucose 6-phosphate dehydrogenase, 381
γ-Glutamyltransferase, serum, 39
Glycosuria, 203
Goiter, neopterin levels in, 115
Goodness-of-fit, 48, 50–51, 54
Growth hormone, 379
Guanosine triphosphate, in protein biosynthesis, 90–92, 94
Guanosine triphosphate cyclohydrolase, 86, 90, 92, 94
Gynecological tumors, neopterin levels in, 119–120

H

Haemophius influenzae, monoclonal antibodies, 337
Hantaan virus, 324
Haptoglobin, serum, 39
HBV, *see* Hepatitis B virus
Health, as concept, 16
Heart allograft, effect on neopterin levels, 103–104
Hemagglutination, as polyalbumin assay, 181
Hematocrit, 40, 114, 120
Hematological neoplasia, neopterin levels in, 118–119
Hemoglobin, 20, 21, 22, 40, 65, 113, 120, 209, 213
  dissociation, and acid–base balance, 237, 240
Hemoglobinuria, 20
Henderson–Hasselbalch equation, 227–228, 230

Hepatitis
  assay, using monoclonal antibodies, 331
  effect on neopterin levels, 105
  α-fetoprotein, as marker for, 359
  test for, in monoclonal antibodies, 325
  treatment, using monoclonal antibodies, 335, 337
Hepatitis B virus
  antigens, 150–151
  biology, 146–147
  coinfection with delta virus, 147
  deoxyribonucleic acid
    in bone marrow, 173
    cloned, 155–156, 168
    correlation with serological markers, 165–166
    dot hybridization, 158–159
    extrahepatic, 170–174
    hepatic, 166–168
      characterization, 168–170
    from human plasma, preparation, 155
    hydridization, 151–153
    integrated, 166–170
    in leukocytes, 171–173
    as marker, 151–174
    melting temperature, 152
    molecular forms, 159–160
    nucleotide sequence, 149
    polymorphism, 149
    probes, 154–158
    RNA-hybrids, 163–164
    in saliva, seminal fluid, and urine, 171
    significance in liver disease, 163–170
    Southern blots, analysis, 159–163
    specimen preparation, 153–154
    structure, 147–150
  detection, 143–199
  DNA polymerase, 146, 151, 174–178
    activity in hepatitis B disease, 177–178
    differential assays, 176–177
    factors affecting activity, 176
    serological assays, 174–176
  epidemiology, 144–145
  genome, 148
  markers of infection, 150–151
  molecular biology, 147–150
  perinatal transmission, 174
  polypeptides, 179–183
    polyalbumin assays, 180–181
    polyalbumin and hepatropism, 182–183
    polyalbumin receptors, 181–182
  pre-S region, 183
  in region X, 183
  protein kinase, 146, 178–179
  replication, 149–150
  surface antigen, immunosensor, 340
Hepatocellular carcinoma, 145, 168–169, 363
Hepatoma, 359
Hepatosplenomegaly, 83
Hepatotropism, 182
Herpes simplex virus, 177
  and monoclonal antibodies, 331, 332, 337, 388–389
High-performance liquid chromatography, in neopterin measurement, 95–96
Histocompatability antigens, 384–386
Histograms, in reference value data, 40–41, 66
HIV, see Acquired immunodeficiency syndrome
Hodgkin's lymphoma, neopterin levels in, 118–119
Hormones, research, monoclonal antibodies in, 376–379
HPLC, see High-performance liquid chromatography
Hybridization, see Hepatitis B virus, deoxyribonucleic acid, hybridization
Hybridoma, cultivation of, 315–316
Hydrochloric acid, as treatment for alkalosis, 243, 246
Hydrogen peroxide, secretion, 122
6-Hydroxylumazine, 83, 91, 92
6-Hydroxymethylpterin, 87, 119
Hyperaldosteronism, 240, 242
Hyperbasemia, 240, 241, 245, 247
Hypercapnia, 228, 231, 232, 239–242, 245, 246, 248
Hyperchloremia, and acid–base balance, 233, 244, 247, 249
Hyperglycemia, 205, 207, 219, 222, 225
Hyperhydration, and natremia, 213–215
Hyperkalemia, 215–216, 224, 236
Hypermineralocorticism, 240, 242
Hypernatremia, see Natremia, hypernatremia
Hyperosmolality, see Osmolality, monitoring
Hypertension, 243
Hyperventilation, and electrolyte balance, 210, 231, 236, 239, 244, 246, 247
Hypoalbuminemia, 235
Hypobicarbonatemia, 244–245
Hypocapnia, 228, 231, 244
Hypokalemia, 215–216, 222, 224, 236, 241, 244, 259

Hyponatremia, *see* Natremia, hyponatremia
Hypoosmolality, *see* Osmolality, monitoring
Hypophosphatemia, 244
Hypoproteinemia, and electrolyte balance, 208, 214, 235, 241, 246
Hypotension, and kidney function, 217
Hypovolemia, 210, 213
Hypoxemia, and acid-base balance, 232, 237, 240, 244
Hypoxia, 237, 239, 241, 244, 246-248

I

Immunoaffinity purification, using monoclonal antibodies, 333-334
Immunoassays, for polyalbumin, 181
Immunodeficiency virus, *see* Acquired immunodeficiency syndrome
Immunoglobulins, *see also* Monoclonal antibodies
　immunosensors, 340
　serum, 21, 22, 113, 367, 369
Immunosensors, monoclonal antibodies as, 338-340
Immunosuppressive therapy, 108
Index of atypicality, 63-65
Induction signal, for neopterin, 93-94
Inflammatory diseases, neopterin levels in, 116-118
Influenza, effect on neopterin levels, 105
Information management, *see* Computers, in organization and management
Inhibitors, of DNA polymerase, 176
Interferons, 83, 93-94, 100-104, 110-116, 122-123, 178-179
Interfractile interval, in reference limits, 42-43
Interleukin-2, role in AIDS, 107, 108, 123
Intermediate filament proteins, in tumor cell typing, 370
Iron, serum, 19, 21, 40
Islet cells, monoclonal antibodies, 337
Isoxanthopterin, 86, 87, 91, 92

K

Keratin, 370
Ketoacidosis, diabetic, 236, 237, 246, 247

Kidney
　function, in electrolyte balance, 203, 214, 217-227
　　clearance parameters, 218-220
　　diuresis, types of, 221-222
　　fractional excretions, 220-221
　　in ICU patients, 222-227
　as hepatitis carrier, 173-174
　transplant
　　monoclonal antibodies and, 383-384
　　neopterin levels, 101-103, 106
Kilham rat virus, 325
Kolmogorov-Smirnov test, 50-51, 54
Kurtosis, in reference values, 52-53, 55, 57

L

Lactate, in acid-base balance, 234, 236-238, 241, 244-247
Lactate dehydrogenase, serum, 20, 40
Lactogen, in human placenta, 67
Lepidoptera, 84-85
Leprosy, neopterin levels in, 111
Leucopterin, 90
Leukemia
　cell surface antigens, 383
　neopterin levels in, 118-119
　treatment, with monoclonal antibodies, 373-374, 381
Leukocytes
　antigens, 382-386
　heptatis B virus DNA in, 171-173
Limit, reference, *see* Reference values, reference limit
Linnaeus, C., 17
Lipids, serum, 18, 19
Lipoproteins, serum, 19, 21
Lithium, effect on electrolyte balance, 221, 235
Liver
　cancer, 358
　hepatitis DNA in, 166-169
　　characterization, 168-170
　transplant, effect on neopterin levels, 101, 103
Lumazine, 84, 85, 87, 91
Lung cancer
　incidence rate, 357
　markers for, 364
　neopterin levels in, 121-122

therapy, 373
tumor localization, 367–368
Lungs, role in acid–base balance, 228, 239
Lupus erythematosus, 112, 113, 384, 390
Lymphocytes, 40, 93
Lymphocytic choriomeningitis, 324–325
Lymphokines, 100, 110, 112, 113, 115, 116, 117
Lysine hydrochloride, 243

M

α_2-Macroglobulin, serum, 21, 39
Macrophages, 91, 93, 94
Magnesium ion, role in acid–base balance, 234, 240, 243
Malaria
  monoclonal antibodies, 337
  neopterin levels in, 109–110
Malignant lymphoma, cell typing, 369
Management, see Computers, in organization and management
Manitol, 214, 218, 220, 222
Measles–mumps vaccine, neopterin levels, 106
Melanoma
  cell typing, 369–370
  markers, 362
  neopterin levels in, 122
Meningitis, and electrolyte balance, 221, 244
Metabolic acidosis, see Acidosis, metabolic
Metabolic alkalosis, see Alkalosis, metabolic
Methanol, 206, 236
Methanopterin, 88
Methodology, in specimen collection, 26–27
Methotrexate, in cancer therapy, 373
Methoxyfluoran, 221
β_2-Microglobulin, 103, 109, 119
Mineralocorticoid, 242–243
Minute virus, 325
Mixed seminoma, 359
Modeccine, in cancer therapy, 372
Molybdopterin, 86
Monapterin, 87, 93, 95
Monoclonal antibodies, 303–415
  bispecific, 340–341
  catalytic, 341–342
  in cell surface antigen studies, 381–387
  clinical applications, 329–342, 355–415
    in Chlamydia infection, 388

    in herpes simplex virus, 388–389
    immunoaffinity purification, 333–334
    in vitro diagnosis, 330–333
    in Neisseria gonorrhoeae, 388
    in neoplastic disease, 357–376
      cell typing, 368–370
      diagnosis, 358–370
      therapy, 334–335, 337, 371–375
      tumor localization, 364–368
    rheumatology, 389–390
  developing areas, 338–342
  heterogeneity, 326–329
  in hormone and protein research, 376–381
  human antibody products, 335–338
  as immunosensors, 338–340
  versus polyclonal antibodies, 356
  production, 306–317
    difficulties, 310–311
    human, 313–315
    hybridoma cultivation, 315–316
    from mouse, 307–311
    quadroma generation, 311–313
    in serum-free media, 316–317
  purification, 317–329
    affinity purification, 320–322
    for medical applications, 322–323
    methods, 318
    removal of viral contaminants, 324–326
  recombinant, 341
  regulatory considerations, 323–324
  structure and diversity, 304–306
Monocytes, 40
Mononucleosis, effect on neopterin levels, 105
Mouse antibody production test, 325
Multiple myeloma, neopterin levels in, 118–119
Multiple sclerosis, monoclonal antibodies and, 384
Multiple sclerosis, neopterin levels in, 117–118
Murine leukemia viruses, 324
Myocardial infarction, 22–23
Myocarditis, 17

N

Natremia, 203, 205, 207–215
  hypernatremia
    and dehydration, 210–213
    and hyperhydration, 214–215

in ICU patients, 222, 224, 225, 243
  and normohydration, 208–209
 hyponatremia
  and acid–base balance, 235
  and dehydration, 210
  and hyperhydration, 213–214
  and normohydration, 207–208
  interpretation of, 207–215
  sodium doses, recommended, 259–261
Necrosis, electrolyte balance in, 221
Neisseria gonorrhoeae, monclonal antibodies and, 388
Neoplastic disease, *see* Cancer
Neopterin, 81–141
 age differences, 97, 98
 biochemistry, 90–91
 catabolism, 91–92
 and cellular immunity, 122–123
 cellular source, 92–93
 in cerebrospinal fluid, 117–118
 chemistry, 88–89
 circadian variations, 98–99
 clinical measurement, 89
 creatinine ratio, 97, 99, 100
 discovery, 84–85
 as disease indicator, 99–123
  allograft rejections, 100–104
  autoimmune thyroiditis, 115
  bacterial infection, 110–112
  celiac disease, 116–117
  Crohn's disease, 114–115
  diabetes mellitus, type I, 115–116
  intracellular protozoa infection, 109–110
  malignancy, 118–123
  multiple sclerosis, 117–118
  rheumatoid arthritis, 112–113
  sarcoidosis, 116
  ulcerative colitis, 114
  viral infections, 104–109
 induction signal, 93–94
 long-term stability, 99–100
 marker for immunity, 81–141
 measurement, 94–97
  by high-performance liquid chromatography, 95–96
  by radioimmunoassay, 96–97
 oxidation, 88
 physiological role, 94
 serum, 97–98, 103, 104, 109, 111, 116, 118, 119

 in synovial fluid, 113
 synthesis, 82
 urinary excretion, 82–83, 87, 97–99, 101–105, 109–122
Nephritis, 204
Nephropathy, and electrolyte balance, 210, 221
Neurofilaments, 370
Neurons, monoclonal antibodies, 337
Neutrophils, serum, 40
Nomenclature, of pteridines, 84
Non-Hodgkins lymphoma
 cell typing, 369
 neopterin levels in, 118–119
Nonparametric method, in estimating confidence intervals, 43–45, 66
Norepinephrine, 85
Normal values, 2–4
Numerical methods, in presentation of observed data, 62–64

O

Observed value, 8, 12–13
Oligoanuria, 218, 220, 224
Oligonucleotides, as DNA probes, 156–157
Oliguria, 209, 210, 218
Opiates, 203, 213
Oral cancer, 358
Orosomucoid, 40, 114
Osmolality, monitoring, in intensive care patients
 computer programs, 255–258
 fractional excretions, 220, 223–226
 hypersomolality, 209–210, 212, 217, 222, 227
 hypoosmolality, 204–206, 208, 210, 214, 227, 235
 observed values, 223, 226
 reference interval, 202, 223
 renal clearance, 219, 223, 225, 226
Osteoarthritis, neopterin levels in, 112–113
Outliers, in reference value data, 40–41
Ovarian carcinoma
 markers for, 360, 363, 364
 neopterin levels in, 119–120
Ovarian carcinomatosis, 87
Ovarian endodermal sinus tumor, 359
Oxytocin, 203, 213

P

Pancreas, as carrier of hepatitis B virus, 173–174
Pancreatic allograft, effect on neopterin levels, 101, 103–104
Pancreatic carcinoma
  markers for, 359–361, 363, 364, 370
  neopterin levels in, 122
Pancreatic oncofetal antigen, as human tumor marker, 361
Pancreatitis, 360
Parametric method, in estimating confidence intervals, 43–48, 66
Parathyroid hormone, 379
Parovirus, 324
Partitioning, of reference individuals, 25–26, 37–40
Perinatal transmission, of hepatitis B virus, 174
Peroxidase–antiperoxidase, in cell typing, 369
Phenylalanine, accumulation, 85, 100
Phenylketonuria, 85–86, 100
Phorbol myristate acetate, 93, 94
Phosphate, serum, 19
Phosphofructokinase, 379–381
Phosphonoformate, as inhibitor of DNA polymerase, 176–177
Physarum polycephalum, 87
Phytohemagglutinin, 92
  role in AIDS, 107
Placenta, as carrier of hepatitis B virus, 174
Plasminogen-activating enzyme, 381
Plasminogen, serum, 21
Plasmodium, see Malaria
Platelets, 40
Pneumonia
  acid–base balance in, 247
  test, 325
Pneumonitis virus, 325
Polyalbumin, as marker for hepatitis B virus, 151, 180–182
Polycythemia vera, neopterin levels in, 118–119
Polydipsia, psychogenic, and electrolyte balance, 221
Polymorphism, in hepatitis B virus deoxyribonucleic acid, 149
Polyoma virus, 324–325
Polypeptides, as markers of hepatitis B virus, 179–183
Polyuria, 204, 219, 222, 224, 225, 241

Population, reference, 9
Posture, in specimen collection, 19–20, 26–27
Potassium
  deficits, calculation, 258
  doses, in ICU patients, 259–261
  effect on DNA polymerase activity, 176
  monitoring, 215–217
  and acid–base balance, 234, 236, 237, 240–243
  fractional excretion, 220–221, 223, 224, 259
  observed values, 223
  reference values, 215, 223
  serum, 20, 40
  urinary, 21
Preanalytical factors, in specimen collection, 17–22
Prediction interval, in reference limits, 42
Pregnancy, acid–base status disorders, 246
Pregnancy tests, using monoclonal antibodies, 331, 332, 363
Preprocessing, see Statistics, in reference values, preprocessing
Probes, for hepatitis B virus DNA, 154–158
Progesterone receptors, 377–378
Progestins, 21
Prostate cancer
  incidence rate, 357
  markers for, 360–361, 380
  neopterin levels in, 120–121
  therapy, 373
Prostate-specific antigen, as human tumor marker, 361–362
Prostatic acid phosphatase
  as human tumor marker, 361–362
  monoclonal antibodies, 380
Protease, in DNA preparation, 153–154
Protein kinase, of hepatitis B virus, 146, 178–179
Protein, serum, 19, 21, 40
Protozoa infection, neopterin levels in, 109–110
Pseudomonas aeruginosa, monoclonal antibodies, 337
Pteridine, 84
Pteridines, see also specific compounds
  biochemistry, 90–91
  biological function and occurrence, 85–88
  cellular source, 92–93
  chemistry, 88–89
  as chemotactic signals, 87
  as cofactors, 87–88

discovery, 84–85
early syntheses, 85
measurement, 94–97
nomenclature, 84
oxidation reactions, 88–89
oxidation states, 88–89
in oxygenase reactions, 86
urinary, 83
Pterin, 84, 88, 89, 91
6-Pterinaldehyde, 86, 87
Pterin deaminase, 91
Pterins
　biosynthesis, 90–91
　measurement, 95, 119
Purification, of monoclonal antibodies, see Monoclonal antibodies, purification
Pyrazino-(2,3-d)-pyrimidine, see Pteridine
Pyruvate kinase, 381

Q

Quadroma, generation of, 311–313
Quality control, 34–36
　computers in, 285–286, 294–296

R

Rabies, treatment, with monoclonal antibodies, 335, 337
Race, effects in specimen collection, 22
Radioimmunoassay
　in neopterin measurement, 95–97
　in tumor localization, 366–368
Radioimmunotherapy, 371–372
Radioiodine, urinary, 66
Radiolabeling, as DNA probe, 156–158
Recombinant monoclonal antibodies, 341
Rectum cancer, effect on neopterin levels, 122
Reference values, 1–79
　alternatives, 58–60
　analytical methods, 34–35
　classes, 12–13
　collection, checklist, 29–30
　concept, 4–14
　definition, 8
　description, 4–6
　distributions, 10–11, 13, 40–41, 43–58, 63–64, 66
　IFCC recommendations, 6–7

quality control, 34–36
reference groups, 14–17
　diseased, 17
　healthy, 16
reference individuals, 7, 9–10, 13, 17–34
　partitioning, 25–26
　population, 9, 13
　sample group, 9, 13
　selection, 23–25
reference interval, 12–13
　for acid–base balance, 227
　alternatives, 65–66
　for bicarbonate, 227
　for fractional excretions, 220
　for osmolality, 202
　for potassium, 215
　for renal clearance, 218
　of renal function, 223
　for sodium, 206
reference limit, 11, 13, 36, 41–58
　calculation, 57–58
　confidence intervals, 41–43, 57–58
　nonparametric method, 45
　parametric method, 45–48
　parametric versus nonparametric methods, 43–44
reference state, 25–26
relation to observed values, 61–67
　data management, 66–67
　graphical presentation, 64–65
　numerical methods, 62–64
specimen collection, 17–35
statistical treatment, see Statistics, in reference values
transfer, 35–36
transformations, 47, 54–57
REFVAL, 37, 41, 50–51, 53, 54, 57
Renal cell carcinoma, neopterin levels in, 121
Renin, 21, 242
Renin–angiotensin system, 210
Reoviruses, 324–325
Reporting, computer-assisted, see Computers, in organization and management
Respiratory acidosis, see Acidosis, respiratory
Respiratory alkalosis, see Alkalosis, respiratory
Retroviruses, 326
Rheumatic disease, monoclonal antibodies in, 389–390
Rheumatoid arthritis, 17
　monoclonal antibodies in, 389–390
　neopterin levels in, 112–113

428 INDEX

Riboflavin, 87
Ricin, in cancer therapy, 372–373
Rotavirus-like agent, 324
Rubella, treatment, with monoclonal antibodies, 335, 337

S

Salicylates, 236, 244, 246
Sample group, reference, 9–10
Sarcoidosis
  electrolyte balance in, 221
  neopterin levels in, 116
Schizophrenia, 17
  electrolyte balance in, 213
Security, in information management, 275–277
Sendai virus, 325
Sepiapterin, 95
Sepsis, 204, 207, 220, 244, 246
  treatment, with monoclonal antibodies, 335
Serological markers, in hepatitis B infection, 150–151, 165–167, 172, 175–176
Serotonin, 85
Serous cystadenocarcinoma, α-fetoprotein as marker for, 359
Sex, effects in specimen collection, 22, 25
Sexual activity, effects in specimen collection, 21
Skewness, in reference values distribution, 52–55, 63
Skin, as carrier of hepatitis B virus, 173–174
Smoking, pharmacological effects, 21, 25
Socioeconomics, effect on specimen collection, 22
Sodium
  deficits, calculation, 258
  doses, for ICU patients, 259–261
  monitoring, 206–215
    and acid–base balance, 232–235, 239–243, 245
    fractional excretion, 220, 223, 224, 259
    natremia, see Natremia
    observed values, 223
    reference intervals, 206, 223
  serum, 19, 20, 40
Southern blots, of hepatitis B virus DNA, 159–164
Specimens, collection
  accessioning, computer-assisted, 280

age effects, 22
altitude effects, 21
biological factors, 26
chronobiological rhythm effects, 20–21
collection, 17–34
computer-assisted management, 281–282
effects of pharmacologicals, 21
fasting effects, 18–19
handling, 27
of hepatitis B virus, preparation, 153–154
labeling, computer-assisted, 280–281
methodologies, 26–27
posture effects, 19–20
preparation of individuals, 26–28
procedures, 28–34
  skin puncture, 31–34
  venous blood, 31
sex effects, 22
sexual organ effects, 21
site of collection, 20
socioeconomic effects, 22
standardization, 27–28
tourniquet effects, 19, 20
Sperm, monoclonal antibodies, 336, 337
Spina bifida, 359
Spironolactone, 214, 243
Standardization schemes, in specimen collection, 27–28
Staphylococcal pneumonia, neopterin levels, in, 111
Statistics, in reference values, 36–58
  coefficient-based tests, 51–53
  correction for data rounding, 53–54
  distribution tests, 48–54
  goodness-of-fit tests, 50–51, 54
  multivariate alternatives, 58–60
  numerical methods, 62–64
  outliers, 40–41
  preprocessing, 37–41
    partitioning, 37–40
  reference limits, estimation, 41–58
  time-specified alternatives, 60
  transformations, 47, 54–57
    rigid and elastic, 54–55
    transforming functions, 56
    two-stage, 55–57
  visual inspection, 40–41
Stenocardia, 17
Steroid hormone receptors, 377–378
Stomach cancer, markers for, 364

T

Tachycardia, 244
T Cells, 93, 100, 102, 104, 107–109, 111–117, 122, 123, 373–375, 382–384, 390
Testicular cancer
  markers for, 363, 364
  neopterin levels in, 120–121
Tetanus toxoid, monoclonal antibodies, 337
5,6,7,8-Tetrahydrobiopterin, 85, 86, 88, 91
Tetrahydrofolate, cofactors of, 87–88
Tetrahydrofolic acid, 91
5,6,7,8-Tetrahydroneopterin, 88, 89, 91–92
5,6,7,8-Tetrahydropterin, 88, 95
Thymine, 87
Thyrotropin, 379
Thyroxine, 40, 340
Thyroxine-binding globulin, serum, 21
Tissue typing, using monoclonal antibodies, 331
Tolerance interval, in reference limits, 41–42
Tourniquet, in specimen collection, 19–20
Toxoplasma gondii, monoclonal antibodies, 340
Transaminase, as marker for hepatitis B virus, 150
Transferrin, serum, 21, 40
Transformations, *see* Statistics, in reference values, transformations
Triglycerides, blood, 40
Trophoblastic disease, markers for, 363
Tropoblastic neoplasm, markers for, 363
Tuberculosis, 17
  electrolyte balance in, 221
  neopterin levels in, 110
Tumors
  cell typing, using monoclonal antibodies, 368–370
  markers, assays for, 359–364
  monoclonal antibodies, 337
  neopterin levels in, 118–123
  registry, computer-assisted, 296
Tyrosine hydroxylase, 381

U

Ulcerative colitis, 360
  neopterin levels in, 114
Urea, 19, 40, 120, 209, 219, 222, 223
  excretion, 203–204, 225
Uremia, 237, 246
Uric acid, serum, 19, 40
Urinary bladder carcinoma, markers for, 360
Urothione, urinary, 86
Uterine carcinoma, markers for, 360

V

Vaccines, immunoaffinity purification, 333
Variation, intra- versus interindividual, 39–40
Varicella–zoster virus, treatment, with monoclonal antibodies, 335
Vasopressin, 203–205
Vimentin, 370
Vindecine, in cancer therapy, 373
Viral infections, effect on neopterin levels, 104–109
Viruses, tests for, 324–326

W

Water, clearance, renal, 219–220, 223–227, 259
White blood cells, count, 40, 113, 120

X

Xanthine oxidase, 91, 381
Xanthopterin, 85, 88, 89, 91

Z

Zymosan, 93